CliffsAP®

Physics B & C

by

James R. Centorino

D1441683

WILEY

Wiley Publishing, Inc.

About the Author

James R. Centorino received B. S. and M. S. degrees in Geophysics from Boston College and B.A. and M.A. degrees in music composition and trumpet from The Boston Conservatory of Music. He has taught secondary school physics in the Boston and Los Angeles areas for more than 30 years.

Author's Acknowledgments

This book is dedicated to Susan and Jimmy, Nickie and Jim, with many thanks to Bob Marks, Jerry Bobrow, Kelly Dobbs Henthorne, and Philip McKinley.

Publisher's Acknowledgments

Editorial

Project Editor: Kelly Dobbs Henthorne

Acquisitions Editor: Greg Tubach

Technical Editor: Philip R. McKinley

Composition

Proofreader: Christine Pingleton

Wiley Publishing, Inc. Composition Services

Cliffs AP® Physics B & C

Published by:
Wiley Publishing, Inc.
111 River Street
Hoboken, NJ 07030-5774
www.wiley.com

Note: If you purchased this book without a cover, you should be aware that this book is stolen property. It was reported as "unsold and destroyed" to the publisher, and neither the author nor the publisher has received any payment for this "stripped book."

Copyright © 2004 James R. Centorino, www.centorino.com

Published by Wiley Publishing, Inc., New York, NY
Published simultaneously in Canada

Library of Congress Cataloging-in-Publication Data
Centorino, James R., 1949-
CliffsAP physics : B & C exams / by James R. Centorino.-- 1st ed.
p. cm.
Includes bibliographical references and index.
ISBN 0-7645-3985-X (pbk.)
1. Physics--Examinations--Study guides. 2. Physics--Examinations, questions, etc. 3. Universities and colleges--United States--Entrance examinations--Study guides. 4. Advanced placement programs (Education)--Examinations--Study guides. I. Title: Cliffs AP physics. II. Title.

QC32.C395 2004
530'.076--dc22

2004000644

ISBN: 0-7645-3985-X

Printed in the United States of America

10 9 8 7 6 5 4 3 2

1B/RQ/QW/QV/IN

No part of this publication may be reproduced, stored in a retrieval system, or transmitted in any form or by any means, electronic, mechanical, photocopying, recording, scanning, or otherwise, except as permitted under Sections 107 or 108 of the 1976 United States Copyright Act, without either the prior written permission of the Publisher, or authorization through payment of the appropriate per-copy fee to the Copyright Clearance Center, 222 Rosewood Drive, Danvers, MA 01923, 978-750-8400, fax 978-646-8600. Requests to the Publisher for permission should be addressed to the Legal Department, Wiley Publishing, Inc., 10475 Crosspoint Blvd., Indianapolis, IN 46256, 317-572-3447, or fax 317-572-4447.

THE PUBLISHER AND THE AUTHOR MAKE NO REPRESENTATIONS OR WARRANTIES WITH RESPECT TO THE ACCURACY OR COMPLETENESS OF THE CONTENTS OF THIS WORK AND SPECIFICALLY DISCLAIM ALL WARRANTIES, INCLUDING WITHOUT LIMITATION WARRANTIES OF FITNESS FOR A PARTICULAR PURPOSE. NO WARRANTY MAY BE CREATED OR EXTENDED BY SALES OR PROMOTIONAL MATERIALS. THE ADVICE AND STRATEGIES CONTAINED HEREIN MAY NOT BE SUITABLE FOR EVERY SITUATION. THIS WORK IS SOLD WITH THE UNDERSTANDING THAT THE PUBLISHER IS NOT ENGAGED IN RENDERING LEGAL, ACCOUNTING, OR OTHER PROFESSIONAL SERVICES. IF PROFESSIONAL ASSISTANCE IS REQUIRED, THE SERVICES OF A COMPETENT PROFESSIONAL PERSON SHOULD BE SOUGHT. NEITHER THE PUBLISHER NOR THE AUTHOR SHALL BE LIABLE FOR DAMAGES ARISING HEREFROM. THE FACT THAT AN ORGANIZATION OR WEBSITE IS REFERRED TO IN THIS WORK AS A CITATION AND/OR A POTENTIAL SOURCE OF FURTHER INFORMATION DOES NOT MEAN THAT THE AUTHOR OR THE PUBLISHER ENDORSES THE INFORMATION THE ORGANIZATION OR WEBSITE MAY PROVIDE OR RECOMMENDATIONS IT MAY MAKE. FURTHER, READERS SHOULD BE AWARE THAT INTERNET WEBSITES LISTED IN THIS WORK MAY HAVE CHANGED OR DISAPPEARED BETWEEN WHEN THIS WORK WAS WRITTEN AND WHEN IT IS READ.

NOTE: THIS BOOK IS INTENDED TO OFFER GENERAL INFORMATION ON ADVANCED PLACEMENT PHYSICS B & C EXAMS. THE AUTHOR AND PUBLISHER ARE NOT ENGAGED IN RENDERING LEGAL, TAX, ACCOUNTING, INVESTMENT, REAL ESTATE, OR SIMILAR PROFESSIONAL SERVICES. ALTHOUGH LEGAL, TAX, ACCOUNTING, INVESTMENT, REAL ESTATE, AND SIMILAR ISSUES ADDRESSED BY THIS BOOK HAVE BEEN CHECKED WITH SOURCES BELIEVED TO BE RELIABLE, SOME MATERIAL MAY BE AFFECTED BY CHANGES IN THE LAWS AND/OR INTERPRETATION OF LAWS SINCE THE MANUSCRIPT IN THIS BOOK WAS COMPLETED. THEREFORE, THE ACCURACY AND COMPLETENESS OF THE INFORMATION PROVIDED HEREIN AND THE OPINIONS THAT HAVE BEEN GENERATED ARE NOT GUARANTEED OR WARRANTED TO PRODUCE PARTICULAR RESULTS, AND THE STRATEGIES OUTLINED IN THIS BOOK MAY NOT BE SUITABLE FOR EVERY INDIVIDUAL. IF LEGAL, ACCOUNTING, TAX, INVESTMENT, REAL ESTATE, OR OTHER EXPERT ADVICE IS NEEDED OR APPROPRIATE, THE READER IS STRONGLY ENCOURAGED TO OBTAIN THE SERVICES OF A PROFESSIONAL EXPERT.

Trademarks: Wiley, the Wiley Publishing logo, Cliffs, CliffsNotes, CliffsAP, CliffsComplete, CliffsTestPrep, CliffsQuickReview, CliffsNote-a-Day, and all related trademarks, logos and trade dress are trademarks or registered trademarks of Wiley Publishing, Inc., in the United States and other countries and may not be used without written permission. All other trademarks are the property of their respective owners. Wiley Publishing, Inc. is not associated with any product or vendor mentioned in this book. For general information on our other products and services or to obtain technical support, please contact our Customer Care Department within the U.S. at 800-762-2974, outside the U.S. at 317-572-3993, or fax 317-572-4002.

Wiley also published its books in a variety of electronic formats. Some content that appears in print may not be available in electronic books.

WILEY

Table of Contents

PART I: INTRODUCTION

PART II: SUBJECT AREA REVIEWS WITH SAMPLE QUESTIONS AND ANSWERS

PART III: PRACTICE TESTS: SAMPLE B & C EXAMS WITH ANSWERS AND COMMENTS

PART I

INTRODUCTION

Introduction

This book is designed to be a study guide for the AP Physics B and C Exams. It contains the basic material that students might expect to encounter on these exams, along with comments and helpful hints. Although derivations and details are omitted, brief refreshers are included at the beginning of each section. Simplicity of style is, in the author's opinion and experience, most beneficial to both learning and review. Attention is also paid to simplifying complex concepts and refreshing problem-identification and problem-solving proficiency. The chapters end with sample problems typical of those found on the exams. The practice exams at the end of the book include typical multiple-choice and free-response problems. Following these practice exams are answers and comments meant to be of assistance in review and preparation.

How You Should Use This Book

This book will assist you in preparing for the AP Physics Exam. Since the Exam is administered in May, you should start your review several weeks beforehand. In the usual course of high school study, the Mechanics portion is generally covered during the first semester and Electricity and Magnetism topics are covered in the second semester.

This book contains

- Review sections of important material
- Review questions after each section, with solutions, explanations, and helpful comments
- Two sample B Exams and two sample C Exams

You may use this book as an adjunct to your regular text and skim through the sections, taking the sample exams included, or you may read through the entire book and review as you read. Use it to reinforce what you have learned in class and as a self-check of your level of understanding. Highlight important items, make notes in the margins, and use tape or sticky-tabs to mark important sections.

When you are working out problems, there may be more than one method of solving a physics problem. Your work or methodology may not be the same as what is given in the solutions, but you may still arrive at the correct answer. Do whatever works best for you. The sample problems included in each section are meant to demonstrate typical problems that you might expect to encounter on the exam itself.

Many students have found that studying for this exam in small groups can be extremely helpful. If you can, gather a few friends who are also preparing for the exam. Review together. Such small group interactions are especially beneficial from the aspect of alternative explanations and additional insight into concepts and problem-solving approaches.

When you take the appropriate practice exams in this book, note the concepts for which you may need extra study. Review those sections in this book and in your regular text. Seek additional texts if necessary.

The AP Exams in Physics: B and C

There are two AP Physics exams: the B exam and the C exam. You may choose to take either exam but not both in the same year.

The B Exam

The B exam covers more material than the C exam, but in much less depth. Mathematical content on the B exam includes geometry, algebra, and trigonometry. Calculus is not part of the B exam.

The C Exam

The C exam is offered in two parts: Mechanics, and Electricity and Magnetism. You may opt to take either one or both for the same fee. The two separate parts cover less material than the B exam, but the questions are more challenging, and the mathematical content includes basic calculus.

Format

Both the B and C exams consist of a multiple-choice section and a free-response section.

Multiple-choice questions have five possible answers, but only one is correct. Each correct answer is worth 1 point and .25 of a point is deducted for each wrong answer, therefore, random guessing is of little value. Although a reference sheet of constants and common trigonometric functions is provided, use of a calculator is not allowed on the multiple-choice section. There are 70 multiple-choice questions on the B exam, with a 90-minute time limit. On the C exam, there are 35 multiple-choice questions in the Mechanics part and 35 multiple-choice questions in the E & M part. The time limit on each part is 45 minutes.

The free response section allows the use of non-qwerty calculators and supplies an additional sheet of equations and mathematical formulas. The B exam includes about 6 multi-part questions, for which you must show your work when finding solutions. The time limit is 90 minutes for this section. On the C exam, there are usually three multi-part questions to the Mechanics part and three for the E & M part, 45 minutes being allotted for each.

Grading

The multiple-choice sections are computer-graded. The free response sections are hand-graded by high school and college physics teachers, who allow partial credit for incorrect or incomplete answers. It is highly recommended that you show *all* work and *do not erase* any part of your response. Even though your final answer may be incorrect, points are awarded for correct processes and steps found in your response.

In both the B and C exams, the multiple-choice and free-response sections are each worth 50 percent of your final score.

AP exam grades are reported as a number from 1 to 5, with 5 being the highest and 3 being a passing score. An approximate score of 75 percent is necessary to receive a 5, approximately 60 percent earns a 4, and approximately 45 percent earns a passing score of 3.

What's on the Exams

The AP Physics exams' content area is specified every year by The College Board in their course description "Acorn" booklet. It lists the following content areas and the percentage of exam inclusion:

	Physics B	***Physics C***
Mechanics (Newtonian)	**35%**	**50%**
Kinematics (includes vectors)	7%	9%
Newton's Laws of Motion (includes friction and centripetal force)	9%	10%
Work, energy, power	5%	7%
Impulse, momentum, particle systems	4%	6%
Circular motion and rotation	4%	9%
Oscillation and gravitation	6%	9%
Fluid Mechanics and Thermal Physics	**15%**	

	Physics B	*Physics C*
Fluid mechanics	5%	
Temperature and heat	3%	
Thermodynamics and kinetic theory	7%	
Electricity and Magnetism	**25%**	**50%**
Electrostatics	5%	15%
Conductors, capacitors, and dielectrics	4%	7%
Electric circuits	7%	10%
Magnetism	4%	10%
Electromagnetism	5%	8%
Waves and Optics	**15%**	
Wave motion (includes sound)	5%	
Physical optics	5%	
Geometric optics	5%	
Atomic and Nuclear Physics	**10%**	
Atomic physics and quantum effects	7%	
Nuclear physics	3%	

Each exam usually includes one or more questions dealing with laboratory or experimental situations, which assess your knowledge of practical experience with laboratory settings, including data collection and analysis. Knowledge of important historical physics experiments and topics may also be encountered.

Hints for Taking the Different Sections

Multiple-Choice Section

Process of elimination plays an important role in successfully navigating the multiple-choice section of the exam. It is helpful to go through and answer the easier questions first. In any given question, there is a 1-in-5 chance of randomly guessing the correct answer. If you can eliminate one or more possible answers, you will increase your probability of getting the correct answer and reduce the possibility of having .25 of a point deducted for a wrong answer.

Free-Response Section

Since each free-response problem has multiple parts, the easiest portions of this type of problem are the first and second parts. In other words, a free-response question may have six parts, **A** through **F**, with **A** and **B** being the easiest. You may wish to answer those parts first, then proceed to the next problem, using the same strategy of answering the preliminary, easier parts first. By answering those first parts in each problem, you can be assured of scanning all the problems for content, rather than becoming bogged down on one problem without even reading the succeeding question(s). Otherwise, you may not do an easier problem, simply because you are unaware of it.

In all of your free-response answers, *always show all of your work* and *never erase anything*. Even if you believe your work to be incorrect and start over, leave your previous work as well. You may draw a line through it, but *do not erase* it, because it may contain correct information, and you can *still* receive credit for it.

If you are certain of your answers, it will help the grader if you underline or circle your final answers. Neatness and legibility will also aid the persons who will grade your free-response answers. It is helpful to jot down the basic equations you will use to solve the problem. Also, do not forget the benefits of units both in your answers and in the solution process. Often, dimensional analysis will aid you in setting up the problem correctly. Also, AP Physics exams do not make an issue of significant figures.

If time allows, go back and check your work on your free-response answers.

PART II

SUBJECT AREA REVIEWS WITH SAMPLE QUESTIONS AND ANSWERS

Vectors

Definition

A **vector** is defined as a quantity having both magnitude and direction. For example, 88 km/hr East or 200N 20° North of East are vectors. (By contrast, a *scalar* is a quantity that has magnitude only. For example, 35 meters, 50 ft/sec, and 100N are all *scalars* . . . numbers with no direction.) Typical vectors encountered in physics studies include displacement, velocity, acceleration, force, momentum, magnetic fields, and electric fields.

Displacement is the straight line connecting starting and ending points.

Velocity is speed *with direction.*

Acceleration is the change in velocity with respect to time.

A **force** is a push or a pull in a particular direction.

Momentum is an object's mass multiplied by its velocity.

Magnetic and **electric** fields both have directional orientations.

All are *vectors* because they have *magnitude and direction*.

Addition and Subtraction

Geometric Addition of Vectors

Add "head to tail" as follows:

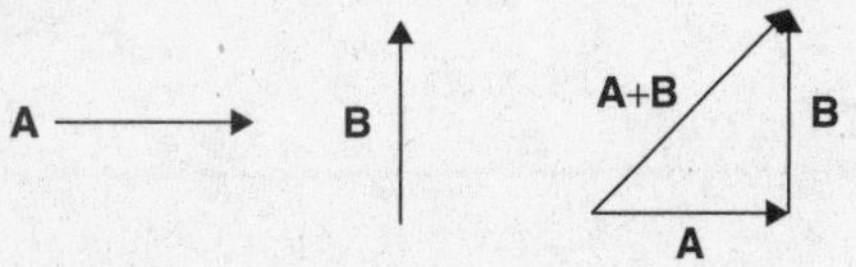

Following the commutative law:

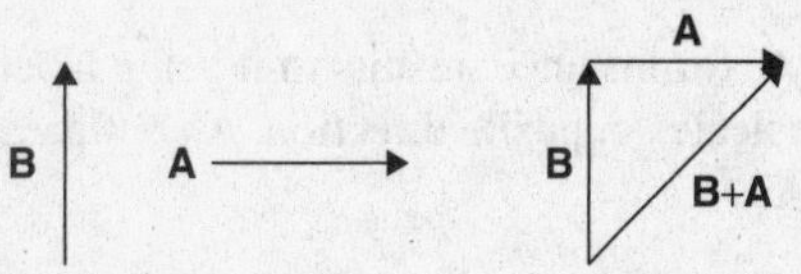

$$\mathbf{A} + \mathbf{B} = \mathbf{B} + \mathbf{A}$$

Magnitude of $\mathbf{A} + \mathbf{B} = \sqrt{\mathbf{A}^2 + \mathbf{B}^2}$ (Pythagorean Theorem)

Geometric Subtraction of Vectors

Subtraction of vectors is accomplished by adding the opposite of the vector to be subtracted.

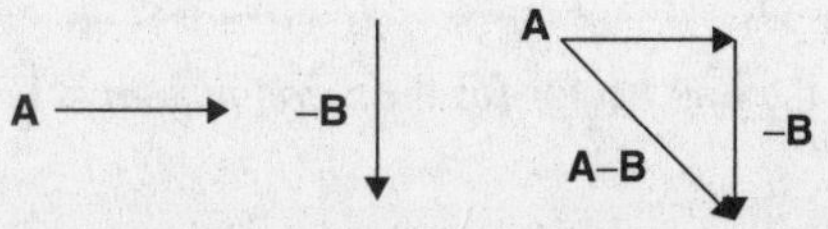

$$\mathbf{A} - \mathbf{B} = \mathbf{A} + (-\mathbf{B})$$

Example

Using the following three vectors, find:

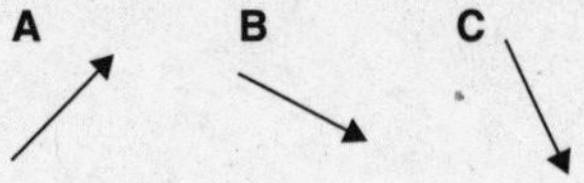

(a) A + B
(b) A − B
(c) A + B + C
(d) A + B − C

Solution

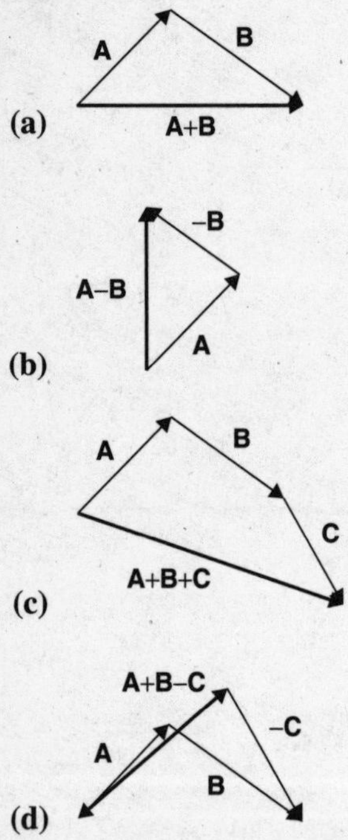

Simple Multiplication of Vectors

Doubling the size of vector **A**, for instance, results in it being labeled 2**A** and being twice as long as **A** and in the same direction. A negative sign indicates opposite direction. An antiparallel vector is an oppositely directed vector. For example, −2**A** is antiparallel to 2**A**.

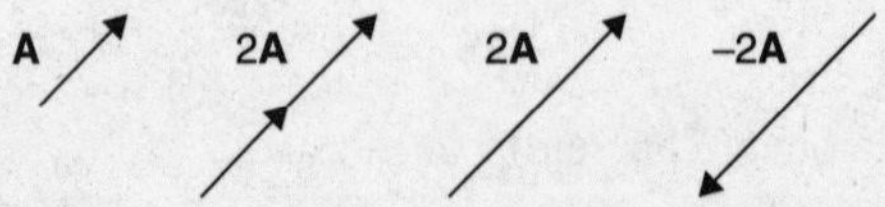

Vector Components in Two and Three Dimensions (C Exam)

In the following diagrams, the unit vector for the *x component* is **i**, for the *y component* is **j**, for the *z component* is **k**.

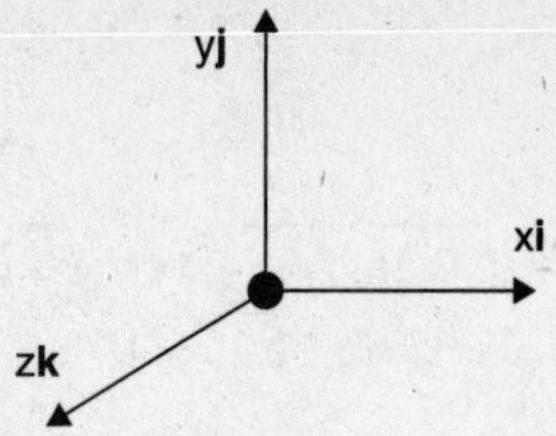

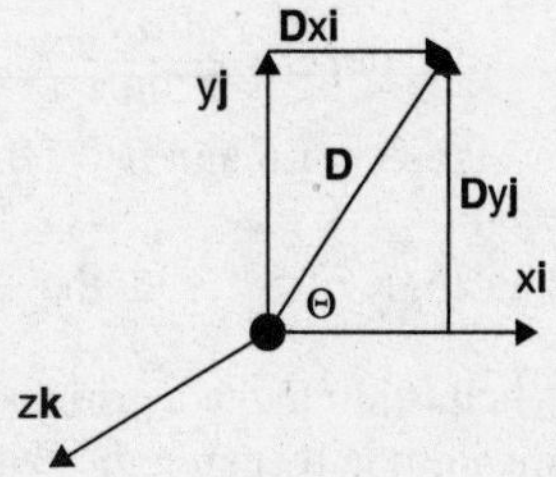

$\mathbf{D_x i}$ and $\mathbf{D_y j}$ are the components of vector **D**.

$$\mathbf{D} = \mathbf{D_x i} + \mathbf{D_y j}$$

The magnitude of **D** is calculated as follows:

$$\mathbf{D} = \sqrt{(\mathbf{D_x i})^2 + (\mathbf{D_y j})^2}$$

In order to find the angle the resultant vector **D** makes with the +x-axis, use the following equation:

$$\text{Tan } \Theta = \frac{\mathbf{D_y j}}{\mathbf{D_x i}}$$

For a three-dimensional problem, the magnitude of the resultant of the **i**, **j**, and **k** component vectors follows the "triple Pythagorean":

$$\mathbf{R}^2 = [(x\mathbf{i})^2 + (y\mathbf{j})^2 + (z\mathbf{k})^2]$$

The angles can be determined relative to the x-, y-, and z-axes.

Example

Determine the resultant vector **C** of two vectors acting on a common point.

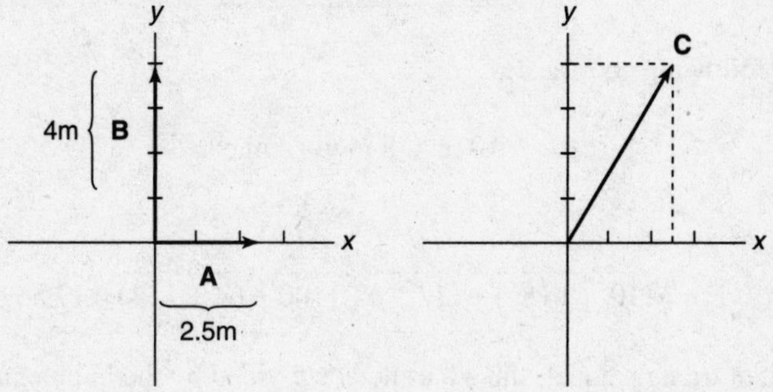

Vector **A** = 2.5 m x **i**

Vector **B** = 4 m y **j**

Solution

First, find the magnitude:

$$C = \sqrt{(2.5\text{m})^2 + (4\text{m})^2} = \sqrt{6.25 + 16}$$
$$C = \sqrt{22.25} = 4.7 \text{ m}$$

Next, find the angle that **C** makes with the *x*-axis:

$$\tan \Theta = \frac{4\text{m y}}{2.5\text{m x}}$$
$$\tan \Theta = 1.6 \text{ and } \Theta = 58°$$

The resultant vector **C** is 4.7 m long and makes an angle of 58° with the +x-axis.

When working with more than two vectors, it is helpful to make a grid and add up the x components and the y components. Then you can find the resultant vector's magnitude by using the Pythagorean Theorem and the vector's angle is found by using trigonometry.

Example

Find the resultant vector **E** of four vectors (with coordinates given from the origin):

A = (3 m x, 2 m y, 5 m z)
B = (1 m x, 4 m y)
C = (2 m y, 3 m z)
D = (6 m x, 4 m z)

Solution

First make a grid:

	X	Y	Z
A	3	2	5
B	1	4	0
C	0	2	3
D	6	0	4
SUM:	10	8	12

The resultant vector **E** has the following coordinates:

10 m x, 8 m y, 12 m z

E has a magnitude:

$$E = \sqrt{(10^2) + (8^2) + (12^2)} = \sqrt{(100 + 64 + 144)} = 17.55 \text{ m}$$

The cosine of any angle in a right triangle equals the adjacent side divided by the hypotenuse, and the hypotenuse is the magnitude of the resultant vector (17.55 m). Therefore, dividing 17.55 into the *x*, *y*, or *z* components given in the SUM row yields the cosines of the angles α, β, or γ, respectively.

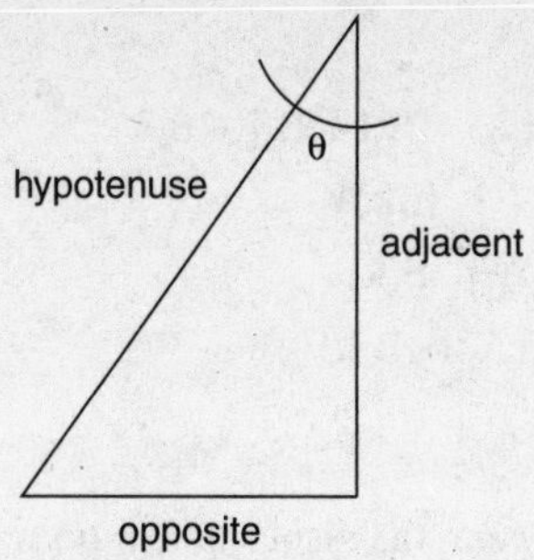

$$\cos \alpha = \frac{10}{17.55} = 0.5698 \text{ and } \alpha = 55.3°$$

$$\cos \beta = \frac{8}{17.55} = 0.4558 \text{ and } \beta = 62.9°$$

$$\cos \gamma = \frac{12}{17.55} = .06838 \text{ and } \gamma = 46.9°$$

Resultant vector **E** has a magnitude of 17.55 m and makes an angle of 55.3° with the x-axis, 62.9° wih the y-axis, and 46.9° with the z-axis.

Vector Multiplication: Dot Product and Cross Product (C Exam)

Dot Product (·)

The **dot product** or **scalar product** yields a scalar by multiplying two vectors. This is useful in physics when computing *work, electric flux,* and *magnetic flux.*

In computing the **dot (·) product**, we multiply the magnitudes of two vectors by the **cosine** of the angle that separates them:

$$\mathbf{A} \cdot \mathbf{B} = \mathbf{AB} \cos \Theta$$

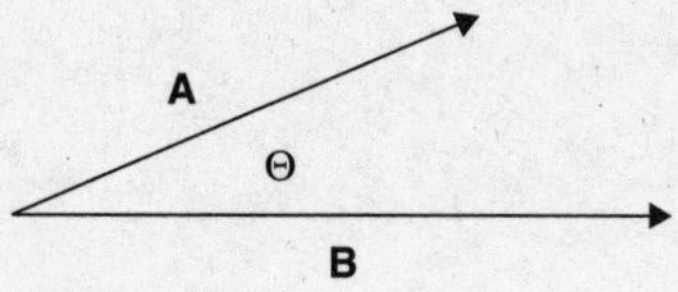

Dot Products and Unit Vectors

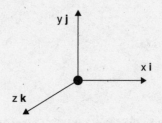

$\mathbf{i} \cdot \mathbf{i} = (\mathbf{i})(\mathbf{i}) \cos 0° = 1$

$\mathbf{j} \cdot \mathbf{j} = (\mathbf{j})(\mathbf{j}) \cos 0° = 1$

$\mathbf{k} \cdot \mathbf{k} = (\mathbf{k})(\mathbf{k}) \cos 0° = 1$

$\mathbf{i} \cdot \mathbf{j} = (\mathbf{i})(\mathbf{j}) \cos 90° = 0$

$\mathbf{i} \cdot \mathbf{k} = (\mathbf{i})(\mathbf{k}) \cos 90° = 0$

$\mathbf{j} \cdot \mathbf{k} = (\mathbf{j})(\mathbf{k}) \cos 90° = 0$

When in Unit Vector Notation

$$\mathbf{A} \cdot \mathbf{B} = \mathbf{AB} \cos \Theta = (A_x\mathbf{i} + A_y\mathbf{j} + A_z\mathbf{k}) \cdot (B_x\mathbf{i} + B_y\mathbf{j} + B_z\mathbf{k})$$
$$= (A_x\mathbf{i} \cdot B_x\mathbf{i}) + (A_x\mathbf{i} \cdot B_y\mathbf{j}) + (A_x\mathbf{i} \cdot B_z\mathbf{k}) +$$
$$(A_y\mathbf{j} \cdot B_x\mathbf{i}) + (A_y\mathbf{j} \cdot B_y\mathbf{j}) + (A_y\mathbf{j} \cdot B_z\mathbf{k}) +$$
$$(A_z\mathbf{k} \cdot B_x\mathbf{i}) + (A_z\mathbf{k} \cdot B_y\mathbf{j}) + (A_z\mathbf{k} \cdot B_z\mathbf{k})$$
$$= 1 + 0 + 0 + 0 + 1 + 0 + 0 + 0 + 1 = 3$$

Work, for instance, is the product of **F** and **D**, or *force* times *displacement*, the displacement occurring in the same direction as the applied force. If vector **F** represents the applied force and vector **D** represents the displacement, the *work* done by the force on the object that has moved a distance **D** is equal to the *part of the applied force* that actually moves the object in the direction of **D**.

$$W = \mathbf{F} \cdot \mathbf{D} = \mathbf{FD} \cos \Theta$$

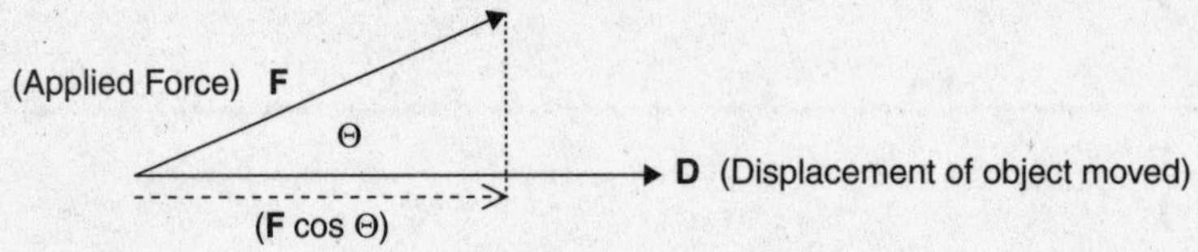

The dotted line, **F** cos Θ, is the part of the applied force that acts in the direction of motion.

Example

A force of 25 N pulls a 10 kg box a distance of 5 meters along a level floor. Compute the work done by the force if the angle between the force and the floor is

(a) 0°
(b) 30°
(c) 60°
(d) 90°

For each case, assume a constant speed.

Solution

Since work done is $W = \mathbf{F} \cdot \mathbf{D} = \mathbf{FD} \cos \Theta$,

(a) (25 N)(5 m)(cos 0°) = 125 Nm or **125 J**
(b) (25 N)(5 m)(cos 30°) = **108.3 J**
(c) (25 N)(5 m)(cos 60°) = **62.5 J**
(d) (25 N)(5 m)(cos 90°) = **0**

It is important to note that since no part of the applied force at 90° is actually in the direction of motion, *no work is done* by that force.

Cross Product (×)

The **cross product** or **vector product** yields a vector by multiplying two vectors. One vector is multiplied by the **component** of the second vector that is **perpendicular** to the first vector. The resultant vector is **perpendicular to both** of the vectors multiplied. This is useful in physics when computing *torque*, *angular momentum*, and *magnetic force*.

In computing the cross (×) product, we multiply two vectors by the sine of the angle that separates them, as follows:

$$\mathbf{A} \times \mathbf{B} = \mathbf{AB} \sin \Theta$$

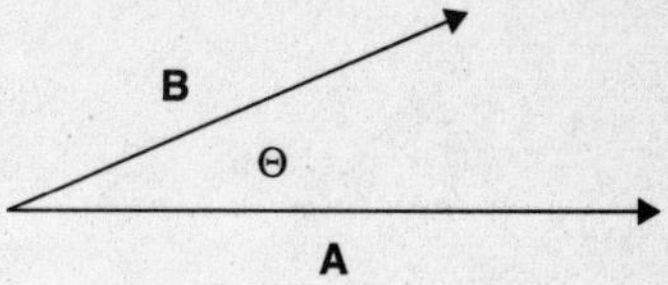

The computation $\mathbf{A} \times \mathbf{B} = \mathbf{AB} \sin\Theta$ is made with the two vectors placed tail to tail.

It is easier to visualize the cross product by placing the vectors head to tail: *torque*, for instance, is $\mathbf{r} \times \mathbf{F}$, or radius times the applied force.

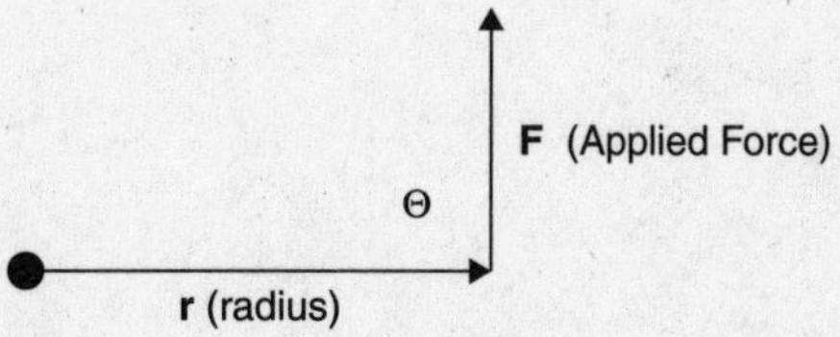

Let **r** be the radius of a wheel. If the applied force **F** is perpendicular to **r,** then the **maximum torque** is applied and the wheel will turn in the **counterclockwise** direction. If, however, the applied force is *not* perpendicular to the radius, then a torque is still applied, but it is less than maximum. Forces applied at 0° or 180° produce zero torque.

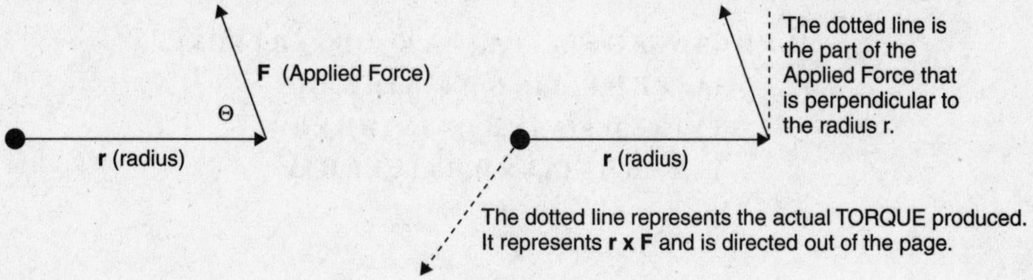

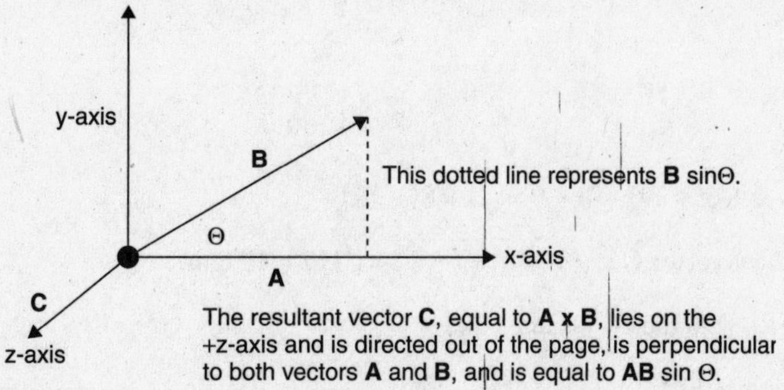

When obtaining the cross product, it is helpful to apply the **right hand screw rule**: Extend your *right* hand. It is the **A** vector. Pivot your right hand at the wrist and "slap" your open palm toward the second vector **B**. Your thumb points in the direction of the resultant vector **C** and is perpendicular to both vectors **A** and **B**.

Cross Products and Unit Vectors

Use the Right Hand Screw Rule: Two unit vectors in the same direction have nowhere to cross to and yield a zero resultant cross product. Two unit vectors at 90° yield a complete cross product resultant that is perpendicular to both.

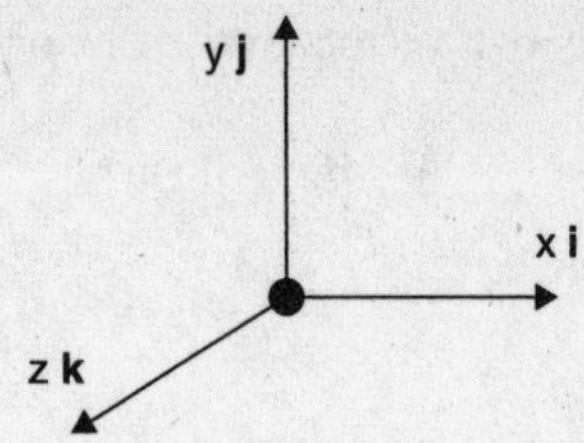

$\mathbf{i} \times \mathbf{i} = (\mathbf{i})(\mathbf{i}) \sin 0° = 0$

$\mathbf{j} \times \mathbf{j} = (\mathbf{j})(\mathbf{j}) \sin 0° = 0$

$\mathbf{k} \times \mathbf{k} = (\mathbf{k})(\mathbf{k}) \sin 0° = 0$

$\mathbf{i} \times \mathbf{j} = (\mathbf{i})(\mathbf{j}) \sin 90° = (1)(\mathbf{k}) = \mathbf{k}$

$\mathbf{i} \times \mathbf{k} = (\mathbf{i})(\mathbf{k}) \sin 90° = (-1)(\mathbf{j}) = -\mathbf{j}$

$\mathbf{j} \times \mathbf{i} = (\mathbf{j})(\mathbf{i}) \sin 90° = (-1)(\mathbf{k}) = -\mathbf{k}$

$\mathbf{j} \times \mathbf{k} = (\mathbf{j})(\mathbf{k}) \sin 90° = (1)(\mathbf{i}) = \mathbf{i}$

$\mathbf{k} \times \mathbf{i} = (\mathbf{k})(\mathbf{i}) \sin 90° = (1)(\mathbf{j}) = \mathbf{j}$

$\mathbf{k} \times \mathbf{j} = (\mathbf{k})(\mathbf{j}) \sin 90° = (-1)(\mathbf{i}) = -\mathbf{i}$

$\mathbf{A} \times \mathbf{B} = -\mathbf{B} \times \mathbf{A}$, therefore $\mathbf{i} \times \mathbf{j} = -\mathbf{j} \times \mathbf{i}$, and so on

When in Unit Vector Notation

$$\begin{aligned}\mathbf{A} \times \mathbf{B} = AB \sin \Theta &= (A_x\mathbf{i} + A_y\mathbf{j} + A_z\mathbf{k}) \times (B_x\mathbf{i} + B_y\mathbf{j} + B_z\mathbf{k}) \\ &= (A_x\mathbf{i} \times B_x\mathbf{i}) + (A_x\mathbf{i} \times B_y\mathbf{j}) + (A_x\mathbf{i} \times B_z\mathbf{k}) + \\ &\quad (A_y\mathbf{j} \times B_x\mathbf{i}) + (A_y\mathbf{j} \times B_y\mathbf{j}) + (A_y\mathbf{j} \times B_z\mathbf{k}) + \\ &\quad (A_z\mathbf{k} \times B_x\mathbf{i}) + (A_z\mathbf{k} \times B_y\mathbf{j}) + (A_z\mathbf{k} \times B_z\mathbf{k})\end{aligned}$$

Example

Vector **A** is (2 m **i**, 3 m **j**, 4 m **k**), and Vector **B** is (3 m **i**, 1 m **j**, 2 m **k**).

Find the resultant of $\mathbf{A} \times \mathbf{B}$.

Solution

$\mathbf{A} \times \mathbf{B} = 0 + 2\mathbf{k} - 4\mathbf{j} - 9\mathbf{k} + 0 + 6\mathbf{i} + 12\mathbf{j} - 4\mathbf{i} + 0 = (2\mathbf{i} + 8\mathbf{j} - 7\mathbf{k})$

The magnitude of the resultant vector **C** is $\sqrt{2^2 + 8^2 + (-7)^2} = \sqrt{117} = 10.8$ m.

It makes the following angles with the x-, y-, and z-axes:

$\cos \alpha = \frac{2}{10.8} = 0.185; \alpha = 79.3°$

$\cos \beta = \frac{8}{10.8} = 0.741; \beta = 42.2°$

$\cos \gamma = \frac{-7}{10.8} = -0.648; \gamma = -130.4°$

Review Questions and Answers

B & C Exam Question Types

Multiple Choice

1. This vector cannot be a resultant of

(A) 4m E and 2m W.
(B) 4m W, 2m E, 4m E, 2m NE, 2m SW.
(C) 5m W, 3m S, 7m E, 2m N, 1m N.
(D) 5m NW, 5m SE, 10m W, 12m E.
(E) 6m N, 5m S, 4m E, 2m W.

2. A displacement vector is a

(A) change in position.
(B) velocity.
(C) scalar.
(D) distance without direction.
(E) dimensionless quantity.

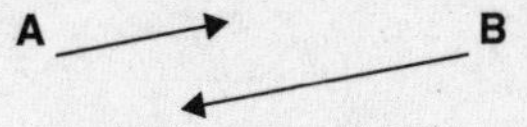

3. Vectors **A** and **B** cannot be

(A) antiparallel.
(B) opposite in magnitude.
(C) coplanar with each other.
(D) multiples of one another.
(E) resultant vectors of other vectors.

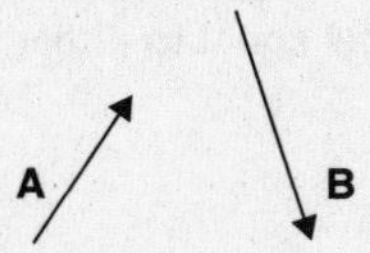

4. The resultant of **A** – **B** is most nearly:

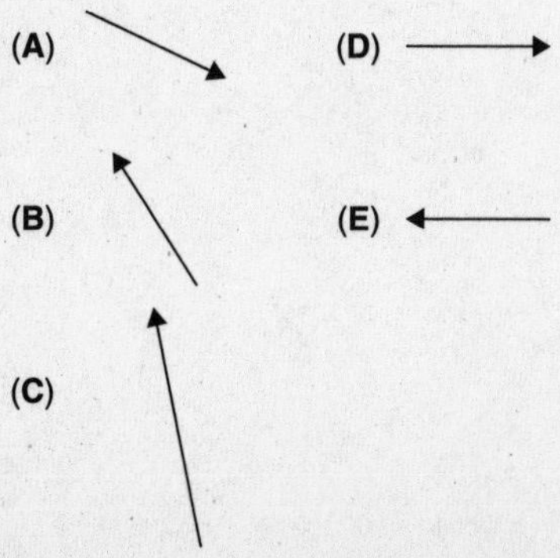

5. Vectors **A** and **B**, at right angles to each other,

(A) cannot be component vectors.
(B) can be antiparallel.
(C) cannot form a third vector.
(D) can be part of a three-dimensional system.
(E) can have no components.

6. Vector **A**, 5m long, and vector **B**, 19m long, cannot have a resultant vector of

(A) 14 m.
(B) 14.5 m.
(C) 18 m.
(D) 20 m.
(E) 25 m.

7. A vector in three-dimensional space

(A) has no coordinates.
(B) is directionless.
(C) is dimensionless.
(D) may possess only 2 coordinates.
(E) cannot be the resultant of 4 component vectors.

8. Of two vectors placed head to tail, which statement is false?

(A) They can be added.
(B) They can be components of another vector.
(C) They can produce a third vector that is longer than either one.
(D) They can produce a third vector that is shorter than either one.
(E) They cannot produce a third vector that is equal to either one in magnitude.

C Exam Question Types

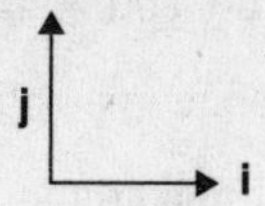

9. Vectors **i** and **j**

(A) can have a dot product equal to 1.
(B) can have a dot product equal to 0.
(C) can have a dot product **k**.
(D) can have a negative dot product.
(E) cannot be coplanar.

10. Vectors **i** and **j**

(A) can have a cross product with a magnitude equal to 1.
(B) can have a cross product equal to 0.
(C) cannot have a cross product +**k**.
(D) can have a cross product –**j**.
(E) can both lie in the same plane as their cross product.

Answers to Multiple-Choice Questions

1. **(E)** It is the only answer whose components do not add up to 2m East.
2. **(A)** By the definition of displacement, it is a change in position.
3. **(B)** Vectors may be opposite in direction but not magnitude.
4. **(C)** The simple vector sum of **A** + (–**B**) yields choice **C**.
5. **(D)** Any two vectors at right angles may be oriented in an infinite number of planes and may indeed be part of a three-dimensional system.
6. **(E)** The longest possible resultant is 24 m, the shortest possible is 14 m.
7. **(D)** Yes, it can lie in the x-y, y-z, or x-z planes.
8. **(E)** They may form an equilateral triangle.
9. **(B)** Since $\mathbf{i} \cdot \mathbf{j} = \mathbf{ij} \cos 90° = 0$ and $\mathbf{j} \cdot \mathbf{i} = 0$
10. **(A)** Since $\mathbf{i} \times \mathbf{j} = \mathbf{ij} \sin 90° = (\mathbf{i})(\mathbf{j})(1)$

Free Response

B & C Exam Question Type

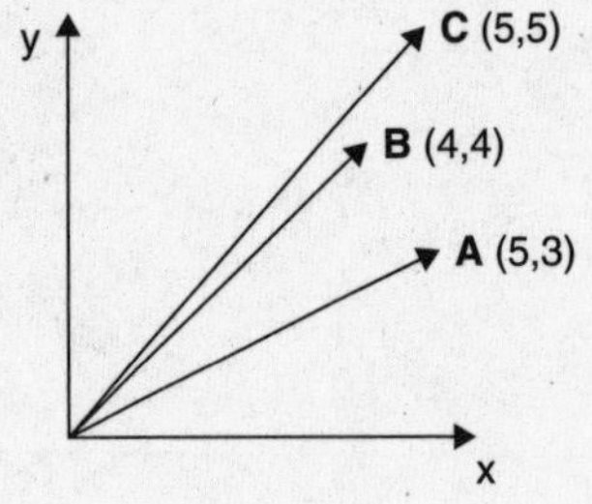

1. Given three vectors, all from the origin, **A** (5m, 3m), **B** (4m, 4m), and **C** (5m, 5m):

   (a) Find the magnitude of their resultant vector.

   (b) Find the angle their resultant vector makes with the y-axis.

   (c) Repeat (a) and (b) if **A** is made twice as long.

   (d) If a 4th vector **D** (2m, 2m) is added to the original three, repeat (a) and (b).

   (e) If the original three vectors are replaced by their opposites, repeat (a), (b), and (d).

C Exam Question Type

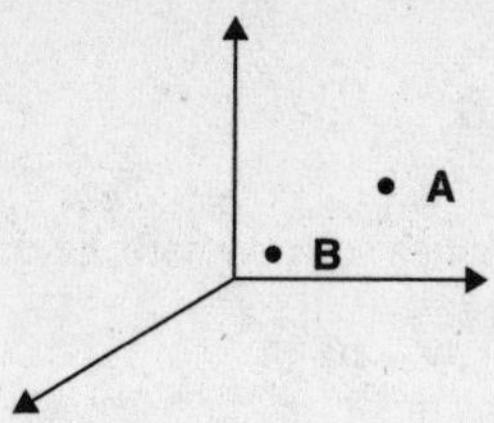

2. Given vectors from the origin **A** (5, 3, 2) and **B** (3, 3, 5):

(a) Find their resultant vector's magnitude.
(b) Find the angles the resultant vector makes with the x-axis and the y-axis.
(c) Find the dot product of the two vectors.
(d) Find the magnitude of the cross product of the two vectors.
(e) Find a third vector **C** that, when added to **A** and **B**, gives a zero resultant.

Answers to Free-Response Questions

1. **(a)** The magnitude is the Pythagorean:

$$\left[\left(\overset{\Sigma x}{5+4+5}\right)^2+\left(\overset{\Sigma y}{3+4+5}\right)^2\right]^{1/2}=\mathbf{18.4\ m}$$

(This is the hypotenuse of the right triangles formed with Σx and Σy.)

(b) $\cos\Theta=\dfrac{\text{adj}}{\text{hyp}}=\dfrac{\Sigma y}{18.4}=\dfrac{12}{18.4}=0.652$ and $\Theta=\mathbf{49.3°}$.

(c) Since **A** is now (10, 6), the new Pythagorean is $\sqrt{19^2+15^2}=24.2$ m.

For the angle: $\cos\Theta=\dfrac{\text{adj}}{\text{hyp}}=\dfrac{\Sigma y}{24.2}=\dfrac{15}{24.2}=0.6198$ and $\Theta=51.7°$

The new vector is 24.2 m long and makes an angle of 51.7° with the y-axis.

(d) $\Sigma x=5+4+5+2=16$; $\Sigma y=3+4+5+2=14$

The hypotenuse is the Pythagorean, $\sqrt{(16)^2+(14)^2}=21.3$ m.

For the angle: $\cos\Theta=\dfrac{\text{adj}}{\text{hyp}}=\dfrac{\Sigma y}{21.3}=\dfrac{14}{21.3}=0.6573$

This gives an angle of 48.9°.

The new vector is 21.3 m long and makes an angle of 48.9° with the y-axis.

(e) –**A** = (–5, –3), –**B** = (–4, –4), –**C** = (–5, –5)

$\Sigma x=-5-4-5=-14$, $\Sigma y=-3-4-5=-12$.

i. Since the Pythagorean involves squaring negatives, the magnitude is still **18.4 m**.

ii. $\cos\Theta\ \dfrac{\Sigma y}{18.4}=\dfrac{-12}{18.4}=-0.652$ and $\Theta=\mathbf{131°}$

iii. a. Resultant vector: Magnitude $=\sqrt{(-5-4-5+2)^2+(-3-4-5+2)^2}=\mathbf{15.6\ m}$

b. Angle with y-axis: $\cos\Theta=\dfrac{\Sigma y}{15.6}=\dfrac{-10}{15.6}=-0.6410$ $\Theta=\mathbf{130°}$.

2. **(a)** $\left[\left(\overset{\Sigma x}{5+3}\right)^2+\left(\overset{\Sigma y}{3+3}\right)^2+\left(\overset{\Sigma z}{2+5}\right)^2\right]^{1/2}=\mathbf{12.2\ m}$

(b) $\cos\alpha=\dfrac{\Sigma x}{12.2}=\dfrac{8}{12.2}=0.6557$ and $\alpha=\mathbf{49.0°}$

$\cos\beta=\dfrac{\Sigma y}{12.2}=\dfrac{6}{12.2}=0.4918$ and $\beta=\mathbf{60.5°}$

(c) $\mathbf{A}\cdot\mathbf{B}=\mathbf{AB}\cos\Theta=(\mathbf{A_xi}+\mathbf{A_yj}+\mathbf{A_zk})\cdot(\mathbf{B_xi}+\mathbf{B_yj}+\mathbf{B_zk})=(5\mathbf{i}+3\mathbf{j}+2\mathbf{k})\cdot(3\mathbf{i}+3\mathbf{j}+5\mathbf{k})$

$=15\cos 0°+15\cos 90°+25\cos 90°$

$+9\cos 90°+9\cos 0°+15\cos 90°$

$+6\cos 90°+6\cos 90°+10\cos 0°=15+0+0+0+9+0+0+0+10=\mathbf{34m}$

(d) $\mathbf{A}\times\mathbf{B}=\mathbf{AB}\sin\Theta=(\mathbf{A_xi}+\mathbf{A_yj}+\mathbf{A_zk})\times(\mathbf{B_xi}+\mathbf{B_yj}+\mathbf{B_zk})=(5\mathbf{i}+3\mathbf{j}+2\mathbf{k})\times(3\mathbf{i}+3\mathbf{j}+5\mathbf{k})$

$=(0+15\mathbf{k}-25\mathbf{j}-9\mathbf{k}+0+15\mathbf{i}+6\mathbf{j}-6\mathbf{i}+0)\text{m}=(9\mathbf{i}-19\mathbf{j}+6\mathbf{k})\text{m}=\mathbf{21.9m}$

(e) This means that **C** = –(**A** + **B**). If (**A** + **B**) = (5, 3, 2) + (3, 3, 5) = (8m, 6m, 7m), then the opposite or antiparallel vector is **(–8m, –6m, –7m)**.

Mechanics: Forces

Definition

A force is generally defined as a push or a pull. Because pushes and pulls are directed, all forces are vectors. A *net force* is the single force that represents the sum of all forces acting on an object. Hence, if a net force exists and acts on an object, by Newton's Second Law, that object will be accelerated.

For the C exam, the identity $\mathbf{F(x)} = -\frac{dU(\mathbf{x})}{dx}$ unites the concepts of *force*, *displacement*, and *potential energy*.

Example 1 (B & C Exams)

A force of 10 N East and another force of 5 N West act on an object. If both forces are constant, what is the net force?

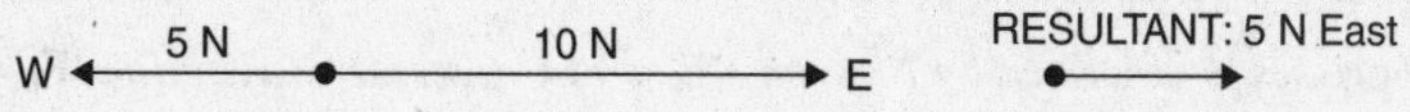

Solution

The *net force* is **5 N East**.

Example 2 (B & C Exams)

A force of 10 N East and another force of 5 N North act on an object. If both forces are constant, what is the net force?

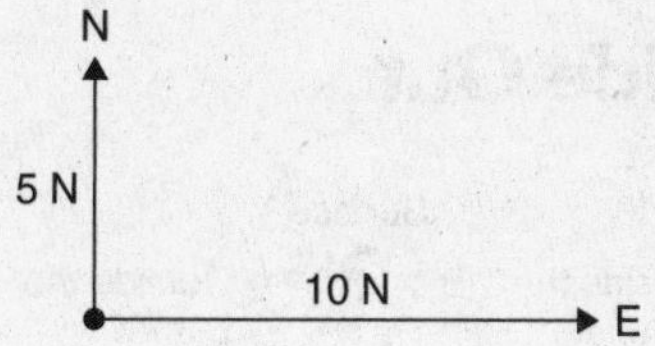

Solution

The net force is $\mathbf{F}^2 = (10\text{N})^2 + (5\text{N})^2$. The **magnitude** of the *resultant vector* is $\sqrt{100+25}$ or **11.2 N**. Because 11.2 N is a vector, the direction is also necessary and can be computed by designating 5 N North as the *y*-component of a right triangle and 10 N East as the *x*-component of the triangle. Using trigonometry, the tangent of the resultant angle = $^5/_{10}$, making the resultant angle **26.6° North of East**.

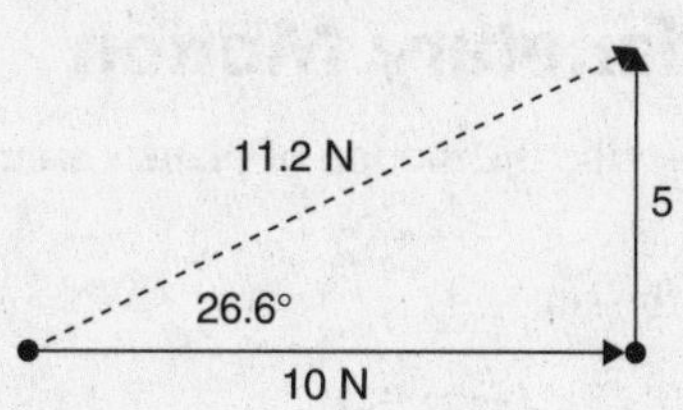

Weight and Mass

Mass is defined as the amount of material comprising an object. *Weight* is defined as the effect of the local gravitational acceleration on that object.

Example (B & C Exams)

A mass of 10 grams has what weight?

Solution

Because weight is a constant force and the mass is given in metric units, the metric unit of weight is the Newton, which is measured in kg m/s^2. Therefore, change the grams to kg and multiply by **g** to get the object's weight: $\mathbf{F_{wt}} = \text{mg} =$ (0.01 kg)(9.8 m/s^2) = **0.098 N**.

Gravitation

The interaction of any two masses is described by *Newton's Law of Universal Gravitation*: $\mathbf{F_G} = \frac{-Gm_1 m_2}{\mathbf{r}^2}$, where $G = 6.67 \times 10^{-11} \frac{Nm^2}{kg^2}$, and the negative sign indicates an attractive force. The two masses, m_1 and m_2, are separated, center-to-center, by distance *r*. The gravitational acceleration near a mass can be represented by $\mathbf{a} = -G\frac{m}{\mathbf{r}^2}$. [Note the application of Newton's Second Law: $\mathbf{F_G} = m\mathbf{a} = (m_1)(\frac{Gm_2}{\mathbf{r}^2})$.] *Gravitational Potential Energy* is represented through the Work-Energy Theorem ($W = F \cdot d = \Delta E$) and is applied as $E_G = |\mathbf{Fd}| = |(\frac{-Gm_1 m_2}{\mathbf{r}^2})\,(\mathbf{r})| = \frac{Gm_1 m_2}{\mathbf{r}}$.

If work is done on or by a mass in a gravitational field, the work may be positive or negative depending on direction — the absolute value of work done is used to signify a certain amount of energy involved in the process.

Gravitation from the Inside Out

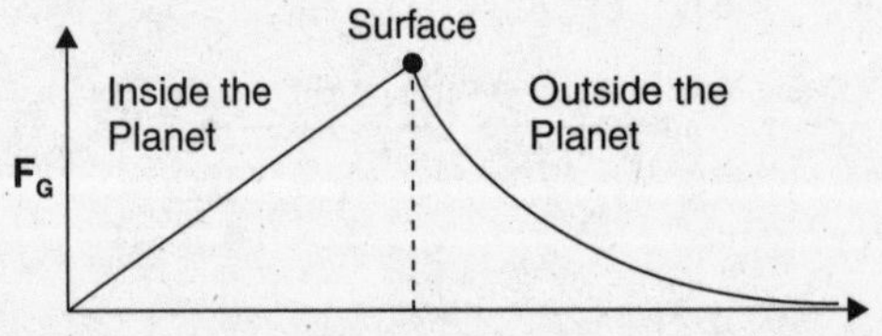

The force of gravitation at the center of the earth or any planet is zero, as all force vectors cancel. The gravitational force experiences a direct linear relationship between the center and the surface. Traveling away from the surface, the inverse-square law takes effect.

Kepler's Three Laws of Planetary Motion

Johannes Kepler developed three laws that describe the motion of planets around the sun:

1. Planets travel in elliptical paths around the sun.

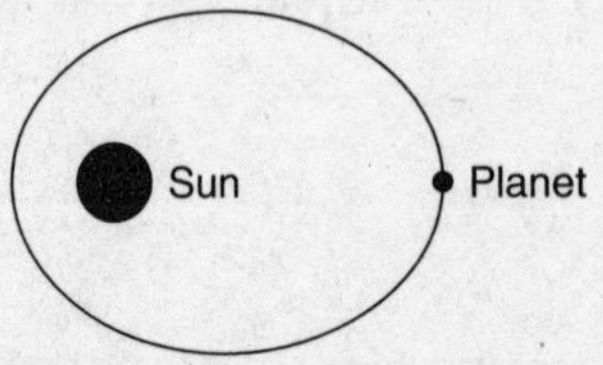

2. Orbiting planets sweep out equal areas in equal amounts of time. Area 1 = Area 2 when $T_1 = T_2$.

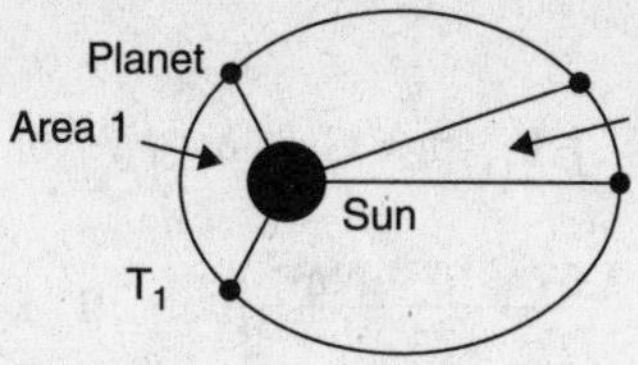

3. $\frac{T^2}{R^3}$ is the same for all planets, where T is the period of rotation and R is the mean distance from the sun.

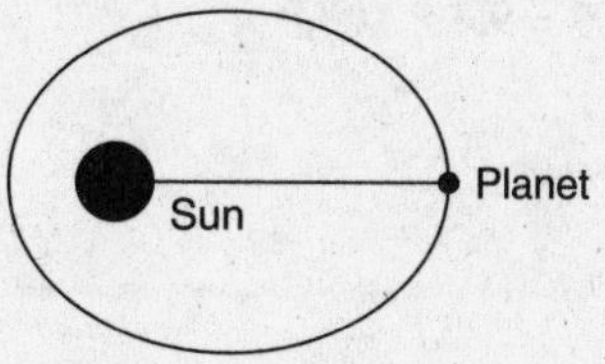

Example 1 (B & C Exams)

What is the gravitational force of attraction between two trucks, each of mass 10,000 kg, separated by a distance of 20 m?

Solution

Using $\mathbf{F_G} = \frac{-\mathbf{Gm_1 m_2}}{\mathbf{r^2}}$, $\mathbf{F_G} = \frac{\left(\frac{-6.67 \times 10^{-11}\,\text{Nm}^2}{\text{kg}^2}\right)\left(10^4\,\text{kg}\right)\left(10^4\,\text{kg}\right)}{(20\text{m})^2} = \mathbf{-1.67 \times 10^{-5}\,N}$

Example 2 (B & C Exams)

Two planets, separated by a distance *r*, exert a certain gravitational force on one another. What happens to that force if

(a) the mass of both planets is doubled?
(b) their distance of separation is doubled?
(c) both their masses are doubled as well as their distance of separation?

Solution

(a) The new force: $\mathbf{F} = \frac{\text{G2m}_1\,\text{2m}_2}{\mathbf{r^2}}$ = **4 times the original force**

(b) The new force: $\mathbf{F} = \frac{\text{Gm}_1\,\text{m}_2}{(\mathbf{2r})^2} = \frac{1}{4}$ **the original force**

(c) The new force: $\mathbf{F} = \frac{\text{G2m}_1\,\text{2m}_2}{(\mathbf{2r})^2}$ = **Equal to the original force**

Example 3 (C Exam)

How much work is done launching a satellite of mass 1000 kg into an orbit above the Earth, at an elevation equal to the radius of the Earth? (Earth's mass = 6×10^{24} kg, Earth's radius = 6.4×10^6 m.)

Assume that all the work applied to the satellite is transformed into potential energy.

Solution

Since work is $\mathbf{F} \cdot \mathbf{d}$,

$$\mathbf{F} \cdot \mathbf{d} = \int_{r_1}^{r_2} \mathbf{F}_{(r)}\, dr = Gm_1 m_2 \int_{r_1}^{r_2} r^{-2}\, dr$$

$$= \left(\frac{6.67 \times 10^{-11}\, Nm^2}{kg^2}\right)\left(6 \times 10^{24}\, kg\right)\left(10^3\, kg\right)\left[-r^{-1}\right]_{r_1}^{r_2}$$

$$\frac{-40 \times 10^{16}\, Nm^2}{12.8 \times 10^6\, m} - \frac{-40 \times 10^{16}\, Nm^2}{6.4 \times 10^6\, m}$$

$$= \mathbf{9.4 \times 10^{10} J}$$

Equilibrium

There are two types of equilibrium:

TRANSLATIONAL equilibrium exists when all the linear forces acting on an object cancel, leaving no net force to accelerate the object in a linear fashion.

ROTATIONAL equilibrium exists when all the turning forces (TORQUES) that act on an object are equal and opposite. Torque, the product of the spin radius and the part of the turning force that is perpendicular to the radius, equals *Fr* or $r \times F$. Rotational motion will be discussed further in the next chapter.

Example 1 (B & C Exams)

Linear forces of 5 N north and 5 N south act on a point. Describe the resultant force.

Solution

The resultant force equals zero. Because no net force exists, no linear acceleration exists, and the object is in *Translational* equilibrium.

Example 2 (B & C Exams)

Linear forces of 10 N north and 8 N SE act on an object. What is the equilibrant force?

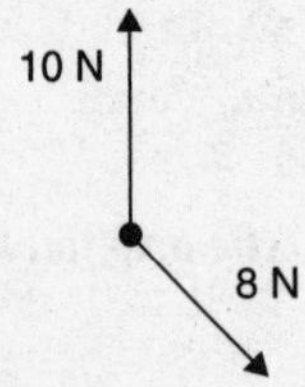

Solution

Make a grid; designate E as $+x$ and N as $+y$ and name the 10N force **A** and the 8N force **B**:

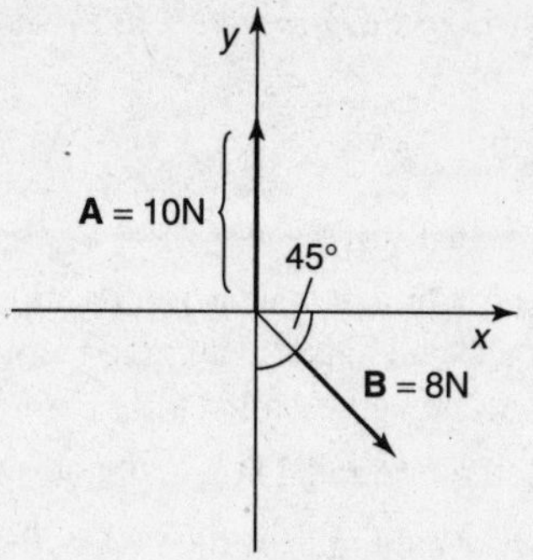

	x	y
A	0	10N
B	5.7N	–5.7N
Total	5.7N	4.3N

The magnitude of the resultant vector is

$$\sqrt{(5.7\text{N})^2 + (4.3\text{N})^2} = \sqrt{32 + 18} = 7.1\text{N}$$

The direction of the resultant vector is

$$\tan \alpha = \frac{4.3\text{N}}{5.7\text{N}} = 0.754,\ \alpha = 37°$$

The dotted line is the resultant vector and is 7.1N in magnitude and makes an angle of 37° with the x-axis, placing it 37° N of E.

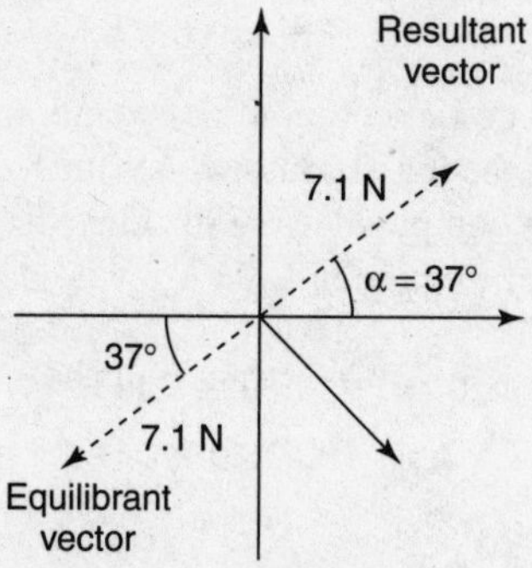

The Equilibrant vector has the magnitude of the resultant vector and the opposite direction: 7.1N at an angle of 37° S of W.

Example 3 (B & C Exams)

A wheel of radius 0.5 m and a mass of 10 kg sustains forces of 5 N (perpendicular to the radius in a counterclockwise direction) and 10 N (perpendicular to the radius in a clockwise direction). Describe the resulting motion of the wheel about its axle.

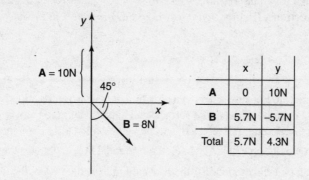

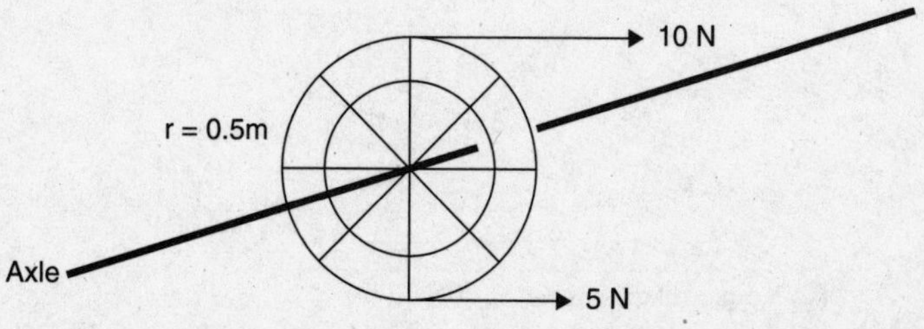

Solution

Two opposing torques act on the wheel, one being stronger than the other; therefore, there will be a net torque on the wheel, causing it to accelerate in a rotational fashion. A *net force* of 5 N *clockwise* will cause **the wheel to rotationally accelerate in the clockwise direction**. *(The rotational aspect of this problem will be discussed in the next chapter.)*

Friction

Friction is a force that opposes motion. Because it acts in a direction, it is a vector. Friction between two solid surfaces is a function of the relative roughness or smoothness of the surfaces in contact. The ratio is given as a decimal value, usually (but not always) less than 1. The Greek letter μ denotes this value and is called the *Coefficient of Friction* and equals the Frictional Force divided by the Normal Force, $\mu = \mathbf{F_f}/\mathbf{F_n}$. $\mathbf{F_f}$, the Frictional Force, denotes the *net force* of friction, which acts in the opposite direction of an object's motion. $\mathbf{F_N}$, the Normal Force, is the part of the object's weight that acts against the surface the object rests or slides on. *Static* friction, the resistance to starting an object moving along a surface, is stronger than *Sliding* or *Kinetic* friction, the resistance to a body already in motion along a surface. The coefficients of both types of friction usually are denoted by μ_S for static friction and μ_K for kinetic friction. (In fluids, the frictional force is as follows: $\mathbf{F} = \frac{1}{2}\mathrm{CA}\rho\mathbf{v}^2$ and depends on the *drag coefficient* C, the object's cross-sectional area A, the fluid density ρ, and the square of the object's velocity $\mathbf{v}^2$.) **(B Exam)**

Example 1 (B & C Exams)

Determine the frictional force acting on a 5 kg box that has an initial speed of 10 m/s and slides to a stop after sliding 15 m along a level floor.

Solution

This is a problem that involves Newton's Second Law of motion, $\mathbf{F} = m\mathbf{a}$. In this case, **F** is the net force (friction) that acts on the object to slow and stop it. Because the mass is known, we must compute the acceleration and multiply it by the mass to get the answer. The acceleration is computed using the facts that the initial speed is 10 m/s, the final speed is 0, and the distance it covers is 15 m. Using the equation $\mathbf{V_f^2 = V_o^2 + 2\,ad}$,

$$0^2 = (10\text{ m/s})^2 + 2\,\mathbf{a}(15\text{ m})$$

$$\mathbf{a} = -100/30\text{ m/s}^2 = -3.3\text{ m/s}^2$$

Completing the solution,

$$\mathbf{F} = (5\text{ kg})(-3.3\text{ m/s}^2) = \mathbf{-16.5\ N}$$

Example 2 (B & C Exams)

Determine the frictional force acting on a 5 kg box that has an initial speed of 10 m/s and slides to a stop after sliding 15 m down an incline, which makes an angle of 30° with the horizontal.

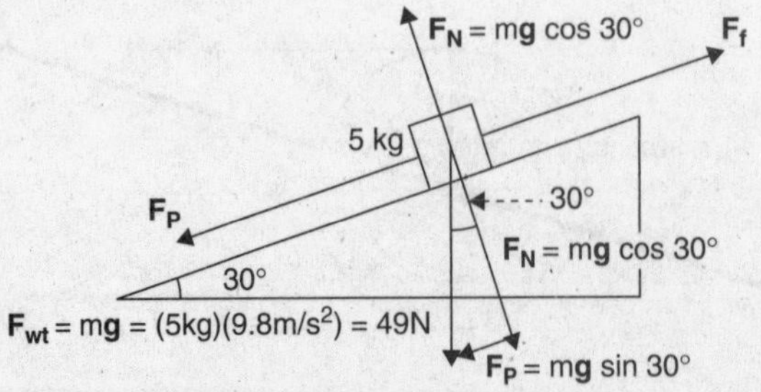

Solution

In this case, the *frictional* force acts against the *parallel* force. The resultant force of one against the other is the *net force* (in **F** = **ma**) that stops the box.

$$\mathbf{F_{net}} = \mathbf{F_p} - \mathbf{F_f} = \text{ma}$$

Here, the acceleration can be obtained by again using the equation

$$\mathbf{V_f^2 = V_o^2 + 2\ ad}$$

and the combination of the two equations will yield the solution. For the acceleration:

$$0^2 = (10\text{ m/s})^2 + 2\,\mathbf{a}\,(15\text{m})$$
$$\mathbf{a} = -100/30\text{ m/s}^2 = -3.3\text{ m/s}^2$$

For the $\mathbf{F_{net}} = \mathbf{F_p} - \mathbf{F_f}$, we need to calculate the $\mathbf{F_p}$. Using the fact that $\mathbf{F_p}$ = mg sin 30°,

$$\mathbf{F_p} = (5\text{ Kg})(9.8\text{ m/s}^2)(0.5) = 24.5\text{ N}$$

Substitutions give

$$24.5\text{N} - \mathbf{F_f} = (5\text{ kg})(-3.3\text{ m/s}^2)$$
$$\mathbf{F_f} = 24.5\text{ N} - (5\text{ kg})(-3.3\text{ m/s}^2) = \mathbf{41.0\ N}$$

Centripetal Force, Centripetal Acceleration

Centripetal force is a center-directed or center-seeking force. It is the force responsible for holding an object in a path of uniform circular motion.

Centripetal force is a vector whose direction is toward the center of the circle in which it acts.

$\mathbf{F_c} = \text{m}\mathbf{v}^2/\mathbf{r}$ also is a statement of Newton's Second Law, where the *centripetal acceleration*, $\mathbf{a_c} = \mathbf{v}^2/\mathbf{r}$.

Example 1 (B & C Exams)

Determine the centripetal force and centripetal acceleration of a 2 kg mass M, traveling with a linear speed of 5 m/s, being whirled in a horizontal circle on a string of length 1 m.

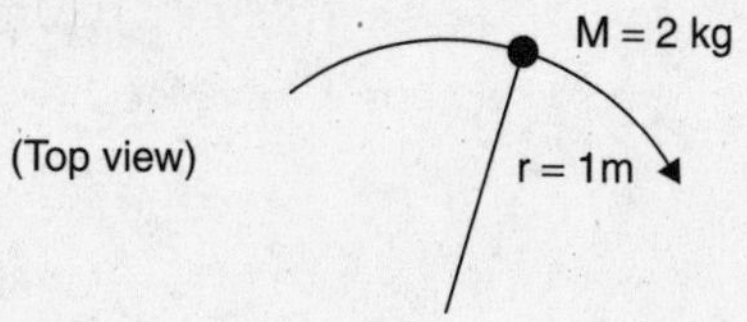

Solution

Since $\mathbf{F_c} = \text{m}\mathbf{v}^2/\mathbf{r} = (\text{m})(\mathbf{a_c}) = (\text{m})(\mathbf{v}^2/\mathbf{r})$, find the centripetal acceleration first and then multiply it by the mass:

$$\mathbf{a_c} = \mathbf{v^2}/\mathbf{r} = (5\text{m/s})^2/\mathbf{r} = 25\text{m}^2/\text{s}^2/1\text{ m} = \mathbf{25\ m/s^2}$$
$$\mathbf{F_c} = \text{m}\mathbf{v}^2/\mathbf{r} = (\text{m})(\mathbf{a_c}) = (2\text{ kg})(25\text{m/s}^2) = \mathbf{50\ N}$$

In problems involving gravitational forces holding planets or satellites in orbit, usually a circular orbit is specified and the gravitational force equals the centripetal force on the planet or satellite.

Example 2 (B & C Exams)

Determine the linear velocity **v** of a communications satellite a distance equal to the Earth's radius above the Earth.

$G = 6.67 \times 10^{-11}\ Nm^2/kg^2$

Earth's radius $= 6.37 \times 10^6$ m

Earth's mass $= 5.98 \times 10^{24}$ kg

Solution

The gravitational force on the satellite equals the centripetal force on the satellite.

$\mathbf{F_G} = Gm_1m_2/\mathbf{r}^2$

$\mathbf{F_C} = m\mathbf{v}^2/\mathbf{r}$

Setting them equal,

$$Gm_Em_S/\mathbf{r}^2 = m_S\mathbf{v}^2/\mathbf{r}$$

Canceling m_S and **r** gives $\mathbf{v} = \sqrt{\frac{Gm_E}{\mathbf{r_E}}}$ and substituting numerical values yields $\mathbf{5.59 \times 10^3}$ **m/s.**

In problems involving a car traveling over a circular-shaped hill and just staying on the hill at the very top, let the car's weight equal the centripetal force.

Example 3 (B & C Exams)

A cart of mass 50 kg rides over a semicircular bridge of radius 10 m. What is the maximum speed the cart can have without lifting off the bridge?

Solution

Set the cart's weight equal to the centripetal force on it. (The inertia of the cart away from the center of the bridge "equals" the centripetal force on it, which in turn equals the cart's weight.)

$$\mathbf{F_{WT}} = \mathbf{F_C}$$
$$mg = m\mathbf{v}^2/\mathbf{r}$$

Canceling the mass results in the following:

$$\mathbf{v} = \sqrt{\mathbf{gr}}$$
$$= \sqrt{(9.8\,\text{m/s}^2)(10\,\text{m})}$$
$$= \mathbf{9.9\ m/s}$$

Note that the mass is inconsequential.

Variable Forces: Elastic Forces and Springs

In considering variable forces, the most common are springs and variations of elastic forces, both being functions of the spring constant, k, and the stretch displacement, **x**. Since **F** = –**kx** (Hooke's Law: Where **F** is the restoring force, k is the spring constant, or springiness of the spring, **x** is the stretch displacement, and the negative sign signifies that the spring pushes back in the opposite direction), the work done *on* or *by* a spring is given by $\mathbf{W = \bar{F}d}$, where **W** is work in Joules, $\bar{\mathbf{F}}$ is the average force exerted on or by the spring, and **d** is displacement.

Variable Forces (C Exam)

(Work is covered in a later chapter, along with momentum, energy, and power.) For work done by a variable force, it is necessary to take the definite integral of the force over the extent of displacement. For a variable force **F**, which acts from point $\mathbf{x_1}$ to point $\mathbf{x_2}$, the work done would be as follows:

$W = \int_{x_1}^{x_2} \mathbf{F(x)\,dx_2}$. The graph of force **F** versus displacement **x** is given by the following:

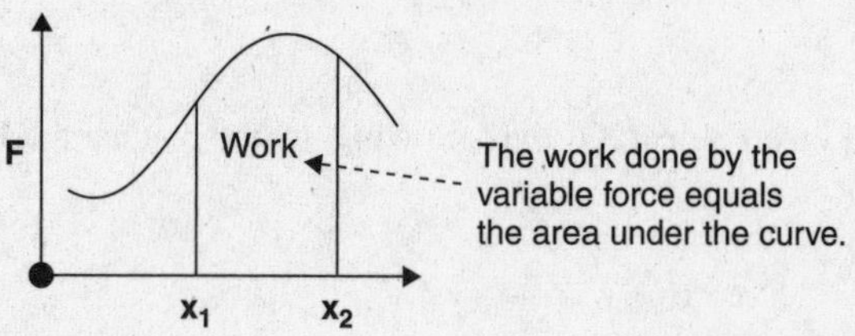

Example 1 (C Exam)

A certain spring of negligible mass has a spring constant k of 10 N/m. How much work is required to stretch it 20 cm from its rest position?

Solution

Use the following formula:

$$\begin{aligned} W &= \int_0^{.2} -kx\,dx \\ &= -(\tfrac{1}{2})k\, x^2 /_0^{.2} \\ &= (-\tfrac{1}{2})(10\text{N/m})(0.2\text{m})^2 \end{aligned}$$

Since the work is done by an applied force, the work is positive:

$$= \mathbf{0.2\ J}$$

Example 2 (B Exam)

A certain spring of negligible mass has a spring constant k of 10 N/m. How much work is required to stretch it 20 cm from its rest position?

Solution

Using Hooke's Law and the work definition, $\mathbf{W = \bar{F}d}$, **W** = The average spring force **(F/2)** times the distance stretched. Since the maximum spring force is **F = –kx**, the average force is **F/2 = –kx/2**. Work is done on the spring and is therefore positive:

$$(-kx/2)(x) = |(-10\ \text{N/m})(0.20\ \text{m}/2)(0.2\ \text{m})| = \mathbf{0.2\ J}$$

Review Questions and Answers

B & C Exam Question Types

Multiple Choice

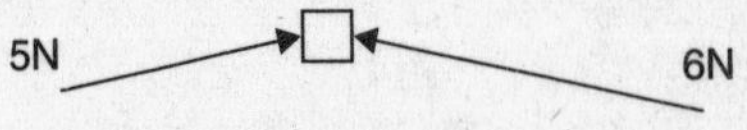

1. Two forces of 5 N and 6 N act on an object. Of the following, which one cannot be the resultant force?

(A) 1 N
(B) 4.5 N
(C) 9.8 N
(D) 11 N
(E) 13 N

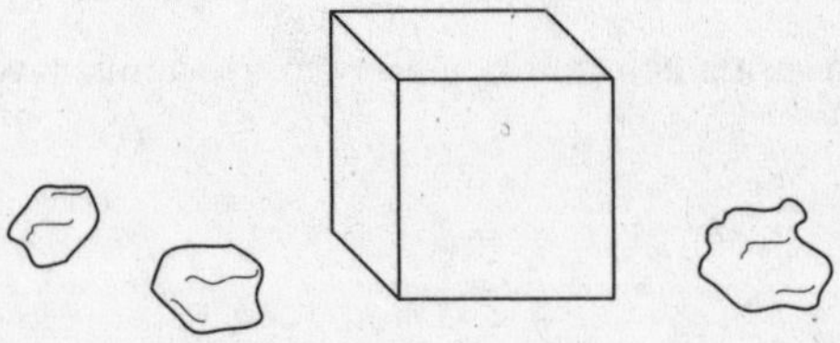

2. Three rock samples, 3.3 kg, 5.1 kg, and 6.6 kg, are put into a box by astronauts before leaving the moon for Earth. If the moon's gravitational acceleration is $\frac{1}{6}$ that of Earth, about how much weight will the rocks add to the return liftoff?

(A) 2.5 N
(B) 25 N
(C) 73.5 N
(D) 147 N
(E) 882 N

3. Four forces, 5 N north, 6 N 20° S of E, 9 N 20° N of W, and 10 N south, act on an object. The equilibrant force equals

(A) 4.9 N 35° W of S
(B) 2.8 N SW
(C) 2.8 N W
(D) 4.9 N 35° E of N
(E) 49 N 35° S of W

4. A certain planet X has a moon $\frac{1}{10}$ its mass with an orbit of average radius R and velocity **v**. If the moon's mass were doubled in the same orbit, its velocity would need to be

(A) v/2
(B) $\frac{\mathbf{v}}{\sqrt{2\mathbf{v}}}$
(C) $\mathbf{v}\sqrt{2}$
(D) $\sqrt{4\mathbf{v}}$
(E) 2v

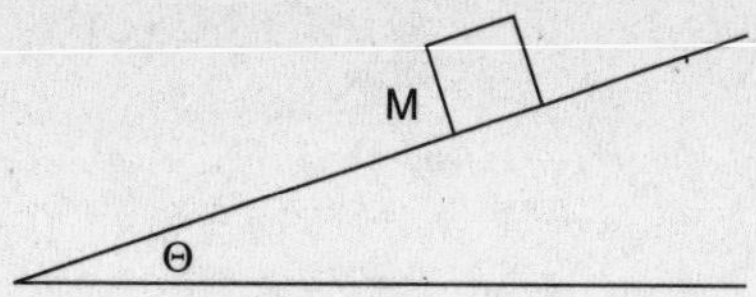

5. A block of mass M sits on a frictionless ramp, which makes an angle Θ with the ground. The block's acceleration is

(A) g /2
(B) g sin Θ
(C) g tan Θ
(D) g cos Θ
(E) g

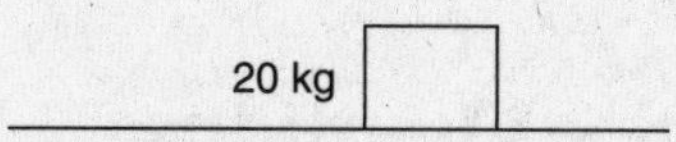

6. A 20 kg box is to be pushed along the floor at constant speed. If the force of friction between the box and the floor is 25 N, the coefficient of friction between their surfaces is

(A) 0.125
(B) 0.250
(C) 0.350
(D) 0.450
(E) 0.700

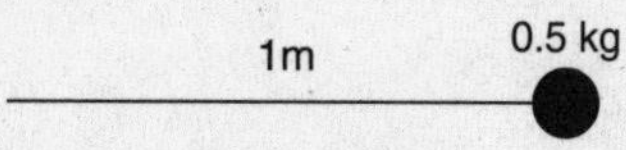

7. A mass of 0.5 kg is swung in a vertical circular path at the end of a 1 m string. If the speed of the mass is 2 m/s at the lowest point, the tension in the string there is most nearly

(A) 2 N
(B) 7 N
(C) 29 N
(D) 45 N
(E) 98 N

8. Its centripetal acceleration is most nearly

(A) 0.2 m/s^2
(B) 0.4 m/s^2
(C) 2.0 m/s^2
(D) 4.0 m/s^2
(E) 20 m/s^2

9. The work done by the string on the mass in the previous problem is

(A) 0
(B) 2 J
(C) 5.4 J
(D) 9.8 J
(E) 98 J

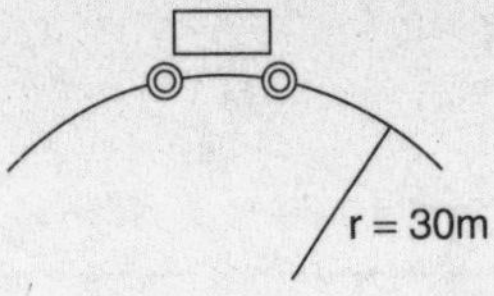

10. A 10 kg cart travels over the top of a semicircular hill of radius 30 m. The closest maximum speed it can attain without leaving the surface is

(A) 10 m/s
(B) 15 m/s
(C) 17 m/s
(D) 18 m/s
(E) 20 m/s

Answers to Multiple-Choice Questions

1. **(E)** The least value can be 1 N; the greatest can be 11 N. 13 N is too large.

2. **(B)** 5.1 + 3.3 + 6.6 = 15.0 kg and **Fw** = (15 kg)(10 m/s^2) = 150 N on Earth. On the moon, 150 N/ 6 = **25 N.**

3. **(D)** Addition of the four vectors gives a resultant force of 4.9 N 35° W of S. The **equilibrant vector**, therefore, is equal and opposite, **4.9 N 35° E of N.**

4. **(C)** $v_x = \sqrt{G\,2m/R}$ :: $v = \sqrt{G\,m/R}$ $v_x = v\sqrt{2}$

5. **(B)** $\mathbf{F_P} = \text{mgsin}\Theta$, $\mathbf{a} = \mathbf{F_{NET}} / m = \text{gsin}\Theta$.

6. **(A)** $\mu = F_f / F_N = 25\text{ N} / mg = 25\text{ N}/ 200\text{ N} = \mathbf{0.125}$

7. **(B)** $\mathbf{T = F_C + F_{WT}}$ $\mathbf{T} = [(mv^2/r) + (mg)] = \mathbf{7.0\ N}$

8. **(D)** $\mathbf{a_C = (v^2 / r) = 4.0\ m/s^2}$

9. **(A)** The mass does not move in the direction of the force.

10. **(C)** $\mathbf{F_{WT} = F_C}$ $mg = mv^2 / r$ and $v = \sqrt{(30\text{m})(10\text{m/s}^2)} = \mathbf{17.3\ m/s}$

Free Response

B & C Exam Question Types

1. **Force Problem:** Four concurrent forces are applied to a 5 Kg mass: 50 N North, 25 N East, 35 N South, and 35N West.

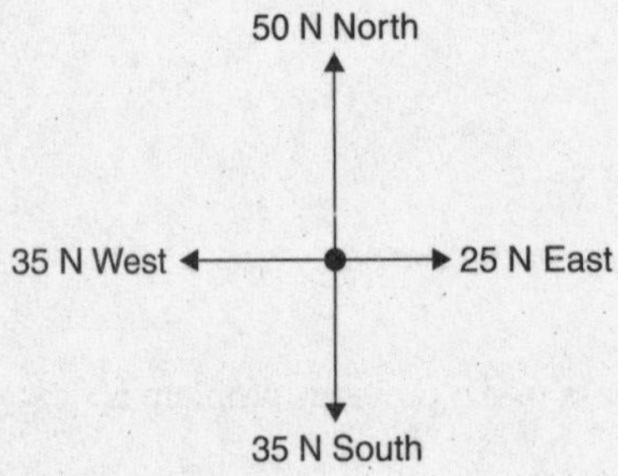

(a) Find the resultant force on the mass.
(b) Find the acceleration of the mass.
(c) What equilibrant force will sustain equilibrium?
(d) What will be the acceleration of the object if its mass doubles?
(e) Find the equilibrant force if all the initial forces are doubled.

2. **Weight and Mass Problem:** A 50 kg experiment package is placed on the Moon, which has a gravitational acceleration of $\frac{1}{6}$ that of the Earth.

   (a) How much less is its weight on the moon than on Earth?

   The package has wheels and is motorized. It is capable of 1 hour of sustained remote-controlled operation, its power supply has a 40-W rating and 5×10^5 J of stored energy. The motor can exert a maximum force of 40 N. Neglect friction. If Power = work/time or energy/time,

   (b) Theoretically, for how long can the package safely operate?
   (c) If it has a top speed of 0.8 m/s, how far from the base camp can it be sent to retrieve surface samples?
   (d) What is the maximum angle of slope that it can climb without losing speed?
   (e) Assuming a frictional force of between 5 N and 10 N on the incline, what would the maximum attainable angle now be?

3. **Equilibrium Problem:** A 25 kg mass is hung at the 0-cm mark on a steel meter stick, which rests on a pivot at its 50 cm mark.

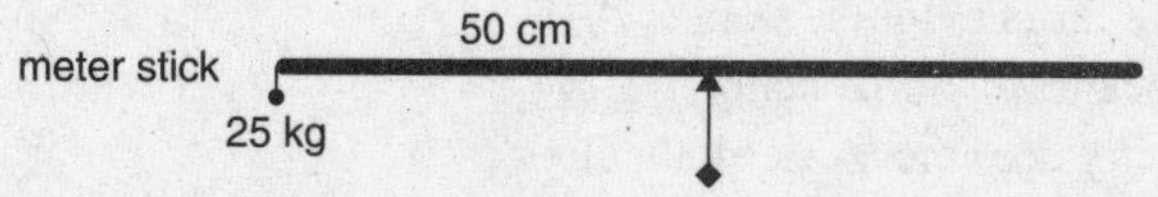

   (a) Where must a 40 kg mass be placed in order to achieve rotational equilibrium?
   (b) If an additional 10 kg mass is added to the original 25 kg mass at the left end, where must the 40 kg mass *now* be placed to achieve rotational equilibrium?

   The support/pivot point is now moved 10 cm to the left.

   (c) Repeat (a) and (b) for this new situation.

4. **Gravitation Problem (find magnitude only):**

   (a) What is the gravitational force of attraction between two masses, (M1) 1.2×10^3 kg and (M2) 5.2×10^4 kg, respectively, separated by a distance (center-to-center) of 30 m?

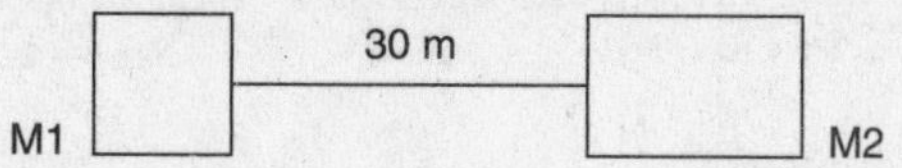

   (b) If a third mass, M3, 8.0×10^4 kg, is placed 20 m away from the 5.2×10^4 kg mass, M2, in line with and on the opposite side of the first one, what is the combined gravitational force on the first mass (M1) now?

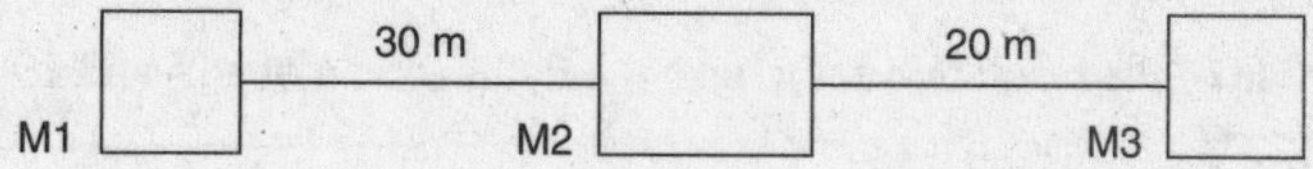

   (c) A fourth mass, M4, 3.5×10^6 kg, now sits 40 m from M2 as shown. Compute the total gravitational force on the first mass (M1).

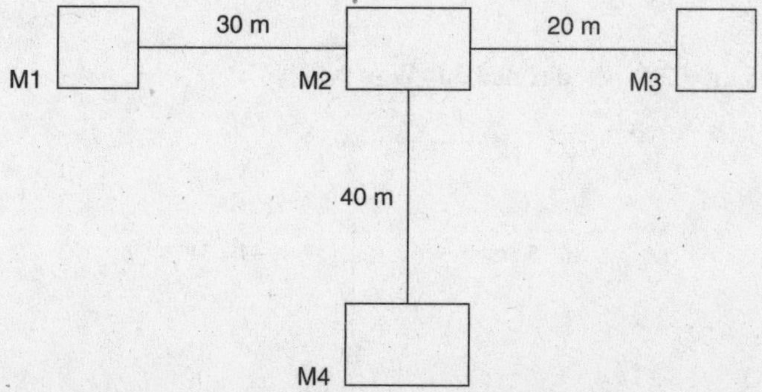

5. **Centripetal Force Problem:** A block of mass m travels in a circular path on the frictionless surface of a table of height 1 m. The circle's radius equals 0.5 m and the small mass is connected to a string that is fed through a hole in the center of the table. A second mass M hangs below.

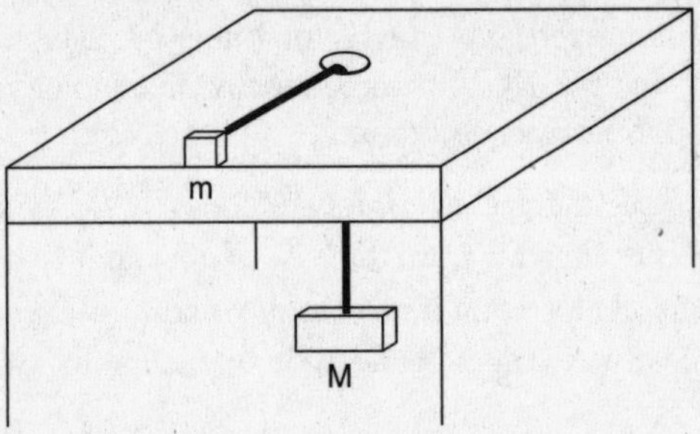

(a) If block m's velocity is 1 m/s and its mass is 2.0 kg, what is the mass of block M?

The string breaks, sending m off the table.

(b) How long does it take for m to hit the floor?

(c) How far does the block travel horizontally?

(d) With what total velocity does the block hit the floor?

(e) The block hits the floor, does not bounce, and slides to a stop. The total sliding time is 0.4 seconds. How far does it slide?

(f) What is the coefficient of friction between the block and the floor?

Answers to Free-Response Questions

1. Add the North and South and then add the East and West to simplify the problem. This gives totals of 15 N North and 10 N West. Assign these totals to an x-y axis system, and this gives a right triangle in the second quadrant. The **x** component is –10 N, and the **y** component is 15 N.

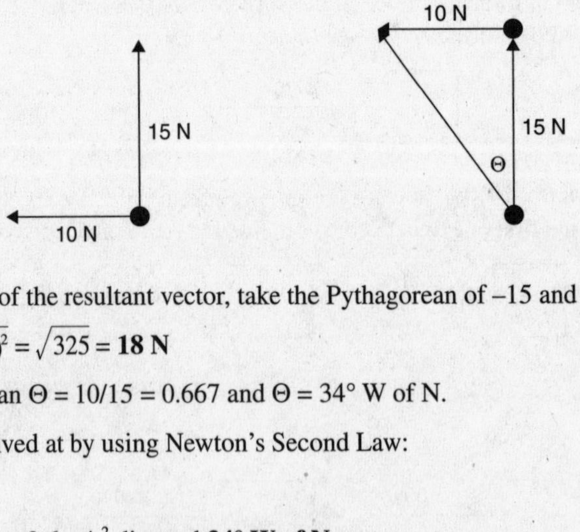

(a) To find the magnitude of the resultant vector, take the Pythagorean of –15 and 10:

$R = \sqrt{(-15N)^2 + (10N)^2} = \sqrt{325} = \mathbf{18\ N}$

For the direction, use tan Θ = 10/15 = 0.667 and Θ = 34° W of N.

(b) The acceleration is arrived at by using Newton's Second Law:

F = ma

a = **F** / m = 18N / 5 kg = 3.6 m/s^2 directed **34° W of N**

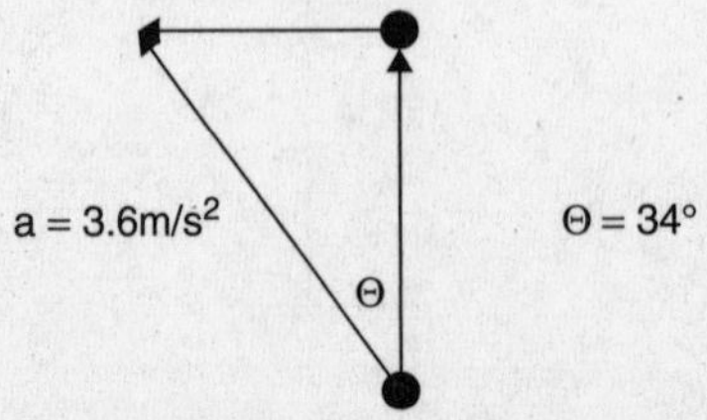

(c) The equilibrant force is the opposite of the resultant force: **18 N, 34° E of S.**

(d) a = F /m = 18 N / 10 kg = 1.8 m/s² directed 34° W of N

(e) Doubling all forces doubles the resultant force, thereby doubling the equilibrant, which is still in the same direction: **36 N, 34° E of S.**

2. **(a)** Fwt (Earth) – Fwt (moon) = (50 kg) (9.8 m/s²) – (50 kg) (9.8 m/s²) / 6 = (490 – 82)N = **408 N** less

(b) (1 sec/40 J)(1 hr/3600 sec)(500,000 J) = **3.47 Hrs**

(c) (12500 sec)(0.8 m/s) = 10,000 m. If it is to return, **5,000 m**

(d) At the maximum slope, $\mathbf{F_P}$ = mg sin Θ = 40 N. Since mg on the moon = 81.7 N, this means that 40 N = 81.7 N sin Θ and Θ = **29.3°**.

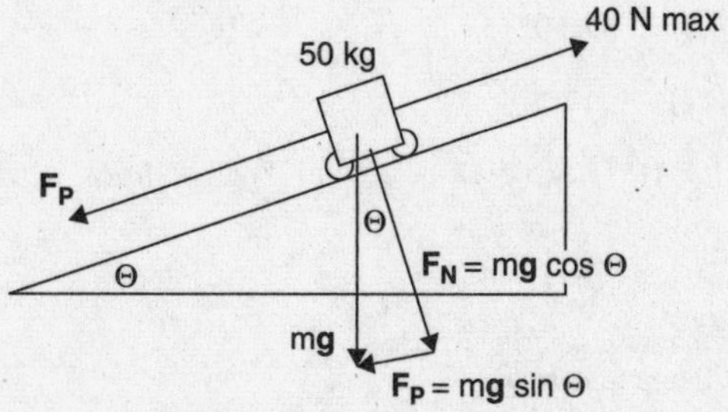

(e) Frictional forces act *down the slope*, in the same direction as $\mathbf{F_P}$.

For a frictional force of 5 N, $\mathbf{F_{NET}}$ = 40N – 5N = 35N = 81.7N sin Θ, Θ = 25.4°.

For a frictional force of 10 N, $\mathbf{F_{NET}}$ = 40N – 10N = 30N = 81.7N sin Θ, Θ = 21.5°

Assuming the maximum frictional force of 10 N, the maximum attainable angle is now 21.5°.

3. **(a)** This is actually a torque problem, in which the force times the distance from the pivot on the left must equal the force times the distance from the pivot on the right. Since the force is the weight, mg, the **g** will cancel, and it is only necessary to multiply the mass times the distance on each side and set them equal: (25kg)(50cm) = (40kg)(X). **X = 31.3 cm to the right of the pivot.**

(b) (35kg)(50cm) = (40kg)(X)

X= 43.8 cm right of the pivot or 6.2 cm from the right end

(c) i. (25kg)(40cm) = (40kg)(X). **X = 25 cm right of the pivot**

ii. (35kg)(40cm) = (40kg)(X). **X = 35 cm right of the pivot**

4.

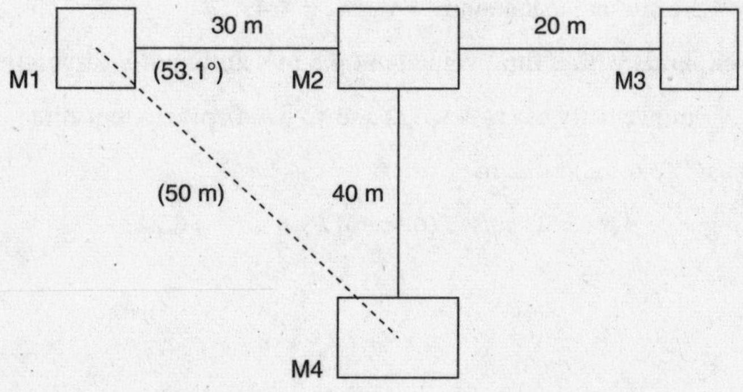

(a) $F_G = Gm_1m_2/r^2 = (6.67 \times 10^{-11} Nm^2/kg^2)(1.2 \times 10^3 kg)(5.2 \times 10^4 kg)/(30\ m)^2$

$= 41.6 \times 10^{-4}\ Nm^2 / 9.0 \times 10^2\ m^2 = \mathbf{4.62 \times 10^{-6}\ N}$

(b) Add the new mass's force to the existing force (4.62×10^{-6} N + F_G3).

$F_G3 = Gm_1m_3 / r^2 = (6.67 \times 10^{-11}\ Nm^2/kg^2)(1.2 \times 10^3 kg)(8.0 \times 10^4 kg)/(50m)^2$

$= 0.0256 \times 10^{-4} N = 2.56 \times 10^{-6} N$

$\mathbf{F_{TOT} = (4.62 + 2.56) \times 10^{-6}\ N = 7.18 \times 10^{-6}\ N}$

(c) Since the fourth mass is located 50 m from the first mass, its force on the first is calculated to be

$F_G4 = (6.67 \times 10^{-11} Nm^2/kg^2)(1.2 \times 10^3 kg)(3.5 \times 10^6 kg) / (50\ m)^2$

$= 1.1 \times 10^{-4} N$

The angle Θ between the line of the first three masses and the fourth mass is calculated from sin Θ = 40m/50m and Θ = 53.1°, which translates into

$\alpha = 53.1°$ and $\gamma = 180° - 53.1° = 126.9°$

Since $C^2 = A^2 + B^2 - 2ab \cos C$

$F_{TOT} = [(7.18 \times 10^{-6})^2 + (1.1 \times 10^{-4})^2 - 2(7.18 \times 10^{-6})(1.1 \times 10^{-4})(\cos 126.9°)]^{\frac{1}{2}}$

$\mathbf{F_{TOT} = 1.2 \times 10^{-4}\ N}$

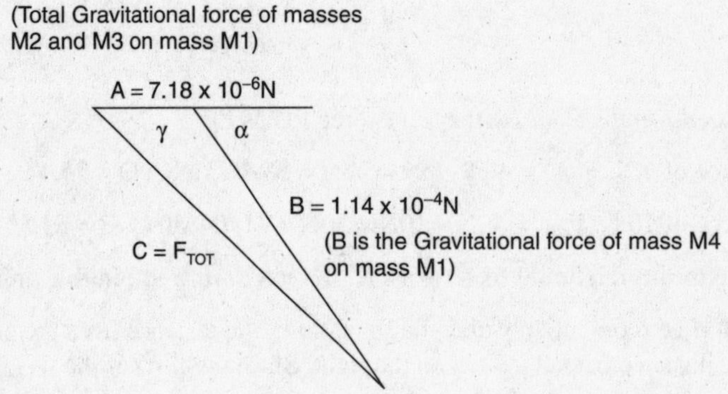

5. **(a)** $F_C = mv^2 / r$ and since the centripetal force is balanced by the weight of M, the mass of M would equal the centripetal force divided by **g**. $M = mv^2 / rg = (2kg)(1m/s)^2 / (0.5)(9.8m/s^2) = \mathbf{0.41\ kg}$

(b) Using the position formula, $s(t) = v_o t + 1/2\ at^2$, $-1m = 0 - 4.9\ t^2$

t = 0.45 sec

(c) $d = v_{AVE}t = (1\ m/s)(0.45\ sec) = \mathbf{0.45\ m}$

(d) $v_F^2 = v_o^2 + 2ad$ gives $v_F = \sqrt{2(-9.8m/s^2)(-1m)} = 4.4$ m/s

$v_{TOT} = \sqrt{v_H^2 + v_v^2} = 4.54$ m/s and using $\tan \Theta = v_H / v_V = 1/4.4$

Θ = 12.7° **The block lands with a final velocity of 4.5 m/s and makes an angle of 12.7° with the vertical.**

(e) Using the average sliding velocity as $(v_F + v_O)/2$ and v_{AVE} = **displacement**/time,

$d = v_{AVE}\ (t) = (1\ m/s/2)(0.4\ sec) = \mathbf{0.2\ m}$

(f) $\mu = F_F / F_N = ma/mg = \Delta v / \Delta t = [(1\ m/s) / (0.4sec)] / 9.8\ m/s^2 = \mathbf{0.26}$

Displacement, Speed, Velocity, and Acceleration

Linear motion is discussed first in this chapter, followed by a discussion of **rotational** motion.

All motion is relative to something, whether it is a fixed point or group of fixed points. A straight line change in position from one point to another is called the **linear displacement**, usually denoted by $\mathbf{\Delta x = x_2 - x_1}$. The linear displacement is a vector, having both magnitude and direction. A change in displacement in a certain amount of time is called *velocity*. **Average velocity** $\mathbf{v}_{AVE}$ usually is denoted by $\frac{\mathbf{x}_2 - \mathbf{x}_1}{t_2 - t_1}$. **Velocity** is a vector. (Velocity without direction is *speed*, a scalar.)

Instantaneous velocity is defined as the instantaneous change of displacement in an instantaneous amount of time, $\Delta x / \Delta t$ for extremely small values or

$$v = \lim_{\Delta t \to 0} \frac{\Delta x}{\Delta t} = \frac{dx}{dt}$$

Accuracy increases as values for time diminish. The closer to zero they come, the more accurate the value for instantaneous velocity becomes. (Derivative and antiderivative notation appears on the C exam only.)

Acceleration is defined as the change in velocity with respect to time,

$$\frac{\mathbf{v}_2 - \mathbf{v}_1}{t_2 - t_1}$$

This is $\frac{\Delta \mathbf{v}}{\Delta t}$, which is **average acceleration**. For extremely small values of time, the **instantaneous** acceleration is defined as $\mathbf{a}(t) = \frac{d\mathbf{v}}{dt}$. Since **acceleration** has a direction, it is a **vector.**

Instantaneous velocity $= \Delta x / \Delta t$.
It is the *derivative* or slope of the tangent to the curve.
$\mathbf{v}(t) = \frac{d\mathbf{x}}{dt}$ at any instant.

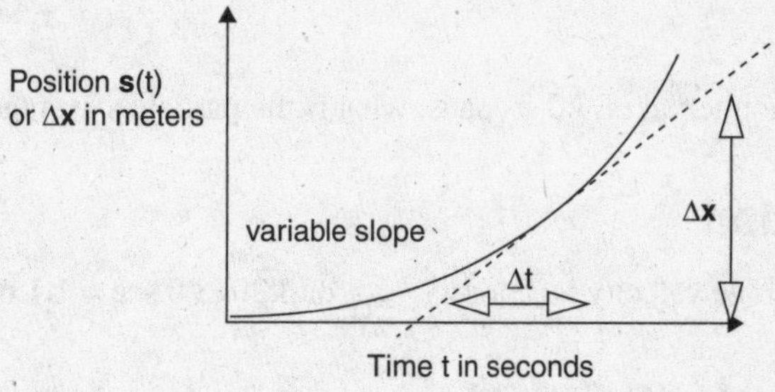

In the event that the acceleration is NOT constant, the graph of velocity versus time is not a straight line. Finding the instantaneous acceleration then would entail finding the slope of a line *tangent* to a particular part of the curve.

Instantaneous acceleration $= \Delta v / \Delta t$.
It is the derivative or slope of this curve and the *second derivative* of the previous curve.
$\mathbf{a}(t) = \frac{d\mathbf{v}}{dt}$ at any instant.

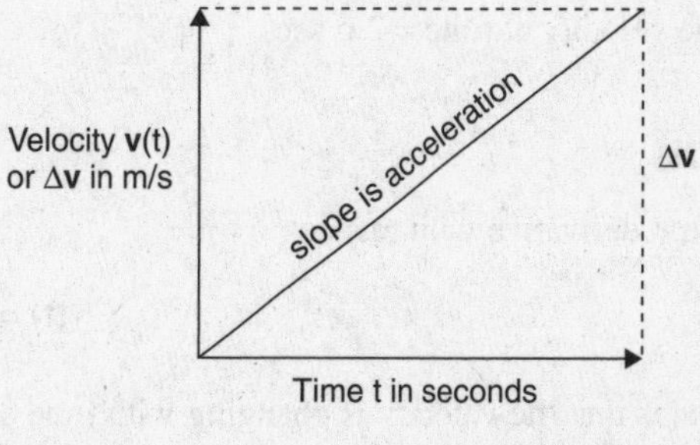

Important Linear Motion Equations

(C Exam)

$$\textbf{Displacement } \mathbf{x}(t) = \int \mathbf{v}(t)\,dt$$

(C Exam)

$$\mathbf{v}_{\text{INSTANTANEOUS}} = \mathbf{v}(t) = {}^{dx}\!/_{dt} = \int \mathbf{a}(t)\,dt$$

(C Exam)

$$\mathbf{a}_{\text{INSTANTANEOUS}} = \mathbf{a}(t) = {}^{dv}\!/_{dt} = {}^{d^2x}\!/_{dt^2}$$

$$\mathbf{a}_{\text{AVE}} = {}^{\Delta v}\!/_{\Delta t}$$

$$\mathbf{v}_{\text{AVE}} = {}^{\Delta x}\!/_{\Delta t}$$

$$\mathbf{v}_F = \mathbf{v}_o + \mathbf{at} \quad \text{[For contant accelerations]}$$

$$\mathbf{d} = \mathbf{s}(t) = \mathbf{v}_o t + \frac{1}{2}\mathbf{a}t^2 \quad \text{[For contant accelerations]}$$

$$\mathbf{v}_F^{\,2} = \mathbf{v}_o^{\,2} + \mathbf{2ad} \quad \text{[For contant accelerations]}$$

Note: All of these linear motion equations also hold for the analogous quantities in rotational motion. "Rotational Motion and Torque" later in this chapter will cover these equations, along with other concepts.

Example (B Exam)

The position of a particle moving along the **x**-axis is given as the following:

$$\mathbf{x}_2 = 8.8\text{m}$$

$$\mathbf{x}_1 = 5.5\text{m}$$

In a time interval of 3.0 seconds, what is the particle's average velocity?

Solution

The average velocity is ${}^{x_2 - x_1}\!/_{t_2 - t_1}$, or 3.3m/3.0 sec = **1.1 m/s**

Example (C Exam)

The position of a particle moving along the **x**-axis is given as the following:

$$\mathbf{x} = 5.4 + 6.8\,t + 10.0\,t^2 \text{ (in meters)}$$

What is the velocity at time = 2.0 sec?

Solution

Take the first derivative with respect to time:

$$\mathbf{v}(t) = \frac{\mathbf{dx}}{dt} = 0 + 6.8 + 20.0t$$

This indicates that the velocity is changing with time and that when t = 2.0 sec, the velocity will be (6.8 m/s) + (20.0 m/s^2)(2.0 sec) = **46.8 m/s.**

Example (B Exam)

The velocity of a particle is given as 10 m/s at t = 0 and 20 m/s at t = 5 sec. What is its average acceleration?

Solution

$$\mathbf{a} = \frac{v_2 - v_1}{t_2 - t_1} = \frac{20\text{m/s} - 10\text{m/s}}{5\,\text{sec}} = \mathbf{2\ m/s^2}$$

Example (C Exam)

The position of a particle is given by $\mathbf{x} = 7 + 16t - t^3$ and t is in seconds, **v** in m/s and **a** in m/s^2. Determine its **instantaneous velocity** **v**(t) and its **instantaneous acceleration** **a**(t).

Solution

$$\mathbf{v}(t) = {}^{dx}\!/_{dt} = 0 + 16 - 3t^2 \text{ or } \mathbf{16 - 3t^2}$$
$$\mathbf{a}(t) = {}^{dv}\!/_{dt} = 0 - 6t = \mathbf{-6t}$$

Newton's Laws

Isaac Newton took information regarding much of the studies of mechanics from Galileo, whose many experiments with bodies in motion laid the groundwork for Newton's famous Three Laws of Motion:

1. **Law of Inertia**: A body at rest will remain at rest, and a body in motion will remain in motion, at the same speed in the same direction forever, unless acted upon by a net external force.

When no net force is applied to a body, it will either remain at rest or will continue to move at the same speed, in the same direction, forever. If more than one force acts on a body that remains at rest, it is in **equilibrium**. If a body is already in motion at a constant velocity and if all forces acting on it add up to zero, the body is also in **equilibrium**. An automobile cruising at a constant velocity (constant speed in a straight line) along a road is a perfect example. The force of the engine turning the wheels and propelling the automobile forward is just equaled and opposed by the sum of all frictional forces acting on the automobile. No *net force* is acting on the automobile, so it does not accelerate, which leads to the second law.

2. **F = ma**: It takes a net force to accelerate a mass; the larger the net force, the greater the acceleration of a given mass.

The most important concept in this beautifully simple equation is that a net force will accelerate a mass. This means that the *sum* of all forces acting on an object will constitute the net force. When two or more objects are attached to each other and the system is influenced by outside forces, again, it is the net force, the *sum* of all the forces acting on the system, which will accelerate the sum of the masses of the system.

3. **Law of Interaction**: For every *action*, there is an equal and opposite *reaction*. For every force there is an equal and opposite force on the other object involved in the interaction. Forces always come in pairs because forces can exist only when two objects interact.

When you touch an object and the object does not move, that object touches you back with the same force with which you touched it. If the object moves at a constant velocity as you push it, no acceleration and no net force exist. The object still pushes you back with the same force with which you push it. When you push an object and it accelerates, the net force on the object is greater than zero. The result is that the net force has accelerated the object. If you are pushing the object against a frictional force and accelerating the object, the object's acceleration is equal to the net force divided by its mass:

$$\mathbf{a} = \mathbf{F}_{NET} / m = \text{(your pushing force – the frictional force) / the object's mass}$$

If the same force acts on two objects of different mass, the less massive object has the greater acceleration, and the larger object has the lesser acceleration. This is demonstrated easily by two ice skaters of different mass who push on each other. They both accelerate in opposite directions. The skater with the larger mass has the lesser acceleration and the smaller skater's acceleration is greater than that of the larger skater. The skaters attain unequal velocities in opposite directions. This involves the Law of Conservation of Momentum, which is covered in the next chapter.

Example 1 (B Exam)

A box of physics textbooks is pulled along a rough floor at a constant speed. If the coefficient of kinetic friction between the box and the floor is 0.35, and the box has a mass of 50 kg, what is the frictional force on the box?

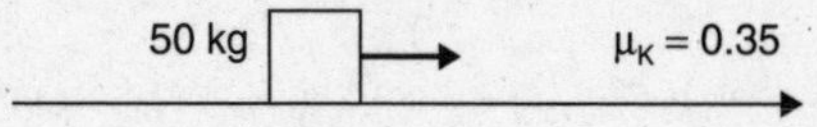

Solution

Because the box is not accelerating, no net force exists. Therefore, the parallel and the frictional forces are equal and opposite. Because $\mu = \mathbf{F_F} / \mathbf{F_N}$,

$$\mathbf{F_F} = \mu\,\mathbf{F_N} = (0.35)(50\text{ kg})(9.8\text{ m/s}^2) = \mathbf{170\text{ N}}$$

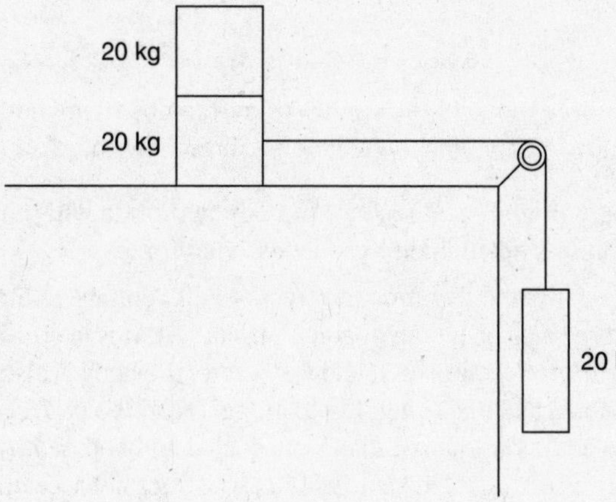

Example 2

Two 20 kg masses rest on a horizontal table. The bottom mass is connected to a string of negligible mass, which passes over a frictionless pulley with negligible mass and supports a third 20 kg mass that hangs over the side of the table. The coefficient of kinetic friction between the table and the stacked masses is 0.15.

(a) What is the tension in the string?

(b) What is the acceleration of the system?

(c) What is the minimum coefficient of friction between the two vertically stacked masses that will prevent the top one from sliding off as they move?

Solution

(a)

$$\begin{aligned}T_{TOT} &= \mathbf{F_{WT}} - \mathbf{F_F} = (20\text{ kg})(\mathbf{g}) - \mu\mathbf{F_N}\\ &= (20\text{ kg})(9.8\text{ m/s}^2) - (0.15)(40\text{ kg})(9.8\text{ m/s}^2) = 196\text{ N} - 59\text{ N}\\ &= \mathbf{137\text{ N}}\end{aligned}$$

(b)

$$a = \frac{\mathbf{F_{NET}}}{\Sigma m} = \frac{137\text{N}}{60\text{kg}} = \mathbf{2.3\text{ m/s}^2}$$

(c) Since the top mass accelerates at 2.3 m/s^2, the force needed to accelerate 20 kg by that amount is $\mathbf{F} = \mathbf{ma} = (20\ \text{kg})(2.3\ \text{m/s}^2) = 46\ \text{N}$. This represents $\mathbf{F_F}$, which means

$$\mu = \frac{\mathbf{F_F}}{\mathbf{F_N}} = \frac{46\ \text{N}}{(20\ \text{kg})(9.8\ \text{m/s}^2)} = \frac{46\ \text{N}}{196\ \text{N}} = 0.23$$

Falling Bodies and Freefall

Galileo's experiments with falling bodies and ramps that diluted the acceleration of gravity led to the understanding that, neglecting air friction, all objects will accelerate at the same steady rate when falling to the earth. It is approximately 9.8 m/s^2 at sea level. When a falling object reaches *terminal velocity*, its weight is balanced and opposed by air friction, and the object ceases to accelerate. In freefall problems, it is assumed that $\mathbf{g} = 9.8$ m/s^2. The usual equations and relationships utilized in freefall problems are as follows.

The position formula

$$\mathbf{d} = \mathbf{s(t)} = \mathbf{v_0 t} + \frac{1}{2}\mathbf{at^2} \text{ (where } \mathbf{a} = \mathbf{g} = 9.8\ \text{m/s}^2)$$

The velocities formulas

$$\mathbf{v_F}^2 = \mathbf{v_O}^2 + 2\ \mathbf{ad} \text{ (where } \mathbf{a} = \mathbf{g} = 9.8\ \text{m/s}^2)$$

$$\mathbf{v_f} = \mathbf{v_o} + \mathbf{at}$$

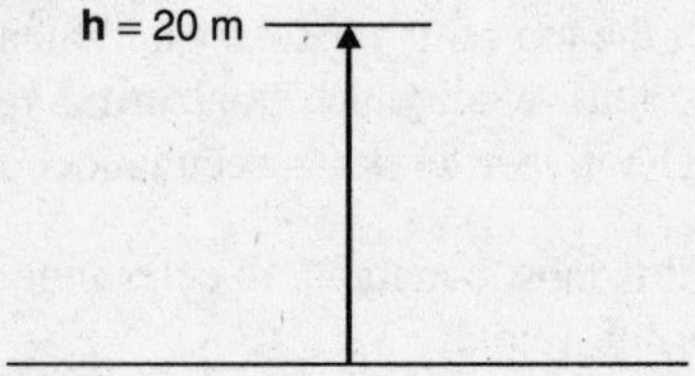

Example (B & C Exams)

A 0.5 kg rocket is launched straight up and reaches a maximum height of 20 m.

(a) What was its initial velocity?
(b) What is its acceleration at the top of its trajectory?
(c) How long is it in the air?

A steady breeze with velocity 2.0 m/s acts on the object while it is in the air. (The breeze is parallel with the ground.)

(d) How far from its launch point does the object land? (projectile motion)
(e) What is the object's velocity when it lands?

Solution

(a) Using $\mathbf{v_F}^2 = \mathbf{v_O}^2 + 2\ \mathbf{ad}$, and taking the final velocity at the top of the trajectory as 0, $0 = \mathbf{v_O}^2 + 2\ (-9.8\ \text{m/s}^2)\ (20\ \text{m})$, which gives the following:

$$\mathbf{v_O} = \mathbf{19.8\ m/s}$$

(b) The acceleration of a projectile *at any time* is **g**, or **–9.8 m/s^2**.

(c) Using $\mathbf{d} = \mathbf{v_0 t} + \frac{1}{2}\mathbf{at^2}$, and since the total displacement of the ball at the end of its flight is 0, $0 = 19.8\ \text{m/s}\ (t) - 4.9\ t^2$. Factoring out t gives the following:

$$0 = 19.8\ \text{m/s} - 4.9\ \text{m/s}^2\ (t) \text{ and } \mathbf{t} = \mathbf{4.0\ sec}$$

(d) $d = v_{AVE}t$ = (2.0 m/s)(4.0 sec) = **8.0 m away.** (Projectile motion problem: Consult the following section if necessary.)

(e) Since the object now has both vertical and horizontal components,

$$v^2 = v_H^2 + v_V^2 = (2.0\text{ m/s})^2 + (19.8\text{ m/s})^2 = 396\text{ m}^2/\text{s}^2$$

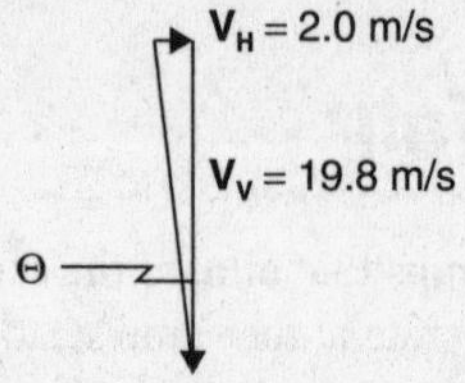

$$\tan \Theta = (2.0\text{ m/s}) / (19.8\text{ m/s}) = 0.101$$

$$\Theta = 5.8°$$

The object lands with a velocity of 19.9 m/s at an angle of 5.8°.

Projectile Motion

When an object is thrown, kicked, hurled, shot, or launched into the air, the *only* acceleration it experiences is that of gravity if the effects of wind resistance are negligible. Its **final vertical velocity $V_{F\ VERT}$** will equal the negative of its **initial vertical velocity $V_{O\ VERT}$** if it lands at the same elevation from which it was launched. Otherwise, its instantaneous vertical velocity diminishes to zero at the top of its trajectory and then increases as it turns and falls back toward its launch elevation. If given an initial horizontal velocity, that **horizontal velocity $V_{O\ HOR}$** will remain constant while airborne because there is no horizontal acceleration after it has been launched.

When solving projectile motion problems, it is most beneficial to determine the *initial horizontal* $V_{O\ HOR}$ and *initial vertical* $V_{O\ VERT}$ components of its initial or **launch velocity V_0**.

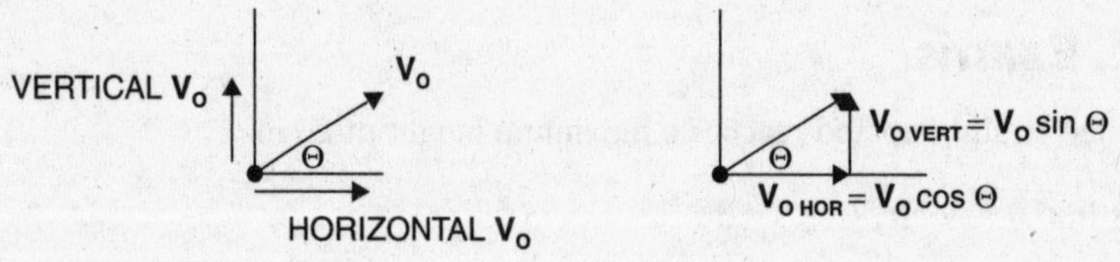

Since the initial velocity and both of its components constitute the parts of a right triangle, the Pythagorean Theorem holds: $V_O^2 = [V_{O\ VERT}]^2 + [V_{O\ HOR}]^2$.

Example (B & C Exams)

Determine the horizontal and vertical components of the velocity of a football kicked at an angle of 30° with the ground with a velocity of 20 m/s.

Solution

$$V_{O\ VERT} = V_O \sin \Theta = (20\text{ m/s})(\sin 30°) = \mathbf{10\text{ m/s}}$$

$$V_{O\ HOR} = V_O \cos \Theta = (20\text{ m/s})(\cos 30°) = \mathbf{17.3\text{ m/s}}$$

When the projectile is launched, either the highest point in its path or the total horizontal distance it travels may be given. The time of flight can be given or calculated as well.

The usual equations useful in solving projectile motion problems are as follows:

$+\frac{1}{2}\mathbf{a}\,t^2$

ad

)(t)

The key to sc … omponents separate. First, find the vertical and horizontal velocity co … and horizontal components *separately*. Also, remember that the initia … ntil the object hits the ground or another object.

0°

E

A soc … ı the ground. Assuming its initial velocity is 15 m/s, answer the following:

(a) H… vertical air resistance.)
(b) How fa… n site will it land?

Solution

First, find the vertical and horizontal components of the initial velocity.

$$\mathbf{V}_{O\ VERT} = \mathbf{V}_O \sin\Theta = (15\text{ m/s})(\sin 20°) = 5.1\text{ m/s}$$
$$\mathbf{V}_{O\ HOR} = \mathbf{V}_O \cos\Theta = (15\text{ m/s})(\cos 20°) = 14.1\text{ m/s}$$

(a) For maximum height reached: $\mathbf{V}_F^2 = \mathbf{V}_O^2 + 2\mathbf{ad}$. Keeping *vertical* values only: $0 = (5.1\text{ m/s})^2 + 2(-9.8\text{ m/s}^2)(\mathbf{d})$, where **d** is the maximum vertical displacement. Solving for **d** yields: **1.3 m for the maximum height.**
(b) For maximum horizontal distance traveled, we first need the total time the ball is in the air. Use $\mathbf{d} = \mathbf{s}\,(t) = \mathbf{V}_O t + \frac{1}{2}\mathbf{a}\,t^2$ with the vertical aspect. The total vertical displacement will be zero once the ball lands on the ground. Therefore, $0 = (5.1\text{ m/s})(t) - 4.9\,(t^2)$. Factoring out (t) gives t = 1.0 sec total flight time. Now use $\mathbf{d} = (\mathbf{v}_{AVE})(t)$:

$$\mathbf{d}_{HOR} = (\mathbf{V}_{O\ HOR})(t) = (14.1\text{ m/s})(1.0\text{ sec}) = \mathbf{14.1\text{ m}}$$

Two- and Three-Dimensional Motion

Most typical B and C exam questions will be limited to projectile motion in two dimensions, as covered in the preceding section. In the event of a three-dimensional projectile motion problem, the same approach for two-dimensional problems can be extended to include the third dimension's velocity component. Because the initial velocity and all three of its components constitute the parts of a right triangle in three-dimensional space, the Pythagorean Theorem applies once again (the "triple Pythagorean"):

$$\mathbf{V}_O^2 = [\mathbf{V}_{O\,VERT}]^2 + [\mathbf{V}_{O\,HOR}]^2 + [\mathbf{V}_{O\,3RD\,DIM}]^2 \textbf{ or } \mathbf{V}_O^2 = [\mathbf{V}_X\mathbf{i}]^2 + [\mathbf{V}_Y\mathbf{j}]^2 + [\mathbf{V}_Z\mathbf{k}]^2$$

In such a problem, the third-dimension velocity vector may be wind blowing in a particular direction at a constant velocity. In solving a problem of this type, use the same basic equations as in two-dimensional projectile motion problems.

Example (B & C Exams)

In the previous soccer ball problem, add a wind blowing into the paper at 5 m/s. How far away from its launch site will it land?

Solution

Since there is 1.0 sec. of total flight time, the ball will land 5.0 m into the page. Taking the 14.1 m to the right and 5.0 m into the page as perpendicular components, the Pythagorean answer is

$$\mathbf{d} = \sqrt{(14.1\,\text{m})^2 + (5.0\,\text{m})^2} = \mathbf{15.0\ m}$$

For the angle:

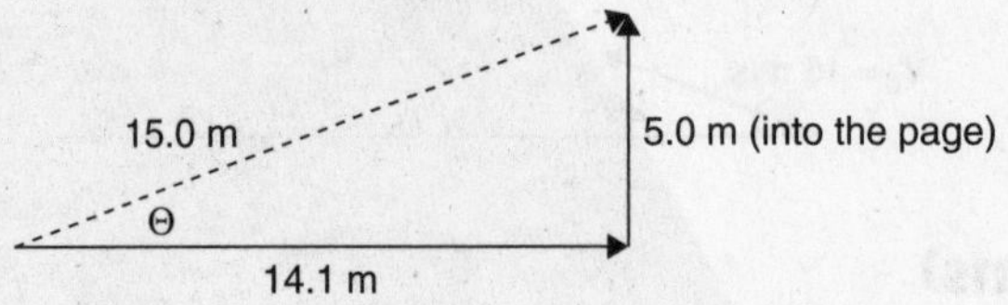

Using $\tan \Theta = \frac{5.0\,\text{m}}{14.1\,\text{m}} = 0.355$, $\Theta =$ **19.5° into the page**

Pendulums and Periodic Motion

There are various types of pendulums and physical harmonic oscillators: simple, conical, spring, torsional, and physical.

The **simple** pendulum consists of a mass swinging in a simple back-and-forth repeating motion. The period of this type of simple pendulum is given as the following:

$$T = 2\pi\sqrt{l/g}$$

Note that the mass has no bearing on the period of this type of pendulum.

The **conical** pendulum is a three-dimensional adaptation of the simple pendulum, since a hanging mass describes a cone when it oscillates in a circular path. The period of this pendulum incorporates the simple pendulum's vertical part of the length by using the cosine of the angle ($l \cos \Theta$). The period of a conical pendulum is given as $T = 2\pi\sqrt{l\cos\Theta/g}$. Note that, once again, the mass has no bearing on the period of this type of pendulum. (The cosine merely gives the pendulum's vertical component.)

The **spring** pendulum may consist of a mass hung vertically on a spring, a mass attached to a fixed spring on a horizontal, often "frictionless" surface, or a semi-rigid metal strip, like a hacksaw blade. The period of this pendulum is given as the following

$$T = 2\pi\sqrt{m/k}$$

Here, the mass *is* significant, as is the spring constant k.

The torsional pendulum involves a disk or other object twisting around the axis of a suspending string or wire. The physical pendulum involves a rigid body such as a stick, fixed at one end and swinging back and forth.

Note: Some texts may denote *period* as T (Greek letter Tau). Here it is denoted by T.

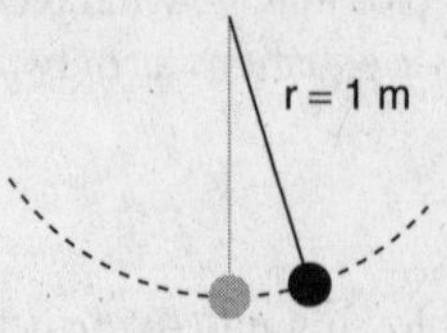

Example (B & C Exams)

What is the period of a 1 m long pendulum at sea level?

Solution

$T = 2\pi\sqrt{1/g} = (6.2832)\sqrt{1\text{m}/9.8\text{m/s}^2} = \mathbf{2.0\ sec}$

Note: At an altitude other than sea level, the value for g is not 9.8 m/s² and must be corrected for in all pertinent calculations.

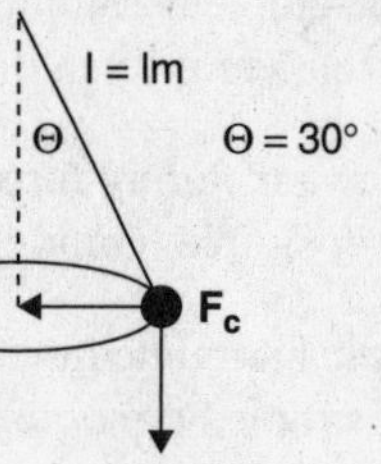

Example (B & C Exams)

(a) What is the period of a 1 m long pendulum at sea level if swung in a circular path with the string making an angle of 30° with the vertical?

(b) If the mass on the end is 1.0 kg, and it has a linear velocity of 0.5 m/s, what is the tension in the string?

Solution

(a) For the **period:**

$$T = 2\pi\sqrt{1\cos\Theta/g} =$$
$$(6.2832)\sqrt{(1\text{m})(.866)/9.8\text{m/s}^2} = \mathbf{1.9\ sec}$$

(b) **Tension** is the result of weight and centripetal force:

$$T = \sqrt{(\mathbf{mg})^2 + (\mathbf{F_c})^2} = \sqrt{(\mathbf{m}^2\mathbf{g}^2) + (\mathbf{mv}^2/0.5\text{m})^2}$$
$$= \sqrt{\left[(1.0\text{kg})(9.8\text{m/s}^2)\right]^2 + \left[(1.0\text{kg})(0.5\text{m/s})^2/(0.5\text{m})\right]^2}$$
$$= \sqrt{(96.0\text{kg}^2\text{m}^2/\text{sec}^4) + (0.25\text{kg}^2\text{m}^4/\text{s}^4\text{m}^2)}$$
$$= \sqrt{96.25\text{kg}^2\text{m}^2/\text{s}^4}$$
$$= \sqrt{96.25\text{N}^2}$$
$$= \mathbf{9.8\ N}$$

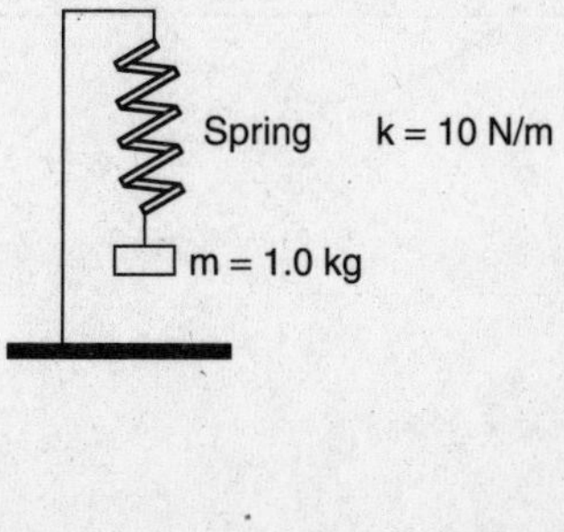

Example (B & C Exams)

What is the period of a vertical spring pendulum if k=10 N/m and the attached mass is 1.0 kg?

Solution

$$T = 2\pi\sqrt{m/k} = (6.2832)\sqrt{1\,kg/10\,N/m} = \mathbf{1.99\ sec}$$

Periodic Motion

Any repetitive motion involving an object either oscillating or repeating the same movement is called **periodic or harmonic motion**. When a body oscillates with simple harmonic motion, a constant interplay exists between kinetic energy and potential energy. (Energy will be discussed fully in the next chapter.)

Periodic or simple harmonic motion is caused by a restoring force proportional to the displacement from equilibrium. Many systems exhibit or approximate periodic or simple harmonic motion conditions.

A marble rolling inside a bowl can exhibit simple harmonic motion, as does an oscillating spring. The linear pendulum and the conical pendulum are also examples of simple harmonic motion, as is the motion of a point on a carousel.

Some Periodic Motion Relationships

Springs:

$$\mathbf{F} = -\mathbf{kx}$$
$$\mathbf{E_P} = \frac{1}{2}\,\mathbf{kx^2}$$

Pendulums:

Linear:

$$T = 2\pi\sqrt{(1/g)}$$
$$\mathbf{F} = -\mathbf{mg}\ \sin\Theta$$

Conical:

$$T = 2\pi\sqrt{1\cos\Theta/\mathbf{g}}$$

Spring Pendulum:

$$T = 2\pi\sqrt{m/k}$$

Simple Harmonic Motion in terms of circular motion:

$$\mathbf{d} = \mathbf{r}\cos\mathbf{\Omega t}$$
$$\mathbf{d} = \mathbf{r}\cos 2\pi\ f(t) = \mathbf{r}\cos 2\pi\ (t)\ /\ T$$

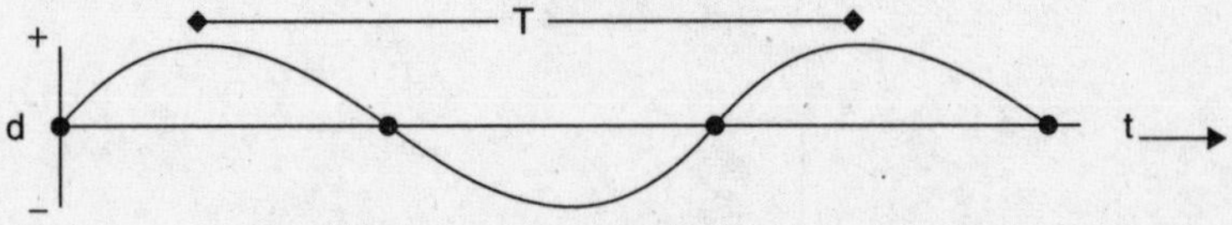

$$\sin \Theta = (r^2 - d^2)^{1/2} / r$$
$$\cos \Theta = d / r$$

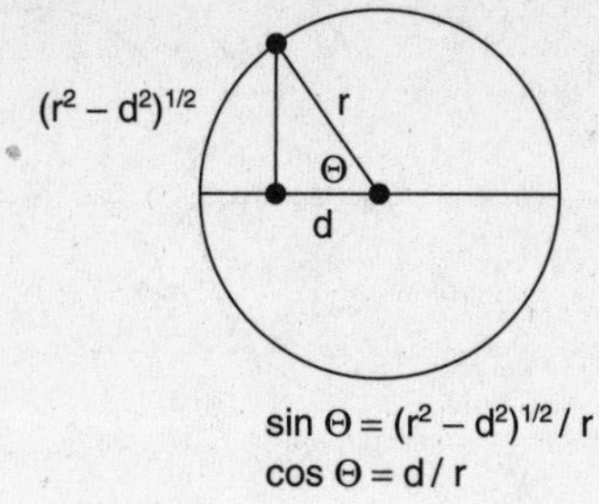

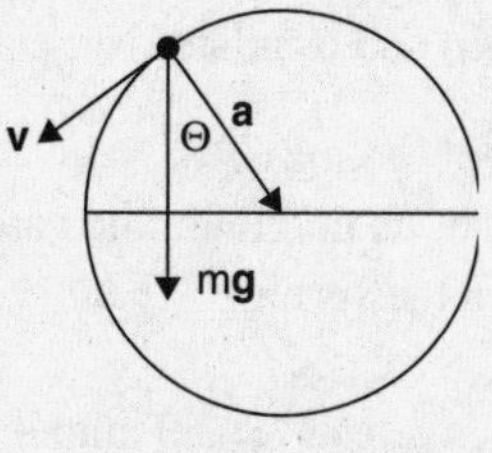

For an object moving in a vertical circular path, its acceleration is toward the center. Its velocity is tangential. If gravity is to be considered, its weight **mg** is directed toward the center of the earth.

Spring Pendulum and Simple Harmonic Motion

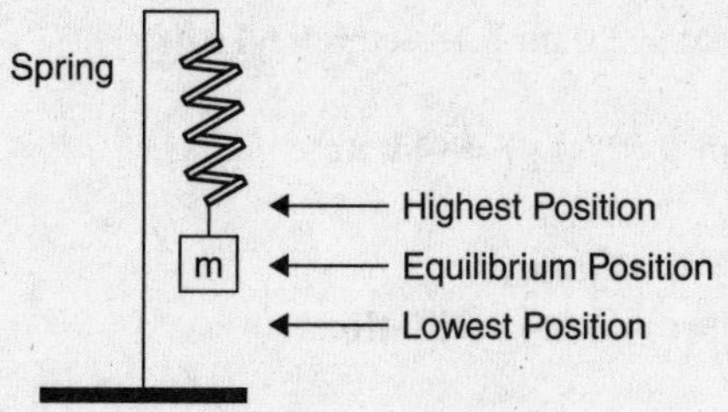

	d	v	a	U
Highest Position	Max	0	Max	Max
Equilibrium Position	0	Max	0	0
Lowest Position	Max	0	Max	Max

At its highest position, the mass has maximum displacement, 0 velocity, maximum acceleration, and maximum potential energy stored in the spring.

At its equilibrium position, the mass has 0 displacement, maximum velocity, 0 acceleration, and 0 potential energy stored in the spring.

At its lowest position, the mass has maximum displacement, 0 velocity, maximum acceleration, and maximum potential energy stored in the spring.

Maximum Displacement: 0 velocity, max. acceleration

Equilibrium Position: Max. velocity, 0 acceleration

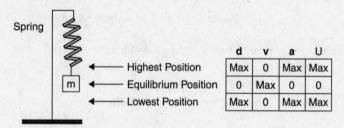

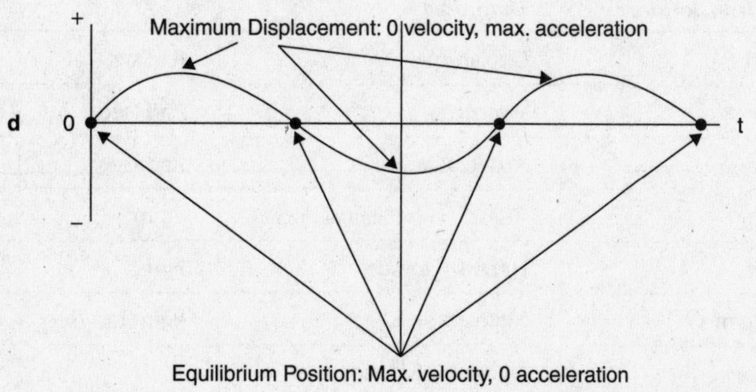

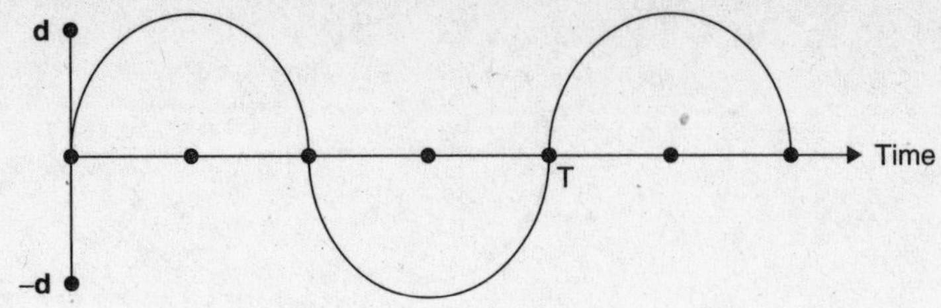

Example (B & C Exams)

A 0.5 kg mass is attached to a spring (k = 2.0 N/m) and oscillates vertically with an amplitude **d** and period T as illustrated.

(a) If **d** = 1.0 m, find the average speed of the mass.
(b) What is the acceleration of the mass at its maximum upward displacement?
(c) What is the frequency of the mass on the spring?
(d) What is the velocity of the mass at t = 1.0 sec?
(e) How many oscillations has the mass made at t = 2.0 minutes, and how far has the mass traveled at that time?

Solution

(a) Using $T = 2\pi\sqrt{m/k}$,

$$T = 2\pi\sqrt{0.5\text{kg}/2.0\text{N/m}} = 3.14 \text{ sec.}$$

$$|\bar{v}| = \Delta x/\Delta t = 2.0 \text{ m}/ 3.14 \text{ sec} = \mathbf{0.64 \text{ m/s}}$$

(b) $\mathbf{a} = \mathbf{F}_{NET}/m = -\mathbf{k}\mathbf{x}/m = (-2.0 \text{ N/m})(1.0 \text{ m}) / (0.5 \text{ kg}) = \mathbf{-4.0 \text{ m/s}^2}$
(c) $f = 1/T = 1/3.14 \text{ sec} = \mathbf{0.32 \text{ Hz}}$
(d) $\mathbf{v}_{1 \text{ sec}} / 1.0 \text{ sec} = (0.64 \text{ m/s}) / (1.57 \text{ sec}) = \mathbf{0.41 \text{ m/s}}$
(e) # oscillations = ft = (0.32 osc/sec)(120 sec) = **38.4 oscillations**

$$\mathbf{v}_{AVE} = 1.0 \text{ m/s and } \mathbf{d} = (\mathbf{v}_{AVE})(t) = (0.64 \text{ m/s})(120 \text{ sec}) = \mathbf{77 \text{ m}}$$

Rotational Motion and Torque

Rotational Motion deals with the manner in which solid objects rotate or revolve about an axis. All concepts and equations dealing with linear mechanics can be translated easily to the language of rotational mechanics.

Linear	*Units (Translational)*	*Rotational (Angular)*	*Units*	*Relationship*
d	m	Displacement Θ	Radians	$\Theta = d/r = \int \omega\, dt$
v	m/s	Velocity ω	Rad /sec	$\omega = v/r = d\Theta/dt = \int \alpha\, dt$
a	m/s^2	Acceleration α	Rad /sec^2	$\alpha = a/r = d\omega/dt$
m	kg	"Mass" = I (rotational inertia)	kg m^2	$I = \Sigma mr^2$
F	N	Force τ = torque	Nm	$\tau = r \times F$
p	kg m/s	Momentum **L**	kgm^2 rad /sec	$L = I\omega$
F = ma = **dp**/dt	→	(Newton's Second Law)	→	$\tau = I\alpha = dL/dt$

Linear	*Units (Translational)*	*Rotational (Angular)*	*Units*	*Relationship*
p is constant	→	(Law of Conservation of Momentum)	→	L is constant
W = $\int$ **F(x) dx**	J	Work **W**	J	**W** = $\int \tau \, d\Theta$
P = **W**/t = **Fv**	W	Power P	W	P = $\tau \omega$
K = 1/2 m**v**^{2}	J	Kinetic Energy K	J	$K = \frac{1}{2} I\omega^2$
W = ΔK	→	(Work-Energy Theorem)	→	**W** = ΔK
$\mathbf{d} = \mathbf{v}_o t + \frac{1}{2}\mathbf{a} t^2$	→	**(Displacement)**	→	$\Theta = \omega_o t + \frac{1}{2}\alpha t^2$
$\mathbf{v} = \mathbf{v}_o + \mathbf{a}t$	→	**(Velocity)**	→	$\omega = \omega_o + \alpha t$
$\mathbf{v}_f^2 = \mathbf{v}_o^2 + 2\mathbf{a}\mathbf{d}$	→	**(Velocity, Acceleration, Displacement)**	→	$\omega_f^2 = \omega_o^2 + 2\alpha\Theta$

Helpful Comments

1. The rotational character often equals the linear character divided by the radius.
2. There are 2π radians per revolution.
3. One radian equals 360°/2π = 57.3° and represents the angle described when the radius of a circle sweeps through a length on the circumference that is equal to the length of the radius. For this reason, radius and radian may be interchanged for purposes of unit cancellation.
4. Torque τ is a turning force.
5. Rotational Inertia **I** is the rotational representation of an object's inertia or mass and is a function of the mass and the radial distance squared: $I = \Sigma\, mr^2 = \int r^2\, dm$.
6. The total amount of kinetic energy possessed by a rolling object equals the sum of the linear and rotational kinetic energies: $K\text{ tot} = \frac{1}{2}mv^2 + \frac{1}{2}I\omega^2$.

- For purposes of dimensional analysis (correct unit notation and cancellation), you may label radius (**r**) in **m/Radian**.
- Commonly used rotational inertias:
  1. Simple Pendulum, Hoop, or Ring: $I = mr^2$
  2. Solid Disk: $I = \frac{1}{2} mr^2$
  3. Solid Sphere: $I = \frac{2}{5} mr^2$
  4. Hollow Sphere: $I = 2/3\ mr^2$

Common Rotational Problems

Example 1 (B & C Exams)

A flywheel starts from rest and has a constant angular acceleration of 10.2 rad/s^2.

(a) How long will it take the wheel to attain an angular velocity of 20.0 rev/s?
(b) How many revolutions will it make in this time?
(c) What angular velocity will it attain in this time?
(d) How many revolutions will it make during the tenth second?

Solution

(a) Using $\alpha = \frac{\Delta\omega}{\Delta t}$, $\Delta t = \frac{\Delta\omega}{\alpha}$

$$\frac{(20.0\text{ rev/s})(2\pi\text{ rad/rev})}{10.2\text{ rad/s}^2} = \mathbf{12.3\ sec}$$

(b) Using $\Theta = \omega_o t + \frac{1}{2}\alpha t^2$

$$= 0 + \frac{1}{2}(10.2\text{ rad/s}^2)(12.3\text{ s})^2(1\text{rev}/2\pi\text{ rad}) = \mathbf{123\ rev}$$

(c) Using $\omega_f^2 = \omega_o^2 + 2\alpha\Theta$

$$\omega_f^2 = (2)(10.2\text{ rad/s}^2)(123\text{ rev})(2\pi\text{ rad/rev}) = 15766\text{ rad}^2/\text{s}^2$$

$$\omega_f = \mathbf{126\ rad/s}$$

(d) Using $\Theta = \omega_o t + \frac{1}{2}\alpha t^2$:

$$\Theta \text{ During tenth second} = \Theta_{10} - \Theta_9$$

$$= [\tfrac{1}{2}(10.2\text{ rad/s}^2)(100\text{ sec}^2)] - [\tfrac{1}{2}(10.2\text{ rad/s}^2)(81\text{ sec}^2)]$$

$$= 510\text{ rad} - 413\text{ rad} = 97\text{ rad }(1\text{ rev}/2\pi\text{ rad}) = \mathbf{15.4\ rev}$$

Example 2 (C Exam)

A gyroscope is spun on a table and has an initial angular velocity and angular acceleration given by $\omega_o = 6$ rad/s and $\alpha = 6t^3 - 5t$.

(a) Find the angular velocity after 2 seconds.
(b) How many revolutions has it made after 2 seconds?
(c) How many revolutions did it make during the 2nd second?

Solution

(a) Since $\alpha = \frac{d\omega}{dt}$, $d\omega = \alpha\, dt \ldots$ and $\ldots \omega = \int \alpha\, dt$

$$\omega = \int (6t^3 - 5t)dt = (6/4)t^4 - (5/2)(t^2) + 6\Big|_0^2,$$

$$= 24 - 10 + 6 = \mathbf{20\ rad/sec}$$

(b) $\Theta = \int_0^2 \omega dt = \int_0^2 \left(\frac{3}{2}\right)t^4 - \left(\frac{5}{2}\right)t^2 + 6dt$

$$= \left(\frac{3}{10}\right)t^5 - \left(\frac{5}{6}\right)t^3 + 6t\Big|_0^2$$

$$= 14.9\text{ rad}\left(\frac{1\text{ rev}}{2\pi\text{ rad}}\right) = \mathbf{2.4\ revolutions}$$

(c) $\Theta_{2nd\ sec} = \Theta_2 - \Theta_1$, and since we already know that $\Theta_2 = 14.9$ rad,

$$\Theta_1 = \left(\frac{3}{10}\right)(1) - \left(\frac{5}{6}\right)(1) + 6 = 5.5\text{ radians}$$

$$\Theta_2 - \Theta_1 = (14.9 - 5.5) = 9.4\text{ radians}\left(\frac{1\text{ rev}}{2\pi\text{ rad}}\right) = \mathbf{1.5\ revolutions}$$

Common Torque Problems

Torque problems commonly involve solid objects that are subject to various forces, which may cause the object to rotate in a given plane. If two forces acting on an object are antiparallel and not colinear, the object will experience a **torque**, or a **turning force**.

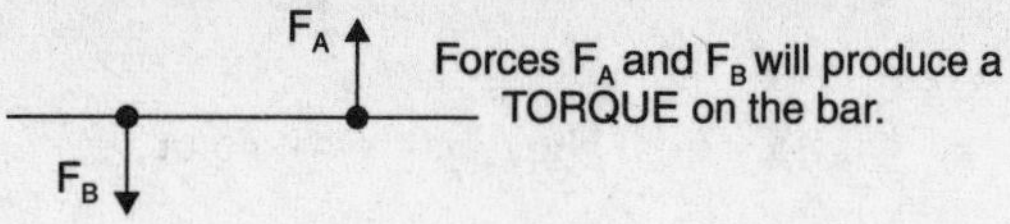

Since the mathematical definition of *torque* is $\boldsymbol{\tau} = \mathbf{r} \times \mathbf{F} = \mathbf{Fr} \sin \Theta$, it is important to note that most torque problems involve forces acting perpendicularly to an object, as shown by both forces $\mathbf{F_A}$ and $\mathbf{F_B}$ in the diagram.

Two basic steps are involved in solving torque problems. For an object to be in both **translational** and **rotational** equilibrium, the following will be true:

1. The sum of the forces UP must equal the sum of the forces DOWN.

   The sum of the forces LEFT must equal the sum of the forces RIGHT.

   The sum of the forces OUT OF THE PAGE must equal the sum of the forces INTO THE PAGE.

   In this case,

$$\Sigma \mathbf{F}_{UP} = \Sigma \mathbf{F}_{DOWN}$$

2. The sum of all CLOCKWISE TORQUES must equal the sum of all COUNTERCLOCKWISE TORQUES.

$$\Sigma \tau_{CW} = \Sigma \tau_{CCW}$$

You may have to pick an arbitrary pivot point around which to calculate the torques. The selection of the pivot point location will not change the outcome of the answer, but a prudent choice of origin can make the analysis easier. Generally a choice of origin where one or more forces acts eliminates some of the terms in the torque equation.

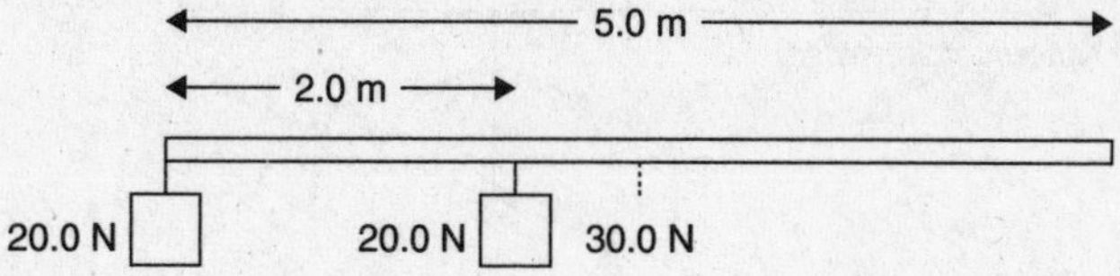

Example (B & C Exams)

A uniformly constructed beam is 5.0 m long and weighs 30.0 N. A mass weighing 20.0 N is hung from the left end, and another 20.0 N mass is hung 2.0 m to the right of the left side. What force must be applied, and where, to maintain translational and rotational equilibrium?

Solution

Follow the two steps:

1. $\Sigma \mathbf{F}_{UP} = \Sigma \mathbf{F}_{DOWN}$

$$\mathbf{F}_{UP} = 20.0 \text{ N} + 20.0 \text{ N} + 30.0 \text{ N} = 70.0 \text{ N}$$

An upward force of 70 N is needed.

2. $\Sigma \tau_{CW} = \Sigma \tau_{CCW}$

Select the midpoint as the pivot point (any point will do, but selecting the midpoint eliminates the bar's weight as a torque-producing force):

$$(\mathbf{F}_{UP})(\mathbf{x}) = (20.0 \text{ N})(2.5 \text{ m}) + (20.0 \text{ N})(0.5 \text{ m})$$
$$= 60.0 \text{ Nm}$$

Substitute 70.0 N for $\mathbf{F}_{UP}$:

$$\mathbf{x} = 60.0 \text{ Nm} / 70.0 \text{ N} = 0.86 \text{ m}$$

The upward force must be placed 0.86 m to the left of center.

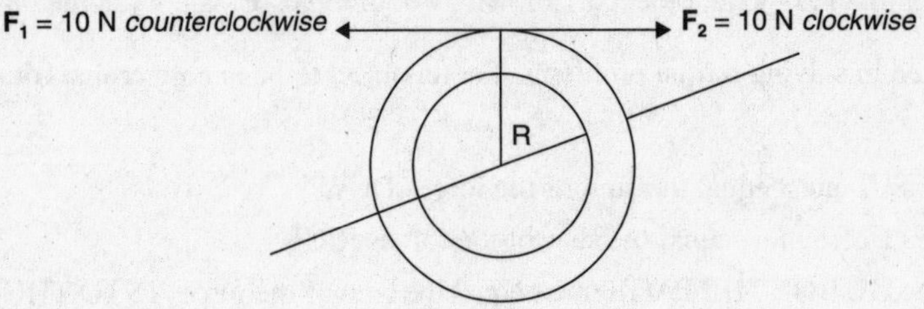

Example (B & C Exams)

A thin wheel is acted upon by two external forces, each acting perpendicularly to the wheel's radius. One force, 10 N, is applied to the wheel so as to give the wheel a clockwise torque. The other force, also 10 N, is applied so as to give the wheel a counterclockwise torque. Describe the resultant situation.

Solution

Because the torques are equal and opposite, the wheel is in *rotational* equilibrium.

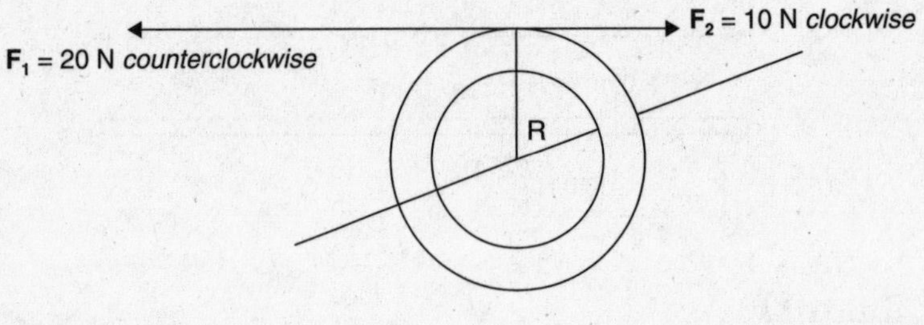

Example (B & C Exams)

In the preceding example, the 10 N force in the clockwise direction remains, but the counterclockwise force is now doubled. Describe the resultant situation.

Solution

Now there is a net force of 10 N in the counterclockwise direction, which causes a resultant counterclockwise **torque**. The wheel accelerates in the counterclockwise direction. Its angular acceleration α would be found by knowing the mass and radius of the wheel, using the relationship $\tau = I\alpha = Fr$.

It is important to note that an object can be in one type of equilibrium but not the other. A car, for example, may be accelerating in a linear direction but not spinning. It is in **rotational** equilibrium only. On the other hand, a wheel on the car may be headed in a straight-line path and accelerating while spinning. It is in **translational** equilibrium only.

A commonly used example of an object in translational equilibrium is a car that is traveling in a straight line at constant speed. Even though it is moving, it is not accelerating; therefore, no net force is acting on it, and the car is experiencing **translational** equilibrium. Conversely, a ball that is dropped and is not spinning accelerates toward the Earth. It is *not* in **translational** equilibrium because the net force of gravity is acting on it, but the ball is in **rotational** equilibrium.

Example (B & C Exams)

A wheel of radius 0.5 m and mass of 10 kg sustains forces of 5 N (perpendicular to the radius in a counterclockwise direction) and 10 N (perpendicular to the radius in a clockwise direction). Describe the resulting motion of the wheel about its axle.

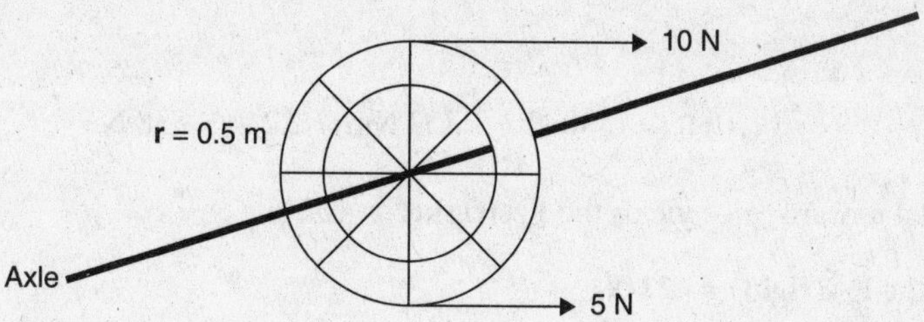

Solution

Two opposing torques are acting on the wheel, one being stronger than the other; therefore, a net torque is acting on the wheel, causing it to accelerate in a rotational fashion. **There is a net force of 5 N clockwise, which will cause the wheel to rotationally accelerate in the clockwise direction.**

The amount of the acceleration can be calculated as follows:

$$\mathbf{Fr = I\alpha}$$

$$\mathbf{\alpha = Fr / I} \text{ (I for a wheel of this type is mr}^2\text{)}$$

This gives the following:

$$\alpha = (5\text{ N})\left(\frac{0.5\text{m}}{\text{rad}}\right) / (10\text{ kg})\left(\frac{0.5\text{m}}{\text{rad}}\right)^2 = \mathbf{1\text{ radian / sec}^2\text{ clockwise}}$$

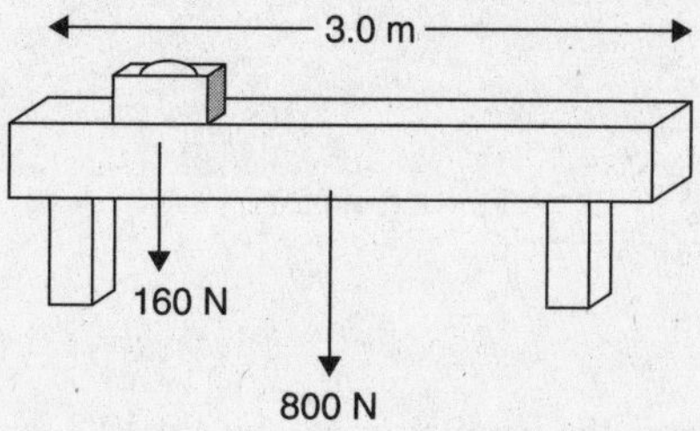

Example (B & C Exams)

A uniform wooden beam is 3.0 m long and weighs 800 N. Its two supports are built 0.4 m in from each end. A box of tools weighing 160 N is placed 0.7 m from the left end. How much weight does each support hold?

Solution

Follow the two steps:

1. $\mathbf{\Sigma F_{UP} = \Sigma F_{DOWN}}$

$$\mathbf{F_{UP} = F_{UP}(left) + F_{UP}(right)} = 800\text{ N} + 160\text{ N} = 960\text{ N}$$

A total upward force of 960 N is needed.

2. $\Sigma\tau_{CW} = \Sigma\tau_{CCW}$

Selecting the support on the right, or the $\mathbf{F_{UP}}$**(right)**, as the pivot point eliminates one variable:

$$[\mathbf{F_{UP}(left)}](2.2 \text{ m}) = (800 \text{ N})(1.1 \text{ m}) + (160 \text{ N})(1.9 \text{ m})$$

This yields:

$$\mathbf{F_{UP}(left)} = (880 \text{ Nm} + 304 \text{ Nm}) / 2.2 \text{ m} = 538 \text{ N}$$

Subtracting this from the total upward force yields the $\mathbf{F_{UP}}$**(right)** as 422 N.

The $\mathbf{F_{UP}}$(left) = 538 N and the $\mathbf{F_{UP}}$(right) = 422 N.

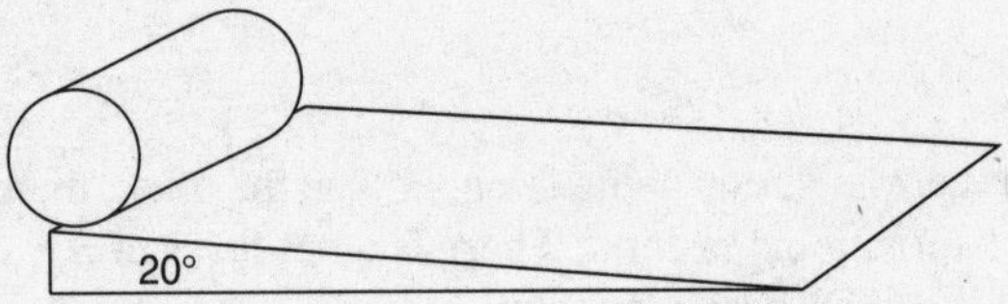

Example (B & C Exams)

A hollow metal tube of mass 1.5 kg and diameter 14.0 cm rolls down an inclined plane from rest. The plane makes an angle of 20° with the horizontal. The tube's rotational inertia is $I = mr^2$.

- **(a)** What is the net force acting on the tube as it rolls down the plane?
- **(b)** What is the tube's linear velocity after rolling1.0 m down the plane?
- **(c)** What is the tube's rotational kinetic energy after rolling 1.0 m down the plane?
- **(d)** What is the tube's angular acceleration as it rolls down the plane?

Solution

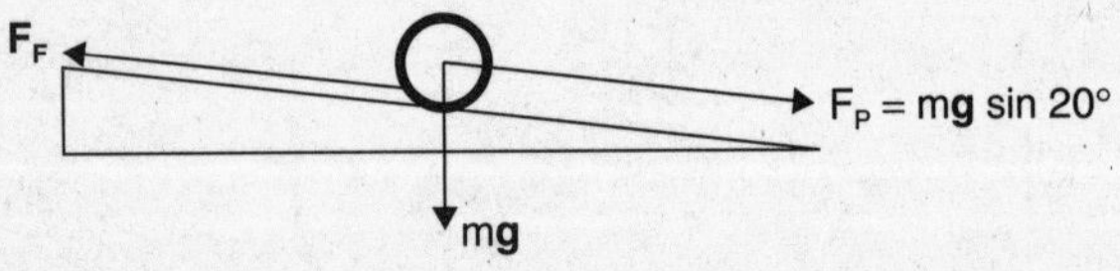

(a) The net force acting on the tube is the parallel force, **mg** sin 20°.
$\mathbf{F_P} = \mathbf{mg} \sin 20° = (1.5 \text{ kg})(9.8 \text{ m/s}^2)(0.342) = \mathbf{5.0\ N}$

(b) $\mathbf{E_P} = \mathbf{E_{K\ TOTAL}} = \mathbf{E_{K\ LINEAR}} + \mathbf{E_{K\ ROTATIONAL}}$

$$\mathbf{mgh} = \left(\frac{1}{2}\right)(\mathbf{mv^2}) + \left(\frac{1}{2}\right)\mathbf{I\omega^2} \quad (\mathbf{h} = 1.0 \text{ m} \sin 20° = 0.34 \text{ m})$$

$$\mathbf{mgh} = \left(\frac{1}{2}\right)(\mathbf{mv^2}) + \left(\frac{1}{2}\right)(\mathbf{mr^2})(\mathbf{v^2/r^2})$$

$$\mathbf{gh} = \left(\frac{1}{2}\right)\mathbf{v^2} + \left(\frac{1}{2}\right)\mathbf{v^2}$$

$$\mathbf{v} = \sqrt{\mathbf{gh}} = \sqrt{(9.8 \text{m/s}^2)(0.34)} = \mathbf{1.8\ m/s}$$

(c) $\mathbf{mgh} = \text{E}_{\text{K LINEAR}} + \text{E}_{\text{K ROTATIONAL}}$

$$\mathbf{mgh} = \left(\frac{1}{2}\right)(\mathbf{mv^2}) + \text{E}_{\text{K ROTATIONAL}}$$

$$\text{E}_{\text{K ROTATIONAL}} = \mathbf{mgh} - \left(\frac{1}{2}\right)(\mathbf{mv^2})$$

$$= \text{m}(\mathbf{gh} - \mathbf{v^2}/2)$$

$$= (1.5 \text{ kg})[(9.8 \text{ m/s}^2)(0.34 \text{ m}) - (1.8 \text{ m/s})^2/2)]$$

$$= \mathbf{2.6\ J}$$

(d) Using $V_F^2 = V_O^2 + 2ad$,
$(1.8\text{ m/s})^2 = 0 + (2)(\mathbf{a})(1.0\text{ m})$
$\mathbf{a} = (3.24\text{ m}^2/\text{s}^2)/2.0\text{ m} = 1.6\text{ m/s}^2$
$\alpha = \mathbf{a/r} = (1.6\text{ m/s}^2)/(0.07\text{ m}) = \mathbf{23\ rad/s^2}$ or
$(23\text{ rad/sec}^2)(1\text{ rev}/2\pi\text{rad}) = \mathbf{3.7\ rev/s^2}$

Review Questions and Answers

B & C Exam Question Types

Multiple Choice

1. A toy remote-controlled car moves from rest with constant acceleration of 2 m/s^2. Its average speed after 3 sec is

(A) 2 m/s
(B) 3 m/s
(C) 4 m/s
(D) 5 m/s
(E) 6 m/s

2. After 2 seconds, the toy car has traveled

(A) 2 m
(B) 3 m
(C) 4 m
(D) 5 m
(E) 6 m

For Questions 3–4 (C Exam): An object moving in a straight line has a velocity **v** m/s that varies with time t sec according to the relationship **v** = 3 + 0.6 t^2.

3. The instantaneous acceleration of the object at t = 2 sec is

(A) 0.6 m/s^2
(B) 1.2 m/s^2
(C) 2.4 m/s^2
(D) 3.6 m/s^2
(E) 4.4 m/s^2

4. The object's total displacement at time t = 3 seconds is most nearly

(A) 0 m
(B) 2 m
(C) 6 m
(D) 8 m
(E) 14 m

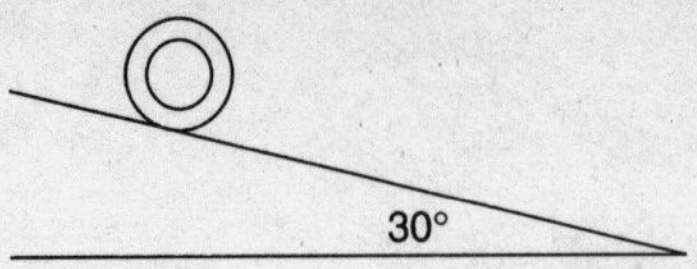

5. **(B & C Exams):** A tire rolls from rest down a 30° hill. It has a mass of 25 kg and a diameter of 0.5 m. Its rotational inertia is mr^2. The net force acting on the tire is approximately

(A) 25 N
(B) 50 N
(C) 75 N
(D) 100 N
(E) 125 N

6. The tire in problem 5 rolls 10 m, where its linear speed is most nearly

(A) 2 m/s
(B) 7 m/s
(C) 15 m/s
(D) 20 m/s
(E) 25 m/s

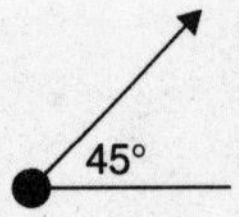

7. A ball is kicked at an angle of 45° with the ground. Its initial speed is 10 m/s. It reaches a maximum height of

(A) 0.5 m
(B) 1.5 m
(C) 2.5 m
(D) 3.5 m
(E) 4.5 m

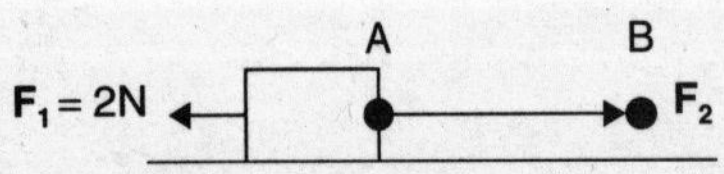

8. A 4.5 kg block is pulled along a surface by a string of mass 0.5 kg. A force $\mathbf{F_2}$ acts on the end of the string. The block is opposed by a frictional force $\mathbf{F_1}$ of 2 N. If the block accelerates at a rate of 1 m/s², the force acting at point A is most nearly

(A) 1 N
(B) 3 N
(C) 5 N
(D) 7 N
(E) 9 N

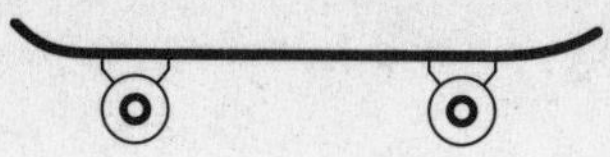

9. A 6 kg skateboard rolls away from its owner at a speed of 5 m/s. If the coefficient of rolling friction between the wheels and the pavement is 0.2, it will come to a stop in

(A) 0.5 second.
(B) 1.0 second.
(C) 1.5 seconds.
(D) 2.0 seconds.
(E) 2.5 seconds.

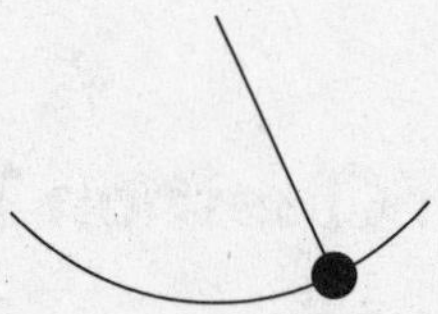

10. A pendulum with a period of 2 seconds at sea level has its length doubled. Its new period is now most nearly

(A) 1 second.
(B) 2 seconds.
(C) 3 seconds.
(D) 4 seconds.
(E) 5 seconds.

Answers to Multiple-Choice Questions

1. **(B)** $\mathbf{v_F = v_O + at}$ = (2 m/s²)(3 sec) = 6 m/s final velocity.

For the average velocity: $\mathbf{v_{AVE}}$ = (0 + 6 m/s)/2 = **3 m/s**

2. **(C)** Again, $\mathbf{v_F = v_O + at}$ = (2 m/s²)(2 sec) = 4 m/s final velocity.

The average velocity is 2 m/s and (2 m/s)(2 sec) = **4 m**

3. **(C)** $\mathbf{a} = (\frac{\mathbf{dv}}{\mathbf{dt}}) = (1.2)\ t\ m/s^2$ and $(1.2\ m/s^2)(2) = \mathbf{2.4\ m/s^2}$

4. **(E)** $\mathbf{v} = \frac{\mathbf{dx}}{\mathbf{dt}}$, and integrate **v**(t) to find **x**:

$\int_0^3 \mathbf{v}(t) = \mathbf{x}(t) \qquad \int_0^3 (3 + 0.6\ t^2)\ dt = 3t + (0.6)(1/3)t^3\big/_0^3 = 9 + (0.6)(9)$

= **14.4 m**

5. **(E)** The net force on the tire is $\mathbf{F_{NET}}$ = mg sin 30°

= (25 kg)(10 m/s²)(0.5) = **125 N**

6. **(B)** $E_P = E_{K\ LINEAR} + E_{K\ ROT}$

$\mathbf{mgh} = \frac{1}{2}\mathbf{mv}^2 + \frac{1}{2}\mathbf{I\omega}^2 \qquad h = 10\ m\ \sin 30° = 5m$

$\mathbf{v} = \sqrt{\mathbf{gh}} = \sqrt{(10m/s^2)(5m)} = \sqrt{50} \approx \mathbf{7\ m/s}$

7. **(C)** At its maximum height, the ball's vertical speed is 0.

Using $\mathbf{v_F^2 = v_O^2 + 2ad}$, $\mathbf{v_{O\,VERT}} = 10 \sin 45° = 10\ (\frac{2^{1/2}}{2}) = 5(2^{1/2})$

$0 = [5(2^{1/2})]^2 + (2)(-10\ m/s^2)(\mathbf{d})$

$\mathbf{d} = \frac{-50m/s^2}{-20m/s^2} = \mathbf{2.5\ m}$

8. **(D)** The net force at point A is the sum of all forces acting on point A.

This means that $\mathbf{F_2 - F_1 = \Sigma ma}$

Therefore, $F_2 = \Sigma ma + F_1 = (5.0\text{ kg})(1\text{m/s}^2) + 2\text{ N} = \mathbf{7\ N}$

9. **(E)** $\mathbf{a = F/m = F_F/m = \mu F_N /m} = (0.2)(6\text{ kg})(-10\text{ m/s}^2) / (6\text{ kg}) = -2\text{ m/s}^2$

$\mathbf{v_F = v_O + at}$ and $t = \mathbf{(v_F - v_O)/a} = (-5\text{ m/s})/(-2\text{m/s}^2) = \mathbf{2.5\ sec}$

10. **(C)** Make a ratio between periods of the pendulum with both lengths:

$$\frac{2\text{sec} = 2\pi\left(1/g_{ALT}\right)^{1/2}}{T_{NEW} = 2\pi\left(2\ 1/g_{ALT}\right)^{1/2}}$$

$T_{NEW} = \sqrt{2}(2\text{ sec})$ yields $2s<T<4s$: **3 sec**

Free Response: B & C Exam Question Types

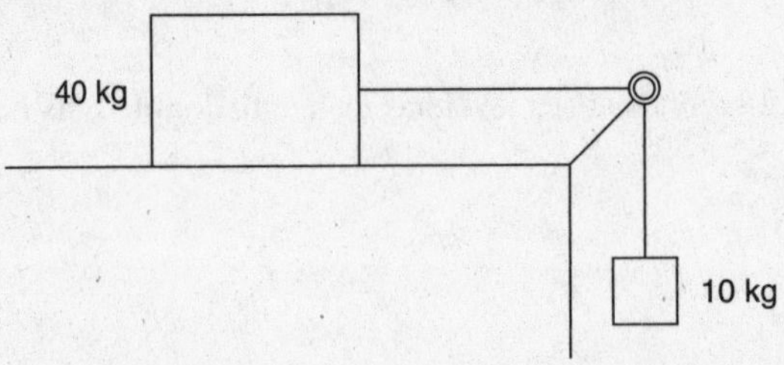

1. A 40 kg mass sits on a rough surface, attached to a string of negligible mass, which passes over a frictionless pulley and suspends a second mass of 10 kg.

(a) If the system does not move, what is the frictional force between the 40 kg block and the surface?
The rough surface is sanded smooth.

Now the system is released, and the 10 kg block falls a distance of 0.5 m in 2.2 seconds.

(b) What is the acceleration of the blocks?
(c) What is the coefficient of kinetic friction between the 40 kg block and the smooth surface?
(d) How far have the blocks moved after 1 second?
(e) What is the speed of both blocks after 0.2 second?

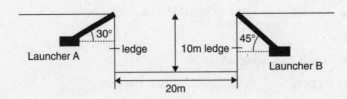

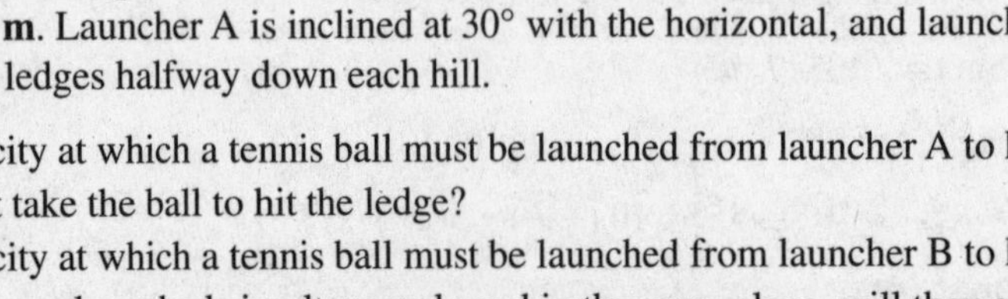

2. Two tennis ball launchers are positioned on adjacent hills, both having a height **10 m**, and are separated by a horizontal distance **20 m**. Launcher A is inclined at 30° with the horizontal, and launcher B is inclined at a 45° angle. There are small ledges halfway down each hill.

(a) What is the velocity at which a tennis ball must be launched from launcher A to hit the opposing ledge?
(b) How long does it take the ball to hit the ledge?
(c) What is the velocity at which a tennis ball must be launched from launcher B to hit the opposing ledge?
(d) If the tennis balls are launched simultaneously and in the same plane, will they collide in mid-air?

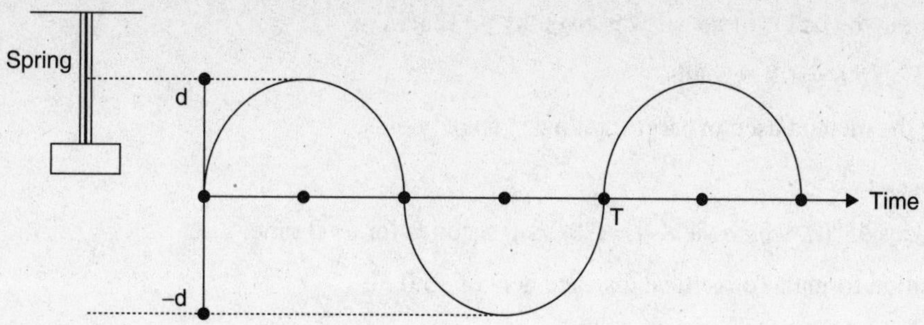

3. **(B & C Exams):** A 0.1 kg mass is attached to a spring (k = 0.4 N/m) and oscillates with an amplitude **d** and period T as illustrated. Let **d** = 0.5 m and T = 3.1 second.

(a) Find the velocity **v** of the mass at time $(\frac{3}{4})$ T.

(b) Find the acceleration **a** of the mass at time $(\frac{3}{4})$ T.

(c) How far has the mass traveled at that point?

(d) The mass is now doubled, and the spring's length is doubled. (An identical spring is attached to one end.) Solve part (b) for this new situation.

(e) The double mass has a second, identical spring placed beside the first spring; solve part (b) again for this situation.

Answers to Free-Response Questions

1. **(a)** Since the system is in equilibrium, this means that the force of friction just balances the weight of the 10 kg mass.

$\mathbf{F} = (10 \text{ kg})(9.8 \text{ m/s}^2) = \mathbf{98\ N}$

(b) Using $d = \mathbf{v_O t} + (\frac{1}{2})\ \mathbf{at^2}$

$(0.5 \text{ m}) = 0 + (\frac{1}{2})(\mathbf{a})(2.2 \text{ sec})^2$ and $\mathbf{a = 0.21\ m/s^2}$

(c) Since the acceleration equals the *net force* divided by the sum of the masses, $\mathbf{a} = \mathbf{F_{NET}} / \Sigma m = (mg_{10} - \mathbf{F_F}) / \Sigma m$

$(0.21 \text{ m/s}^2) = [(10 \text{ kg})(9.8 \text{ m/s}^2) - \mathbf{F_F}]/\ 50 \text{ kg}$, which yields $\mathbf{F_F} = 87.5$ N

Now solving for the coefficient of friction:

$\mu = \mathbf{F_F} / \mathbf{F_N} = 87.5 \text{ N} / (40 \text{ kg})(9.8 \text{ m/s}^2) = \mathbf{0.22}$

(d) Using the position formula $\mathbf{s(t) = d = v_O t} + (\frac{1}{2})\mathbf{at^2}$

$\mathbf{d} = 0 + (\frac{1}{2})(0.21 \text{ m/s}^2)(1 \text{ sec})^2 = \mathbf{0.11\ m}$

(e) Using $\mathbf{v_f = v_O + at}$

$\mathbf{v_f} = 0 + (0.21 \text{ m/s}^2)(0.2 \text{ sec})^2 = \mathbf{0.042\ m/s}$

2. **(a)** The ball must travel a horizontal distance of 20 m with no horizontal acceleration once it is launched.

$\mathbf{d_{HOR}} = \mathbf{v_{O\ HOR}}(t) + 0$

$20 \text{ m} = \mathbf{v_O} \cos 30°(t)$, which yields $t = (23/\mathbf{v_O})$ seconds for time in the air.

Now using the position formula for total vertical displacement –5.0 m:

$(-5.0 \text{ m}) = \mathbf{v_{O\ VERT}}\ (t) + (\frac{1}{2})\mathbf{a}t^2$

$(-5.0 \text{ m}) = \mathbf{v_O} \sin 30°(t) - 4.9(t^2)$

Substituting $(23/\mathbf{v_O})$sec for t yields $\mathbf{v_O}$ **Launcher A = 12.5 m/ s**

(b) The total time the ball is in the air was given by $t = (23/\mathbf{v_O})$

$t = 23\text{ m}/12.5\text{ m/s} = $ **1.8 seconds**

(c) Repeating the method used in part (a) for a 45° angle yields:

$\mathbf{d_{HOR}} = \mathbf{v_{O\,HOR}} + 0$

$20\text{ m} = \mathbf{v_O}\cos 45°(t)$, which yields $t = (28.2/\mathbf{v_O})$ seconds for total time.

Using position formula for vertical displacement of –5.0 m:

$(-5.0\text{ m}) = \mathbf{v_{O\,VERT}}\,(t) + (1/2)at^2$

$(-5.0\text{ m}) = \mathbf{v_O}\sin 45°(t) - 4.9(t^2)$

Substituting $(28.2/\mathbf{v_O})$ sec for t yields **$\mathbf{v_O}$ Launcher B = 12.5 m/ s**

(d) The ball launched from Launcher B takes $t = (28.2/12.5)$ seconds or 2.3 seconds.

Launcher B's ball takes a higher route than Launcher A's ball. **They do not collide.**

3. **(a)** **v** at (3/4) T is at a point of maximum displacement, where **v = 0**

(b) $\mathbf{a} = \mathbf{F}/m = -\mathbf{k}\mathbf{x}/m = (-0.4\text{ N/m})(-0.5\text{ m})/(0.1\text{ kg}) = 2.0$ **m/s²**

(c) $\mathbf{d} = \mathbf{v}_{AVE}(t) = \left(\dfrac{2.0\text{ m}}{3.1\text{ sec}}\right)(0.75)(3.1\text{ sec}) = $ **1.5 m**

(d) Doubling the spring length makes no difference in k. Only the mass being doubled makes a difference.
$\mathbf{a} = \mathbf{F}_{NET}/m = -k\mathbf{x}/\mathbf{m}$

$= (-0.4\text{ N/m})(-0.5\text{ m})/(0.2\text{ kg}) = 1.0$ **m/s²**

(e) Two tandem springs doubles the force per length, giving 2k:

$\mathbf{a} = -2k\mathbf{x}/m = (-0.8\text{ N/m})(-0.5\text{ m})/(0.2\text{ kg}) = $ **2.0 m/ s²**

Momentum, Energy, Work, and Power

Momentum

Momentum is defined as the mass of an object multiplied by its velocity. Because velocity is a vector, momentum is also a vector. Commonly expressed as **p**, the equation for momentum is

$$\mathbf{p} = m\mathbf{v}$$

Closer inspection illustrates that the units of momentum are kgm/s. Because an object with a given momentum usually possesses a constant, unchanging mass and a definite speed and direction, by Newton's First Law, it will continue to have the same velocity, and its momentum will remain constant if no external forces act on the object. If either mass or velocity undergoes change, the momentum will also change accordingly. If an object with momentum **p** collides with another object, it will impart a force (a push or a pull) to the other object. That force may cause the other mass to accelerate. The interaction will impart a force back on the first object, which will cause a change in momentum of the first object by changing its velocity. When objects interact, a number of outcomes are possible. All obey laws of physics.

Law of Conservation of Momentum

In a collision involving two or more objects, the sum of the total momenta *before* the collision equals the sum of the total momenta *after* the collision. A special case of this Law is Newton's Third Law of Action/Reaction.

$$\text{Before} \qquad \text{After}$$

$$\Sigma m\mathbf{v}_O = \Sigma m\mathbf{v}_F$$

$$m_1\mathbf{v}_{1\,o} + m_2\mathbf{v}_{2\,o} + \ldots = m_1\mathbf{v}_{1\,F} + m_2\mathbf{v}_{2\,F} + \ldots$$

Impulse and Momentum

When a force acts on an object for a certain amount of time, we say that it imparts an **impulse** to the object. The **impulse, J**, in turn, imparts **momentum, p,** to or changes the existing **momentum** of the object.

Impulse Produces Momentum

$$\mathbf{J} = \mathbf{p}$$

$$\mathbf{F}\Delta t = m\Delta\mathbf{v}$$

Impulse **Momentum**

UNITS: Ns kgm/s

Even though the units of **impulse** are different than those of **momentum**, they are equivalent. An impulse of 20 Ns imparts 20 kgm/s of momentum. An interesting relationship is that dividing both impulse and momentum by Δt yields Newton's Second Law: $(\mathbf{F}\Delta t)/\Delta t = (m\Delta\mathbf{v})/\Delta t$ = yields $\mathbf{F} = m\Delta\mathbf{v}/\Delta t = d\mathbf{p}/dt = d/dt\,(m\mathbf{v}) = m\,(d\mathbf{v}/dt) = \mathbf{F} = m\mathbf{a}$. (This assumes a constant mass.)

Momentum and Collisions

Two types of collisions exist, **Elastic** and **Inelastic.**

In an elastic collision, the objects rebound from each other completely, without any energy loss or deformation. Collisions between gas molecules and atomic particles generally are regarded as being perfectly elastic. Physics problems involving billiard balls often negate the small amount of heat produced in their collisions and commonly are used as models for elastic collision problems.

In an inelastic collision, objects commonly stick together or become deformed. Two railroad cars that collide and interlock commonly are used as models for inelastic collision problems, as are clay or putty masses that stick together and objects that become embedded in target objects. Most automobile accidents are inelastic collisions.

In considering collisions in two dimensions, it is important to reiterate that the **vector sum** of the momenta of the interacting particles will yield the **vector resultant** of the momenta after the interaction. When an object strikes another object and both travel at different angles after the collision, the **vector sum** of the momenta of the colliding particles before the collision will equal the **vector sum** of the momenta of the particles after the collision.

It is important to note that in *any* type of interaction, momentum is always conserved. Mechanical energy is NOT always conserved in such interactions. The relationship between momentum and energy is covered later in this chapter.

Example (B & C Exams)

A force of 10 N acts for 0.2 seconds on a 10 kg object initially at rest. Compute the object's final velocity.

Solution

$$\mathbf{F\Delta t = m\Delta v}$$

$$\Delta \mathbf{v} = \mathbf{F}\Delta t / m = (10\text{ N})(0.2\text{ sec})/(10\text{ kg}) = \mathbf{0.2\ m/s}$$

Example (B & C Exams)

A 30 kg mass travels east along a frictionless surface with a velocity of 2 m/s. It strikes a stationary 10 kg mass. After the collision, the 10 kg mass moves east with a velocity of 4 m/s. What is the final velocity of the 30 kg mass?

Solution

$$\text{Before} \qquad\qquad \text{After}$$

$$\Sigma m\mathbf{v}_O = \Sigma m\mathbf{v}_F$$

$$m_1\mathbf{v}_{1\,o} + m_2\mathbf{v}_{2\,o} + \ldots = m_1\mathbf{v}_{1\,F} + m_2\mathbf{v}_{2\,F} + \ldots$$

$$(30\text{ kg})(2\text{ m/s}) + 0 = (30\text{ kg})(\mathbf{v}_{F\,30}) + (10\text{ kg})(4\text{ m/s})$$

$$(60 - 40)/30 = \mathbf{v}_{F\,30} \text{ and } \mathbf{v}_{F\,30} = \mathbf{0.67\ m/s\ east}$$

Center of Mass (C Exam)

In any interaction of objects or systems of objects, the center of mass of an object or system of objects is taken to be the interaction point. In other words, we consider the center of mass to be the point about which any calculations are made

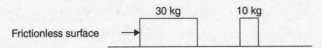

regarding forces, impulses, or momenta. The center of mass of a system of objects, whether in two- or three-dimensional space is given as the following:

$$\mathbf{d_{CM} = \Sigma\, m\, r / M}$$

where $\mathbf{d_{CM}}$ is the distance of the center of mass from the origin or reference point; **m** is the mass of the individual particle; **r** is its distance from the origin or reference point; and **M** is the total mass of all of the particles in the system. If you are finding the center of mass of particles in a two-dimensional system, you find $\mathbf{d_{CM}}$ for $\boldsymbol{x}$ and then $\mathbf{d_{CM}}$ for $\boldsymbol{y}$. Using the Pythagorean Theorem, you will obtain the magnitude of the vector from the origin or reference point to the center of mass of the system. For a two-dimensional system, the angle made with the x-axis by the center of mass of a system of particles is found by letting the ratio of the y-component to the x-component equal the tangent of the angle. This will yield the vector location of the angle that the center of mass makes with the positive $\boldsymbol{x}$-axis. For three-dimensional systems, add $\mathbf{d_{CM}}$ for **z** and use the triple Pythagorean to find the magnitude of the resultant vector. To find the angle that the resultant vector makes with the $\boldsymbol{x}$-axis, for example, use the dot product method discussed in the first chapter.

Example (C Exam)

Find the center of mass for the following three particles: A: 5 kg at (1, 2, 4) cm; B: 4 kg at (–2, 0, 5) cm; C: 3 kg at (–3, 2, 6) cm.

Solution

Particle	*Mass (kg)*	*x (cm)*	*y (cm)*	*z (cm)*
A	5	1	–2	4
B	4	–2	0	5
C	3	–3	2	6
Σ:	12	–4	0	15

$$\mathbf{d_{CM}x} = \frac{\Sigma \mathbf{mx}}{\Sigma \mathrm{m}} = \frac{(5\,\mathrm{kg})(1\,\mathrm{cm}) + (4\,\mathrm{kg})(-2\,\mathrm{cm}) + (3\mathrm{kg})(-3\mathrm{cm})}{12\,\mathrm{kg}} = -1\ \mathrm{cm}\ \mathbf{x}$$

$$\mathbf{d_{CM}y} = \frac{\Sigma \mathrm{my}}{\Sigma \mathrm{m}} = \frac{(5\,\mathrm{kg})(-2\,\mathrm{cm}) + (4\,\mathrm{kg})(0\,\mathrm{cm}) + (3\,\mathrm{kg})(2\,\mathrm{cm})}{12\,\mathrm{kg}} = -0.33\ \mathrm{cm}\ \mathbf{y}$$

$$\mathbf{d_{CM}z} = \frac{\Sigma \mathbf{mz}}{\Sigma \mathrm{m}} = \frac{(5\,\mathrm{kg})(4\,\mathrm{cm}) + (4\,\mathrm{kg})(5\,\mathrm{cm}) + (3\,\mathrm{kg})(6\,\mathrm{cm})}{12\,\mathrm{kg}} = 4.8\ \mathrm{cm}\ \mathbf{z}$$

The center of mass of the three particles is located at (–1, –0.33, 4.8) cm.

Magnitude of the center of mass vector: $[(-1)^2 + (-0.33)^2 + (4.8)^2]^{\frac{1}{2}} = \mathbf{4.9\ cm}$

Example (B & C Exams)

Object A has a mass of 5.0 kg and is moving horizontally with velocity 5.0 m/s. Object A is hit by another object, which causes it to reverse direction. If its new velocity is –10 m/s and the objects were in contact with each other for 0.02 seconds, find the average force, **F**, exerted on object A.

Solution

Take the initial direction of object A to be positive.

Impulse produces momentum as shown in the following equation.

$$F\Delta t = m\Delta \mathbf{v}$$

$$0.02\text{ sec} = (5.0\text{ kg})(\mathbf{v}_F - \mathbf{v}_o)$$

$$(-10\text{ m/s} - 5.0\text{ m/s})$$

$$\mathbf{F} = \frac{m\Delta \mathbf{v}}{\Delta t} = \frac{(5.0\text{ kg})(-15.0\text{ m/s})}{0.02\text{ sec}} = \mathbf{-3750\ N}$$ (The negative sign indicates that the force is applied opposite the initial momentum of object A.)

Angular Momentum

As in linear momentum, **angular** momentum is the product of an object's effective mass and velocity. In this case, however, the effective mass is a result of the *distribution* of mass as an object spins. Also, the velocity is now the **angular** velocity of the object. A rotating object's angular momentum will remain constant unless acted upon by an outside torque. The relationship between **linear** momentum and **angular** momentum, **L,** is described mathematically as follows:

Linear Impulse	Linear Momentum	Angular Impulse	Angular Momentum
$F\Delta t$	$\mathbf{p} = m\mathbf{v}$	$\tau\Delta t$	$\mathbf{L} = \mathbf{I}\boldsymbol{\omega}$

Common representations of **rotational inertia, I,** are

Point mass, hoop, thin ring, hollow cylinder $I = mr^2$ *

Solid disk or solid cylinder(flywheel) $I = \left(\frac{1}{2}\right)mr^2$ *

Solid sphere $I = \left(\frac{2}{5}\right)mr^2$ *

Hollow sphere $I = \left(\frac{2}{3}\right)mr^2$

Thin rod of mass, m, and length, l, through center $I = \left(\frac{1}{12}\right)ml^2$

Thin rod of mass, m, and length, l, at end $I = \left(\frac{1}{3}\right)ml^2$

Thick ring or washer, with central hole of radius $\mathbf{r_1}$ and washer radius $\mathbf{r_2}$ $I = \left(\frac{1}{2}\right)m\,(r_1^2 + r_2^2)$

Solid disk with axis through diameter $I = \left(\frac{1}{4}\right)mr^2$

*The most commonly used rotational inertias

Example (B & C Exams)

A child's toy merry-go-round has a mass of 32 kg, and its plastic platform has a diameter of exactly 1.0 m. What constant unbalanced force, **F**, is required to increase its speed from 0.5 rev/sec to 1.5 rev/sec in 3.0 sec?

Solution

$$\tau = I\alpha = Fr$$

$$\left(\frac{1}{2}\right)mr^2 \quad \frac{(\omega_F - \omega_o)}{\Delta t} \quad \frac{d}{2}$$

$$F = \frac{\left[\left(\frac{1}{2}\right)mr^2\right]\left[\left(\frac{\omega_F - \omega_o}{\Delta t}\right)\right]}{0.5\text{ m/rad}} = \frac{\left[\left(\frac{1}{2}\right)(32\text{ kg})(0.5\text{ m})^2\right]\left[(1.5\text{ rev/s} - 0.5\text{ rev/s})(2\pi\text{ rad/rev})\right]}{0.5\text{ m/rad}} = \mathbf{50.3\ N}$$

Energy

Energy is defined as the ability to do work. Work is the exertion of a force on an object that results in the object being moved through a distance. Because it takes energy to do work (**The Work-Kinetic Energy Theorem: $W = \Delta E_K$** or **Work done equals the change in Kinetic Energy**) and because energy is considered in terms of the work it can do, the units of work and energy are the same. A **Joule**, or Newton-meter, is the unit of energy and of work.

Law of Conservation of Energy

Energy can neither be created nor destroyed but can be changed into other forms of energy. In any closed system (such as the Universe), where no energy is added or removed, the total amount of energy is constant. It may change from one form to another or other forms, matter included, but the sum total amount of energy will remain constant.

An easy way to remember the forms of energy is to remember

I S C R E A M

in which each letter stands for a form of energy:

Internal (B Exam)

The kinetic theory of matter holds that all matter is composed of tiny particles, which are constantly in motion. The motion arises from the internal energy possessed by atoms and molecules in their various states. Any increase in temperature due to the addition of radiant energy increases the motion, while loss of radiant energy (cooling) reduces the motion.

Sound (B Exam)

Caused by a vibratory or compressional disturbance, it is a transfer of mechanical energy through a transmitting substance, or any solid, liquid, or gas. A longitudinal wave disturbance, sound exerts pressure which, by definition, is a force on an area.

Chemical (B & C Exams)

A certain amount of energy can either be absorbed or released by a chemical reaction. The energy may be in radiant, electrical, or mechanical form. A prime example is a common dry cell or battery of cells in which chemical energy is stored until the chemicals react and transfer this stored energy into electrical energy.

Radiant (B Exam)

Commonly and mistakenly limited to thermal phenomena, radiant energy encompasses the entire electromagnetic spectrum. Transverse in nature, radiant energy waves range from approximately 10^{-16} m to 10^8 m in wavelength. All radiant energy travels at the speed of light, 2.9979246×10^8 m/s in a vacuum. The value of 3×10^8 m/s is commonly used for the speed of light in air.

Electrical (B & C Exams)

Electrical energy, or electricity, is the flow of electrons. Metals easily give up their loosely held electrons to nonmetals. When a force is applied to electrons, they will flow, either through a conductive material or through air. A conductive material, such as a metal, will allow electrons to flow when one side is heated or compressed. If electrical pressure, in the form of electromagnetic force (**emf**), is applied, as in an electrical cell, the pressure causes electron flow. If electron repulsion from an electron-rich metal or cloud is great enough, electrons will spark through air.

Atomic or Nuclear (B Exam)

Atomic or nuclear energy arises from the strong force that binds stable atomic nuclei together or the weak force that contains the particles in a radioactive material.

Mechanical (B & C Exams)

Mechanical energy is possessed by a moving object or substance in motion (kinetic). It is also energy that is mechanically stored (potential) and has the capacity to be released and transferred into kinetic energy.

Kinetic Energy

E_K **= (1/2) mv²**, where m is the mass of the object and **v** is its velocity. It is the energy of motion. Since it is dependent on the square of the velocity, it is always positive. The units of kinetic energy are $(kg)(m^2/s^2)$, Nm, or Joules.

Potential Energy

Different types of potential energy examples include:

Gravitational Potential Energy: E_P = mgh, where **m** is the mass of the object, **g** is the gravitational acceleration of the object in a particular location, and **h** is the height above a reference altitude. The units of gravitational potential energy are $(kg)(m/s^2)(m)$, Nm, or Joules.

Elastic Potential Energy: $E_P = (1/2)kx^2$, where **k** is the **spring constant** of the stretched or compressed object and **x** is the distance that the stretched or compressed object is displaced from its equilibrium position. Common examples of elastic potential energy as encountered in physics problems are springs, rubber bands, and bungee cords. Since **k** is the spring constant, which has units of N/m, the units of elastic potential energy are $(N/m)(m^2)$, Nm, or Joules.

Electrical Potential Energy: Measured in Joules or electron volts. An electron volt is defined as the amount of energy required to move an electron through a potential difference of 1 volt. A volt is defined as one Joule of energy per Coulomb of charge. $1\ eV = 1.602 \times 10^{-19}$J. (Electricity is covered in the chapter, "Electric Fields and Forces.")

Magnetic Potential Energy: When a magnetic field is present, a ferromagnetic substance experiences a force. This force follows the inverse-square law, as do all force fields (gravitational, magnetic, electrical, sonic, and radiant). A ferromagnetic substance in a magnetic field is another example of potential energy that can change to kinetic. A steel ball sitting on a table with a magnet nearby feels an attractive force. The force will accelerate the ball. The magnetic potential energy becomes kinetic energy in this case. (Magnetism is covered in the last chapter of this book, "Magnetic Fields and Forces.")

Energy Curve: A marble rolling inside a bowl has both potential and kinetic energy. At the top of its roll, all of its energy is potential (E_P). At the bottom of its roll, all of its energy is kinetic (E_K). At any point between, the total energy equals the sum of the potential and kinetic energies.

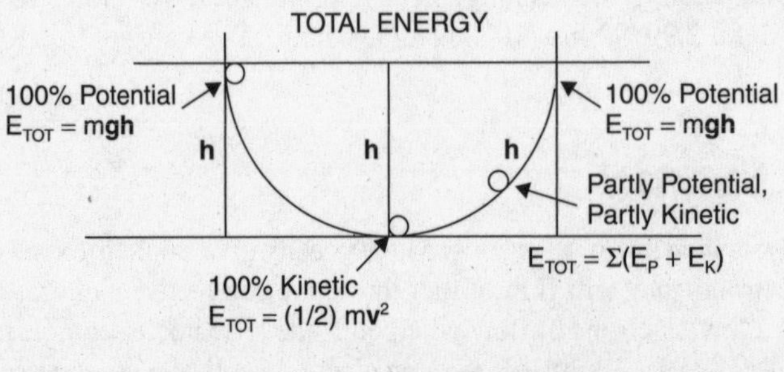

Another example might be a vertically oscillating spring, which experiences 100% potential energy at the top or at the bottom of its path, given by $E_P = 1/2\ (k)(\mathbf{x})^2$. At its midpoint, the potential energy in the spring is 0, and it possesses 100% kinetic energy. A book falling off a table is another example.

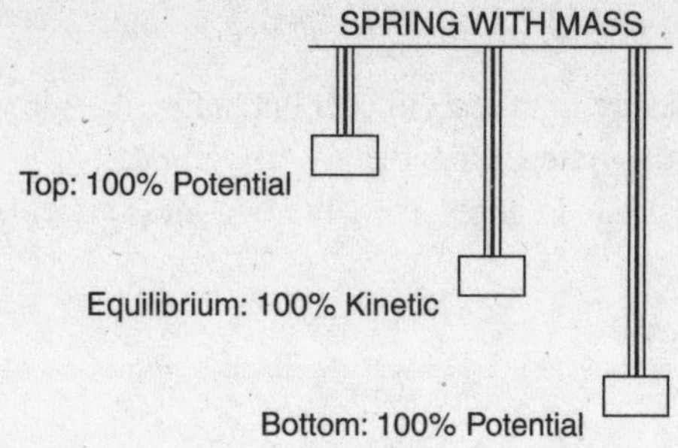

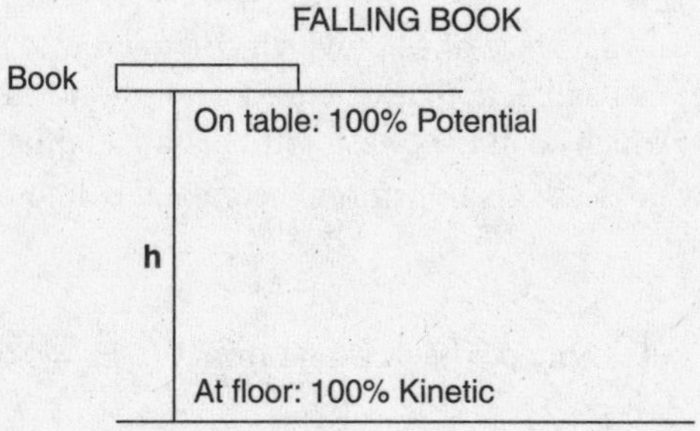

Example (B & C Exams)

A 2 kg box sits on a shelf that is 2 m high. If the box falls, what is its velocity when it hits the floor?

Solution

Since the box possesses all potential energy before it falls, when it hits the floor, the potential energy has all changed to kinetic energy. The following has occurred:

$$E_{P\,GRAV} = E_K$$

$$\mathbf{mgh} = \left(\frac{1}{2}\right)\mathbf{m}\mathbf{v}^2$$

$$(2\text{ kg})(9.8\text{ m/s}^2)(2\text{ m}) = \left(\frac{1}{2}\right)(2\text{ kg})(\mathbf{v}^2)$$

Note that the mass will cancel from both sides of the equation. This is a statement of the fact that all bodies fall at the same rate (in the absence of the outside force of air friction), as Galileo stated.

Cancellation yields $\mathbf{gh} = \mathbf{v}^2/2$ and $\mathbf{v} = \sqrt{2\mathbf{gh}}$

$$\mathbf{v} = \sqrt{(2)(9.8\text{ m/s}^2)(2\text{ m})} = \mathbf{6.3\text{ m/s}}$$

Example (B & C Exams)

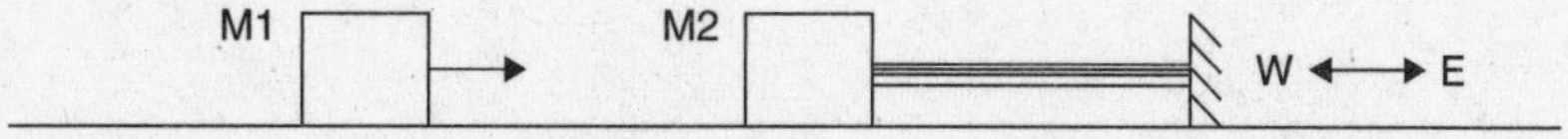

A 10 kg mass (M 1) moves east with a velocity of 0.5 m/s along a frictionless surface toward another 10 kg mass (M 2) attached to a spring with a spring constant of 2.0 N/m.

(a) What is the total mechanical energy of the system after impact?
(b) Describe the action after M 1 collides elastically with M 2.
(c) How much does M 2 compress the spring before stopping?

The process is repeated. This time, M 1 and M 2 now stick together at impact.

(d) What is their combined velocity right after impact but before the spring becomes compressed?
(e) What is the momentum of the system immediately after impact?
(f) How much work does the spring do in reversing the two-mass direction? How much *total energy* is involved in doing so?

Solution

(a) In the absence of friction, the total mechanical energy after collision is equal to the mechanical energy before the collision. $E_K = \left(\frac{1}{2}\right)mv^2 = \left(\frac{1}{2}\right)(10\text{ kg})(0.5\text{ m/s})^2 = \mathbf{1.3\ J}$

(b) M 1 hits M 2, stops, and takes the place of M 2, which continues east into the spring, compressing it to its maximum, stops, and reverses its direction, accelerates west toward M 1 until it collides with M 1, stops, and takes the place of M 1, which continues west with the same initial speed, 0.5 m/ s, as it had initially.

(c) The kinetic energy initially possessed is transformed completely into potential energy stored in the spring.

$$E_K = E_P$$

$$\left(\frac{1}{2}\right)mv_o^2 = \left(\frac{1}{2}\right)kx^2 \text{ and from (a), } E_K = 1.3\text{ J}$$

$$1.3\text{ J} = \left(\frac{1}{2}\right)(2\text{ N/m})(x^2) \text{ and } x = \sqrt{1.3}\text{ m} = \mathbf{1.1\ m}$$

(d)

BEFORE AFTER

$$(10\text{ kg})(0.5\text{ m/s}) + 0 = (20\text{ kg})(V_{FIN\ M1+M2})$$

$$\mathbf{V_{FIN\ M1+M2} = 0.25\ m/s\ east}$$

(e) $p = mv = (20\text{ kg})(0.25\text{ m/s}) = \mathbf{5.0\ kg\ m/s\ east}$

(f) 1.3 J to stop and –1.3 J to reverse direction and motion of the double mass: **0 J**. The total energy involved is **2.6 J.**

Rotational Energy

When a mass rolls, it possesses linear kinetic energy:

$$E_K\text{ LINEAR} = \left(\frac{1}{2}\right)mv^2$$

It also possesses rotational kinetic energy:

$$E_K\text{ ROTATIONAL} = \left(\frac{1}{2}\right)I\omega^2$$

As a round object rolls down an incline, **gravitational potential energy** changes into **linear kinetic energy** and **rotational kinetic energy**.

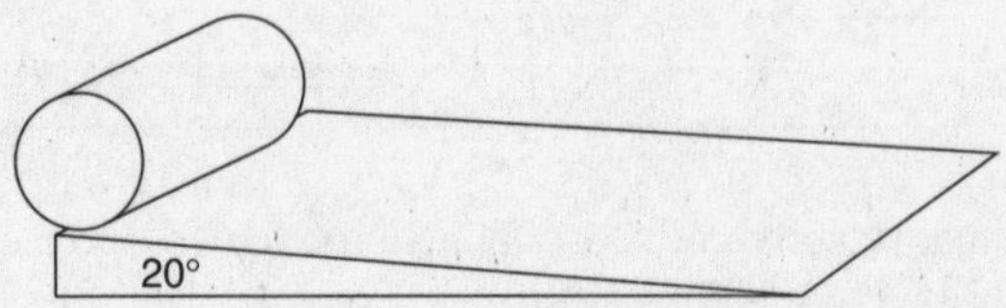

Example (B & C Exams)

A 2.0 kg solid cylinder rolls down a 20° incline from rest. The cylinder has a diameter of 5.0 cm. If it starts from a vertical height of 2.0 cm, and since $I = \left(\frac{1}{2}\right)mr^2$, calculate

(a) its linear velocity at the bottom of the incline.
(b) its angular velocity at the bottom of the incline.
(c) its total kinetic energy at the bottom of the incline.
(d) what percent of its total kinetic energy is rotational.

Solution

(a) E_KTOTAL = E_K LINEAR + E_K ROTATIONAL
Since all kinetic energy is a result of the gravitational potential energy, **mgh** = E_K LINEAR + E_K ROTATIONAL.

$$\mathbf{mgh} = \left(\frac{1}{2}\right)\mathbf{mv^2} + \left(\frac{1}{2}\right)\mathbf{I\omega^2}$$

$$\mathbf{mgh} = \left(\frac{1}{2}\right)[\mathbf{mv^2} + \left(\frac{1}{2}\right)\mathbf{mr^2(v/r)^2}]$$

$$\mathbf{gh = (v^2/2) + (v^2/4) = 3\,v^2/4}$$

$$(9.8\ \text{m/s}^2)(0.02\ \text{m})(4) / 3 = \mathbf{v^2}$$

$$\mathbf{v = 0.5\ m/\ s}$$

(b) $\boldsymbol{\omega} = \mathbf{v/r}$ = (0.5 m/s) / (0.025 m/rad) = **20 rad/ s**

(c) $E_K\text{TOTAL} = \left(\frac{1}{2}\right)\mathbf{mv^2} + \left(\frac{1}{2}\right)\mathbf{I\omega^2} = \left(\frac{1}{2}\right)[\mathbf{mv^2} + \left(\frac{1}{2}\right)\mathbf{mr^2(v/r)^2}]$

$$= \left(\frac{1}{2}\right)\mathbf{mv^2} + \left(\frac{1}{2}\right)\left(\frac{1}{2}\right)\mathbf{mr^2(v/r)^2}$$

$$= \left(\frac{3}{4}\right)\mathbf{mv^2}$$

(Equal to **mgh**,the potential energy at the top) $= \left(\frac{3}{4}\right)(2.0\ \text{kg})(0.25\ \text{m}^2/\text{s}^2) = \mathbf{0.38\ J}$

(d) $E_K\text{ ROTATIONAL} / E_K\text{TOTAL} = \dfrac{\frac{1}{2}\mathbf{I\omega}^2}{\frac{3}{4}\mathbf{mv}^2} = \dfrac{\frac{1}{2}\left(\frac{1}{2}\mathbf{mr}^2\right)\left(\frac{\mathbf{v}^2}{\mathbf{r}^2}\right)}{\frac{3}{4}\mathbf{mv}^2} = [\left(\frac{1}{4}\right)\mathbf{mv^2}] / [\left(\frac{3}{4}\right)\mathbf{mv^2}] = \mathbf{33.3\%}$

Work

In simple terms, work is **force** times **distance**.

$$\mathbf{W = Fd}$$

If a force is applied to an object and the force moves the object, **work** has been done. The amount of work done is the product of the part of the applied force in the direction of the motion times the total distance the object has moved:

$$\mathbf{W = F \cdot d = Fd}\cos\Theta$$

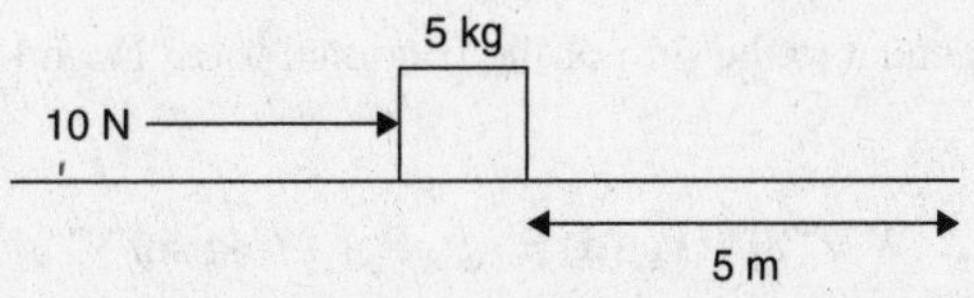

Example (B & C Exams)

A force of 10 N is applied to a 5-kg crate, pushing it a distance of 5 m at a constant speed. How much work is done by the force?

Solution

$W = Fd$ since the force is in the direction of motion. The work done is $W = (10\ N)(5\ m) = \mathbf{50\ J}$

Example (B & C Exams)

If the applied force in the preceding problem is 10 N and the coefficient of kinetic friction between the crate and the surface is 0.2, what is the work done by the friction?

Solution

The work done is the frictional force times the distance. Since $\mu_K = F_F / F_N$, $W = (\mu_K)(F_N)(5\ m) = (0.2)(5\ kg)(9.8\ m/s^2)(5\ m) = \mathbf{-49J}$. (The negative sign indicates work done by friction opposes motion.)

If a force is applied to an object and it does not move in the direction of the applied force, **no work is done by that force.**

Example (B & C Exams)

How much work is done by a planet keeping a satellite in circular or orbit around it?

Solution

Since the force of gravitation is toward the planet and the motion of the satellite is perpendicular to that force, it does *no work.*

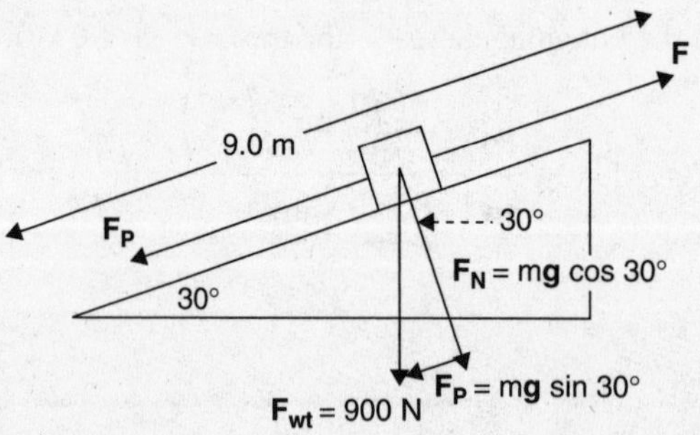

Example (B & C Exams)

A box weighing 900 N is pushed up a ramp a distance of 9.0 m at a constant speed. The ramp is inclined at an angle of 30° with the horizontal, and the coefficient of kinetic friction between the box and the ramp is 0.23. Calculate the work that is done in pushing the box.

Solution

The work done is **Fd**. The force **F** is equal to the sum of the frictional force F_F and the parallel force F_P. The distance is 9.0 m. This yields the following:

$$W = (F_F + F_P)(d) = [(\mu_K F_N) + (mg \sin 30°)](d)$$
$$=[(0.23)(900N)(\cos 30°) + (900\ N)(\sin 30°)](9.0\ m)$$
$$= [179\ N + 450\ N)(9.0\ m) = \mathbf{5700\ J}$$

Work Done by a Variable Force

A force that is not constant may move an object through a certain displacement. Such forces are provided by springs, variable braking systems, and certain types of propulsion systems. The work done by such forces is equal to the area under the curve of variable force, **F**, versus distance or displacement, **x**.

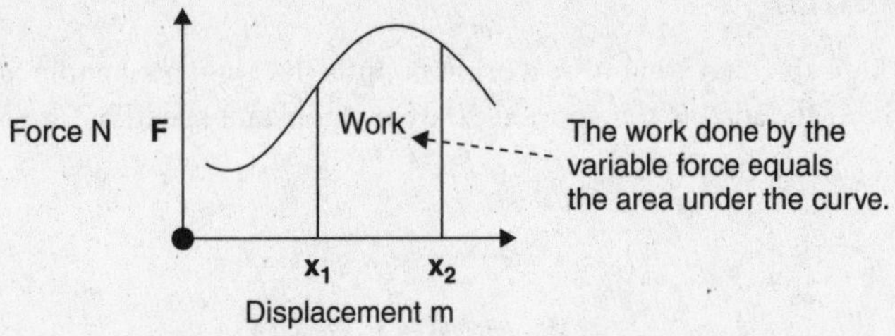

For work done by a variable force, it is necessary to take the definite integral of the force over the extent of displacement. For a variable force, **F**, which acts from point x_1 to point x_2, the work done would be as follows:

$$W = \int_{x_1}^{x_2} F(x)\,dx$$

Example (C Exam)

A certain spring of negligible mass has a spring constant k of 5.0 N/m. How much work is required to stretch it 0.3 m from its rest position?

Solution

Using $W = \int_0^{.3} -kx\,dx = -\left(\frac{1}{2}\right)k\,x^2\Big|_0^{3} = |(-1/2)(5.0\text{N/m})(0.3\text{m})^2| = \mathbf{0.2\ J}$

Example (B Exam)

A certain spring of negligible mass has a spring constant k of 5.0 N/m. How much work is required to stretch it 30 cm from its rest position?

Solution

Using Hooke's Law and the work definition, **W = Fd, W** = the average spring force (**F**/2) times the distance stretched. Since the maximum spring force is **F** = –**kx,** the average force is **F**/2 = –**kx**/2 . Therefore, the work done is (–**kx**/2)(**x**) = |(–5.0 N/m)(0.3 m/2)(0.3 m)| = **0.2 J**

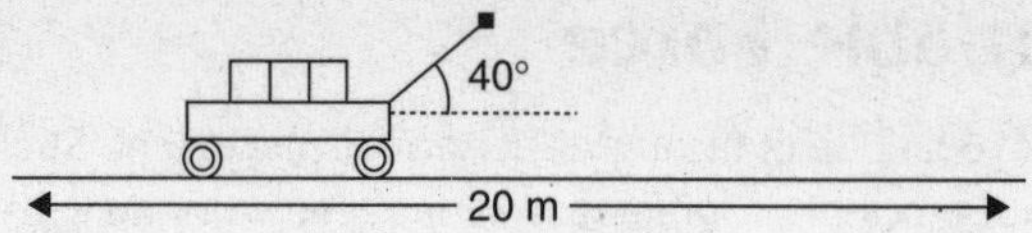

Example (B & C Exams)

A stack of bricks weighing 300 N sits on a wagon. A workman pulls the wagon's handle at an angle of 40° with the horizontal with a force of 15 N. If the load is transported 20 m at a constant speed before stopping, how much work is done by the workman?

Solution

$$\mathbf{W} = \mathbf{F} \cdot \mathbf{d} = \mathbf{Fd} \cos \Theta$$
$$= \text{(Pulling force)(distance)(cos 40°)}$$
$$= (15\text{ N})(20\text{ m})(0.766)$$
$$= \mathbf{230\ J}$$

Power

The definition of **power** is the rate at which work is done. The units are Joules per second or Watts.

$$\text{Average power: } P = \frac{W}{t} \text{ or } \frac{\Delta W}{\Delta t}$$

$$\text{Instantaneous power: } P = \frac{dW}{dt}$$

Inspection reveals that $P = \frac{W}{t} = \frac{Fd}{t} = \mathbf{Fv}$.

Example (B & C Exams)

A demolished car weighing 15,000 N is picked up by an electromagnet on a crane in a junkyard. It lifts the car 10.0 m in 4.0 seconds. What is the output power rating of the crane?

Solution

$$P = \frac{W}{t} \text{ or } \frac{\Delta W}{\Delta t} = \frac{Fd}{t}$$

The force of weight of the car is lifted through a definite displacement in a specified amount of time:

$$P = (15{,}000\text{ N})(10.0\text{ m})/(4.0\text{ sec}) = \mathbf{37{,}500\ W}$$

Energy/Work/Power/Forces and Simple Machines

Simple machines *multiply force*. They do this by spreading the applied force over a larger distance.

These devices are used in our everyday life. They make doing work easier. The amount of mechanical aid they offer is measured in **Mechanical Advantage** (MA).

A machine's MA is the number of times it multiplies the input force. The multiplicative factor may be greater or less than one, as output forces may be less than input forces.

Simple machines *cannot increase the amount of work done*, they merely multiply the force or, in the case of a single fixed pulley, change the direction of the force to a more convenient application angle.

The six simple machines are **inclined plane**, **wedge**, **wheel and axle**, **lever**, **screw**, and **pulley**.

Six Simple Machines: Definition, Illustration, and Mechanical Advantage

1. **INCLINED PLANE:** A ramp or surface elevated at one end, forming a right triangle.

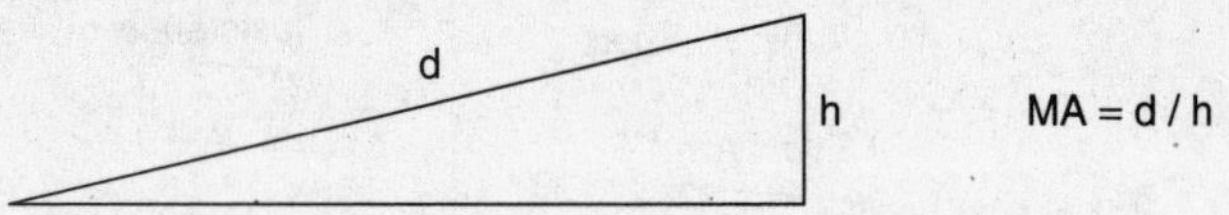

Examples: Wheelchair ramp, truck roller-ramp, freeway on-ramp.

2. **WEDGE:** Two inclined planes back-to-back with all angles less than 90°.

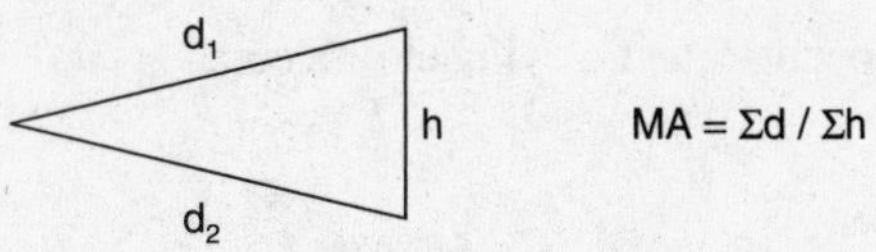

Examples: Knife blade, scissor blade, log-splitter, screwdriver edge.

3. **WHEEL and AXLE:** Two cylinders of different radii attached through a common axis of rotation.

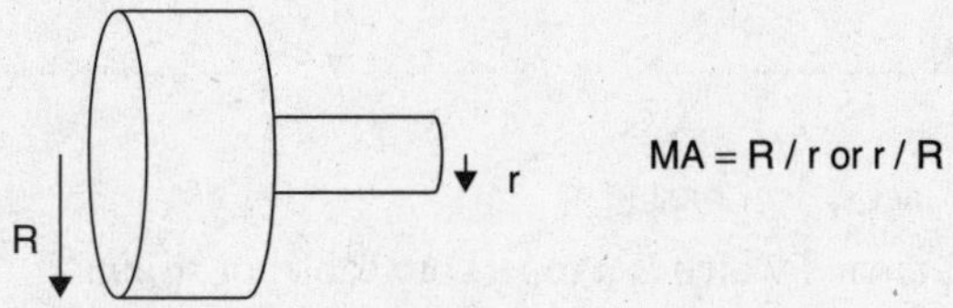

Examples: Doorknob, steering wheel, auto rear wheel, screwdriver.

4. **LEVER:** A pole or board that pivots on a fulcrum.

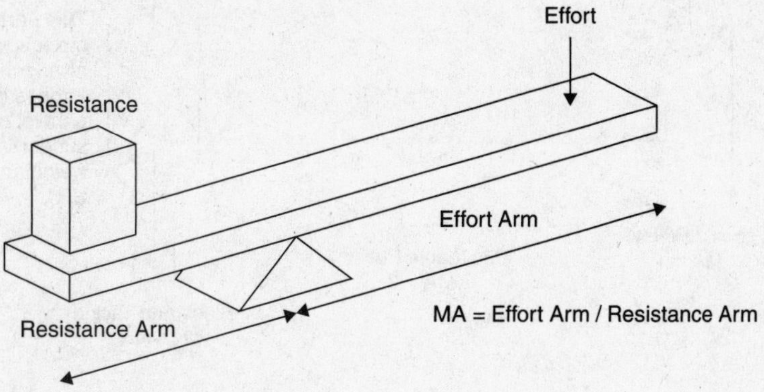

Examples: Crowbar, claw end of a hammer.

There are three classes of levers:

Class 1 has the fulcrum between the effort arm and resistance arm (above).

Class 2 has the fulcrum at one end, effort at the other end and the resistance between them:

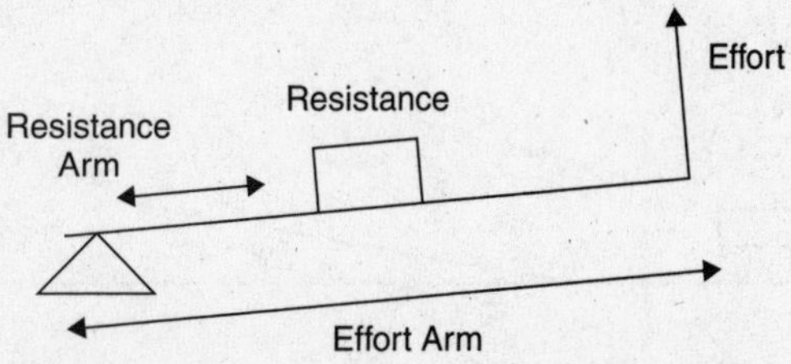

Examples: Wheelbarrow, nutcracker.

Class 3 has the fulcrum at one end, resistance at the other end, and the effort between them:

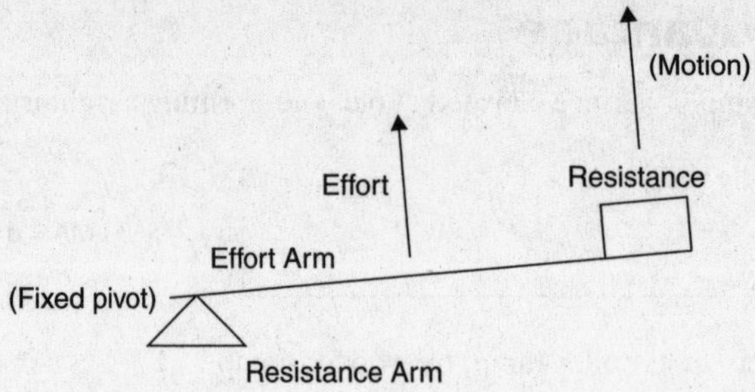

Example: Human forearm.

5. **SCREW:** An inclined plane wrapped around a cylinder or a cone.

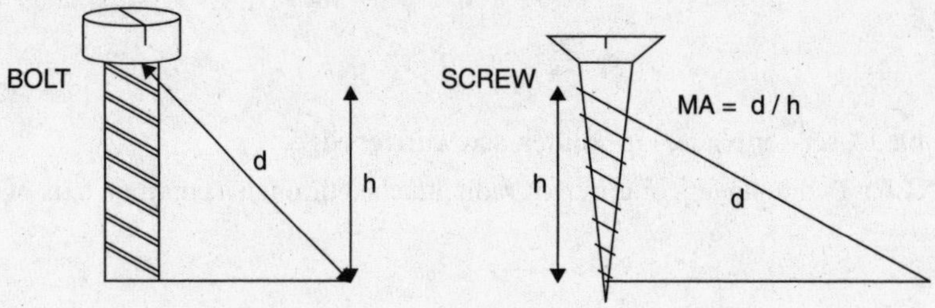

Examples: Bolts, screws, car jacks, drill bits.

6. **PULLEY:** A grooved wheel, around which is a rope, cord cable, or string.

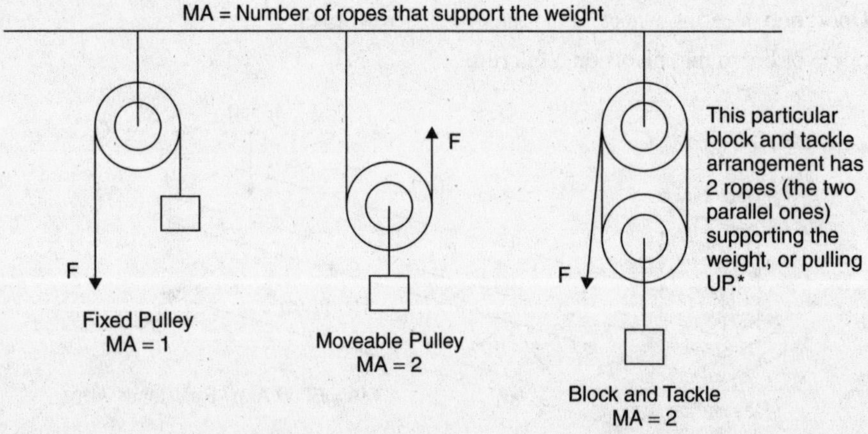

The fixed pulley does not multiply force, but redirects the force to a more convenient angle, thus making it easier to do the work.

The moveable pulley attaches directly to the load. When the upward force is applied, it moves the load *and* the pulley.

*Block and tackle arrangements may have many combinations of turns of rope around both pulleys, hence many Mechanical Advantage possibilities.

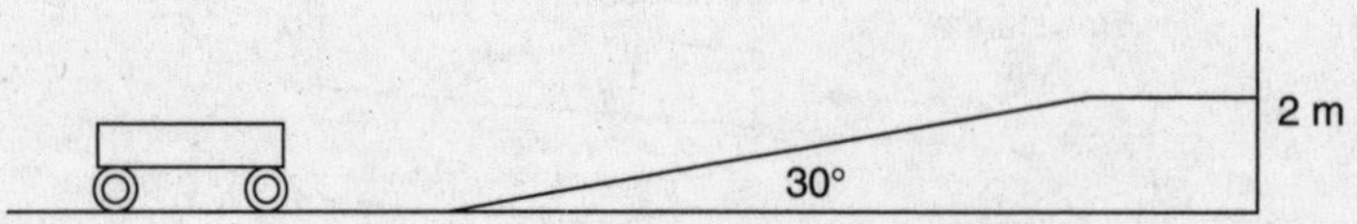

Example (B & C Exams)

An inclined plane makes an angle of 30° with the horizontal as shown. A cart must be pushed, at constant speed, up the plane onto a loading platform 2 m high. The cart's mass is 55 kg. (Neglect friction.)

(a) What percentage of the force needed to lift the cart is eliminated by using the ramp?
(b) How much work is done? (negate friction)

Solution

(a) We need to find the force necessary to *lift* the cart 2 m and the force actually needed to push the cart up the ramp, and then find the difference and divide that amount by the lifting force alone.
Force to lift the cart: **F** = **mg** = (55 kg)(9.8 m/s^2) = 540 N
Force to push the cart up the ramp: Using the parallel force $\mathbf{F_P}$ as the force needed at that angle,
$\mathbf{F_P}$ = mg sin 30° = (540 N)(0.5) = 270 N
% of force eliminated = (540 N – 270 N / (540 N) = 0.5 or **50%**

(b) Work = **Fd** sin Θ or simply the force applied times the distance moved: The actual length of the ramp is **d** = 2 m/sin 30° or 4 m. The work done becomes **W** = (270 N)(4 m) = **1080 J**
(To lift the cart straight up: **W** = (540 N)(2 m) = 1080 J)

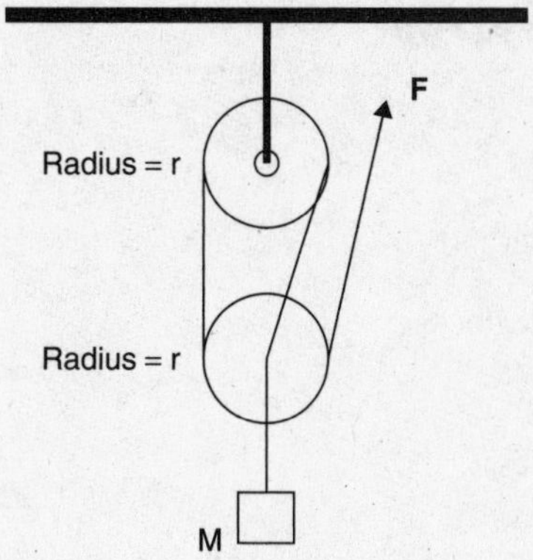

Example (B & C Exams)

A block and tackle pulley system supports a 500 kg mass as shown. (Assume no losses through friction.)

(a) What is the mechanical advantage of the system?
(b) What is the force, **F**?
(c) How much rope must be pulled through the system in order to lift the mass, **M**, a distance of 1.5 m?
(d) How much work is done by the applied force, **F**, in part **(c)**?
(e) If the mass, **M**, is lifted the specified distance in 2.5 seconds, what is the power output of the system?
(f) What is the power input of the system?
(g) The system is constructed and is found to be 80% efficient. Find
i. the actual applied force **F**.
ii. the actual power input.

Solution

(a) Since two strands within the system support the weight and the upward force, **F**, pulls in support, **MA = 3.**
(b) Since the upward force, **F**, has been multiplied 3× to lift the weight, **F** = (500 kg)(9.8 m/s^2)/3 = **1633 N.**

(c) Since work is force times distance and the same amount of work is done by the input force, **F**, as is done on the weight, the force times the distance l_R for the input force, **F**, equals the force times the distance lifted for the weight.

$$(1633\text{ N})(l_R) = (500\text{ kg})(9.8\text{m/s}^2)(1.5\text{ m})$$

$$\text{length of rope } l_R = \mathbf{4.5\ m}$$

(d) **W = Fd** = **(F)**(l_R) = (1633 N)(4.6 m) = **7500 J**

(e) Power = Work out/time = Mgd/t = (500 kg)(9.8 m/s^2)(1.5 m)/(2.5 s) = **3000 W**

(f) Power input = Work in/time = **(F)**(l_R)/t = (1633 N)(4.6 m)/(2.5 s) = **3000 W**

(Note: for parts **(e)** and **(f)**, the machine does not change the amount of work to be done, so the WORK IN = WORK OUT.)

(g) i. If 1633 N is 80% of the new force, the new force is 1633 N/0.8 = **2000 N**
ii. New power input = **(F)**(l_R)/t = (2000 N)(4.6 m)/(2.5 s) = **3700 W**

Review Questions and Answers

B & C Exam Question Types

Multiple Choice

1. As a spinning ice skater extends his arms,

(A) his mass distribution is unchanged.
(B) his angular velocity increases.
(C) his rotational inertia decreases.
(D) his angular momentum increases.
(E) his rotational velocity decreases.

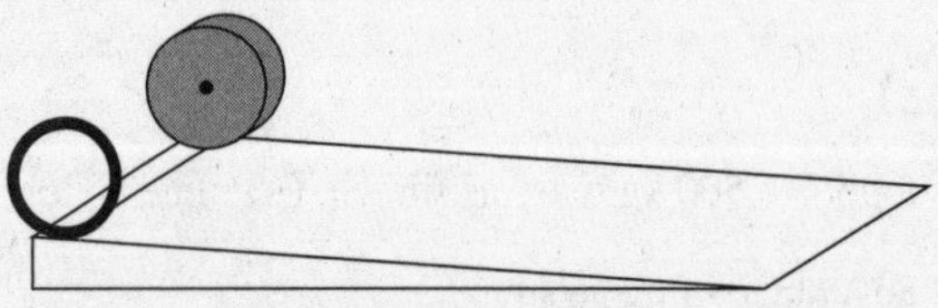

2. A ring and a solid disk of equal mass and equal diameter begin to roll down a ramp simultaneously.

(A) The disk reaches the bottom first.
(B) The ring reaches the bottom first.
(C) They reach the bottom together.
(D) Both have equal linear velocities.
(E) Both have equal mass distribution.

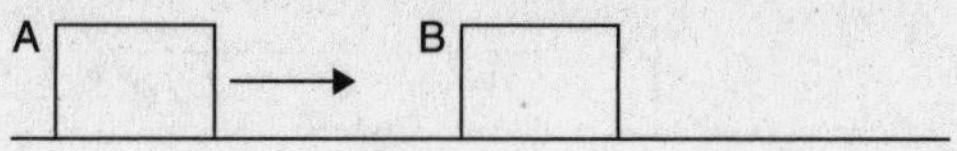

3. Two masses, A and B, of equal mass are on a frictionless horizontal table. Mass A has a velocity of 10 m/s. Mass B is at rest. Mass A collides with Mass B elastically. Their velocities are now

(A) A = 0 m/s, B = 10 m/s
(B) A = 5 m/s, B = 5 m/s
(C) A = 10 m/s, B = 0 m/s
(D) A = 10 m/s, B = 5 m/s
(E) A = 10 m/s, B = 10 m/s

4. The time needed for a net force of 5 N to change the velocity of a 10 kg mass by 2 m/s is

(A) 2 s
(B) 4 s
(C) 6 s
(D) 8 s
(E) 10 s

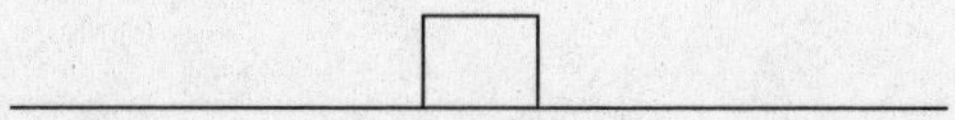

5. If 1450 J of work is done on a mass of 10 kg, it can be raised

(A) 1 m
(B) 1.5 m
(C) 15 m
(D) 21 m
(E) 150 m

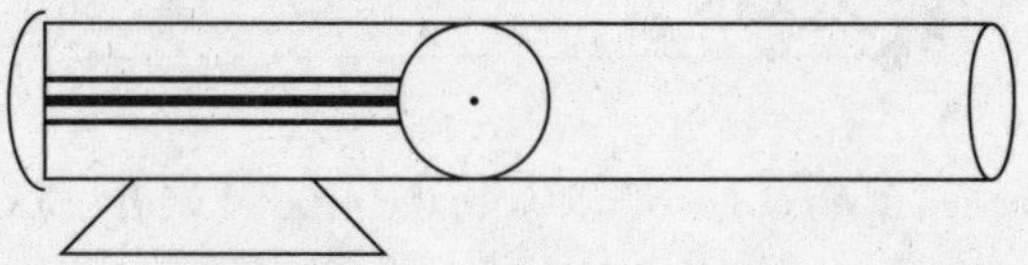

6. A spring-loaded toy cannon has a spring with k = 5 N/m, which is compressed 20 cm. When it is released, a 0.2 kg plastic ball will attain a muzzle velocity of

(A) 0.2 m/s
(B) 0.4 m/s
(C) 0.6 m/s
(D) 0.8 m/s
(E) 1.0 m/s

7. If the cannon in problem 6 fires the ball vertically, it will rise

(A) 0.05 m
(B) 0.10 m
(C) 0.15 m
(D) 0.20 m
(E) 0.25 m

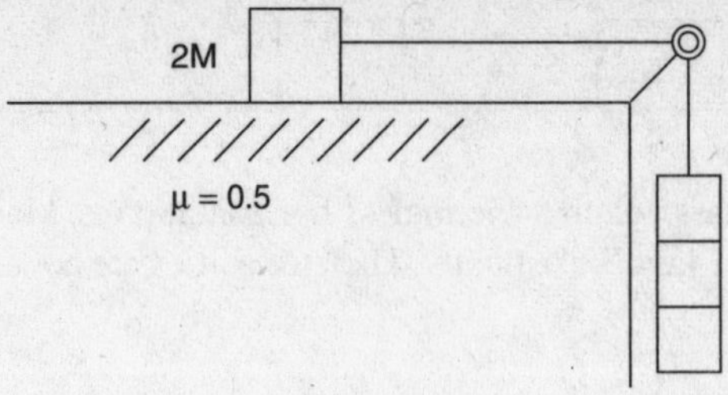

8. A mass of 2M sits on a table and a massless cord runs from it over a frictionless pulley also with negligible mass. Attached to the hanging cord is a mass of 3M. If the $\mu_K = 0.5$, the net force on the cord is

(A) $3Mg - F_F$
(B) $3Mg - \mu F_N$
(C) $3Mg - \mu(2Mg)$
(D) $9Mg/5$
(E) $2Mg$

9. The acceleration of the system in the preceding problem is

(A) $g/4$
(B) $2g/5$
(C) g
(D) $2g$
(E) $4g$

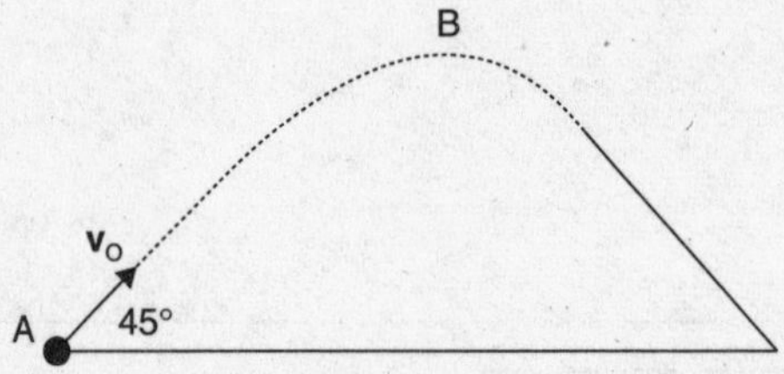

10. A projectile of mass M is launched at an angle of 45° with the ground with velocity v_O. Its momentum at point B is

(A) $\frac{Mv_o}{2}$
(B) $\frac{Mv_o\sqrt{2}}{4}$
(C) $\frac{Mv_o\sqrt{2}}{2}$
(D) $2Mv_o$
(E) $\frac{4Mv_o}{\sqrt{2}}$

Answers to Multiple-Choice Questions

1. **(E)** Increasing mass distribution decreases angular or rotational velocity.

2. **(A)** The disk reaches bottom first, because its mass is concentrated closer to the center, giving it a greater linear inertia.

3. **(A)** A gives all of its momentum to B, since they are the same mass.

4. **(B)** $F\Delta t = m\Delta v$ and $\Delta t = m\Delta v / F = (10\text{kg})(2\text{ m/s})/(5\text{ N}) = \mathbf{4\text{ s}}$

5. (C) $W = Fd$ and $d = W/F = 1450\ \text{J}/(10\ \text{kg})(10\ \text{m/s}^2) = \mathbf{15\ m}$

6. (E) $\left(\frac{1}{2}\right)kx^2 = \left(\frac{1}{2}\right)mv^2$ and $v = \sqrt{kx^2/m} = (5\ \text{N/m})(0.2\text{m})^2/(0.2\ \text{kg}) = \mathbf{1.0\ m/s}$

7. (A) $\left(\frac{1}{2}\right)kx^2 = mgh$ and $h = kx^2/2mg = (5\ \text{N/m})(0.2\ \text{m})^2/(2)(0.2\ \text{kg})(10\ \text{m/s}^2) = (0.1)/(2) = \mathbf{0.05\ m}$

8. (D) The net force on the cord equals the tension due to the weight of the 3M mass minus its accelerating force downward all added to the tension due to friction on the 2M mass plus its accelerating force:
$\mathbf{Mg + 2Ma = 3Mg - 3Ma}$

Solving for **a**, $a = \frac{2g}{5}$ and equating both sides, since the tensions are equal, yields a net force of 9mg/5.

9. (B) $a = \frac{2g}{5}$

10. (C) Since velocity at point B is *all horizontal*, momentum at that point is $(m)(v_o \cos 45°) = m\, v_o \frac{\sqrt{2}}{2}$

Free Response

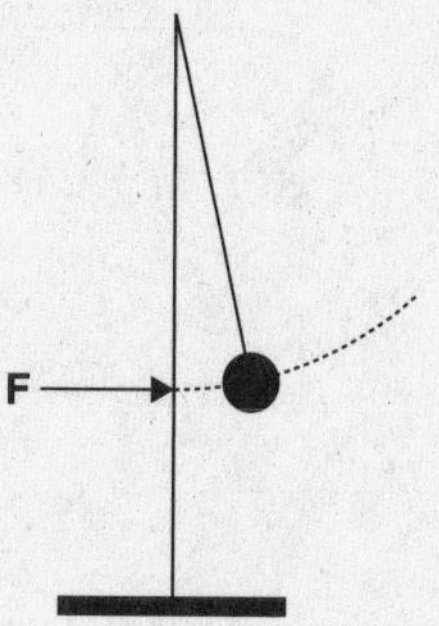

1. A 0.8 kg tetherball hangs on the end of a 3.0 m cord. It is hit by a child and rises 2.1 m.

(a) What is the maximum gravitational potential energy of the ball?
(b) What was the ball's velocity immediately after being hit?
(c) What was the ball's kinetic energy immediately after being hit?
(d) What is the ball's momentum immediately after being hit?
(e) If the child's hand is in contact with the ball for 0.09 seconds, what is the force with which the child's hand hits the ball?
(f) How high will the ball go if the child applies 10 more Newtons of force during the same contact time?

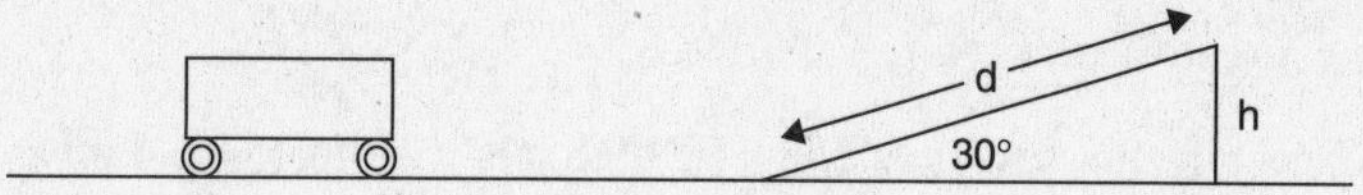

2. A 3.0 kg frictionless cart moves with velocity 2.5 m/s toward a 30° ramp.

(a) How far up the ramp does the cart travel before stopping?
(b) What is the cart's gravitational potential energy once it travels half of its maximum distance up the ramp?
(c) What is the kinetic energy of the cart halfway up the ramp?
(d) The cart is now removed from the ramp and hung on a spring with k = 90 N/m. How far does it stretch the spring?

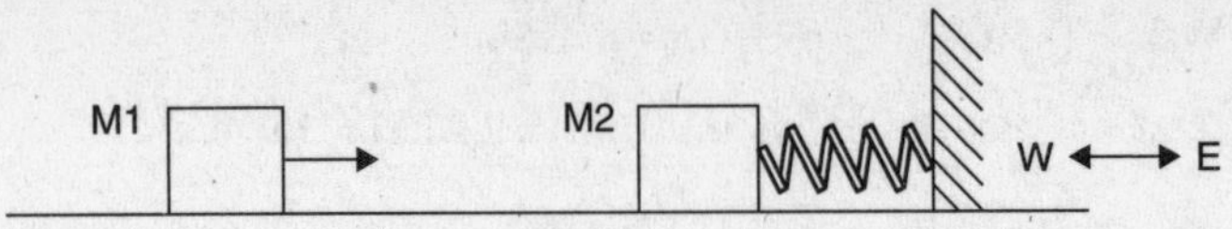

3. Two equal masses of 20 kg each, M 1 and M 2, sit on a frictionless air track. M 1 is moving east at 0.8 m/s and collides with stationary M 2, which is connected to a wall by a massless spring with k = 2.5 N/m.

(a) If the masses collide elastically, what is the total energy of the system immediately after impact?
(b) Describe the motion after M 1 collides elastically with M 2.
(c) How far does M 2 compress the spring before stopping?
Now the masses are reset and adhesive is applied. M 1 sticks to M 2 upon collision.
(d) What is the combined velocity of the masses at the moment the spring begins to compress?
(e) What is the momentum of the system immediately after impact?
(f) How much work does the spring do in reversing the direction of the combined masses?

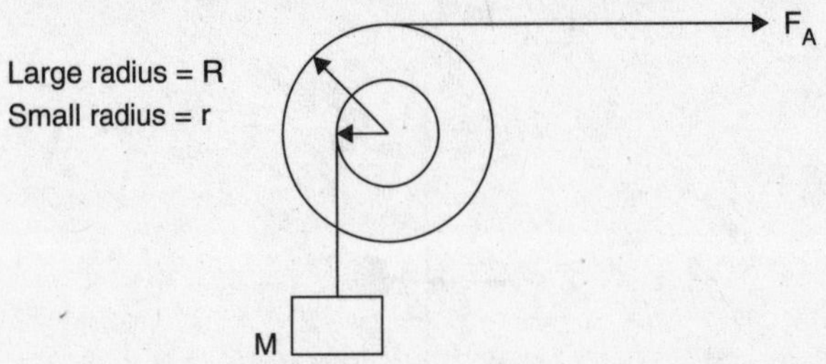

4. Two massless, frictionless pulleys are attached rigidly at and are free to rotate about their central axes. The larger pulley has a radius, **R**, and the smaller one has a radius, **r**. A mass, M, hangs from the smaller pulley, and a force, $\mathbf{F_A}$, is applied at a right angle to the top of the larger pulley. Describe all values in terms of M, **g**, **F**, I, **R** and **r**. (I is the system's rotational inertia.)

(a) What force, $\mathbf{F_A}$, is necessary to lift M at constant speed?
(b) If $\mathbf{F_A}$ disappears, what is the angular acceleration of the system?
(c) If $\mathbf{F_A}$ from part **(a)** is replaced by a new force of $\mathbf{F_A + f}$, find the angular acceleration of the system.
(d) A frictional force, $\mathbf{F_F}$ is added to the system. Repeat part **(c)** for this case.

Answers to Free-Response Questions

1. **(a)** $\mathbf{mgh} = (0.8\text{ kg})(9.8\text{ m/s}^2)(2.1\text{ m}) = \mathbf{16\ J}$

(b) $\mathbf{mgh} = \left(\frac{1}{2}\right)\mathbf{mv^2}$ Cancel the masses: $\mathbf{v} = \sqrt{\mathbf{2gh}} = [(2)(9.8\text{ m/s}^2)(2.1\text{ m})]^{\frac{1}{2}} = \mathbf{6.4\ m/\ s}$

(c) $E_K = \left(\frac{1}{2}\right)(m)(\mathbf{v}^2) = \left(\frac{1}{2}\right)(0.8\text{ kg})(6.4\text{ m/s})^2 = \mathbf{16\ J}$. Note that $E_K = \mathbf{mgh}$ from part **(a)**.

(d) $\mathbf{p} = \text{mv} = (0.8\text{ kg})(6.4\text{ m/s}) = \mathbf{5.1\ kgm/\ s}$

(e) $\mathbf{F\Delta t} = \text{m}\mathbf{\Delta v}$ Solve for force: $\mathbf{F} = \text{m}\,\frac{\mathbf{\Delta v}}{\Delta t} = \frac{(0.8\text{ kg})(6.4\text{ m/s})}{0.09\text{ sec}} = \mathbf{57\ N}$

(f) $\mathbf{F\Delta t} = \text{m}\mathbf{\Delta v}$ Solve for velocity: $\mathbf{\Delta v} = \mathbf{F\Delta t}/\text{ m} = \frac{(57\text{N})(0.09\text{ sec})}{0.8\text{ kg}} = 7.5\text{ m/s}$

$\mathbf{mgh} = \left(\frac{1}{2}\right)\mathbf{mv^2}$ Solve for **h**: $\mathbf{h} = \mathbf{v^2/2g} = \frac{(7.5\text{ m/s})^2}{(2)(9.8\text{ m/s}^2)} = \mathbf{2.9\ m}$

2. **(a)** When the cart stops, all of its kinetic energy has become potential energy.

$\left(\frac{1}{2}\right)mv^2 = mgh$ and $h = v^2 / 2g = (2.5\ m/s)^2 / (2)(9.8\ m/s^2) = 0.32$ m high

Since sin 30° = 0.5 = **h / d**, this yields **d** = 0.32 m/0.5 = **0.64 m**

(b) $E_P = mgh/2 = (3.0\ kg)(9.8\ m/s^2)(0.32m)/2 =$ **4.7 J**

(c) The cart's kinetic energy is equal to its potential energy halfway up the ramp. To check: $E_K = (1/2)mv^2/2 = (3.0\ kg)(2.5\ m/s)^2/4 =$ **4.7 J ANS**

(d) The cart's weight extends the spring to a maximum distance **x**. The force exerted on the cart is –**kx**. This gives $mg = -kx$ $x = mg/k = (3.0\ kg)(9.8\ m/s)^2/90\ N/m =$ **0.3 m**

3. **(a)** The total energy of the system after the collision is the same as that before the collision. Since M 1 possesses all the system's energy before colliding, the total energy of the system remains equal to $E_{K\,1} = (1/2)(M\,1)(v_O^2) = (1/2)(20\ kg)(0.8\ m/s)^2 =$ **6.4 J**

(b) M 1 hits M 2, stops in place, takes the place of M 2, which continues into the spring at 0.8 m/s east, is slowed down and momentarily stopped by the spring, reverses direction, hits M 1 with velocity 0.8 m/s west, stops in place, takes the place of M 1, which continues moving west at 0.8 m/s.

(c) The kinetic energy of the system before impact equals the potential energy ultimately stored in the spring.

E_K of Mass 1 $= \left(\frac{1}{2}\right)kx^2$.

$$x = [(2E_K)/k]^{\frac{1}{2}}$$

From part (**a**)'s answer, $x = [(2)(6.4\ J)/(2.5\ N/m)]^{1/2} =$ **2.3 m**

(d) Total momentum BEFORE collision equals total momentum AFTER: (20 kg)(0.8 m/s east) = (40 kg)(v_{FINAL}) yields v_{FINAL} = **0.4 m/ s east**

(e) $p = mv = (40\ kg)(0.4\ m/s\ east) =$ **16 kgm/ s east**

(f) All work done compressing the spring is equal and opposite the work done decompressing it. The sum is zero.

4. **(a)** Since the system is not accelerating, it is in equilibrium and the sum of the clockwise torque and counterclockwise torque equals zero. This means that $F_AR = Mgr$ and $F_A = Mgr/R$

(b) Using the general relationship, $\tau = Fr = I\alpha$

$$(Mg + Ma)r = I\alpha$$
$$mgr - m\alpha r^2 = I\alpha$$
$$I\alpha + M\alpha r^2 = Mgr$$
$$\alpha(I + Mr^2) = Mgr$$
$$\alpha = \frac{Mgr}{(I + Mr^2)}$$

(c) Again, using the general relationship, $\tau = Fr = I\alpha$

$$(F_A + f)R + Mgr = I\alpha$$
$$\alpha = \frac{(F_A + f)R + Mgr}{I}$$

(d) The general relationship is used again:

$$\tau = Fr = I\alpha$$
$$(F_A + f + F_f)R + Mgr = I\alpha$$
$$\alpha = \frac{(F_A + f + F_f)R + Mgr}{I}$$

Thermodynamics (B Exam)

Heat and Temperature

Thermodynamics is the study of thermal energy and its relationships with heat and other forms of energy. It involves all forms of energy transformations and relationships, beginning with (but not limited to) **radiant energy.**

Forms of Energy ("I S C R E A M")

Internal
Sound
Chemical
Radiant
Electrical
Atomic
Mechanical

Radiant Energy describes the entire electromagnetic spectrum from low-energy power and radio waves to high-energy gamma rays. All radiant energy travels at the speed of light, approximately $\mathbf{3 \times 10^8}$ **m/s** in a vacuum.

Thermal Energy (internal energy) is the total of all potential and kinetic energy of the atoms and molecules of a substance.

Heat is thermal energy transferred from one location to another, either by absorption or emission by conduction, convection, or radiation.

Temperature describes the hotness or coldness of a substance. It is most closely associated with the average kinetic energy of the particles in a substance. Temperature differs from thermal energy in three ways:

1. Thermal energy actually is energy and is therefore measured in energy units such as Joules or calories while temperature is not a form of energy and therefore is measured in its own units.
2. Thermal energy is a measure of total energy while temperature is a measure of average energy. That is why a 1kg block and a 2kg block in the lab might have the same temperature but different amounts of thermal energy within.
3. Temperature is associated only with kinetic energy, while thermal energy is the the total of the kinetic and potential energies that exist within a substance. These ideas can set the stage for clearer descriptions of what is actually happening as an object is heated within a phase or heated through a phase change as described later in this chapter.

Absolute Zero, –273.15°C, which is 0 Kelvins on the Kelvin temperature scale, is theoretically the lowest possible temperature attainable. At 0 Kelvins, gross molecular vibration would stop. The Kelvin is 1/273.16 of the temperature of the triple point of water. The triple point of water is the point where the pressure and temperature conditions can allow its existence in all three of its phases in equilibrium.

A **calorie (c)** is the amount of energy needed to raise the temperature of one gram of water by one degree Celsius. It has the mechanical equivalent of 4.19 Joules of energy.

Specific Heat and Latent Heat

Specific Heat, or Specific Heat Capacity, **c,** is the degree to which a substance can absorb or emit heat as its temperature changes. It is different for different materials and is related to the density and atomic and molecular arrangement of a substance. The specific heat of water is 1 calorie /gram C°.

Latent Heat, L, is the thermal energy that is stored as the potential energy of bonds that exist between particles in solids and liquids. Heat is absorbed by a substance in changing from a solid to a liquid or a liquid to a gas. The heat is then hidden or latent in the internal energy of the molecules. This latent heat then is released when the substance cools and changes state from a gas to a liquid or from a liquid to a solid.

Heat Transfer and Thermal Expansion

Heat transfer or **heat flow** constantly takes place between substances at different temperatures. The objects may be in the same state or in different states. Transfer of thermal energy occurs through conduction (directly touching), convection (currents of thermal origin in fluids), or radiation (electromagnetic waves). As substances absorb heat, their internal energy increases, and their molecules vibrate faster. For most substances this increased vibration translates into an increase in length, area, and volume.

Solids that undergo heating and cooling expand and contract. The length of a metal rod or a length of railroad track, for instance, obey the following relationship:

$$\Delta l = \alpha l_o \Delta T$$

where Δl is the change in length due to heating, l_0 is the original length of the object and ΔT is the change in temperature in C°.

When solids and liquids are heated or cooled, their volumes in turn increase or decrease as their materials expand or contract. For volumetric expansion of solids and liquids, note the following relationship:

$$\Delta V = \beta V_o \Delta T$$

where ΔV is the change in volume due to heating, β is the coefficient of volumetric expansion, V_O is the original volume, and ΔT is the change in temperature in C°. Also, $\beta = 3\alpha$.

Example

A section of steel railroad track (α for steel = 10.5×10^{-6} /C°) is 4.0 m long. If the air temperature increases from 0° C to 15° C, by how much does the length increase?

Solution

Using **$\Delta l = \alpha l_O \Delta T$:**

$$\Delta l = (10.5 \times 10^{-6}\ /\text{C}°)(4.0\ \text{m})(15°) = \mathbf{6.3 \times 10^{-4}\ m\ linear\ increase}$$

Example:

A yellow brass cube with sides of 10 cm is cooled from 100°C to 20°C. By how much does its volume change? (α for yellow brass = 20.3×10^{-6} /°C)

Solution

Using $\Delta V = \beta V_O \Delta T$ and $\beta = 3\alpha$:

$$\Delta V = (3)(20.3 \times 10^{-6})(0.1\ \text{m})^3(-80°\text{C}) =$$
$$\mathbf{4.9 \times 10^{-3}\ m^3\ or\ 0.049\ cm^3\ reduction\ in\ volume}$$

Mechanical Equivalent of Heat

Count Rumford made observations and James Prescott Joule did specific experiments regarding the relationships between thermal energy and mechanical energy. As a result, the unit of mechanical energy (the Joule) and the calorie are related numerically by:

$$4.19\ \text{J} = 1\ \text{cal}$$

When an object is placed beside, near, or into another substance that is not at the same temperature, heat flows from the warmer to the cooler location until thermal equilibrium is reached and the objects are at the same temperature. (This assumes no loss of thermal energy to the surroundings.)

In solving problems involving the addition of a relatively warmer substance to a relatively cooler substance, it is important to keep in mind that the **heat lost by the warmer substance is gained by the cooler substance until they both reach thermal equilibrium.** This can be represented by: **Q Lost = Q Gained.**

When a substance is heated or cooled and **remains in the same state**, it obeys the relationship:

$$\mathbf{Q = mc\Delta T}$$

Q is the amount of heat in calories or joules; m is the mass of the substance; c is the substance's specific heat; and Δ**T** is the change in the substance's temperature.

In metric units, Q may be in calories or Joules, m in g or kg, c either in cal/g°C, cal/kg°C, J/g°C or J/kg°C, and ΔT in °C or K. A degree celsius is numerically equal to a Kelvin.

When a substance is heated or cooled and **changes state**, it obeys the relationship:

$$\mathbf{Q = mL}$$

Q is the amount of heat in calories or joules; m is the mass of the substance; and L is the latent heat of the substance. For solids and liquids, L is the **heat of fusion**.

For liquids and gases, L is the **heat of vaporization** and may be in units of J/g or J/kg.

Note that during either process, **fusion** (freezing or melting) or **vaporization** (vaporizing or condensing), *the temperature of the substance remains constant.* All of the energy being added or extracted goes directly into the breaking apart or the formation of molecular bonds for that particular state. Once all of the bonds have been broken or formed and the process is completed, only then will the temperature of the substance rise or fall.

Example

How many calories must be added to 5.0 kg of water to raise its temperature from 20° C to 80°C? How many Joules would it take?

Solution

Using $Q = mc\Delta T$:

$$Q = (5.0\text{ kg})(1\text{ cal/g°C})(80°\text{C} - 20°\text{C}) = \mathbf{300\text{ cal}}$$
$$= (300\text{ cal})(4.19\text{ J/cal}) = \mathbf{1300\text{ J}}$$

Example

How many Joules of energy are necessary to melt a 1 kg iron ingot initially at 20°C? (Heat of fusion for Fe is 33.0 J/g; specific heat for Fe is 0.449 J/g°C; melting point for Fe is 1540°C.)

Solution

There are two steps in this problem: first, to bring the solid temperature from 20°C to 1540°C; secondly, to change the state of the iron *at its melting point*. Hence, we will use $\mathbf{Q = mc\Delta T}$ first, then $\mathbf{Q = mL}$.

$$\mathbf{Q = mc\Delta T} = (1000\text{ g})(0.449\text{ J/g°C})(1540°\text{C} - 20°\text{C}) = 6.82 \times 10^5\text{ J to heat it.}$$
$$\mathbf{Q = mL} = (1000\text{ g})(33.0\text{ J/g}) = 3.3 \times 10^4\text{ J or } 0.33 \times 10^5\text{ J to melt it.}$$
$$\text{Adding energies gives a total of: } \mathbf{7.15 \times 10^5\text{ J}}$$

Example

A 2.0 kg mass of aluminum at 80°C is immersed in a thermally insulated water bath. If there is a mass of 5.0 kg of water initially at 25°C, what is the final temperature of the system? (Specific heat for Al is 0.897 J/g°C.)

Solution

The warmer aluminum cools to a final temperature, T_F. The cooler water warms to the same final temperature. Therefore, the heat lost by the aluminum is gained by the water until the temperature of both aluminum and water is equal. Because there is no change of state for either substance, $\mathbf{Q = mc\Delta T}$ is used for both.

$$\mathbf{Q\text{ Lost}}\text{ by aluminum} = \mathbf{Q\text{ Gained}}\text{ by water}$$
$$\mathbf{mc\Delta T}\text{ aluminum} = \mathbf{mc\Delta T}\text{ for water}$$
$$(2{,}000\text{g})(0.897\text{ J/g°C})(80°\text{C} - T_F) = (5{,}000\text{ g})(4.19\text{ J/g°C})(T_F - 25°\text{C})$$
$$143{,}520\text{ J} - 1{,}794\text{ J/C° } T_F = 20{,}950\text{ J/C° } T_F - 523{,}750\text{ J}$$
$$667{,}270\text{ J} = 22{,}744\text{ J/C° } T_F$$
$$\mathbf{29°C = T_F}$$

Kinetic Theory

The Kinetic Theory of Matter holds that all matter is comprised of tiny particles that vibrate in constant motion. As matter absorbs heat, its internal kinetic energy increases, and its tiny particles (molecules) vibrate faster. As matter is cooled, heat is extracted from it, causing the kinetic energy of its molecules to decrease and the molecular vibration to decrease.

Just as the kinetic theory has been applied in the previous sections regarding solids and liquids, it is especially important in the study of gases. Gases are unlike solids and liquids due to the fact that in a gas, the molecules are not attached or bonded together as they are in solids and liquids.

Solids occupy a definite volume and assume a definite shape due to the fact that their molecules are rigidly attached or bonded together.

Liquids occupy a definite volume but an indefinite shape due to the fact that their molecules are only weakly bonded to one another and can slide past one another easily.

Gases

Since gas molecules are not attached or bonded to one another, they are free to move about. Therefore, gases occupy an indefinite shape and an indefinite volume. Their motion is controlled by their temperature and pressure. When gas molecules collide with one another, their collisions generally are viewed as being perfectly elastic. The study of gases addresses relationships between **energy, pressure, temperature, and volume.**

When a gas is confined to a fixed volume, such as a closed container, the molecules collide with each other and the sides of the container as well, exerting a force on the interior. This force per unit area is the **pressure** of the gas.

$$P = F / A$$

The metric (SI) units for pressure are N/m^2 or Pascals ($1\ Pa = 1\ N/m^2$). Gas pressures are a result of the size of the container (gas volume) and the gas temperature. Since the motion of the gas molecules increases with temperature, the number of collisions per unit of time is directly related. Increased gas temperature means increased pressure on the interior of the container.

Gas Laws

The study of gases incorporates a number of physical facts about their basic nature. To start with, any equation dealing with gas temperature utilizes Kelvins because there is a direct correlation between gas volume and Kelvins.

For gases at constant pressure: **Charles' Law:** $V_1 / T_1 = V_2 / T_2$

Where volume is in liters and temperature is in Kelvins.

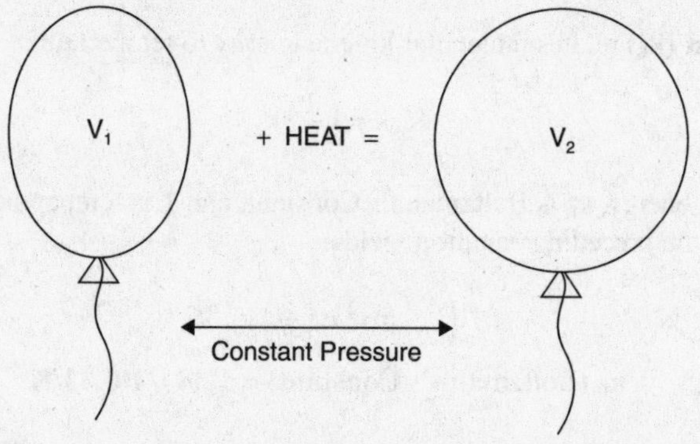

A simplified example is a half-inflated balloon at a certain temperature. If the balloon is tied at the bottom and heated by a hair dryer, the balloon will get larger as the increased temperature will cause more collisions and, therefore, more interior force. This increases the impacted area as the interior pressure remains constant. It also decreases the density of the gas in the balloon.

For gases at a constant pressure: **Boyle's Law:** $P_1V_1 = P_2V_2$,

where pressure is in Pascals (N/m^2) and volume is in liters.

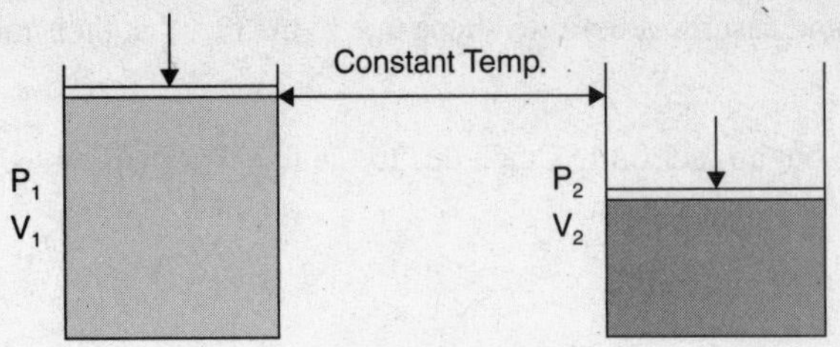

An example is a moveable piston in a cylinder. At a beginning pressure, $\mathbf{P_1}$, the gas occupies a given volume, $\mathbf{V_1}$. At a greater pressure, $\mathbf{P_2}$, the volume decreases to $\mathbf{V_2}$. This also has the effect of increasing the density of the gas in the cylinder.

The elements of both Charles' Law and Boyle's Law are incorporated in the **Combined Gas Equation:**

$$\mathbf{P_1V_1/\,T_1 = P_2V_2\,/\,T_2.}$$

Since pressure, temperature, and volume are so important and interrelated in the study of gases, it is necessary to identify *how much* gas is present at a given temperature and volume agreed upon and specified for scientific reference.

At standard temperature and pressure (STP), 0°C and 1 atmosphere pressure, one mole of gas molecules (6.02×10^{23}, Avogadro's Number) occupies 22.4 Liters.

Another important relationship involves pressure, temperature, volume, and number of moles in a sample of gas. This is the

Ideal Gas Law: PV = nRT

Where P is gas pressure; V is gas volume; n is the number of moles present (**n also equals m/M**, mass divided by the gram-molecular mass); T is the gas temperature in Kelvins; and R is the Universal Gas Constant:

$$R = 8.21 \times 10^{-2} \text{ Lit-Atm/ mole K or}$$
$$R = 8.31 \text{ J/ mole K}$$

For a gas, **Boltzmann's Constant ($\mathbf{k_B}$)** relates molecular kinetic energy to temperature:

$$K_{AVE} = \left(\frac{3}{2}\right) k_B T$$

where K_{AVE} is the average kinetic energy, k_B is Boltzmann's Constant, and T is temperature. Since kinetic energy is $E_K = (1/2)\,mv^2$, substitution into the preceding equation yields:

$$\left(\frac{1}{2}\right) \mathbf{mv^2} = \left(\frac{3}{2}\right) \mathbf{k_B\, T}$$

$$\mathbf{k_B \text{ (Boltzmann's Constant)} = 1.38 \times 10^{-23} \text{ J/K}}$$

Solving for the **v**, the average velocity of the gas molecules, called the **root-mean-square velocity:** $\mathbf{v_{rms} = (3k_BT/m)^{\frac{1}{2}}}$, which can also be expressed as:

$$v_{rms} = (3RT/\,M)^{\frac{1}{2}}$$

where m is the mass of the molecule and M is the gram-molecular mass.

Laws of Thermodynamics

The ZEROTH LAW of Thermodynamics: Two systems in thermal equilibrium with a third system are in thermal equilibrium with each other. When an object is placed beside, near, or into another substance that is not at the same temperature, heat flows from the warmer to the cooler location until **thermal equilibrium** is reached and the objects are at the same temperature. (This assumes no loss of thermal energy to the surroundings.)

The FIRST LAW of Thermodynamics states that heat is a form of energy. The change in internal energy equals the heat minus work done (change in **mechanical energy**).

$$\Delta E_{INTERNAL} = Q - W$$

In a system, the quantity of energy that cannot be converted into mechanical work refers to the **entropy** of the system. The difference in entropy or the *change in entropy* as heat is added to the system is given as

$$\Delta S = \frac{\Delta Q}{T}$$

where ΔS is the change in entropy in units of Joules per Kelvin, ΔQ is the heat energy in Joules, and T is the temperature in Kelvins.

If heat is added to the system, ΔS is positive. If heat is removed, ΔS is negative.

The second law of thermodynamics relates directly to entropy. Natural processes take place in ways that increase the entropy of the universe. For isolated systems, the entropy of the system increases. Even though the total amount of energy in the universe remains constant, the availability of the energy present constantly decreases or is lost in the conversion processes.

Heat Engines

An ideal or perfect theoretical engine extracts heat (INPUT) and converts it all into work (OUTPUT). Such a machine has never been built. A real engine extracts heat (INPUT) and converts a portion of it into work (OUTPUT), with the remainder of the heat expelled into a reservoir (EXHAUST) at lower temerpature (T_{COLD}).

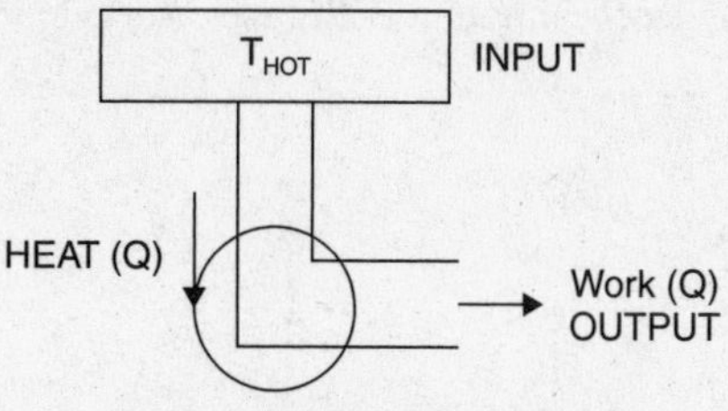

Perfect Theoretical Engine
100% Energy Conversion

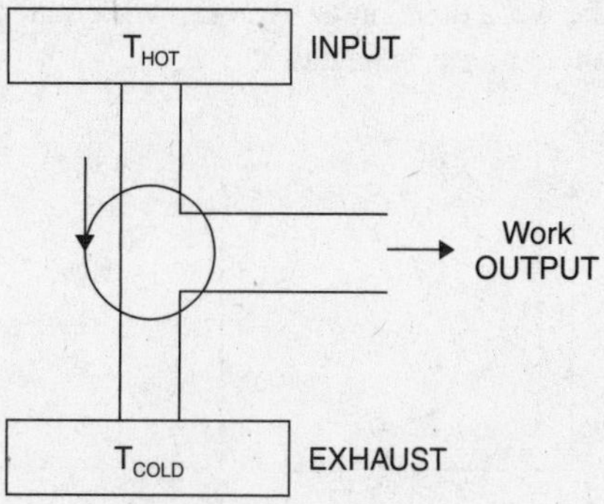

Real Engine

The SECOND LAW of Thermodynamics states that when no other change takes place, heat cannot be completely changed into work. The **efficiency** of a heat engine, for instance, is always less than 100 percent and that efficiency is calculated using:

$$\text{eff} = 1 - \frac{T_{COLD}}{T_{HOT}}$$

Where T_{COLD} is the output heat (exhaust) and T_{HOT} is the input heat (intake). (In order to be 100% efficient, all the input heat T_{HOT} would need to be utilized by the engine, with *no exhaust heat*, leaving, **T_{COLD} equal to zero.** This never is the case due to friction and thermal energy leakage.)

Work Done on or by a Gas

The First Law of Thermodynamics is concerned with work, pressure, and volume relating to gases:

$$W = P\Delta V$$

In terms of units (dimensional analysis), work is in Joules (N-m) or kg m^2/s^2; pressure is in units of Pa (N/m^2) or kgm/s^2m^2, which equals kg/s^2m; volume is in units of m^3.

A dimensional analysis gives kg $m^2/s^2 = (kg/s^2m)(m^3)$, which is correct.

(An important hint in solving AP Exam questions in general is to be mindful of units. A quick check of the units expected in your answer is often helpful in setting the problem up correctly.)

Gas Graphs (P – V Graphs)

Common graphs involving gases usually will pertain to work, volume, and pressure. Many will pertain to gases under *isothermal conditions*, that is, at a constant temperature.

Also, a process that involves no heat being added to or removed from a system is called an **adiabatic** process.

In an **isobaric** process, the system is kept at constant pressure. Calculating work is done using $W = P\Delta V = P(V_2 - V_1)$.

In an **isochoric** process, the system is kept at constant volume. Since $\Delta V = 0$, $W = 0$.

In an **isothermal** process, the system is kept at constant temperature. In these cases, the work is generally not zero and is computed using $W = P\Delta V$. Detailed calculations with this formula can be difficult, however, because the pressure keeps changing as the volume changes.

Some common examples of graphs involving gases and work include the following. Since work equals pressure times volume change, the work done on or by a gas will equal the area under the curve in each case. Work done *by* a gas is positive. Work done *on* a gas is negative.

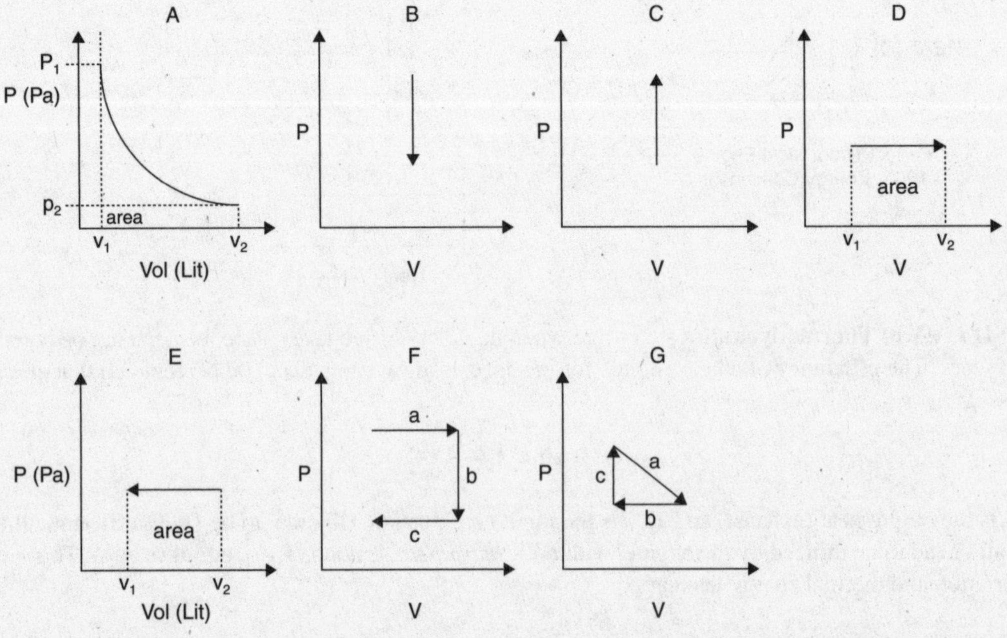

A. Graph of P versus V: Work done on or by the gas kept at constant temperature equals the area under the curve. (Graphical picture of Boyle's Law) $W = P\Delta V = P(v_2 - v_1)$. Since the pressure keeps changing, this needs to be broken up into smaller segments representing smaller areas of $P\Delta V$ and each small value of $P\Delta V$ added up to give a more accurate representation of the work.

B. Graph of decreasing pressure at constant volume and temperature. Since $\Delta V = 0$, $W = P\Delta V = 0$.

C. Graph of increasing pressure at constant volume. Since $\Delta V = 0$, $W = P\Delta V = 0$. Unless you are pumping in additional gas, the temperature would have to be increasing to increase the pressure.

D. Graph of constant pressure and increasing volume. Work done BY THE GAS = W = $P\Delta V$ = area under the curve.

E. Graph of constant pressure and decreasing volume. Work done ON THE GAS = W = $P\Delta V$ = area under the curve.

F. Graph of work being done by a gas at higher pressure (a), then a decrease of pressure with no work involved (b), then work done on the gas at a lower pressure (c). End result: Total $W = P\Delta V$ for paths (a) + (b) + (c):

Path (a) – The gas does work.

Path (b) – No work is done.

Path (c) – Work is done on the gas (but less work than was done by the gas in (a).

END RESULT: Work is done by the gas in the amount of **W**(a) – **W**(b).

G. (a) Pressure decreases as volume increases. Work is done by the gas in the amount of the area under the curve of (a).

(b) Pressure remains constant as volume decreases: Work is done on the gas in the amount of the area under the curve of (b).

(c) Volume remains constant as pressure increases. No work is done.

END RESULT: (a) to (b) to (c): The area enclosed within the triangle is the net work done by the gas, which must also be equal to the net heat absorbed by the gas.

Example:

A sample of gas occupies a volume of 1.0 liter at a pressure of 1.0×10^5 Pa (1 atm of pressure) at a temperature of 300 K. If the pressure is increased to 3.5×10^5 Pa, and the temperature is held constant, what will the volume become?

Solution

Using the combined gas equation: $\mathbf{P_1V_1/T_1 = P_2V_2/T_2}$, we find that the temperature remains constant and $\mathbf{P_2 = P_1V_1/V_2}$.

$$P_2 = \frac{(1.0 \times 10^5 \text{ Pa})(1.0 \text{ Lit})}{3.5 \times 10^5 \text{ Pa}} = \mathbf{0.29\ liters}$$

Example:

A sample of carbon dioxide gas (CO_2) at STP has a mass of 20.0 g. What is its volume?

Solution

Using the Ideal Gas Law: **PV = nRT** and **n = m/ M** and rounding to 273 K, since the gram-molecular mass for $CO_2 = (12) + 2(16) = 44$ g:

$$V = nRT/P = mRT/MP = \frac{(20.0\text{ g})(8.21 \times 10^{-2}\text{ L atm/mole K})(273\text{ K})}{(44\text{ g})(1\text{ atm})} = 10.2\text{ Liters}$$

Example

If CO_2 has a gram-molecular mass of 44 g/mole, how fast do carbon dioxide molecules travel at STP?

Solution

Using $v_{rms} = (3RT/M)^{\frac{1}{2}}$:

$$v_{rms} = \left[\frac{(3)\left(8.31\frac{J}{mole\,K}\right)(273\,K)}{\left(44\frac{g}{mole}\right)\left(\frac{1\,kg}{1000\,g}\right)}\right]^{\frac{1}{2}} = \mathbf{390\ m/s}$$

(Note the conversion of g to kg. This is due to the fact that a JOULE is a kgm^2/s^2.)

Example

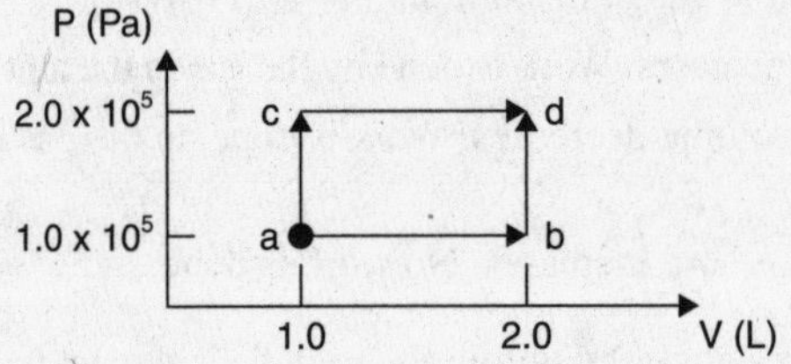

A sample of gas at 1 atm pressure (1.0×10^5 Pa) and 1.0 L volume at point a is brought to points b, c and d in separate exercises. Determine the work involved

(a) from point a to point b.
(b) from point b to point d.
(c) from point a to point c.
(d) from point c to point d.
(e) from point a to point d via point b.
(f) from point a to point d via point c.

Solution

Using $W = P\Delta V$ and the fact that a Pa = 1 N/m^2,

(a) ab: $W = (1.0 \times 10^5\ Pa)(2.0\ L - 1.0\ L) = 1.0 \times 10^5\ (N/m^2)(1.0\ L)\left(1m^3/10^3 L\right) = 1.0 \times 10^2\ Nm = \mathbf{1.0 \times 10^2\ J}$
(b) bd: W = 0: No change in volume, therefore **no work is done.**
(c) ac: W = 0: No change in volume, therefore **no work is done.**
(d) cd: $W = (2.0 \times 10^5\ Pa)(2.0\ L - 1.0\ L) = 2.0 \times 10^5\ (N/m^2)(1.0\ L)\left(1m^3/10^3 L\right) = 2.0 \times 10^2\ Nm = \mathbf{2.0 \times 10^2\ J}$
(e) abd: W = Work for ab + bd = $\mathbf{1.0 \times 10^2\ J}$
(f) acd: W = Work for ac + cd = $\mathbf{2.0 \times 10^2\ J}$

Review Questions and Answers

B Exam Question Types

Multiple Choice

1. A substance's specific heat is a function of its

(A) mass.
(B) weight.
(C) volume.
(D) color.
(E) molecular structure.

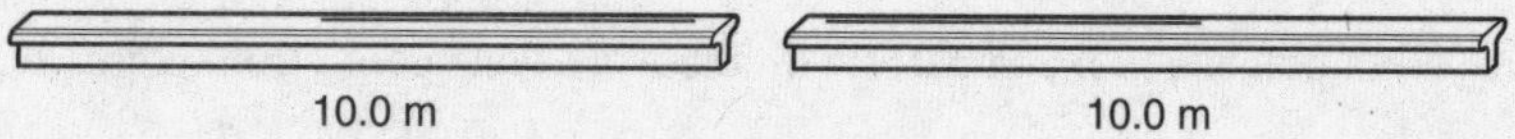

2. Two iron 10.0 m long rails are separated by 1.0 cm. The value of α for iron is $10.5 \times 10^{-6}/C°$. If the given temperature is 0° C, the temperature at which the rails will touch is most nearly

(A) 10°C
(B) 30°C
(C) 60°C
(D) 100°C
(E) 140°C

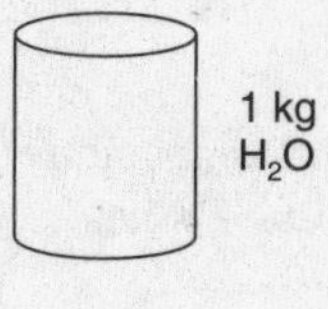

3. The amount of heat liberated when 1 kg of water cools from 100°C to 0°C is

(A) 4.2×10^2 J
(B) 4.2×10^5 J
(C) 4.2×10^8 J
(D) 4.2×10^{11} J
(E) 4.2×10^{14} J

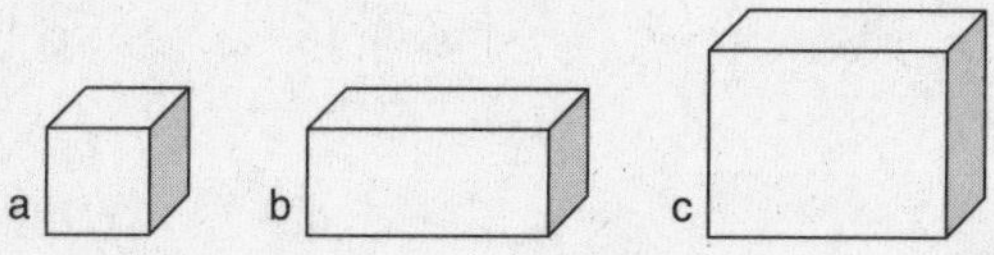

4. Tank a contains 10 L of water at 20°C. Tank b contains 20 L of water at 10°C. When they are emptied into 30 L tank c, the equilibrium temperature is

(A) 11°C
(B) 13°C
(C) 15°C
(D) 17°C
(E) 19°C

5. The number of Joules it takes to raise the temperature of 500 g of water from 20°C to boiling is closest to

(A) 0.4×10^5 J
(B) 1.6×10^5 J
(C) 4.0×10^5 J
(D) 1.6×10^6 J
(E) 4.0×10^6 J

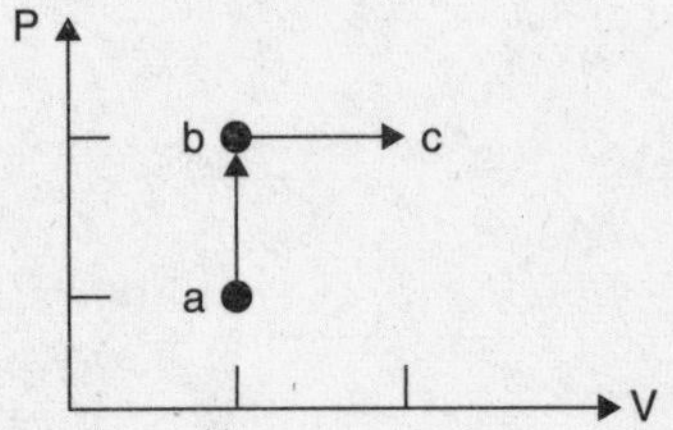

6. The work done going from points **a** to **b** compares with the work done going from points **b** to **c** such that

(A) W **a** to **b** < W **b** to **c**
(B) W **a** to **b** = W **b** to **c**
(C) W **a** to **b** > W **b** to **c**
(D) W **a** to **b** + W **b** to **c** = 0
(E) W **a** to **b** – W **b** to **c** = 0

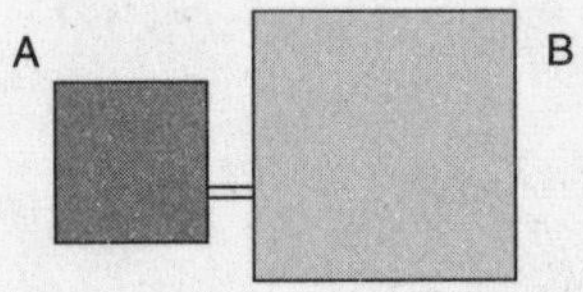

7. Enclosure A contains gas at two atm pressure and temperature T. The gas is then pumped into enclosure B, which has 4× the volume of enclosure A. If the temperature is held constant, the new gas pressure is

(A) 5×10^2 Pa
(B) 5×10^3 Pa
(C) 5×10^4 Pa
(D) 5×10^5 Pa
(E) 5×10^6 Pa

8. The gas in the previous problem is kept at –173°C. The number of moles of this gas in the initial 1L container is closest to

(A) 0.25
(B) 2.5
(C) 25
(D) 250
(E) 2500

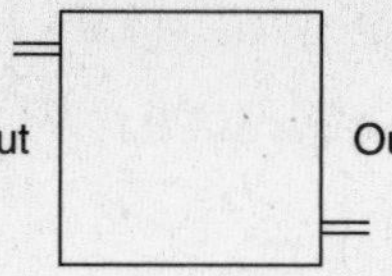

9. The maximum efficiency of a heat engine which operates between the measured temperatures of 227°C and 727°C is

(A) 10%
(B) 20%
(C) 30%
(D) 40%
(E) 50%

10. A $\frac{1}{4}$ L cup of hot water at 90°C is set on a table. By the time it has cooled to 60°C it has lost

(A) 7.5 cal
(B) 75 cal
(C) 750 cal
(D) 7,500 cal
(E) 75,000 cal

Answers to Multiple-Choice Questions

1. **(E)** Molecular structure is the only choice that can determine specific heat or heat capacity.

2. **(D)** Using $\Delta l = \alpha l_O \Delta T$, $\Delta T = \Delta l/\alpha l_O = (0.01\text{ m})/(10\text{ m})(1 \times 10^{-5}/\text{C°}) = \mathbf{100°C}$

3. **(B)** Using $Q = mc\Delta T = (1000\text{g})(4.19\text{ J/g°C})(100°\text{C}) = \mathbf{4.19 \times 10^5\ J}$

4. **(B)** Q lost by tank **a** water = Q gained by tank **b** water

$(10\text{L})(10^3\text{g/L})(1\text{ cal/g°C})(20° - T_F) = (20\text{L})(10^3\text{g/L})(1\text{ cal/g°C})(T_F - 10°\text{C})$

Canceling the following from both sides: $(10^3\text{g/L})(1\text{ cal/g°C})$ gives:

$$200°\text{C} - 10T_F = 20T_F - 200°\text{C}$$
$$40°\text{C} = 3T_F \text{ and } \mathbf{T_F = 13°C}$$

5. **(B)** Using $Q = mc\Delta T = (500\text{g})(4.19\text{ J/g°C})(80°\text{C})$ = approx $\mathbf{1.6 \times 10^5\ J}$

6. **(A)** Using $W = P\Delta V$, W **a** to **b** = 0 since $\Delta V = 0$

W **b** to **c** > 0 so **W a to b < W b to c**

7. **(C)** Using Boyle's Law: $P_1V_1 = P_2V_2$, $(2\text{Atm})(V_1) = (P_2)(4V_1)$

$P_2 = 0.5$ atm, which translates into 0.5×10^5 Pa or $\mathbf{5 \times 10^4}$ **Pa**

8. **(C)** Using $PV = nRT$, $n = PV/RT$

$= (2\text{ atm})(1\text{ L})/(8.2 \times 10^{-2}\text{ L atm/mol K})(100\text{ K})$ = **approx 0.25 moles**

9. **(E)** Using $e = 1 - T_{COLD}/T_{HOT}$,

$e = 1 - \dfrac{227 + 273}{727 + 273} = 1 - 500/1000 = 0.5$ or **50%**

10. **(D)** Using $Q = mc\Delta T$

$= (0.25\text{L})(1\text{ cal/g°C})(1\text{g/ml})(1000\text{ ml/L})(90°\text{C} - 60°\text{C})$

$= 30{,}000$ cal/4 = **7500 cal**

Free Response

1. A brick of mass M = 3.00 kg slides along a surface with an initial velocity 1.50 m/s. While sliding, it begins to compress a spring attached to an immobile baffle before coming to a complete stop, leaving the spring compressed 10.5 cm. The spring has k = 2.50 N/m, and the coefficient of kinetic friction between the brick and the surface is 0.450.

How many Joules of energy are lost to heat before the brick stops?

2. How many Joules and how many calories of heat would it take to heat a 1kg chunk of ice at –20°C to superheated steam at 120°C? (Answer for each step.)

c for ice = 2.22 J/g°C

c for water = 4.19 J/g°C

c for steam = 2.10 J/g°C

L fusion for water = 334 J/g

L vaporization for water = 2260 J/g

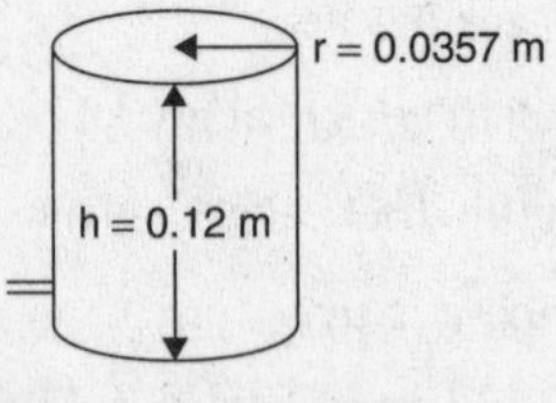

3. A cylindrical container of height 0.12 m and radius 0.0357 m is filled with an ideal gas at 1 atm pressure (1.0×10^5 Pa). The container is sealed and heated to 370 K.

(a) What is the pressure of the heated gas?
(b) How much force is exerted on the round lid by the gas?
(c) The inlet valve is opened on the side of the container, allowing some of the heated gas to escape until the interior of the container is again at 1 atm pressure. The valve is immediately closed, and the container is cooled to 300K. What is the pressure of the cooled gas?
(d) How many moles of the gas are in the container now?

Answers to Free-Response Questions

1.

$$E_{K\,O}\,\text{BRICK} = \text{Work done by friction} + E_P\,\text{SPRING}$$

$$\left(\frac{1}{2}\right)mv^2\ \text{BRICK} = \mathbf{F_F d}\ \text{STOPPING} + \left(\frac{1}{2}\right)kx^2$$

$$\left(\frac{1}{2}\right)(3.00\ \text{kg})(1.50\ \text{m/s})^2 = \mathbf{F_F d}\ \text{STOPPING} + \left(\frac{1}{2}\right)(2.50\ \text{N/m})(0.105\ \text{m})^2$$

$$3.38\ \text{J} = \mathbf{F_F d}\ \text{STOPPING} + 0.014\ \text{J}$$

$$3.38\ \text{J} - 0.014\ \text{J} = \mathbf{F_F d}\ \text{STOPPING}$$

$$3.37\ \text{J} = \mathbf{F_F d}\ \text{STOPPING}$$

$$3.37\ \text{J} = (\mu \mathbf{F_N})(\mathbf{d})$$

$$3.37\ \text{J} = (0.450)(3.00\ \text{kg})(9.80\ \text{m/s}^2)(\mathbf{d})$$

$$0.250\ \text{m} = \text{total sliding distance to stop}$$

$$0.250\ \text{m} = \text{distance to slide without spring} + 0.105\ \text{m with spring}$$

$$0.250\ \text{m} - 0.105\ \text{m} = \text{springless sliding distance only}$$

$$0.145\ \text{m} = \text{springless sliding distance only}$$

Whole sliding distance times *friction force* = Heat Lost + Compression Energy

$$(0.250\ \text{m})[(0.450)(3.00\ \text{kg})(9.80\ \text{m/s}^2)] = \text{Heat Lost} + \text{Compression Energy}$$

$$3.31\ \text{J} = \text{Heat Lost} + 0.015\ \text{J}$$

3.30 J = HEAT LOST or Q Lost

2. **(a)** Heat to warm the ice from –10°C to 0°C:

$Q = mc\Delta T$ = (1 kg)(2.22 J/g°C)(1000g/kg)(20°C) = 44,400 J or 44,400J/4.19 J/cal

= **44,400 J** or **10,600 cal**

(b) Heat to melt the ice:

$Q = mL$ = (1000 g)(334 J/g) = **334,000 J** or **79,700 cal**

(c) Heat to warm the ice water to 100°C:

$Q = mc\Delta T$ = (1000 g)(4.19 J/g°C)(100°C) = **419,000 J** or **100,000 cal**

(d) Heat to vaporize the water:

$Q = mL$ = (1000 g)(2260 J/g) = **2,260,000 J** or **539,000 cal**

(e) Heat to superheat the steam:

$Q = mc\Delta T$ = (1000 g)(2.10 J/g°C)(20°C) = **42,000 J** or **10,000 cal**

Total Heat Needed = 45,100 J **or** 10,730 cal

= **4.51×10^4 J** or **1.07×10^4 cal**

3. **(a)** Since $V_2 = V_1$, $P_1/T_1 = P_2/T_2$ and $P_2 = P_2T_2/T_1$ = $(1.0 \times 10^5$ Pa)(370 K)/273 K

= **1.36×10^5 Pa**

(b) $F = PA$ = $(1.36 \times 10^5$ Pa)(3.1416)(0.0357m)^2 = **545 N**

(c) Since $V_2 = V_1$, $P_1/T_1 = P_2/T_2$ and $P_2 = P_1T_2/T_1$ = $(1.0 \times 10^5$ Pa)(300 K)/370 K

= **8.10×10^4 Pa**

(d) Using PV = nRT, n = PV/RT

The volume is (area of base)(height) = $(3.1416)(0.0357\ \text{m})^2(0.12\ \text{m})$

$= 4.80 \times 10^{-4}\ \text{m}^3$

$$n = \frac{\left(8.10 \times 10^{-4}\ \text{N/m}^2\right)\left(4.80 \times 10^{-4}\ \text{m}^3\right)}{(8.31\ \text{J/moleK})(300\ \text{K})} = 0.0156 = \mathbf{1.56 \times 10^{-2}\ moles}$$

Waves (B Exam)

A wave is a transfer of energy initiated by a vibration, motion, or disturbance. It is the physical embodiment of an energy form in transit from one location to another. A wave does not transmit matter, only energy. Waving a jump rope sends a transverse wave down the rope, but the actual molecules return to their original positions. Pushing a spring parallel to its axis sends a longitudinal wave down the spring, but the actual molecules return to their original positions.

Waves are caused by **vibratory disturbances** in an equilibrium condition. The disturbances may be seen or unseen, felt or unfelt by humans, depending on their frequency of vibration and our physiological reception capabilities. We see ocean waves but not microwaves. We feel infrared rays but not radio waves. We feel earthquake waves above a certain magnitude, yet are unaware of the vast majority of seismic tremors.

The two basic types of waves are **mechanical** and **electromagnetic**.

Mechanical waves need a medium through which to travel. The denser the medium, the faster mechanical waves travel. Sound and earthquake waves are examples. Sound, for instance, travels fastest in solids and slowest in gases. Sound can not travel through a vacuum.

By contrast, *electromagnetic waves* do not need a medium to travel through and travel fastest in a vacuum, slowing as the medium becomes more and more dense, until they are stopped altogether when the medium is sufficiently dense to be described as being opaque.

The electromagnetic spectrum runs from low-energy power waves through radio and TV waves to infrared, visible, and ultraviolet radiation upward through X-rays, gamma rays, and hard gamma rays. All electromagnetic waves travel at the speed of light (**c**), approximately 3×10^8 m/s in a vacuum, the value usually used for the speed of electromagnetic radiation traveling through air. A mechanical wave is a disturbance of an equilibrium condition in matter. An electromagnetic wave is a disturbance of a different type of equilibrium condition, one that involves electric and magnetic fields. The electric field and magnetic field in an electromagnetic wave travel at right angles to each other. This phenomenon is covered in "Electricity and Magnetism" later in this book.

Transverse and Longitudinal Waves

Both mechanical and electromagnetic waves have shape or form.

In a **transverse** wave, the particles of the medium vibrate in directions that are **perpendicular** to the direction of wave travel. Mechanical waves operating in this manner are easily visible traveling along stretched wires, cables, and springs. Mechanical transverse waves are also visible at surfaces where different fluids interface. Ocean waves are a perfect example. L-waves in seismic disturbances, which travel along the surface of the ground, are another example. All electromagnetic waves are two-dimensional and take on sinusoidal form. Sine wave models are used because the sine wave is the most common wave shape in both the natural and designed world.

Transverse Wave

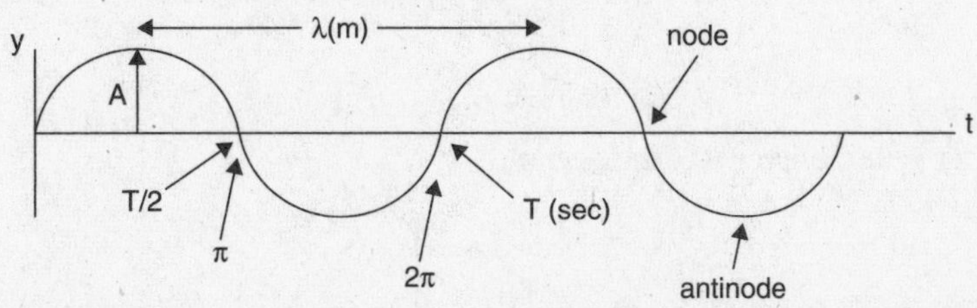

In the preceding figure, a sinusoidal wave is labeled with **A** being the *amplitude*, or wave height, **T** signifying the *period* of the wave, and λ being the wavelength. In the standing wave, the nodes are points having no vertical movement, and antinodes are points of maximum vertical movement. Relationships in a sinusoidal wave are as follows:

$$y(x,t) = A \sin 2\pi ft = A \sin (\omega t + \kappa x)$$

$\kappa = \frac{1}{\lambda}, v = \frac{\omega}{\kappa}$ where in **y** is a function of **x** and **t**, **A** is amplitude, ω is the angular velocity $\left(\frac{2\pi}{T}\right)$, κ is the angular wave number $\left(\frac{2\pi}{\lambda}\text{ rad/m}\right)$ and **f** is the frequency.

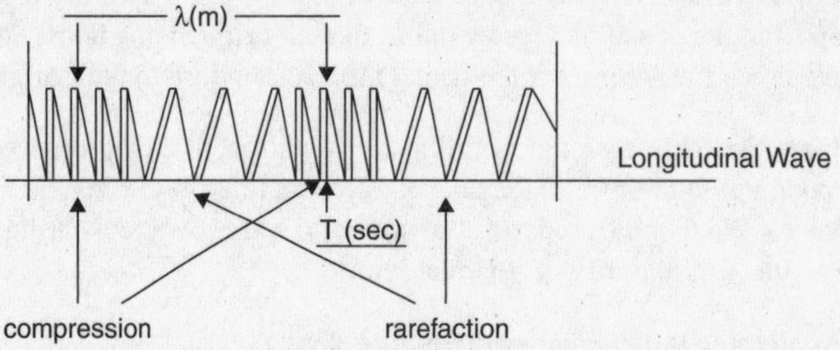

In the preceding figure, the wavelength of the longitudinal or compressional wave is labeled λ, and areas of compression and rarefaction are labeled. A compression area is equivalent to a wave front. In a *longitudinal* wave, the direction of the vibration of the particles of the medium is *parallel* to the direction of wave travel.

Some important wave relationships involve frequency (f), period (T), wavelength (λ), and wave velocity (**v**).

For electromagnetic waves traveling through either vacuum or air, the velocity is considered to be **c**, the "speed of light," 3×10^8 m/s.

For sound waves traveling through air, the velocity is directly related to the air temperature. The velocity of sound at 0°C is 331.5 m/s and increases by 0.6 m/s for each degree Celsius warmer. The speed of sound decreases by 0.6 m/s for each degree Celsius cooler. This is due to the increased motion of air molecules at higher temperatures. The increase in molecular collisions associated with this phenomenon allows faster transfer of sound through the molecules.

There comes a point where the transfer of sound energy from molecule to molecule is too weak to continue due to the fact that the sound energy has spread out in a bubble, and the energy cannot reach the next air molecule. As the sound disturbance grows in distance from the source, the bubble of energy covers a larger and larger area and diminishes in strength by the inverse-square law, $I \propto \frac{1}{r^2}$. (This is true of all force and energy fields that emanate from a point source. . . sound, electromagnetic radiation, electric fields, magnetic fields, and gravitational fields, although only sound waves reach a point where they cannot transfer their energy through the medium, because they are the only type of energy requiring a medium for transmission.) All sound waves eventually turn into radiant energy, heating the molecules of their medium as they pass through.

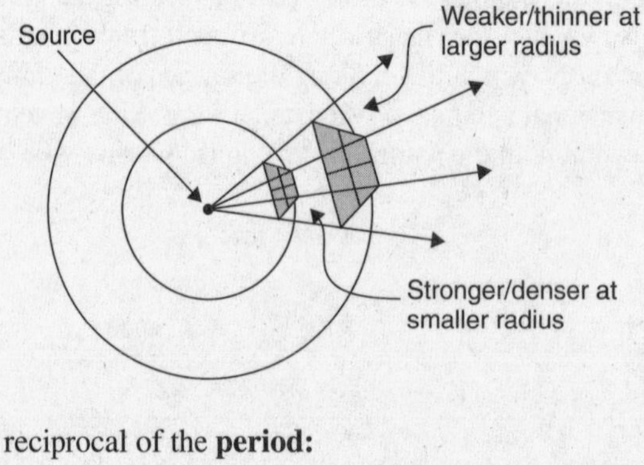

In any wave, the **frequency** is the reciprocal of the **period:**

$$f = 1/T$$

In any wave, the **frequency** is also equal to the **velocity** divided by the **wavelength:**

$$f = \frac{v}{\lambda}$$

Superposition of Waves

Wave amplitude (wave height) can be altered by the interaction or combination of two or more waves (superposition), resulting in an enlarged or reduced amplitude in the resultant wave form.

All waves interact or interfere when combined. If two transverse waves having equal characteristics approach each other from opposite directions, they will combine either positively or negatively. Positive combination of waves is called **constructive interference** and results in the amplitudes of the waves being added to form a third, larger wave. Auditorium acoustics attempt to amplify sounds from the stage through the physical construction of the room.

Negative combination of waves is called **destructive interference** and results in the amplitudes of the waves being subtracted or even canceling each other out.

Pitch

The frequency of a sound wave is interpreted by humans as the **pitch** of the sound. In music, pitch is equated with a specific note. Doubling the frequency of a sound wave changes the pitch of a note by making it one octave higher. Cutting the frequency in half lowers the pitch of the note by one octave. For example, concert A has a frequency of 440 Hz. At 880 Hz, the note A would be heard, but one octave higher. At 220 Hz, the A heard would be one octave lower.

Beats

The difference between two slightly different sound wave frequencies, when combined or superimposed, results in a third sound disturbance called a **beat**.

Beats are recurring, throbbing sounds of wavering loudness. In such cases the listener will hear the average of two frequencies wavering at the the difference of the frequencies. For example, if two violin players play the same note at slightly different pitches (one playing concert C at 126 Hz and the other playing concert C at 128 Hz), the listener will hear a frequency of 127 Hz with beats of frequency 2 Hz. In fine symphonic playing, this is considered poor intonation and is avoided by the careful tuning of the instruments before playing with the ensemble.

Standing Waves

A standing wave is an equally repeating system of wave pulses whose nodes do not move and whose antinodes have no horizontal movement.

When transverse waves of equal amplitude and frequency travel in opposite directions toward one another, superposition produces resultant waves whose amplitudes are larger or smaller than those of either wave.

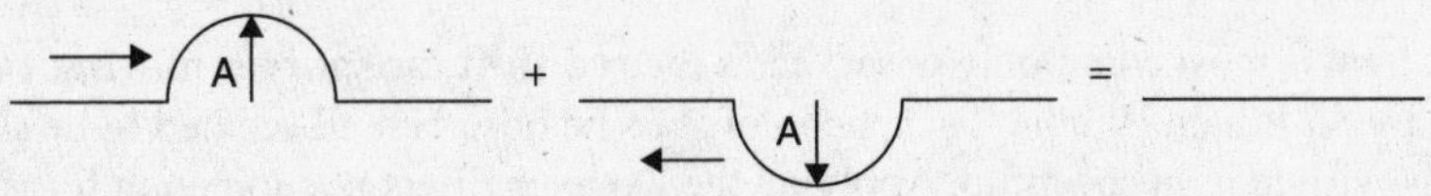

If the wave pulses meet 180° out of phase, and crests from one wave interfere exactly with troughs from the second wave (see the preceding illustration), total cancellation results, and there is no resultant wave. This is complete **destructive interference**.

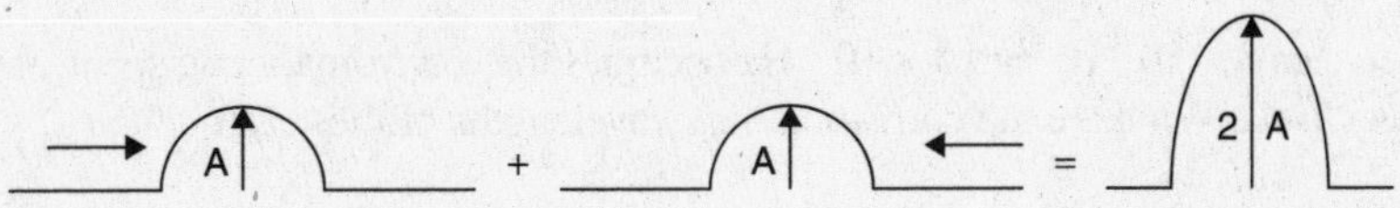

If the wave pulses meet perfectly in phase, and crests and troughs of both waves align perfectly (see the preceding illustration), a standing wave results. This resultant wave has twice the amplitude of either original wave and the same frequency as both original waves.

Referring to the basic transverse wave equation

$$\mathbf{y(x,t) = A \sin(\omega t + \kappa x)},$$

a standing wave can be mathematically described as

$$y(x,t) = A\,[\sin(\omega t - \kappa x) - \sin(\omega t + \kappa x)].$$

In sound waves, sustaining and energizing the standing wave by inputting waves at the same frequency of the standing wave produces the phenomenon called **resonance** and amplifies sound waves. It is the phenomenon that shatters wine glasses whose natural frequency is matched by incoming sound waves of the same frequency, causing the walls of the glass to sustain a standing wave. When the amplitude of the wave exceeds the elastic limits of the glass, it shatters. Resonance is caused by vibrating lips in brass instruments, by vibrating reeds and air columns in woodwinds, by vibrating strings in string instruments, and by vibrating surfaces and solid bars in percussion instruments.

Doppler Effect

The relative motion of the source and observer can distort the perceived frequency and pitch of a sound.

Christian Johann Doppler noted that sound waves approach a stationary listener at an apparently higher frequency and recede at an apparently lower frequency than that emitted at the source. This phenomenon occurs because the sound waves emitted by a moving source are **compressed** as the object moves **into** its own sound waves and are **rarefied** as the object moves **away from** its own sound waves. The resulting wave compression as a sound source approaches causes the sound waves to be closer together, thus raising the apparent frequency of the compressed sound waves. When the sound source passes and moves away from the listener, the resulting wave decompression causes the sound waves to be farther apart, thus lowering the apparent frequency of the rarefacted sound waves.

$$\text{For a moving source: } f_{OBS} = \frac{(f_{SOURCE})(V_{SOUND})}{(V_{SOUND} \pm V_{SOURCE})}$$

(Positive for approaching and negative for receding)

$$\text{For a moving observer: } f_{OBS} = \frac{(f_{SOURCE})(V_{SOUND} \pm V_{SOURCE})}{(V_{SOUND})}$$

(Positive for approaching and negative for receding)

Red Shift

The Doppler Effect for light is most commonly observed as the **red shift for sources moving rapidly away from the observer**. It was noted by Sir Edwin Hubble that spectra of stars he observed all seemed to be shifted into the lower frequency **red** end of the visible light spectrum. Applying the Doppler Effect phenomenon for sound waves to light led to the concept that the observed stars' apparent light frequency shift may be a result of the stars receding from our view. This extension of the Doppler Effect led to the postulation that the universe may be expanding.

Example

Light of frequencies between 3×10^{14} Hz and 3×10^{15} Hz occupies the electromagnetic spectrum from the near-infrared to near-ultraviolet, respectively. What are the corresponding wavelengths of these frequencies?

Solution

Using $f = \frac{v}{\lambda}$, $\lambda = \frac{v}{f}$

$$\lambda_{IR} = \frac{3 \times 10^8 \text{ m/s}}{3 \times 10^{14} \text{ waves/sec}} = 1 \times 10^{-6} \text{ m/wave}$$

$$\lambda_{UV} = \frac{3 \times 10^8 \text{ m/s}}{3 \times 10^{15} \text{ waves/sec}} = 1 \times 10^{-7} \text{ m/wave}$$

Example

How long does it take a thunder clap to travel 5 km if the air temperature is 25°C?

Solution

The velocity of sound at 25°C is

$$331.5 \text{ m/s} + (0.6 \text{ m/s/°C})(25\text{°C}) = 346.5 \text{ m/s}$$

$$\mathbf{v_{AVE} = d/}t \text{ and } t = \mathbf{d/v_{AVE}} = (5000 \text{ m})/(346.5 \text{ m/s}) = \mathbf{14.4 \text{ sec}}$$

Example

A transverse wave is given as

$$y(x,t) = 0.075 \sin (1.02\, t + 55.0\, x),$$

where A is in m, ω is in rad/sec and κ is in rad/m. Find the wave's

- **(a)** amplitude.
- **(b)** wavelength.
- **(c)** frequency.
- **(d)** period.
- **(e)** speed.

Solution

Using $y(x,t) = A \sin (\omega t + \kappa x)$,

- **(a)** amplitude = 0.075 m
- **(b)** $\lambda = (2\pi \text{ rad/cycle})/(55.0 \text{ rad/m}) = 0.11 \text{ m}$
- **(c)** $f = 1/T = (\omega \text{ rad/sec})/(2\pi \text{ rad/cycle}) = (1.02 \text{ rad/sec})/(6.28 \text{ rad/cycle}) = 0.16 \text{ Hz}$
- **(d)** $T = 1/f = 1/(0.16 \text{ cycles/sec}) = 6.25 \text{ sec/cycle}$
- **(e)** $v = \omega/\kappa = (1.02 \text{ rad/sec}) / (55.0 \text{ rad/m}) = \mathbf{0.019 \text{ m/s}}$

Example

A train approaches a crossing gate at a velocity of 10.2 m/s. It sounds its horn, which has a frequency of 400 Hz. What are the observed frequencies of the horn (**a**) as it approaches and (**b**) after it has passed by the crossing gate? (The speed of sound in air at the time is 342 m/s.)

Solution

$$\text{Using } f_{OBS} = \frac{(f_{SOURCE})(f_{SOUND})}{(V_{SOUND} \pm V_{SOURCE})}$$

(Negative for approaching and positive for receding)

(a) Approaching: $f_{OBS} = \frac{(400\text{ Hz})(342\text{ m/s})}{(342\text{ m/s} - 10.2\text{ m/s})} = \mathbf{412\ Hz}$

(b) Receding: $f_{OBS} = \frac{(400\text{ Hz})(342\text{ m/s})}{(342\text{ m/s} + 10.2\text{ m/s})} = \mathbf{388\ Hz}$

Interference

The corpuscular theory of light staunchly held by Isaac Newton was proven to be incorrect by Thomas Young's double slit experiment. Newton had held that light was made of particles because it behaved as particles behave: traveling from one location to another in straight lines, reflecting from objects in its path, and refracting or bending when passing through substances of differing densities.

When Young passed light through two narrow, parallel slits, he observed that the light produced an **interference pattern**, something that particles cannot do.

Wave interference patterns were well-known through studies with water waves and sound waves.

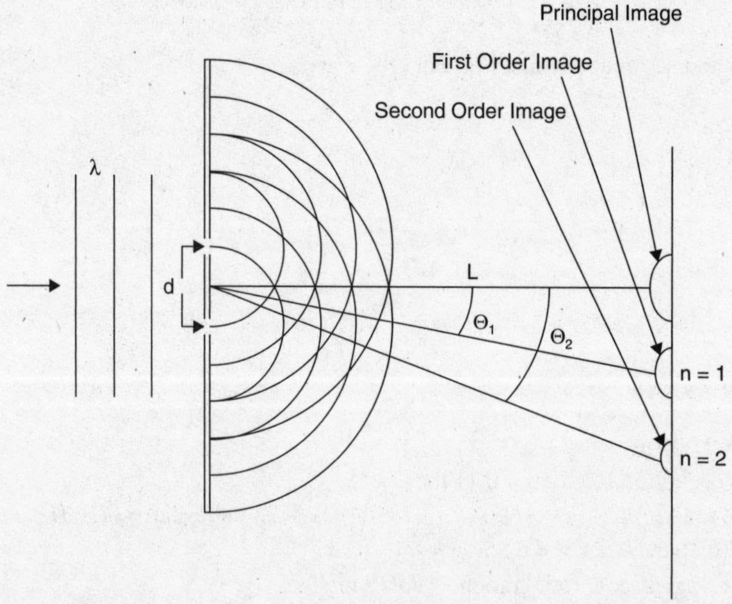

In diffraction pattern problems, waves of wavelength λ and velocity **v** approach a baffle with two slits, which are separated by a distance **d**. The baffle is located a distance **L** from a screen, onto which are projected a series of bright and dark areas. The bright spot directly across from the center of the slits is called the **principal image.** Each successive bright spot becomes smaller and less bright. The first bright spot *after* the principal image is the **first order image (n = 1)**; the second bright spot is the **second order image (n = 2)**; and so on. Each order image lies on a line designated to make an angle Θ with the line perpendicular to the center of the slits as seen in the preceding illustration. The relationship between λ, **d**,**Θn** and **n** is as follows:

$$\lambda = \frac{d \sin \Theta_n}{n}$$

Example

Find the wavelength and frequency of green light whose second order image makes a diffraction angle of 40.2° if the source is a diffraction grating having 6.30×10^3 lines per cm.

Solution

1 cm containing 6.30×10^3 lines yields:

$$\frac{1\,\text{cm}}{6.30 \times 10^3\,\text{lines}}\frac{(1\,\text{m})}{(100\,\text{cm})} = 1.59 \times 10^{-6}\,\text{m/line} = d$$

$$\text{Using } \lambda = \frac{d \sin \Theta_n}{n} \text{ yields } \lambda = \frac{\left(1.59 \times 10^{-6}\right)\left(\sin 40.2^\circ\right)}{2} =$$

$$\mathbf{5.13 \times 10^{-7}\,m}$$

For the frequency, $f = v/\lambda = (3 \times 10^8\text{ m/s})/(5.13 \times 10^{-7}\text{ m/wave}) = \mathbf{5.84 \times 10^{14}\ Hz}$

Dispersion

When white light is passed through a prism, its beam separates or **disperses** into the colors of the spectrum. The visible light spectrum spans only a narrow part of the entire electromagnetic spectrum, from about 3.9×10^{-7}m in the range of violet/ultraviolet to about 7.2×10^{-7}m in the range of red/infrared. (One wonders how the world would look to a human if our eyes could see even 20 percent more of the electromagnetic spectrum, let alone the entire range!) The separation of frequencies of visible light into what we perceive as distinct colors is blurred by a continuous spectral gradient of gradually changing hues. The phenomenon of dispersion is ever-present in our lives, from the multicolored patterns of oil on water to the shimmer of a rainbow to a planar circular musical diffraction grating (compact disc) with our favorite music.

The diffraction patterns discussed in this chapter are mainly concerned with light and dark order images produced by a line spectrum, but the concept of dispersion includes all references to visible light of any certain wavelength, frequency, and our perception of its color.

Review Questions and Answers

B Exam Question Types

Multiple Choice

1. A compressional disturbance has a frequency of 100 Hz. If the speed of the disturbance is 340 m/s, the distance from a compression to the nearest rarefaction is

(A) 0.7 m
(B) 1.0 m
(C) 1.7 m
(D) 3.4 m
(E) 4.7 m

2. The difference between two audible frequencies is 4 Hz. If one frequency is 380 Hz and the speed of sound is 340 m/s, the other frequency might be about

(A) 300 Hz
(B) 325 Hz
(C) 350 Hz
(D) 375 Hz
(E) 400 Hz

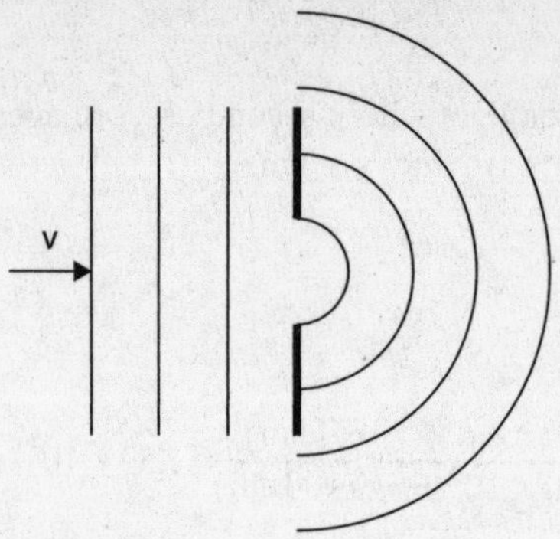

3. Spreading of wave disturbances past the edges of barriers refers to

(A) Amplitude
(B) Damping
(C) Diffraction
(D) Dispersion
(E) Refraction

4. An automobile horn has a frequency of 440 Hz and is traveling at 25 m/s. On a day when the speed of sound is 346 m/s, the apparent frequency of the horn as heard by a person on the sidewalk as the auto approaches is most nearly

(A) 110 Hz
(B) 220 Hz
(C) 440 Hz
(D) 460 Hz
(E) 880 Hz

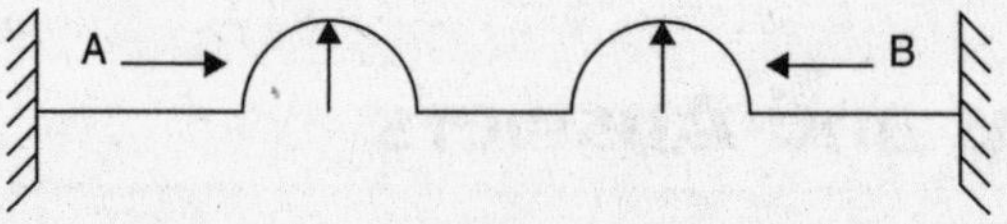

5. Wave A and wave B move toward each other on a string. Which of the following is impossible?

(A) Constructive interference
(B) Destructive interference
(C) Reflection
(D) Superposition
(E) Change of amplitude

6. X-rays having a wavelength of 6×10^{-10} m have a frequency most nearly

(A) 5.5×10^{-10} Hz
(B) 5.0×10^{2} m/s^2
(C) 5.0×10^{17} m
(D) 5.0×10^{17} Hz
(E) 5.5×10^{19} Hz

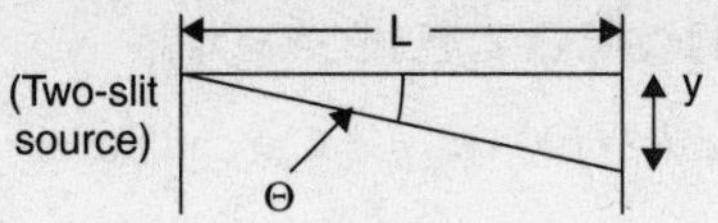

7. A two-slit light source of wavelength λ is located a distance L from a screen. The distance from the principal image to the first order image is given as y. The slit separation d is

(A) λn
(B) $\lambda n / y$
(C) $\lambda n \sin\theta_n$
(D) $\sin\theta_n / \lambda n$
(E) $\lambda n / \sin\theta_n$

8. A wave moving along a spring

(A) cannot be transverse.
(B) transports particles of matter.
(C) has no kinetic energy.
(D) has a definite wavelength.
(E) cannot be reflected.

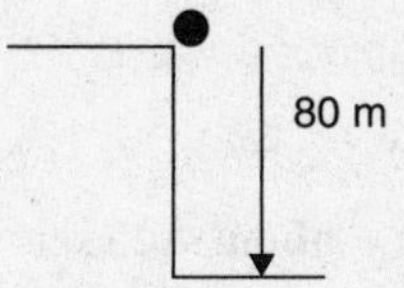

9. A ball falls a distance of 80 m on a day when the speed of sound is 340 m/s. The person who drops the ball will hear the ball hit bottom after approximately

(A) 1.25 sec
(B) 2.00 sec
(C) 2.25 sec
(D) 4.00 sec
(E) 4.25 sec

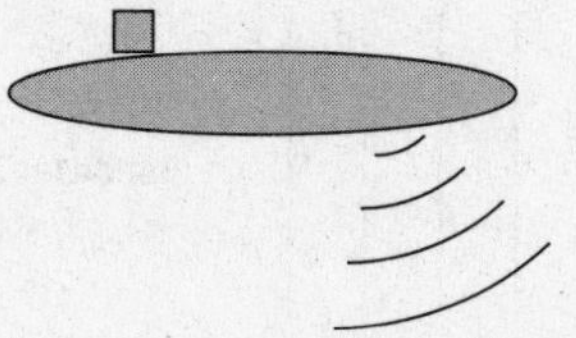

10. A submarine's sonar system generates pulses of 2.0×10^5 Hz. If the speed of sound in water is 1450 m/s and echoes from the bottom take 0.4 sec from transmission to reception, the distance from the sub to the bottom is closest to

(A) 100 m
(B) 200 m
(C) 300 m
(D) 400 m
(E) 500 m

Answers to Multiple-Choice Questions

1. **(C)** Using $f = v/\lambda$, $\lambda = (v/f) = (340 \text{ m/s})/(100 \text{ Hz}) = 3.4$ m and the distance from crest to adjacent trough is $\lambda/2 =$ **1.7 m**

2. **(D)** Beat frequency of 4 Hz means that there is a difference of 4 Hz between the given frequencies. 380 Hz ± 4 Hz is either 384 Hz or 376 Hz. The closest answer is **375 Hz.**

3. **(C)** The definition of wave diffraction is the bending of waves as they pass the edges of a barrier.

4. **(D)** Since the horn is approaching, the apparent frequency must be higher than 440 Hz. This leaves only 460 Hz and 880 Hz. At 880 Hz, the pitch of the horn would be an octave higher, which is impossible at normal automotive speeds, thus leaving **460 Hz.**

5. **(B)** The waves will add amplitudes and interfere constructively. They will then continue on and reflect from the boundaries at each end. **The only impossible answer is destructive interference.**

6. **(D)** Using $f = (v/\lambda)$, $f = \dfrac{3 \times 10^8 \text{ m/s}}{6 \times 10^{-10} \text{ m/wave}} = \mathbf{5.0 \times 10^{17}}$ **Hz**

7. **(E)** Using $\lambda = \dfrac{d \sin \Theta_n}{n}$, $\mathbf{d = \lambda n/\sin\theta_n}$

8. **(D)** **All waves have a definite wavelength.**

9. **(E)** First, the time to fall 80 m: $d = v_0 t + \left(\frac{1}{2}\right) at^2$

   $-80 \text{ m} = 0 - 4.9 \text{ m/s}^2(t^2)$ yields time to fall = 4 sec.

   Next, the time for the sound to travel back up: $t = d/v_{SOUND}$

   $t = 80 \text{ m}/ 340 \text{ m/s} = 4/17$ sec (approximately $\frac{1}{4}$ sec)

   Total time = 4 sec + approximately 0.25 sec = **about 4.25 sec**

10. **(C)** It takes half of 0.4 sec or 0.2 sec for sound to travel from the sub to the bottom. $d = (1450 \text{ m/s})(0.2 \text{ sec}) =$ 290 m **(closest is 300 m).**

Free Response

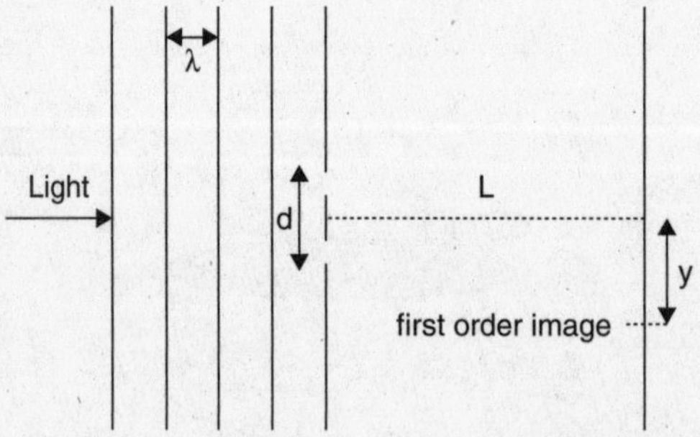

1. Light of wavelength λ equal to 4.9×10^{-7} m strikes a barrier which has two small slits separated by a distance **d**. A screen is placed a distance **L** from the barrier and is parallel to the barrier. The first order image is located at a distance **y** from the principal image as shown. Distance **L** is 0.90 m.

   (a) If **y** is 1.40×10^{-2} m, determine **d**.
   (b) If light of wavelength 5.1×10^{-7} m is now used, determine the new distance **y** for the first order image.
   (c) What are the frequencies of light used in **(a)** and **(b)**?

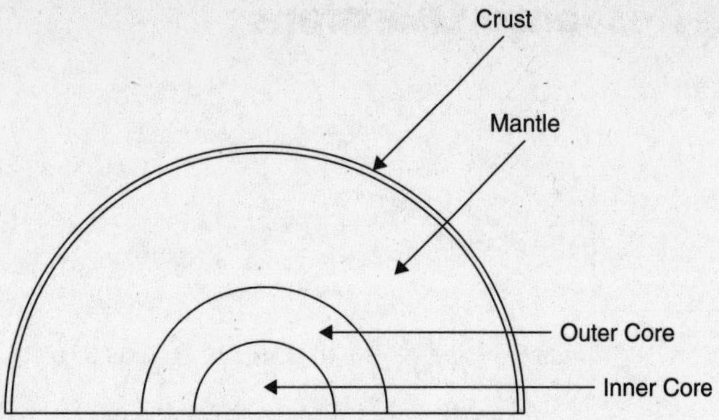

2. A seismic (earthquake) wave of frequency 5 Hz is recorded by a seismograph. The P-wave (primary/pressure) traveled through the Earth's mantle to the crust and was recorded first at the epicenter, directly above the focus or origin of the earthquake. After 2.5 minutes, the S-wave (secondary/sinusoidal) was recorded. The average S-wave velocity in the Earth's mantle is approximately 6200 m/s.

(a) How deep is the focus (origin) of the earthquake?

The Earth's mantle begins at an approximate depth of 30 km and ends at an approximate depth of 3000 km. The P-wave velocities at these depths are approximately 7.7 km/s to 8.2 km/s, respectively.

(b) Calculate the approximate wavelengths of P-waves traveling at these two depths.

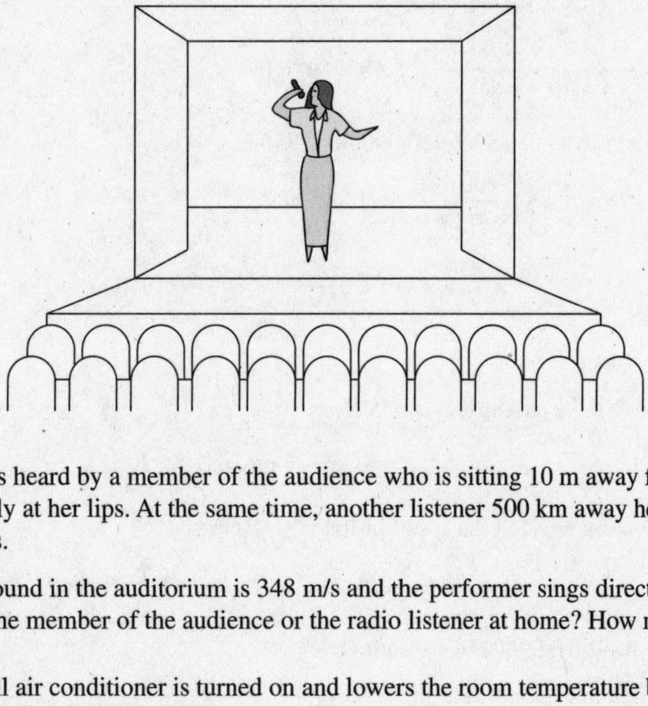

3. A singer at a concert is heard by a member of the audience who is sitting 10 m away from the singer, who holds the microphone directly at her lips. At the same time, another listener 500 km away hears the concert broadcast on his radio earphones.

(a) If the speed of sound in the auditorium is 348 m/s and the performer sings directly into the mike, who hears the singer first, the member of the audience or the radio listener at home? How much sooner does that person hear it?

(b) If the concert hall air conditioner is turned on and lowers the room temperature by 5°C, who hears the concert first and how much sooner?

(c) If the radio broadcast is also carried by another station with a broadcast frequency that is higher by 10 kc/s than that of the first station, which station's signal reaches the home listener first and by how many seconds?

Answers to Free-Response Questions

1. (a)

$$\text{Using } \lambda = \frac{d \sin \Theta_n}{n},\ d = \lambda n / \sin\Theta_n$$

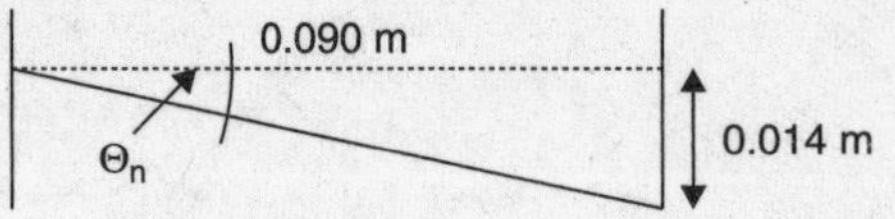

$$\tan \Theta_n = \frac{0.014\text{ m}}{0.090\text{ m}} = 0.0156 \text{ and } \Theta_n = 0.89°$$

$$\text{solving for } \mathbf{d}\text{: } \mathbf{d} = \frac{(4.90\times10^{-7})(1)}{\sin 0.89°} = \frac{(4.90\times10^{-7}\text{ m})}{0.0155} = \mathbf{3.16\times10^{-5}\ m}$$

(b) $$\text{Using } \lambda = \frac{d\sin\Theta_n}{n},\ \sin\Theta_n = \lambda n / d = \frac{(5.1\times10^{-7}\text{ m})}{(3.16\times10^{-5}\text{ m})} = 0.0161 \text{ and } \Theta_n = 0.92°$$

$\tan 0.92° = y / (0.90\text{ m})$ and $y = \mathbf{1.44\times10^{-2}\ m}$

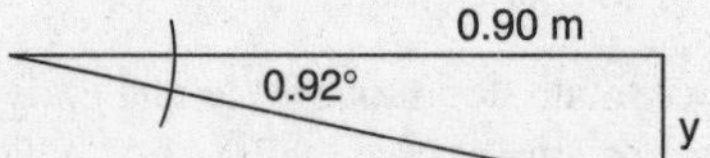

(c) $$\text{For part (a): } f = v/\lambda = \frac{(3\times10^{8}\text{ m/s})}{(4.9\times10^{-7}\text{ m/wave})} = \mathbf{6.12\times10^{14}\ Hz}$$

$$\text{For part (b): } f = v/\lambda = \frac{(3\times10^{8}\text{ m/s})}{(5.1\times10^{-7}\text{ m/wave})} = \mathbf{5.88\times10^{14}\ Hz}$$

2. (a) Using $\mathbf{d = vt}$ = (6200 m/s)(150 sec) = 930,000 m or **930 km**

(b) Using $\mathbf{f = v/\lambda}$, $[\lambda_{30KM} = v/f = \frac{(7700\text{ m/s})}{(5\text{ waves/sec})} = 1540\text{ m}]$

$$\lambda_{3000KM} = v/f = \frac{(8200\text{ m/s})}{(5\text{ waves/sec})} = 1640\text{ m}$$

3. (a) Using $\mathbf{v_{AVE} = d/t}$, $t = \mathbf{d/v_{AVE}}$

Since the speed of sound in the auditorium is 348 m/s:

$$t_{AUDIENCE} = 10\text{ m}/348\text{ m/s} = 0.0287\text{ sec}$$

Since radio broadcast waves travel at the speed of light, 3×10^8m/s,

$$t_{RADIO} = 5\times10^5\text{m}/3\times10^8\text{m/s} = 0.00167\text{ sec}$$

The home listener hears it first, 0.0270 sec sooner.

(b) The speed of sound in the cooled auditorium becomes:

$$348\text{ m/s} - (0.6\text{ m/s/°C})(5\text{°C}) = 345\text{ m/s}$$

This changes the travel time for the live sound to 10 m/345 m/s = 0.0299 sec

The home listener still hears it first, now 0.0282 sec sooner.

(c) Since radio waves all travel at the same velocity, 3×10^8 m/s, increasing or decreasing broadcast frequency has no effect on travel time.

The home listener still hears it first, 0.0282 sec sooner.

Geometric Optics (B Exam)

The study of geometric optics focuses on applications of the physical characteristics of visible light, an electromagnetic radiation that spans a wavelength range of about 4×10^{-7}m (violet) to about 7×10^{-7}m (red). Visible light frequencies fall approximately between 4.0×10^{14} Hz (red) and 7.5×10^{14} Hz (violet).

ELECTROMAGNETIC SPECTRUM

λ (m) approximately		f (Hz) approximately	
10^6	POWER	10^2	AC Electric Generators
$10^2 - 10^3$	AM RADIO	$10^5 - 10^6$	Communications
1 – 10	TV, FM RADIO	$10^7 - 10^8$	Communications
10^{-1}	RADAR	10^9	Communications / Sensing
10^{-1}	MICROWAVES	10^9	Communications / Sensing
10^{-4}	INFRARED	10^{12}	Heat Lamps / Night Photos
7×10^{-7}	VISIBLE LIGHT	4.3×10^{14}	Red Orange Yellow Green Blue Indigo
4×10^{-7}		7.5×10^{14}	Violet
10^{-8}	ULTRAVIOLET	10^{16}	Lasers / Lamps
10^{-10}	X-RAYS	10^{18}	Medical Diagnostics
10^{-12}	GAMMA RAYS	10^{20}	Accelerators
10^{-16}	HARD GAMMA COSMIC RAYS	10^{24}	Accelerators Natural Incoming Radiation

Visible Light

Isaac Newton believed that light obeyed the same natural laws as particles. He held that light consisted of tiny particles. This particle or *corpuscular* theory of the nature of light was drastically altered by Thomas Young's experiments, which demonstrated that light displays definite wavelike qualities as well, which was discussed in the previous chapter.

This chapter reviews how the study of geometric optics applies principles of the characteristics of light, principally reflection and refraction. As previously mentioned, all radiant energy (including all force and energy fields) diminishes in intensity as it propagates away from a point source and obeys the **inverse-square law**:

$$I \propto \frac{1}{r^2}$$

where **I** is intensity and **r** is the linear distance away from the source. In plain language, the **inverse-square law** states the following:

The intensity of the energy or force is inversely related to the square of the distance.

This means that **doubling** the distance from a source of energy or force decreases the intensity to $\frac{1}{4}$ its original value. Tripling the distance away results in the intensity being reduced to $\frac{1}{9}$ its original value, and so forth.

This book has already discussed the inverse-square law with respect to gravity and sound waves. (A discussion of this phenomenon as it refers to electrical and magnetic phenomena is covered in upcoming chapters.)

Since light covers an area when it is projected outward from the source onto the spherical bubble as it travels a certain distance **r**, its strength diminishes by the inverse-square law.

When light emanates from any point source, such as a candle flame, an LED (light-emitting diode), an incandescent lamp, or a star, its illuminative strength diminishes according to the inverse-square law as follows:

$$\mathbf{E} = \frac{\mathrm{I}}{r^2}$$

Where **E** is **illumination** in lumens/m^2, **I** is intensity in **candles** or **candelas,** and **r** is the linear distance from the light source.

The rule to remember is that **candles** are the units **at the source**, and **lumens/m^2** are the units **at some distance away from the source.**

Since the area of a sphere is $4\pi r^2$, a point source of 1 candle produces an illumination rate or *luminous flux* of 4π lumens in a sphere of radius 1m.

4π lumens/candle is the equivalency used for luminous calculations at 1 m.

For example, a 2-candle source at a 2-m radius would produce a luminous flux of (2 cd)(4π lumens/candle)/(2 m)^2 or 2π lumens.

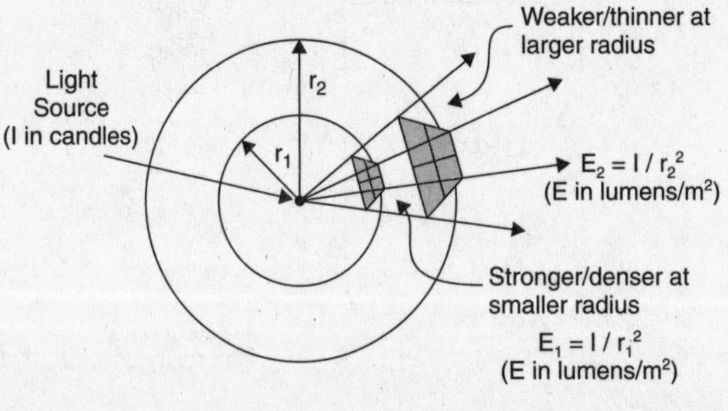

If two unequally bright lamps are separated by some distance **d** and a screen is placed between them, it is easy to determine where the screen should be placed in order for it to be equally illuminated by both lamps.

When illumination from two lamps falling on a screen somewhere between them is the same, set E_1 equal to E_2 as follows:

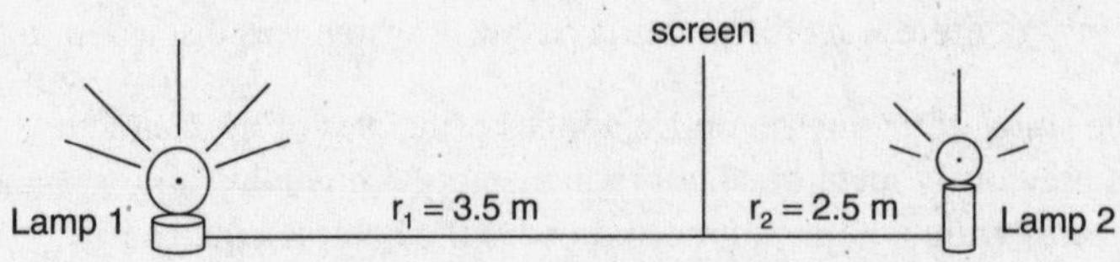

Example

A photosensitive screen receives equal illumination from two lamps, placed 3.5 m and 2.5 m, as shown in the diagram. If the intensity of the first lamp (at 3.5 m distance) is 50 cd, what is the intensity of the second lamp?

Solution

Using $\frac{I_1}{r_1^2} = \frac{I_2}{r_2^2}$

$$\frac{50\,\text{cd}}{(3.5\,\text{m})^2} = \frac{x}{(2.5\,\text{m})^2}$$

$$x = \frac{(50\,\text{cd})(6.25\,\text{m}^2)}{(12.25\,\text{m}^2)}$$

$$x = \mathbf{25.5\ cd}$$

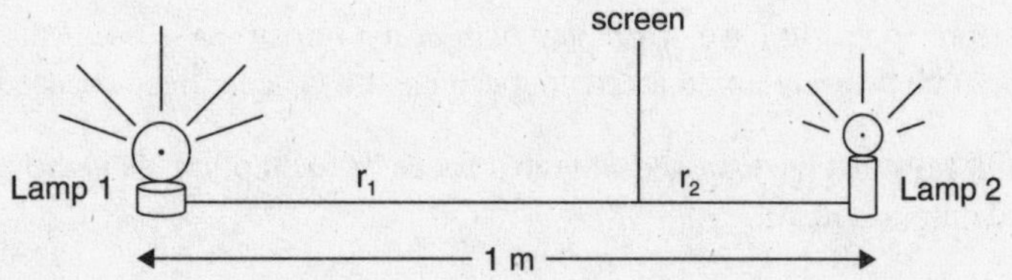

Where should a screen be placed to receive equal illumination from two lamps if **Lamp 1** = 5 cd and **Lamp 2** = 2 cd and if the lamps are separated by 1 m?

Solution

Let the Illumination **E** be equal for both lamps. Then using the illumination relationship for both,

$$E_1 = \frac{I_1}{r_1^2} = E_2 = \frac{I_2}{r_2^2}$$

$$5\frac{\text{cd}}{r_1^2} = 2\frac{\text{cd}}{r_2^2}$$

$$(\text{Let } \mathbf{r_2} = 1\text{ m} - \mathbf{r_1})$$

$$5\frac{\text{cd}}{r_1^2} = \frac{2\text{cd}}{(1\text{m} - r_1)^2}$$

$$5(1 - 2\,\mathbf{r_1} + \mathbf{r_1}^2) = 2\mathbf{r_1}^2$$

$$3\mathbf{r_1}^2 - 10\mathbf{r_1} + 5 = 0$$

Using the quadratic equation $x = \frac{-b \pm (b^2 - 4ac)^{1/2}}{2a}$ where x is $\mathbf{r_1}$ yields $\mathbf{r_1}$ **= 0.61 m and 2.7 m.**

This places equal illumination at 0.61 m *and* 2.7 m to the right of Lamp 1. The screen between the lamps is equally illuminated by both when it is 0.61 m to the right of Lamp 1, *and* at another location of equal illumination 2.7 m to the right of Lamp 1, which is 1.7 m to the right of Lamp 2. The problem yields a solution of r = 0.61 m to the right of Lamp 1, for a screen *between* the lamps.

Reflection

When a ray of light is directed at a polished reflecting surface, the angle of the incoming or incident ray (**i**) is measured from the normal to the reflecting surface, as is the angle for the reflected ray (**r**).

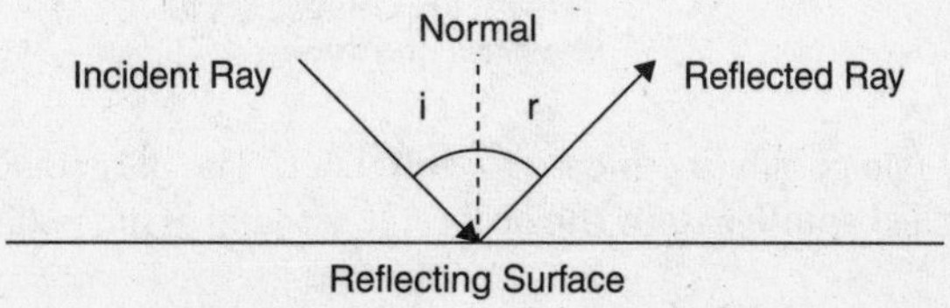

Reflected rays obey these rules:

a. The angle of incidence (i) equals the angle of reflection (r).

angle i = angle r

b. The incident ray, the reflected ray and the normal to the reflecting surface all lie in the same plane.

Images

The image formed by rays of light reflecting from a polished surface can be either virtual or real.

Virtual images made with mirrors appear to have originated *behind* the mirror (as is always the case with plane/flat mirrors). A virtual image cannot be projected onto a screen, because it has never been focused.

A real image is formed by light rays that have passed though a focus or focal point, as is the case with concave mirrors. A real image can be projected onto a screen.

Mirrors

The three basic types of mirrors are **plane**, **convex**, and **concave**.

a. Plane or flat mirror

The image appears to have originated behind the mirror. The distance from the object to the mirror is equal to the apparent or **virtual** distance from the image to the mirror ($\mathbf{S_O = S_I}$). A plane mirror can produce only a virtual image.

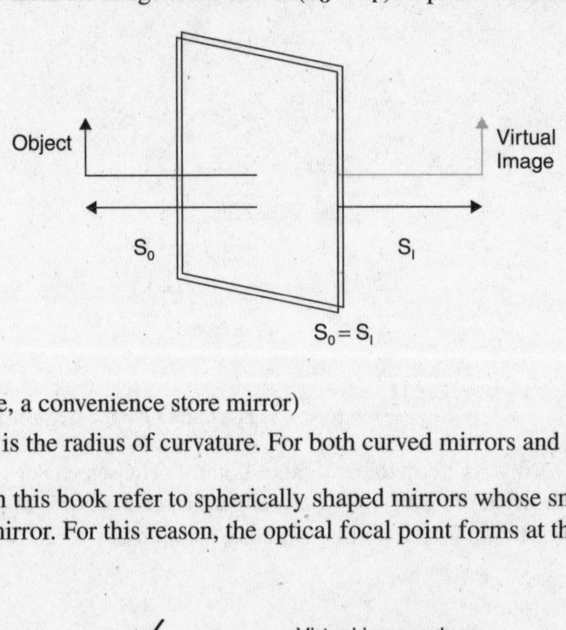

b. Convex mirror (for example, a convenience store mirror)

f is the focal length, and **c** is the radius of curvature. For both curved mirrors and lenses, c = 2f.

Spherical mirrors as discussed in this book refer to spherically shaped mirrors whose small aperture closely approximates the shape of a parabolic mirror. For this reason, the optical focal point forms at the focus of the parabolic shape common to both mirrors.

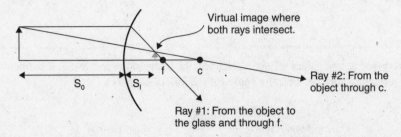

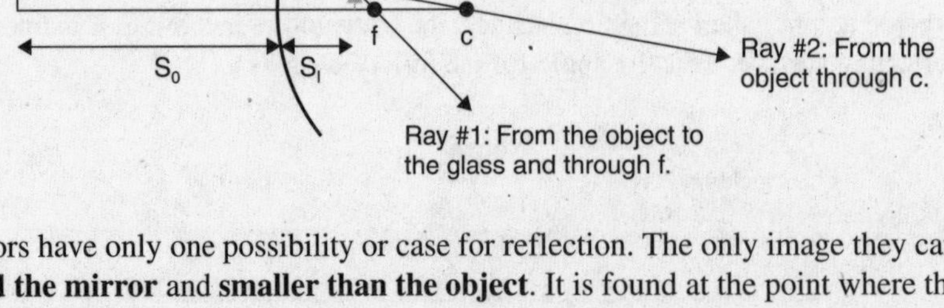

Convex spherical mirrors have only one possibility or case for reflection. The only image they can give is **virtual, right side up** (erect), **behind the mirror** and **smaller than the object**. It is found at the point where the two rays intersect.

Ray Diagrams

When drawing ray diagrams for any type of curved mirror, there are two rays to draw:

1. From the object straight to the glass and then through the focus F
2. From the object directly through the center of curvature C

The image will be found where the two rays **intersect**.

The image is a **virtual image** when it appears behind the mirror.

The image is **right side up** when it appears behind the mirror.

The image is a **real image** when it appears on the same side as the object.

The image is a **real image** when it is inverted.

The distance from the vertex of the mirror to the image (S_I) is negative when the image is behind the mirror.

The image distance, the image height, and the magnification can all be computed using the following formula:

$$\frac{1}{f} = \frac{1}{S_O} + \frac{1}{S_I}$$

where **f** is the focal length, $\mathbf{S_O}$ is the distance from the object to the mirror, and $\mathbf{S_I}$ is the distance from the mirror to the image.

$$\frac{h_i}{h_O} = \frac{S_I}{S_O}$$

where $\mathbf{h_I}$ is the height of the image and $\mathbf{h_O}$ is the height of the object.

h_I is negative when the image is inverted.

$$M = -\frac{S_i}{S_O}$$

where **M** is the magnification (how many times larger or smaller the image is compared to the object).

In the preceding diagram for the convex mirror, notice that the image formed is between the mirror and **f**. The focal distance **f** and the distance to the image $\mathbf{S_I}$ are both **negative** because they are located behind the mirror.

These formulas will hold for both convex and concave spherical mirrors. In every case, the distance from the object to the mirror $\mathbf{S_O}$ will be positive. If the image forms behind the mirror, $\mathbf{S_I}$ will be negative and will be *virtual* and *erect*.

Convex spherical mirrors can form only smaller, erect, virtual images behind the mirror.

c. **Concave mirror** (household enlarging mirror),

c is the distance from the mirror to its center of curvature,

f is the mirror's focal length, and

c = 2f

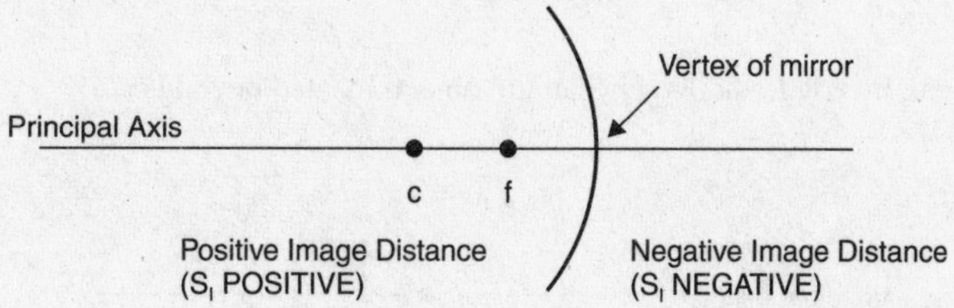

There are six possibilities or cases for concave spherical mirrors.

Case 1. Object at "Infinite" Distance (**very large distance compared with the focal length of the mirror)**

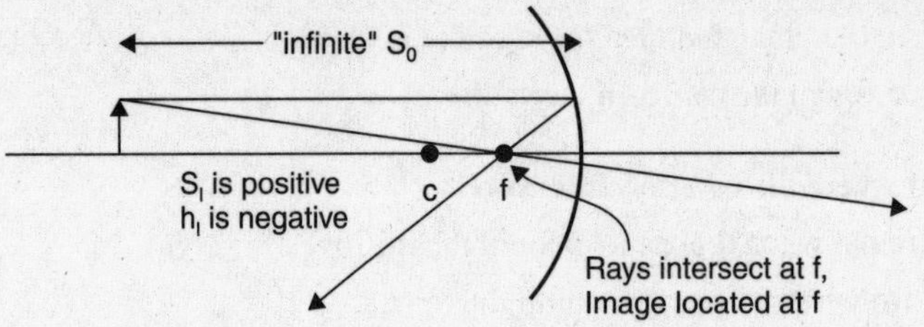

The image formed is real, inverted, and smaller than the object, located at the focus **f.**

Case 2. Object at Finite Distance beyond C

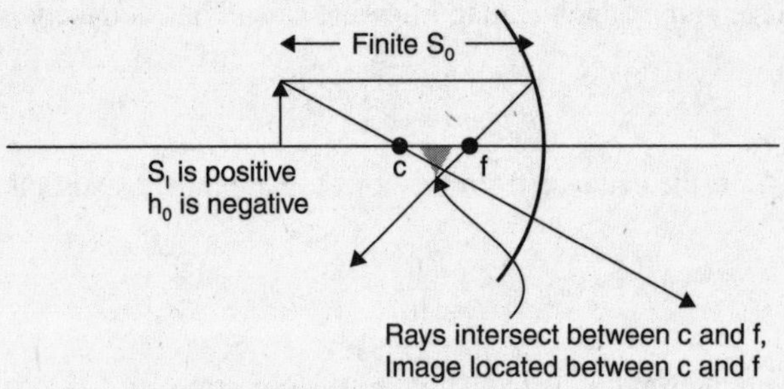

The image formed is real, inverted, and smaller than the object, located between **c** and **f.**

Case 3. Object at **c**

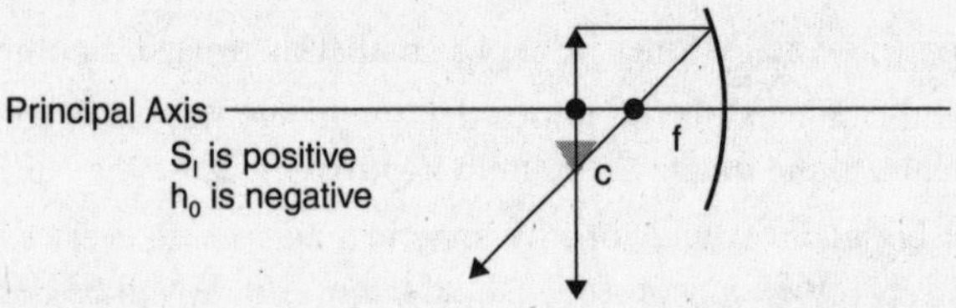

The image formed is real, inverted, and the same size as the object, located at **c**.

Case 4. Object between **c** and **f**

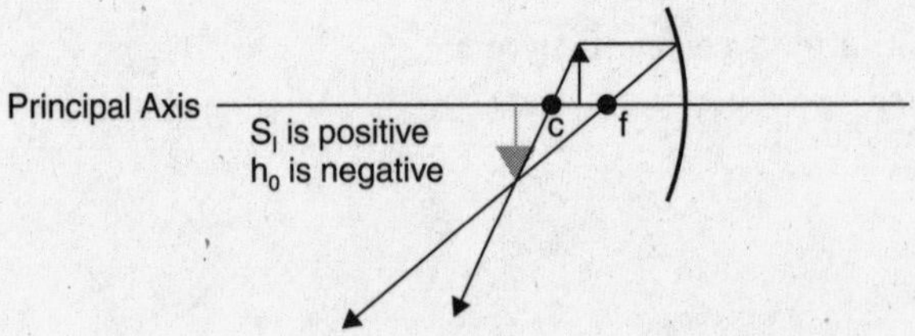

The image formed is real, inverted, and larger than the object, located beyond **c**.

Case 5. Object at **f**

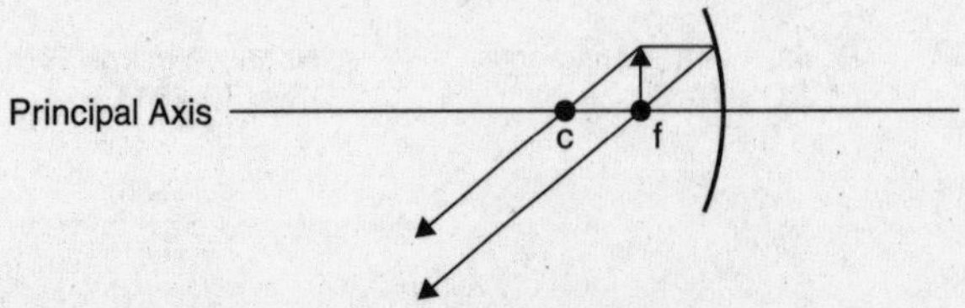

Since the rays emerge parallel and never intersect, **no image is formed.**

Case 6. Object between **f** and Mirror

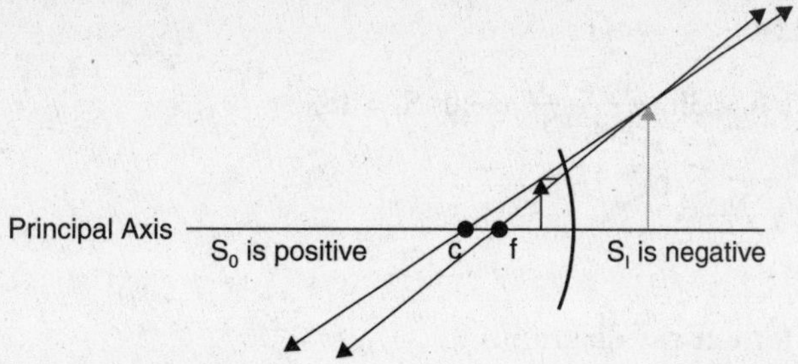

The image formed is virtual, erect, and larger than the object, located behind the mirror.

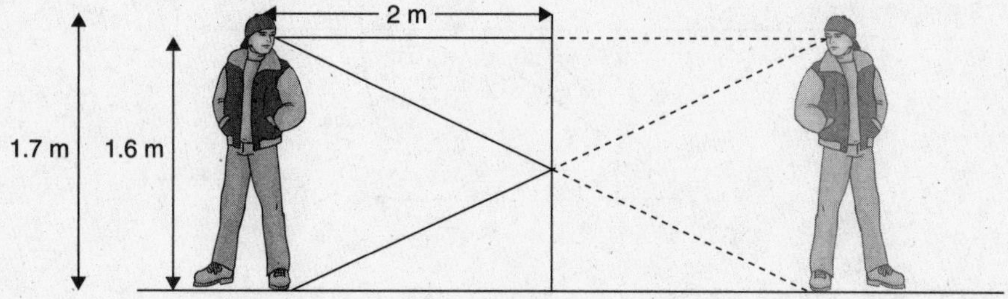

Example

A boy who is 1.7 m tall stands in front of a mirror that is 2 m away.

His eyes are located 0.1 m from the top of his head. If he looks into the mirror, he can just see all of himself from the top of his head to the bottom of his shoes but no further. How large is the mirror and what is its location with respect to the floor?

Solution

Since the angle of incidence equals the angle of reflection, he only needs the mirror to be $\frac{1}{2}$ the distance from his eyes to his feet in order to see his feet (that would be 1.6 m/2 or 0.8 m). Then he needs the mirror to be $\frac{1}{2}$ the distance from his eyes to the top of his head in order to see the top of his head (that would be 0.1 m/2 or 0.05 m).

That makes the size of the mirror 0.8 m + 0.05 m = **0.85 m**

Because he needs to see his shoes, that would make the distance the mirror is up from the floor $\frac{1}{2}$ the distance from his eyes to his shoes, or 1.6 m/2 = 0.8 m.

The mirror is 0.85 m in size and is located **0.8 m up from the floor**.
Note that the mirror needed to be only half the size of the boy.

Example

A concave spherical mirror of small aperture has a 30-cm radius of curvature. Where must an object be placed, relative to the mirror, in order to produce an image that is 3 times the size of the object? Solve the problem and draw the ray diagram.

Solution

(Case 4: Object between c and f)

Since **c** = 30 cm, **f** = 15 cm; since $\mathbf{h_I} = 3\mathbf{h_O}$, $\frac{h_I}{h_O} = \frac{S_I}{S_O}$ yields $\mathbf{S_I} = 3\mathbf{S_O}$.

Using $\frac{1}{f} = \frac{1}{S_O} + \frac{1}{S_I}$: $\left(\frac{1}{15}\text{ cm}\right) = \left(\frac{1}{3S_O}\right) + \left(\frac{1}{S_O}\right)$

$$\mathbf{S_O = 20\ cm}$$

(This makes S_I equal to 60 cm for our ray diagram.)

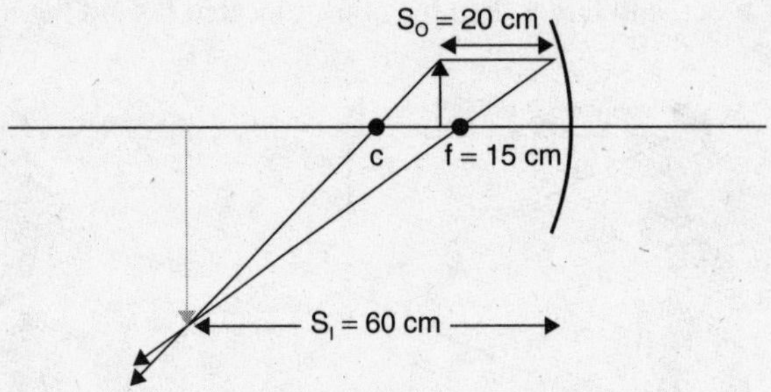

The image formed is real, inverted, three times larger, and 60 cm away.

Note: Since the image is inverted, h_I is negative, which makes the magnification –3 for that reason.

Example

A miniature tree 7 cm high is placed 28 cm away from a concave spherical mirror of radius 20 cm curvature. Describe the image completely and draw the ray diagram.

Solution

(Case 2: Object at a finite distance.)

Using $\frac{1}{f} = \frac{1}{S_O} + \frac{1}{S_I}$: $\left(\frac{1}{10}\text{ cm}\right) = \left(\frac{1}{28}\text{ cm}\right) + \left(\frac{1}{S_I}\right)$

$$S_I = 15.6\text{ cm}$$

(This makes S_I equal to 15.6 cm for our ray diagram.)

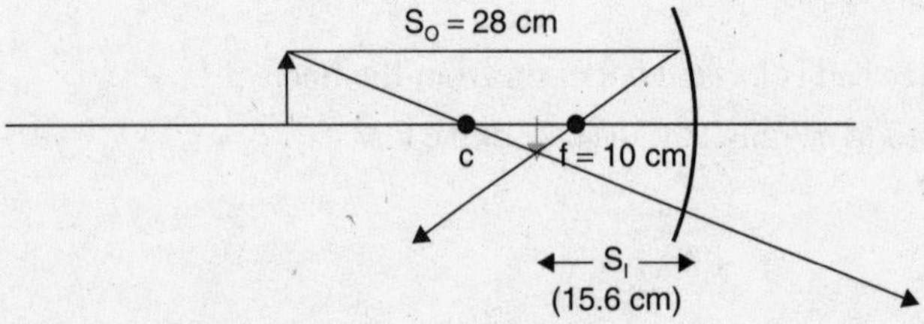

$$\mathbf{M} = -\frac{S_I}{S_O} = \frac{(-15.6\text{ cm})}{(28\text{ cm})} = (-0.56) \textit{ and } h_I = (-0.56)(7\text{ cm}) = -3.9\text{ cm}$$

The image formed is real, inverted, 0.56 times as large, 15.6 cm away (between **c** and **f**) and 3.9 cm in size.

Example

The image of the full moon is formed by a concave mirror having a radius of curvature 40 cm. If the distance to the moon from that location on Earth is 3.83×10^8 m, and the moon's radius is 1.74×10^6 m, describe the image fully and draw a ray diagram.

Solution

(Case 1: Object at an "infinite" distance.)

Using $\frac{1}{f} = \frac{1}{S_O} + \frac{1}{S_I} : \frac{1}{S_I} = \frac{1}{f} - \frac{1}{S_O} = \frac{1}{0.20\,\text{m}} - \left(\frac{1}{3.83 \times 10^8\,\text{m}}\right)$

Since the focal length is insignificant with respect to the object distance, the image is formed approximately at the focus.

(This makes S_I approximately equal to the focal length of 20 cm for our ray diagram.)

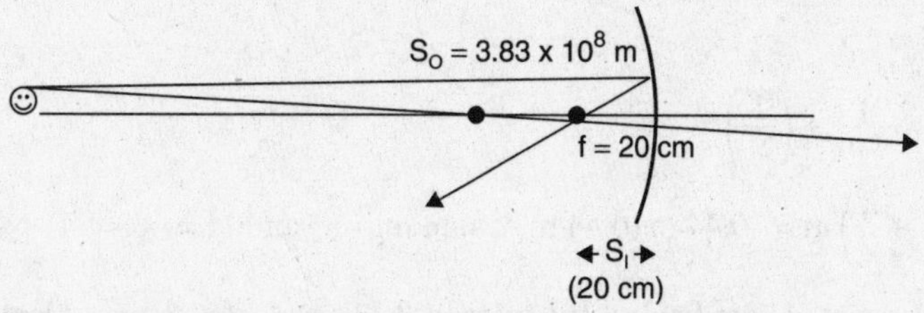

$\mathbf{M} = -\frac{S_I}{S_O} = \frac{-0.20\,\text{m}}{3.83 \times 10^8\,\text{m}} = (-5.7 \times 10^{-10})$ times the size, and $h_I = (-5.7 \times 10^{-10})(0.20\text{ m}) = -1.1 \times 10^{-10}$m or -1.1×10^{-8}cm high.

The image formed is real, inverted, approximately at the focus and -1.1×10^{-8}cm in size.

Example

A spherical concave mirror of radius 38 cm is used in an optical experiment. A 3 cm-high thimble is placed exactly 10 cm from the mirror along its principal axis. Describe the image fully and draw the diagram.

Solution

(Case 6: Object between f and the mirror.)

Using $\frac{1}{f} = \frac{1}{S_O} + \frac{1}{S_I} : \frac{1}{19}\text{ cm} = \frac{1}{10}\text{ cm} + \left(\frac{1}{S_I}\right)$

$$\frac{1}{S_I} = -\frac{9\,\text{cm}}{190\,\text{cm}} \text{ and } \mathbf{S_I} = -21.1\text{ cm}$$

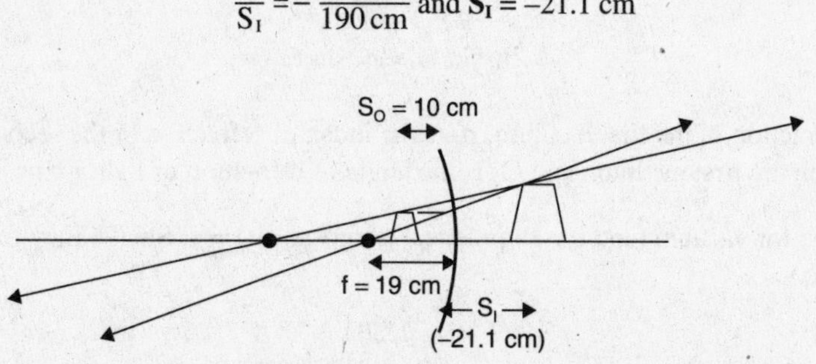

$$\mathbf{M} = -\frac{S_I}{S_O} = -\frac{-21.1}{10\,\text{cm}} = 2.1\times \text{ and } h_I = (2.1)(3\text{ cm}) = 6.3\text{ cm}$$

The image formed is virtual, erect, 2.1× as large as the object (6.3 cm high), and located 21.1 cm behind the mirror.

Example

A convex spherical mirror of focal length 50 cm is used in a convenience store. A curious cat looks at his image and sits 4 m away, staring at it. Describe the image fully and draw the diagram.

Solution

Because this is a convex mirror, the focal distance and the image distance will be negative, since they are located behind the mirror.

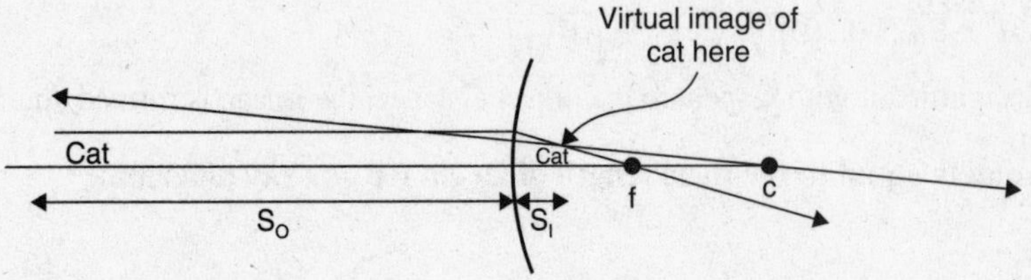

Using $\frac{1}{f} = \frac{1}{S_O} + \frac{1}{S_I}$: $\frac{1}{-0.5\,\text{m}} = \frac{1}{4\,\text{m}} + \left(\frac{1}{S_I}\right)$

$$S_I = -\left(\frac{4}{9}\right)\text{m} = -0.44\text{ m (0.44 m behind mirror) and } M = -0.44/4) = -0.11\times$$

The image formed is virtual, erect, 44 cm behind the mirror, 0.11× as large as the object (cat).

Refraction

When light passes from one medium into another medium of different optical density, it is refracted. This means that its speed changes, and its direction can change if it enters the medium at an angle other than 90°.

Like particles, light waves slow when entering a denser medium and change direction **toward** the normal. However, unlike particles lacking a constant applied force, light speeds up when entering a less dense medium and changes direction **away from** the normal. This phenomenon is due to the partial wavelike characteristics of light. A wave train of light possesses sustaining radiant energy throughout its existence.

The amount of wave refraction depends on the optical densities of the mediums involved. Since the densities dictate the speeds at which light will travel through the mediums, the ratio of the speed of light in a vacuum to the speed of light in a substance is known as that substance's **index of refraction, n.**

Refracted light rays behave in accordance with **Snell's Law:**

$$n_1 \sin \Theta_1 = n_2 \sin \Theta_2$$

where $\mathbf{n_1}$ is the index of refraction of the first medium; $\mathbf{n_2}$ is the index of refraction of the second medium; Θ_1 is the angle of incidence of light in the first medium; and Θ_2 is the angle of refraction of light in the second medium.

Since the index of refraction for vacuum (and for all practical purposes air) = 1, Snell's Law for an air interface with another substance becomes:

$$\mathbf{n} = \frac{\sin i}{\sin r}$$

where **sin i** and **sin r** are the sines of the **incident** and **refracted** angles, respectively. Since **n** also represents the ratio of light's speed through a vacuum as compared to light's speed in a different substance, Snell's Law translates into

$$\frac{n_1}{n_2} = \frac{\text{speed of light in medium 1}}{\text{speed of light in medium 2}} = \frac{\sin \Theta_2}{\sin \Theta_1}$$

When a ray of light is directed at a refracting surface, the angle of the incoming or incident ray (**i**) is measured from the normal to the surface, as is the angle for the reflected ray and the refracted ray (**r**) .

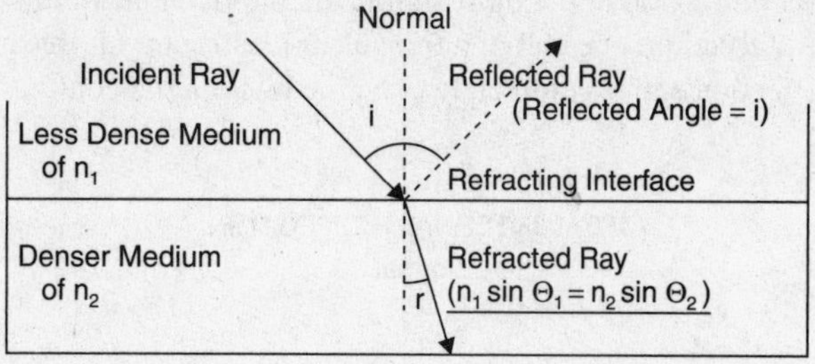

Refracted rays obey these rules:

a. The incident ray, the reflected ray, the refracted ray and the normal to the refracting surface all lie in the same plane.

b. The index of refraction (n) for any particular medium is a constant and is independent of the incident angle.

c. Light rays that pass from substances of lesser density into substances of greater density will bend toward the normal, and light rays that pass from substances of greater density into substances of lesser density will bend away from the normal.

Critical Angle

When a light ray passes from a denser substance into a less dense substance, there is an angle at which there is no refraction into the second substance, but the refracted ray travels along the interface instead. This is known as the substance's **critical angle** (i_C), and it results in a refracted angle of 90°. (At any angle less than the critical angle of incidence, there is both reflection in the denser medium and refraction into the less-dense medium.)

The **critical angle** can be determined using Snell's Law:

$$\mathbf{n_1 \sin \Theta_1 = n_2 \sin \Theta_2}$$

For instance, where the first medium is glass and the second medium is air, light passing from glass into air at the critical angle for glass would be determined as follows (n for glass is 1.5, and n for air is 1):

$$(1.5)(\sin \mathbf{i}_C) = (1)(\sin 90^\circ)$$
$$\sin \mathbf{i}_C = \frac{1}{1.5} = 0.667$$
$$\mathbf{i_C = 41.8^\circ}$$

Since the refracted angle is 90°, for an air interface with a denser medium, the critical angle can be determined easily using $\mathbf{\sin i_C} = \frac{1}{n}$.

CRITICAL ANGLE REFRACTION

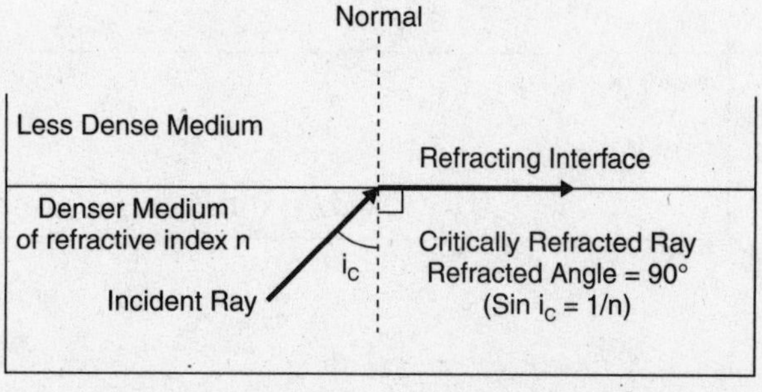

Total Reflection

In addition, light rays that are incident at angles greater than the **critical angle** are entirely **reflected** back into the incident medium (**total internal reflection**). One of the most significant applications of this phenomenon is in the field of fiber optics, in which light is introduced into one end of a thin, clear plastic strand. Because of the tiny diameter, the light experiences total internal reflection and is completely contained within the confines of the strand, entirely transmitted from one end to the other.

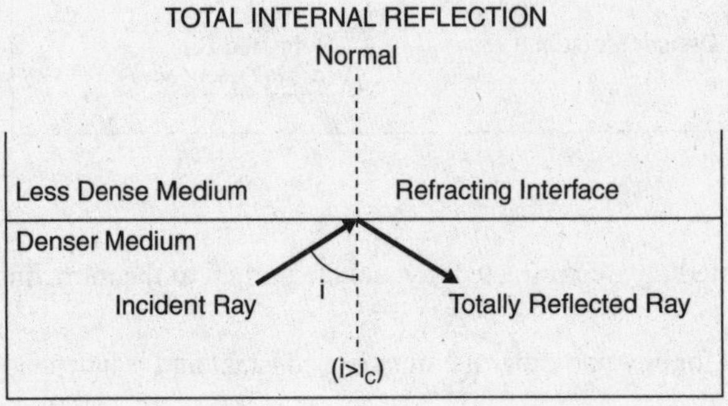

Common refractive indeces include the following:

Air/Vacuum	1.00
Water	1.33
Quartz	1.45
Crown Glass	1.50
Flint Glass	1.59
Diamond	2.42

Example

A beam of light enters a tank of water in a classroom. If the incident angle is 45°, find the refracted angle and draw the diagram.

Solution

Using Snell's Law, $\mathbf{n_1 \sin \Theta_1 = n_2 \sin \Theta_2}$:

$$(1)(\sin 45^\circ) = (1.33)(\sin \Theta_2) \text{ and } \sin \Theta_2 = \frac{(0.707)}{(1.33)} = 0.532$$

This makes the refracted angle, $\Theta_2 = \mathbf{32.1}°$

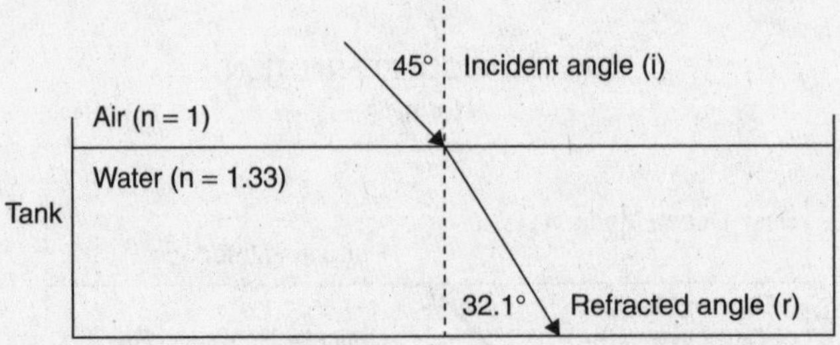

Example

A beam of light travels through water into a slab of crown glass at an incident angle of 60°. Determine the refracted angle and draw the diagram.

Solution

Using Snell's Law, $\mathbf{n_1 \sin \Theta_1 = n_2 \sin \Theta_2}$:

$$(1.33)(\sin 60°) = (1.50)(\sin \Theta_2) \text{ and } \sin \Theta_2 = \frac{(1.33)(0.866)}{(1.50)} = 0.768$$

This makes the refracted angle, $\Theta_2 = \mathbf{50.1°}$

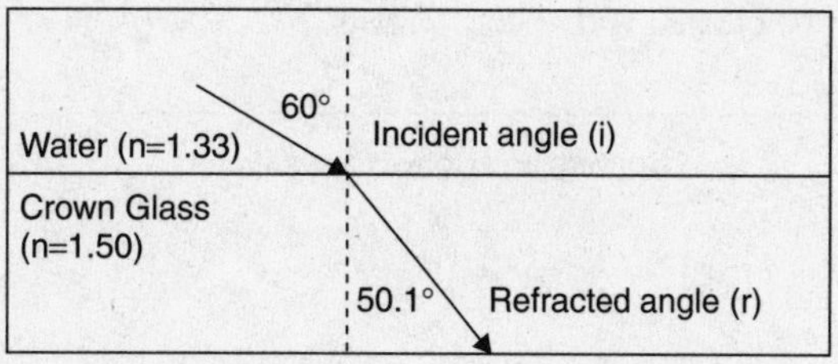

Example

A ray from an underwater pool light strikes the air/water interface at an incident angle of 50°. Is it visible to an observer standing at poolside? Assuming that the water is perfectly still, determine the answer and draw the diagram.

Solution

Use **Sin i_C = 1/n** for the **critical angle.**

Determine the critical angle for water: Sin $i_C = \frac{1}{1.33} = 0.752$.

This makes the critical angle for water i_C = **48.8°**. Since this is less than the incident angle of 50°, **the poolside observer does not see the ray.**

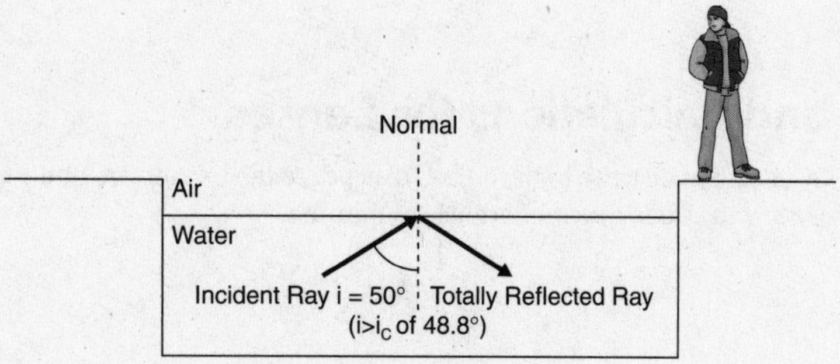

Lenses

Although the complete study of lenses in geometric optics involves more than the following information, the material presented is typical of what the physics student can expect to find on the AP Physics B Exam.

There are basically two types of lenses, **converging** and **diverging.**

Converging or **convex** lenses are thicker in the center and thinner on the edges. They cause light rays to pass through a focal point, become inverted, and form **real** images. Converging lenses can also form virtual images, the most common being the result of an object being placed between the focus and the glass, as occurs when a person wishes to examine an object using a magnifying glass. Other images can also be formed from light rays reflecting from the surfaces of converging lenses, in which case the reflected rays behave as they would when reflected from either type of spherical mirror. Any image that can be created with a concave mirror can be created with a convex lens, and any image that can be created with a convex mirror can be created with a convex lens.

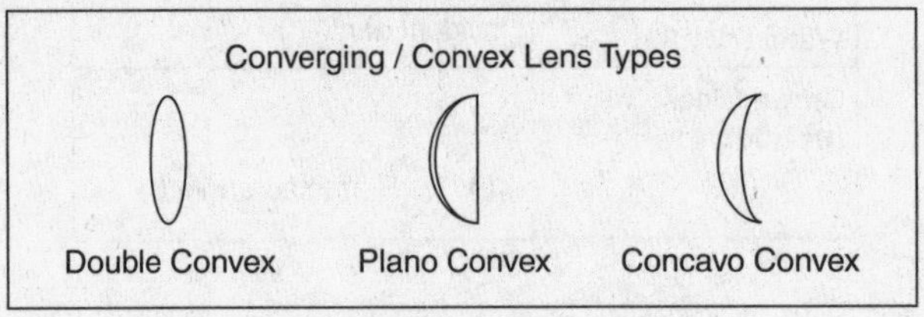

Diverging or **concave** lenses are thicker on the edges and thinner in the center. They cause light rays to pass through and diverge or spread out. Other images can also be formed from light rays reflecting from the surfaces of diverging lenses, in which case the reflected rays again behave as they would when reflected from either type of spherical mirror.

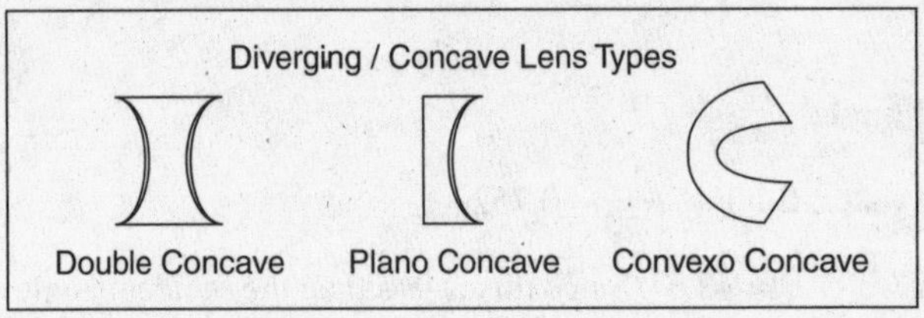

Of these six types, the AP Physics B Exam typically includes only the double convex and double concave lenses in exam problems.

Ray Diagrams and Calculations for Lenses

For calculations involving focal length, distance to the object, distance to the image, size of the object, size of the image, and magnification, all lenses obey the same general formulas as mirrors:

$$\frac{1}{f} = \frac{1}{S_O} + \frac{1}{S_I}$$

$$\frac{h_I}{h_O} = \frac{S_I}{S_O}$$

$$M = \frac{-h_I}{h_O} = \frac{-S_I}{S_O}$$

Convex spherical lenses have only one possibility or case for reflection from their front surfaces. The image they can give is virtual, right side up (erect), behind the lens and smaller than the object, as shown earlier. It is found at the point where the two rays intersect.

Ray Diagrams

As was the case for mirrors, you can draw ray diagrams for any type of lenses using two rays:

1. From the object straight to the lens, and then through the focus **F**
2. From the object directly through the center of the lens

The image will be found where the two rays intersect.

Refracted Images

The image is a **virtual image** when it appears in front of (on the same side as the object) the lens.

The image is **right side up** when it appears in front of (on the same side as the object) the lens.

The image is a **real image** when it appears on the side of the lens opposite of the object.

The image is a **real image** when it is inverted.

The distance from the lens to the image (S_i) **is negative** when the image is located on the same side as the object.

The image distance, the image height, and the magnification can all be computed using the following formulas:

$$\frac{1}{f} = \frac{1}{S_O} + \frac{1}{S_I}$$

where **f** is the focal length, $\mathbf{S_O}$ is the distance from the object to the mirror, and $\mathbf{S_I}$ is the distance from the mirror to the image.

$$\frac{h_I}{h_O} = \frac{S_I}{S_O}$$

where $\mathbf{h_i}$ is the height of the image and $\mathbf{h_O}$ is the height of the object.

$\mathbf{h_i}$ is **negative** when the image is **inverted.**

$$M = \frac{-h_I}{h_O}$$

where **M** is the magnification (how many times larger or smaller the image is compared to the object).

These formulas will hold for both convex and concave lenses. In every case, the distance from the object to the glass $\mathbf{S_O}$ will be positive. If the image forms in front of the lens (on the same side as the object), $\mathbf{S_I}$ will be negative and will be **virtual** and **erect.** Also, as with spherical mirrors, **c = 2 f.**

Convex Lens Ray Diagrams

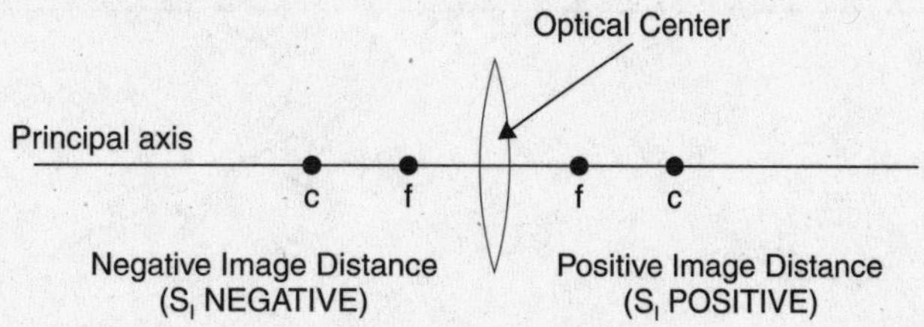

There are six possibilities or cases for convex lenses:

1. Object at "Infinite" Distance

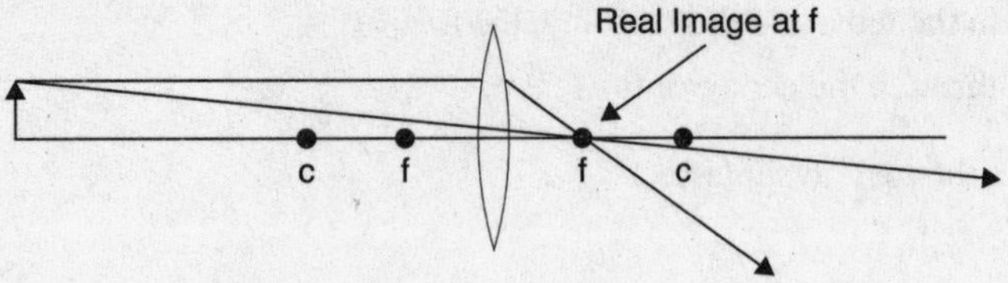

The "image" formed is a point located at the focus f.

2. Object at Finite Distance

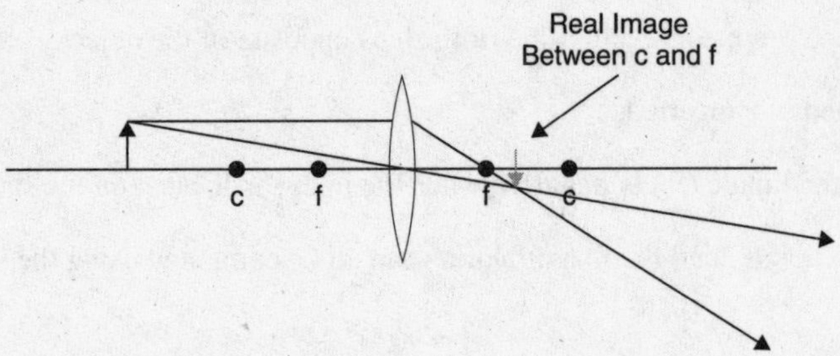

The image formed is **real, inverted, smaller** than the object, and located between **c** and **f**.

3. Object at **c**

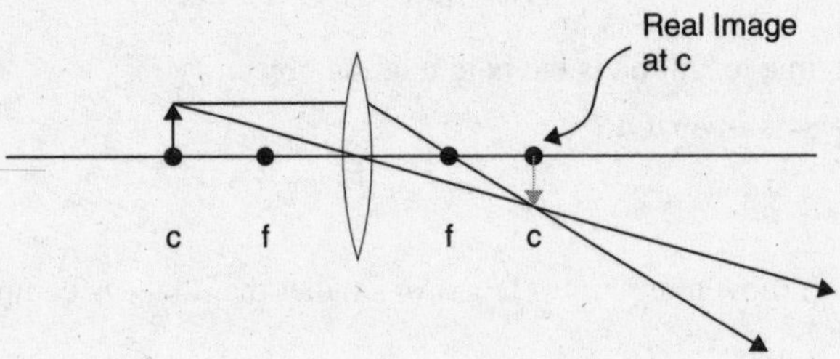

The image formed is real, inverted, the same size as the object, and located at **c**.

4. Object between **c** and **f**

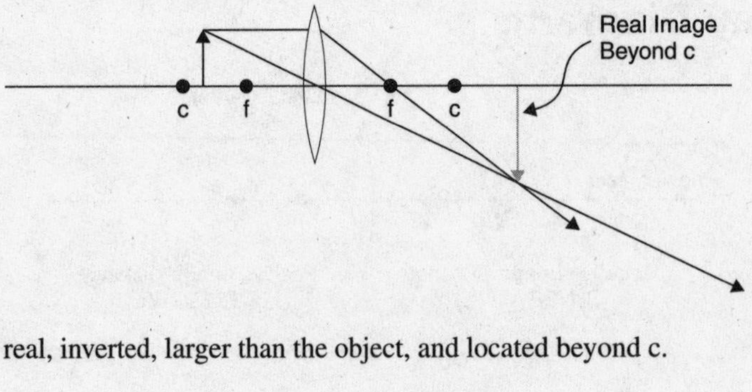

The image formed is real, inverted, larger than the object, and located beyond c.

5. Object at **f**

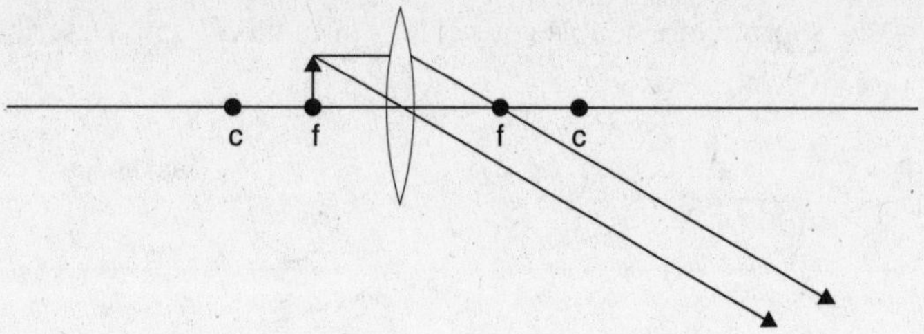

No image is formed.

6. Object between **f** and Lens

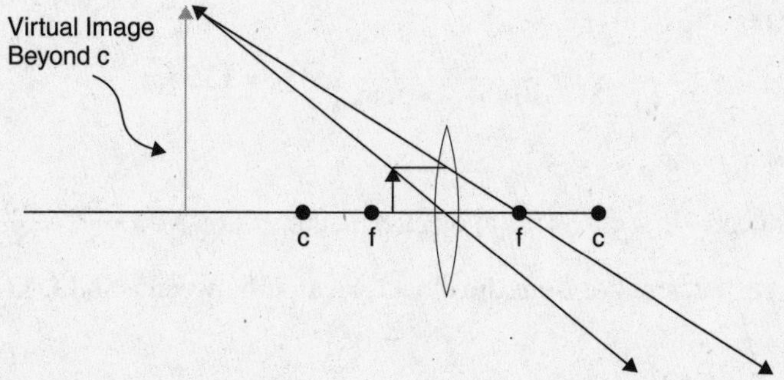

The image formed is virtual, erect, larger than the object, and located beyond **c** on the same side as the object.

Example

An adventurous physics student places a piece of paper behind a symmetrical double convex lens of focal length 12cm. She is able to cast an image of the sun on the paper briefly. If the mean distance from the earth to the sun is 1.50×10^{11}m, describe the image fully and draw the ray diagram.

Solution

Case 1. Object at Infinite Distance

Use $\frac{1}{f} = \frac{1}{S_O} + \frac{1}{S_I}$: $\frac{1}{0.12\,\text{m}} = \frac{1}{S_i} + \frac{1}{1.50 \times 10^{11}\,\text{m}}$. Since the focal length is insignificant in comparison with the distance of the sun from the glass, we can approximate the image distance with the focal length ($S_I = f$).

Also, the magnification $M = \frac{-S_I}{S_O} = \frac{-0.12\,\text{m}}{1.50 \times 10^{11}\,\text{m}} = -8.0 \times 10^{-12} \times$ (The negative sign indicates that the image is inverted.)

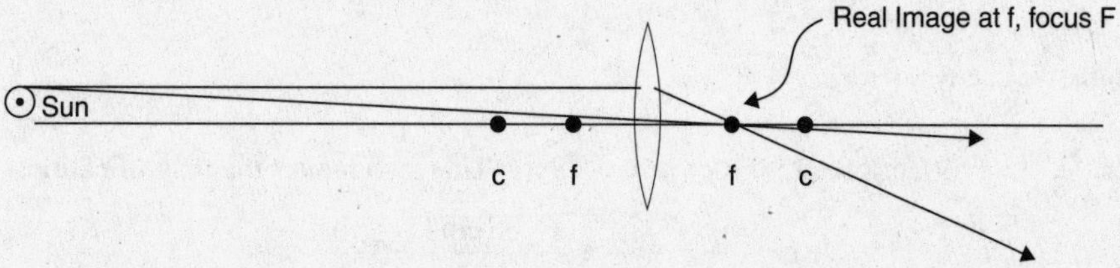

The image formed is **real, inverted, smaller** than the object, **located at the focus F, 8.0×10^{-12} the size of the sun.**

Example

A glass 15 cm high is placed 40 cm in front of a double convex lens of radius 20 cm. Describe the image fully and draw the diagram.

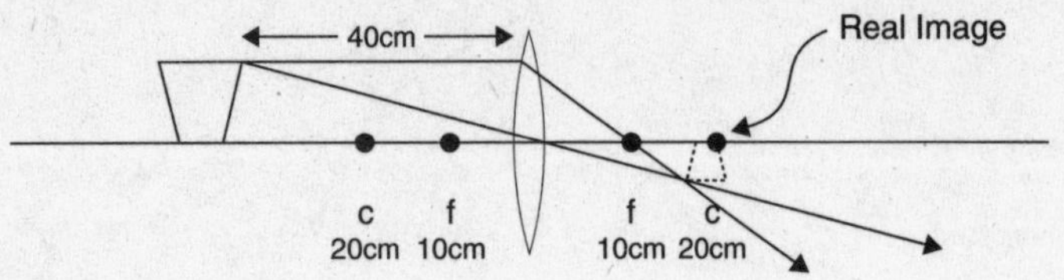

Case 2. Object at Finite Distance

Using $\frac{1}{f} = \frac{1}{S_O} + \frac{1}{S_I}$ yields $\frac{1}{10\,\text{cm}} = \frac{1}{S_I} + \frac{1}{40\,\text{cm}}$.

$$\frac{1}{S_I} = \frac{1}{10\,\text{cm}} - \frac{1}{40\,\text{cm}} \text{ and } \mathbf{S_I = 13.3cm}$$

For the image height h_I, we use $\frac{h_I}{h_O} = \frac{S_I}{S_O}$.

$h_I = \frac{S_I h_O}{S_O}$ = (13.3cm)(15cm)/(40cm) = 5.0 cm. (This makes the magnification $M = \frac{-h_I}{h_O} = \frac{-S_I}{S_O} = \frac{-13.3\,\text{cm}}{40\,\text{cm}} = \frac{-1}{3}$

The image formed is real, inverted, smaller than the object, located between c and f, 13.3 cm behind the glass and $\frac{1}{3}$ the size of the object.

Example

A piece of chocolate candy is placed 10 cm from a converging lens of focal length 15cm. If the chocolate is 4.3 cm high, describe the image fully and draw the diagram.

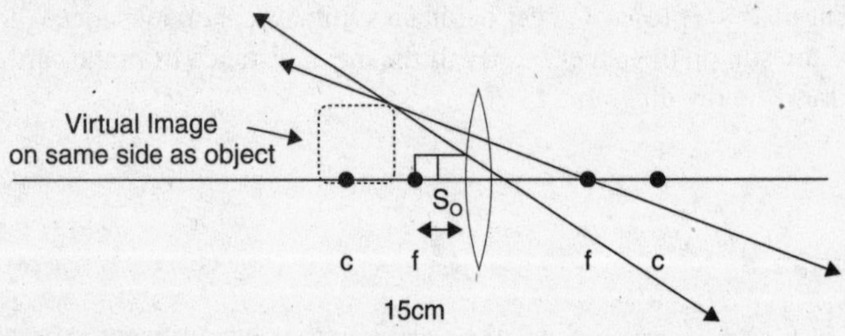

Case 6. Object between f and Lens

Using $\frac{1}{f} = \frac{1}{S_O} + \frac{1}{S_I}$ yields $\frac{1}{15\,\text{cm}} = \frac{1}{S_I} + \frac{1}{10\,\text{cm}}$.

$$\frac{1}{S_I} = \frac{1}{15\,\text{cm}} - \frac{1}{10\,\text{cm}} \text{ and } \mathbf{S_I = -30cm}$$

For the image height h_I, we use $\frac{h_I}{h_O} = \frac{S_I}{S_O}$.

$h_I = \frac{S_I h_O}{S_O}$ = (–30cm)(4.3cm)/(10cm) = –12.9cm. This also makes the magnification:

$$M = \frac{-h_I}{h_O} = \frac{-S_I}{S_O} = -\frac{(-30\,\text{cm})}{10\,\text{cm}} = 3\times$$

The image formed is virtual, erect, larger than the object, located at c on the same side as the object, 30 cm away from the lens and 3× the size of the object, h_I = 12.9 cm high.

Example

Where must an object 20 cm high be placed if a diverging/double concave lens of focal length 22 cm is to form a virtual image 18 cm from the lens on the same side as the object? Find S_O and draw the diagram.

Solution

The case for diverging lenses. (The lens surface reflects some light and acts similarly to a concave mirror.)

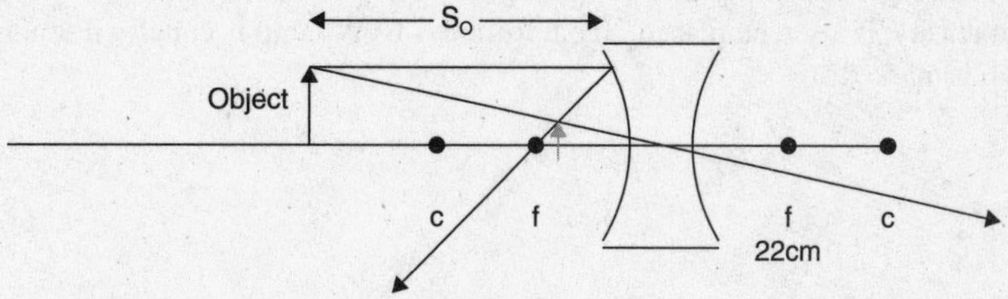

Since the image is to be formed in front of the lens, S_I is negative, as is **f**.

$$\text{Using } \frac{1}{f} = \frac{1}{S_O} + \frac{1}{S_I} \text{ yields } \frac{1}{-22\text{ cm}} = \frac{1}{-18\text{ cm}} + \frac{1}{S_O}.$$

$$\frac{1}{S_O} = \frac{1}{18\text{ cm}} - \frac{1}{22\text{ cm}}$$

$$\mathbf{S_O = 99cm}$$

$$M = \frac{-S_I}{S_O} = \frac{-99\text{ cm}}{18\text{ cm}} = -5.5$$

The object is located 99 cm from the lens on the same side as the image, which is virtual, erect, 5.5× smaller than the object.

Review Questions and Answers

B Exam Question Types

Multiple Choice

1. An object doubles its distance from a light source. The resulting illumination on the object is

(A) quartered.
(B) halved.
(C) doubled.
(D) quadrupled.
(E) not able to be determined.

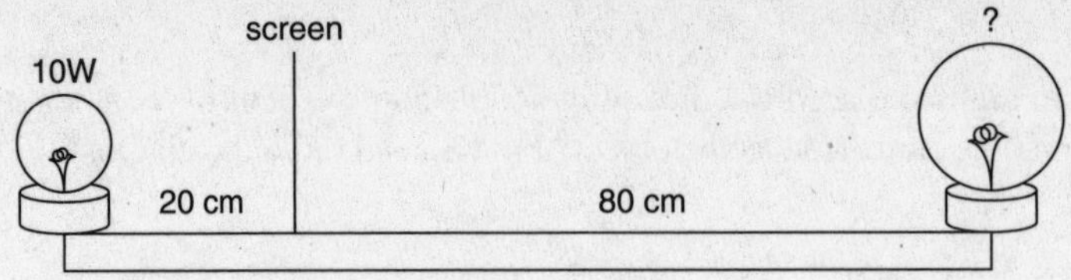

2. A lamp with a 10 W bulb is placed at one end of a meter stick. At the other end is placed a second lamp with a bulb of unknown wattage. Assuming that the wattage rating for the power consumed is proportional to the bulb's luminous intensity, if a screen placed 20 cm from the 10 W lamp is equally illuminated by both lamps, the unknown lamp is marked

(A) 60 W
(B) 80 W
(C) 100 W
(D) 160 W
(E) 200 W

3. I. Plane mirrors
II. Concave mirrors
III. Convex mirrors

Given the preceding choices, virtual images can be formed by

(A) I only
(B) I and II
(C) I and III
(D) II only
(E) I, II and III

4. Given the choices in question 3, real images can be formed by

(A) I only
(B) I and II
(C) I and III
(D) II
(E) I, II and III

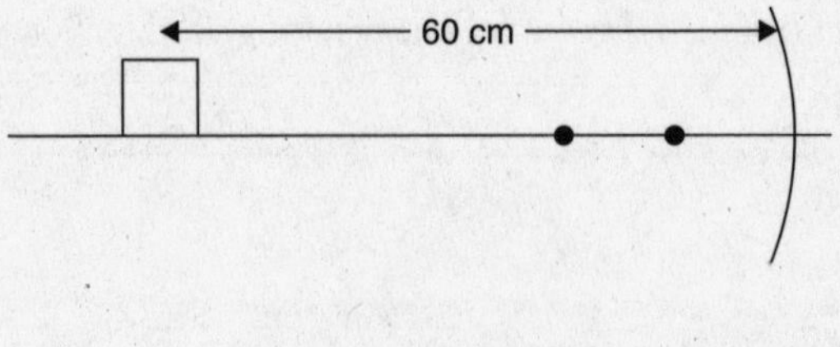

5. A box placed 60 cm away from a concave spherical mirror forms a real image 15 cm from the mirror if the mirror's focal length is

(A) 6 cm
(B) 10 cm
(C) 12 cm
(D) 14 cm
(E) 15 cm

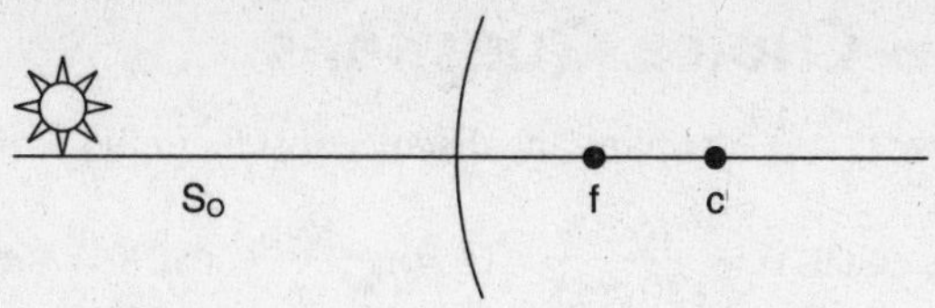

6. An object is placed 100 cm in front of a convex mirror of radius 50 cm. The image formed is

(A) virtual and 20 cm in front of the mirror.
(B) real and 20 cm in front of the mirror.
(C) virtual and 20 cm behind the mirror.
(D) real and 20 cm behind the mirror.
(E) virtual and 40 cm behind the mirror.

7. **I.** Plane mirrors
II. Concave mirrors
III. Convex mirrors
IV. Convex lenses
V. Concave lenses

Given the preceding choices, virtual images can be formed by

(A) I, II and IV only
(B) I, II and V only
(C) II, III and IV only
(D) II, IV and V only
(E) I, II, III, IV and V

8. Given the choices in question 7, real images can be formed by

(A) I, II and IV only
(B) I, II and V only
(C) II, III and IV only
(D) II, IV and V only
(E) I, II, III, IV and V

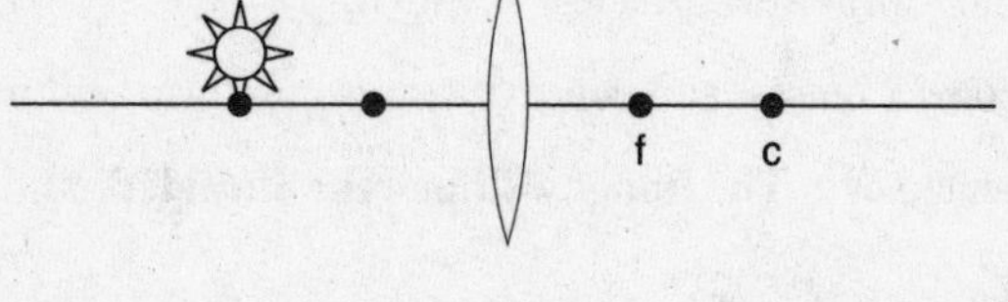

9. An object is placed at the center of curvature of a double convex lens of focus 12 cm. The image is

(A) real, erect and located at **f**.
(B) virtual, erect and located at **c**.
(C) real, erect and located at **c**.
(D) virtual, inverted and located at **c**.
(E) real, inverted and located at **c**.

10. In the previous problem, if the object is placed 12 cm away from the lens,

(A) no image is formed.
(B) a real, half-sized image is formed.
(C) a virtual half-sized image is formed.
(D) a real equal-sized image is formed.
(E) a virtual equal-sized image is formed.

Answers to Multiple-Choice Questions

1. **(A)** By the inverse-square law $\mathbf{E} = \frac{I}{r^2}$, doubling the distance results in the illumination being divided by 2^2.

2. **(D)** Since the illumination is equal, $E_1 = \frac{10\,W}{(20\,cm)^2} = E_2 = \frac{X}{(80\,cm)^2}$ and **X = 160 W.**

3. **(E)** All three mirrors can produce virtual images.

4. **(D)** Only concave mirrors can produce a real image.

5. **(C)**

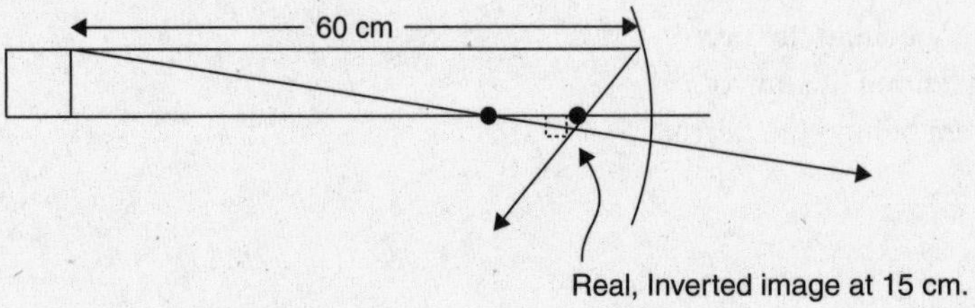

Using $\frac{1}{f} = \frac{1}{S_O} + \frac{1}{S_I}$

$\frac{1}{f} = \left(\frac{1}{60\,cm}\right) + \left(\frac{1}{15}\right)$ cm yields **f = 12 cm.**

6. **(C)**

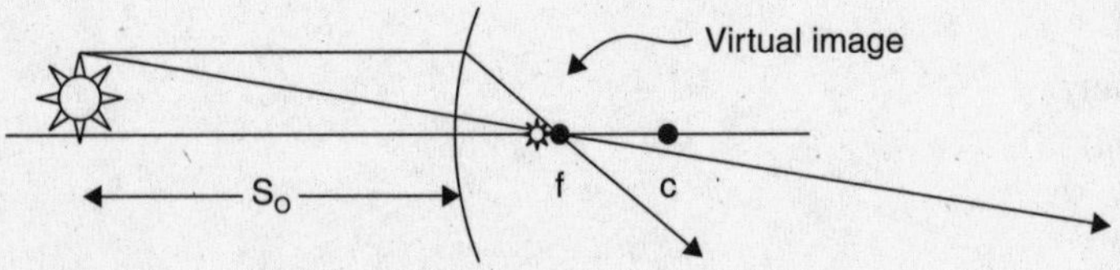

Using $\frac{1}{f} = \frac{1}{S_O} + \frac{1}{S_{I'}} \left(\frac{1}{-25\,cm}\right) = \left(\frac{1}{100\,cm}\right) + \left(\frac{1}{S_I}\right)$. Since this is a convex mirror, both the focus and image distance will be negative because they are found behind the mirror.

$\frac{1}{S_I} = -\left(\frac{1}{100\,cm}\right) - \left(\frac{1}{25\,cm}\right)$ yields $S_I = -20$ cm

The image is virtual and 20 cm behind the mirror.

7. **(E)** All mirrors and lenses can form virtual images.

8. **(D)** Only concave mirrors, convex lenses, and concave lenses can form *real* images.

9. **(E)** This is Case 3, object located at **c**. **The image will be real, inverted, and the same size as the object and located at c.**

10. **(A)** This is Case 5, object located at **f**. **No image is formed.**

Free Response

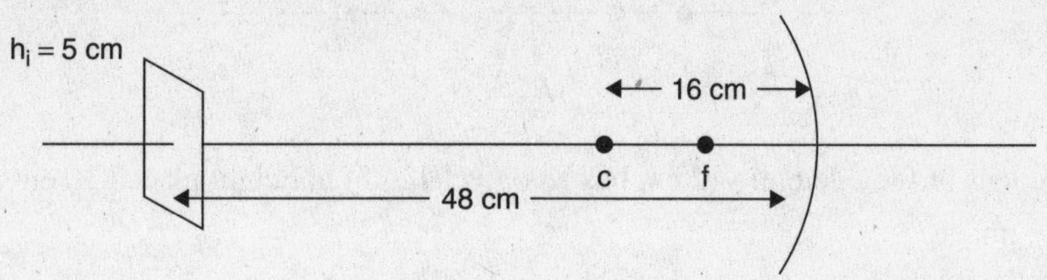

1. A screen 5 cm high is placed a distance 48 cm from a concave mirror of radius 16 cm.

(a) Where must an object be placed in order to produce a real image on the screen?
(b) If the image is exactly 5 cm high, how large will the object be?
(c) Draw the ray diagram and label all parts.
(d) If the concave mirror is replaced with a convex mirror of equal dimensions, and the object remains in the same location, where is the image now located?
(e) Using the convex mirror, where should the object be placed in order to produce a real image on the screen?

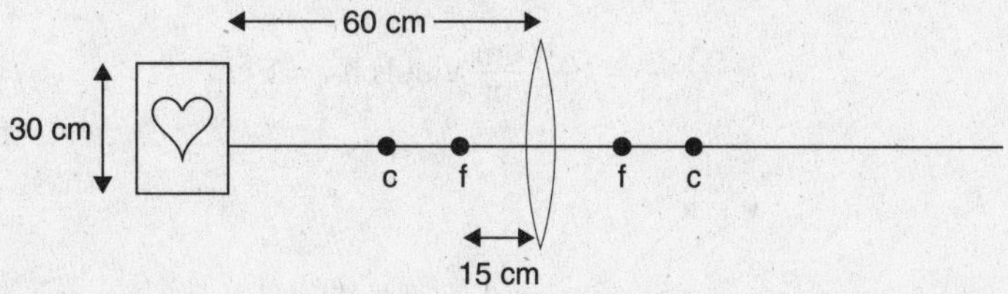

2. A valentine poster 30 cm tall is located 60 cm away from a converging lens of focal length 15 cm.

(a) Where is the image located?
(b) i. Is it real or virtual?
ii. Is it erect or inverted?
iii. How high is the image?
(c) Draw the diagram and label all parts.

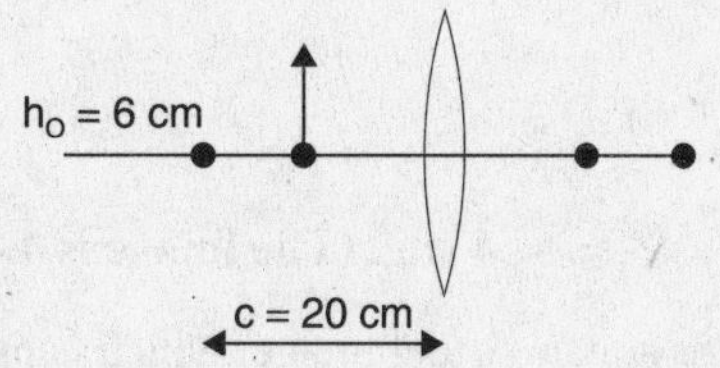

3. An object 6 cm high is placed at the focus of a double convex lens of focal length 10 cm.
(a) Describe the image formed fully.
(b) Draw the ray diagram.
(c) Repeat **(a)** if the object is now placed 20 cm from the lens.
(d) Draw the ray diagram for the situation in **(c)**.

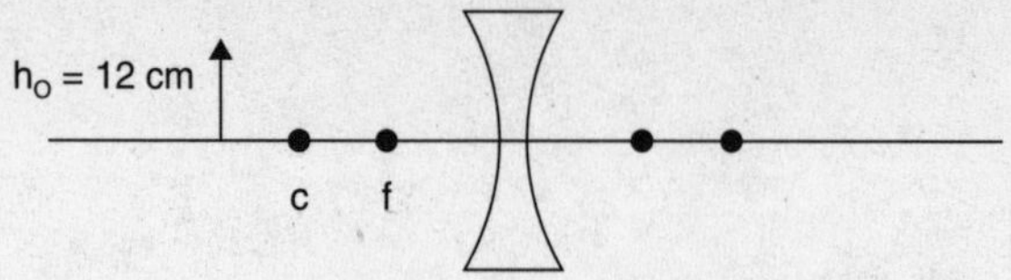

4. A double concave lens of focal length –12 cm has an object 12 cm in height placed 30 cm away as shown.
   (a) Where is the image?
   (b) How high is the image?
   (c) Draw the diagram.
   (d) Describe the image fully.

Answers to Free-Response Questions

1. **(a)** Using $\frac{1}{f} = \frac{1}{S_O} + \frac{1}{S_I}$:

$$\frac{1}{8}\text{ cm} = \frac{1}{S_O} + \frac{1}{48\text{ cm}} \text{ yields } \frac{1}{S_O} = \frac{5}{48} \text{ and } \mathbf{S_O = 9.6\text{ cm}}$$

(b) Using $\frac{h_I}{h_O} = \frac{S_I}{S_O}$:

$$\frac{-5\text{ cm}}{h_O} = \frac{-48\text{ cm}}{9.6\text{ cm}} \text{ yields } \mathbf{h_O = 1\text{ cm}}$$

(c)

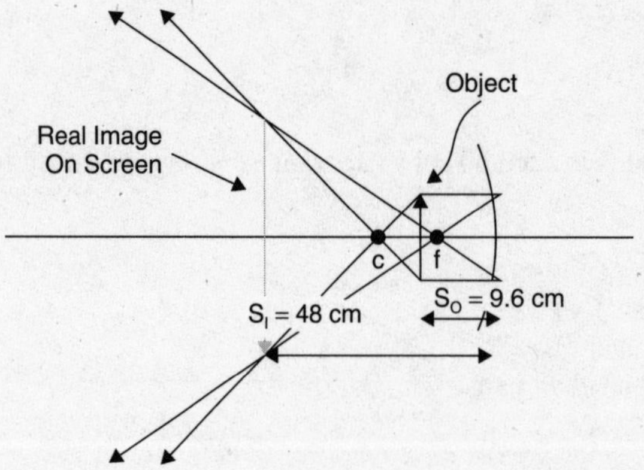

(d) Using $\frac{1}{f} = \frac{1}{S_O} + \frac{1}{S_I}$:

$\left(\frac{1}{-8\text{ cm}}\right) = \left(\frac{1}{9.6\text{ cm}}\right) + \left(\frac{1}{S_I}\right)$ yields $\mathbf{S_I = -4.4}$ **cm . (The image is 4.4 cm behind the mirror.)**

(e) Since a convex mirror can produce only a virtual image, which cannot be projected onto a screen, **there can be no image projected onto the screen.**

2. **(a)** Using $\frac{1}{f} = \frac{1}{S_O} + \frac{1}{S_I}$:

$$\frac{1}{S_I} = \frac{1}{f} - \frac{1}{S_O} = \left(\frac{1}{15\text{ cm}}\right) - \left(\frac{1}{60\text{ cm}}\right) \text{ yields } \mathbf{S_I = 20\text{ cm.}}$$

(b) i. **Real image**

ii. **Inverted**

iii. Using $\frac{h_I}{h_O} = \frac{S_I}{S_O}$, $h_I = (S_I)(h_O)/(S_O)$ = (20 cm)(30 cm)/(60 cm) = **10 cm**

(c)

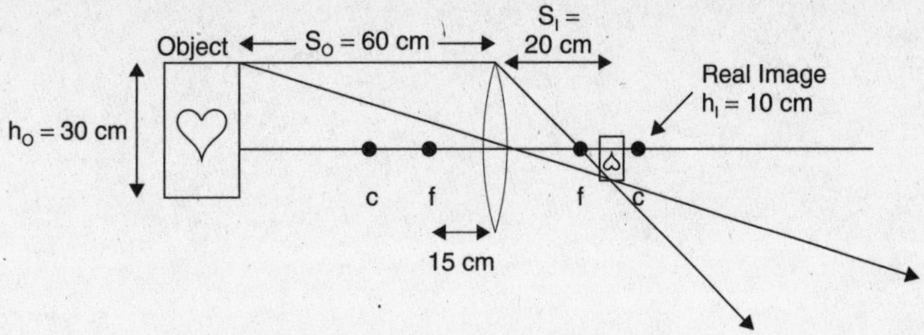

3. **(a)** If the object is placed at **f, no image is formed.**

(b)

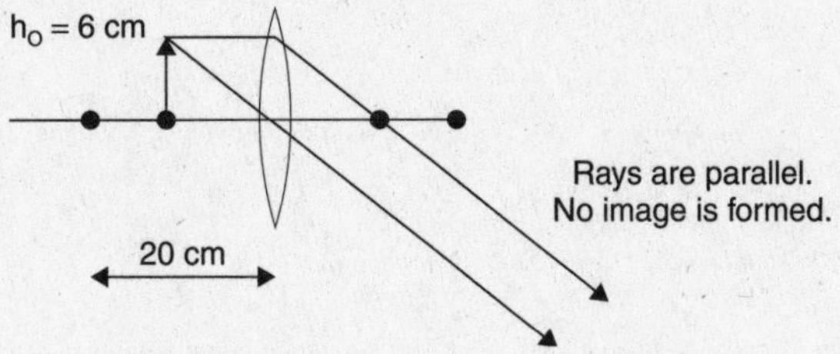

(c) Using the situation for an object placed at the center of curvature, the image formed is **real, inverted, the same size as the object, located opposite of c.**

(d)

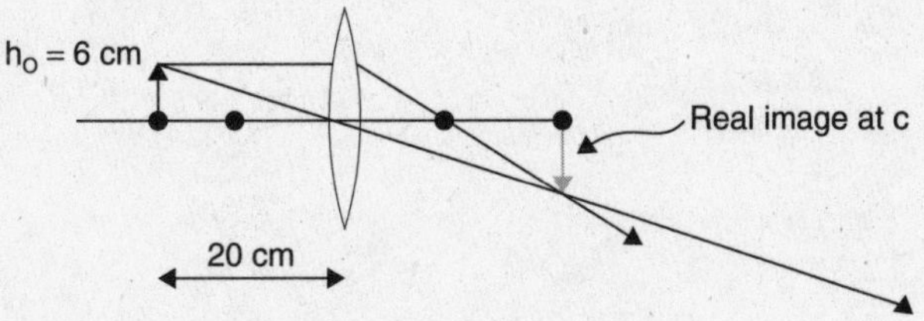

4. In this situation, there is a real image formed, albeit a ghostly one, due to rays reflected *from the surface* of the lens. Lens surfaces do reflect. When reflected rays pass through a focal point, a real image is formed. In other words, *it acts like a concave mirror*.

(a) Using $\frac{1}{f} = \frac{1}{S_O} + \frac{1}{S_I}$

$$\left(\frac{1}{-12\,\text{cm}}\right) = \left(\frac{1}{30\,\text{cm}}\right) + \left(\frac{1}{S_I}\right)$$

$$\left(\frac{1}{S_I}\right) = \left(\frac{-7}{60}\right) \text{ and } \mathbf{S_I = -8.6\ cm}$$

(b) Using $\frac{h_I}{h_O} = \frac{S_I}{S_O}$:

$$h_I = (S_I)(h_O)/(S_O) = (-8.6\ \text{cm})(12\ \text{cm})/(30\ \text{cm}) = \mathbf{-3.4\ cm}$$

(c)

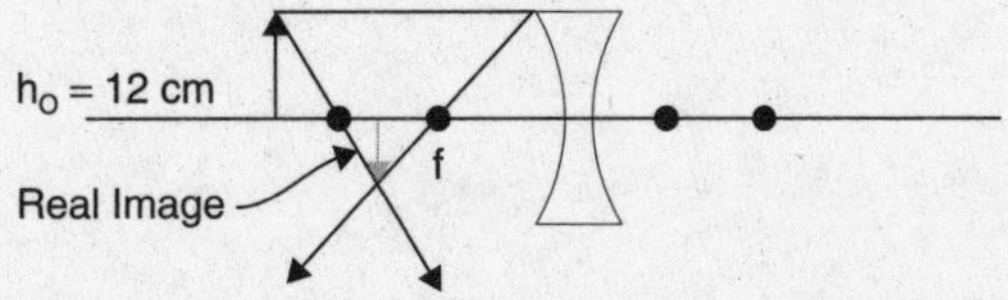

(d) The image is **real, inverted, smaller than the object** (8 cm HIGH), and 20 cm **from the lens** on the same side as the object.

Fluid Mechanics (B Exam)

Fluid mechanics includes the study of fluid dynamics, fluids in motion or objects moving through a fluid medium, and fluid statics, stationary fluids or objects stationary within a fluid. Fluids are the states of matter that will flow and include liquids and gases. These states of matter have no definite shape and take on the shape of their containers.

Fluid Properties

Liquids have a definite volume but no definite shape. Gases, due to their intermolecular separations, have an indefinite shape and indefinite volume, as they can be compressed.

Unlike solids, gases are not capable of resisting shearing force. Liquids can resist shearing forces to some degree, depending on viscosity.

Fluids can exert forces on objects immersed in them (buoyant force and hydrostatic pressure), moving through them (drag), or by flowing past objects (opposing force).

Density is a property of all substances. For fluids, density and viscosity are major determinants of fluid flow rate. The less dense or less viscous a fluid is, the weaker the intermolecular bonds are and the faster the fluid can move. Density is the mass of a substance divided by its volume:

$$\rho = \mathbf{m} / \mathbf{V}$$

where ρ is the density, **m** is the mass, and **V** is the volume. Density can be expressed in units of g/cm^3, g/ml, kg/m^3, and kg/L, among others.

Specific gravity of a substance is the name given to the ratio of the density of the substance to the density of water. It is also the ratio of the mass of an object to the mass of an equal volume of water, as well as the ratio of the weight of the object to the weight of an equal volume of water.

$$\text{sp.gr.} = \frac{\rho_{\text{SUBSTANCE}}}{\rho_{\text{WATER}}} = \frac{m_{\text{SUBSTANCE}}}{m_{\text{EQUAL VOL OF WATER}}} = \frac{F_{\text{WT OF SUBSTANCE}}}{F_{\text{WT EQUAL VOL OF WATER}}}$$

Hydrostatic Pressure

An object immersed in a fluid undergoes a force on its surface due to the surrounding fluid. In the presence of gravity, the greater the amount of overbearing fluid, the greater the force. The pressure exerted by a fluid is called the **hydrostatic pressure** and is characterized by the general formula for pressure:

$$\mathbf{p} = \frac{\mathbf{F}}{\mathbf{A}}$$

where p is pressure exerted on the object, F is the force exerted on the object, perpendicular to an area, and A is the area upon which the force acts. For a relatively small object, the pressure on its surface is the same in all directions. As an object rises in a fluid, the hydrostatic pressure on it is reduced. The taller the object, the lower the pressure becomes on areas of higher elevation in the fluid. This is demonstrated by the relationship:

$$\mathbf{p} = \rho \mathbf{g} \mathbf{h}$$

where **p** is the pressure in N/m^2 or Pa (pascal), ρ is the fluid density, **g** is the acceleration of gravity in that location, and **h** is the height of the fluid column. Dimensionally, the units prove out as follows:

$$Pa = N/m^2 = (kgm/s^2)/(m^2) = \underbrace{(kg/m^3)}_{\rho}\underbrace{(m/s^2)}_{g}\underbrace{(m)}_{h}$$

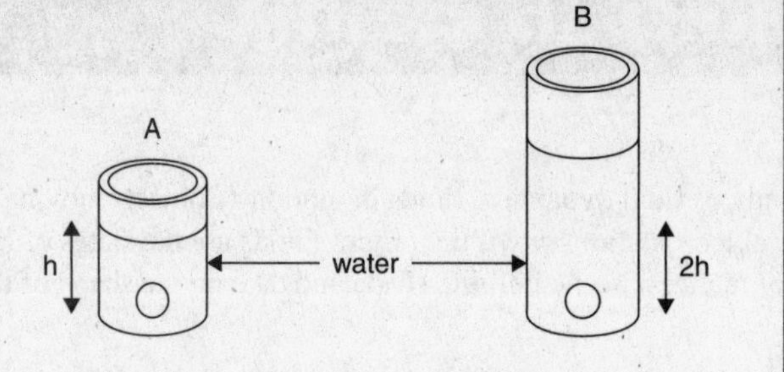

The hydrostatic gauge pressure on the ball in can B is twice the pressure on the same ball in can A since $p_A = \rho g h_A$ and $p_B = \rho g 2h_A$.

Any change in the hydrostatic pressure on a confined fluid results in an equal change in pressure on all parts of the fluid (Pascal's Principle). The ball in can A sustains the same hydrostatic pressure anywhere on its surface. The ball in can B sustains twice the pressure as the ball in can A. Again, anywhere on its surface, the pressure is equal. The **hydrostatic pressure** is *only* due to the submerging fluid and excludes atmospheric pressure. The **total pressure** or **absolute pressure** equals the **hydrostatic pressure** plus the **atmospheric pressure.**

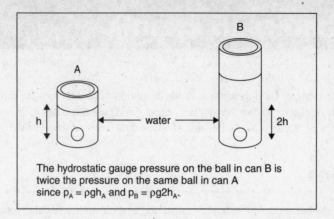

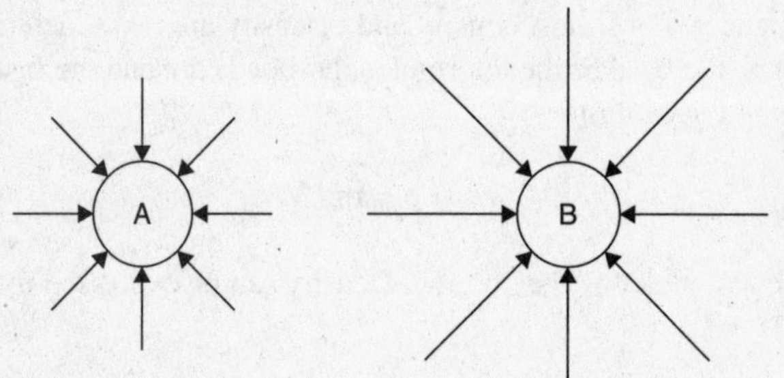

From the preceding image regarding Can A and Can B:
Hydrostatic gauge pressure on ball B is twice that on ball A.

Units of pressure are N/m^2 or Pa (1 Pa = 1 N/m^2). At sea level on the Earth's surface, one atmosphere of pressure has the average value of

$$1 \text{ atm} = 1.01 \times 10^5 \text{ Pa } (= 14.7 \text{ lb/in}^2 \text{ in the English system})$$

Example

What is the hydrostatic pressure on an object submerged in water at a depth of 15 m at or near sea level? ($\rho_{WATER} = 1.00 \times 10^3$ kg/m^3)

Solution

Using **$p = \rho g h$**, $p = (1.00 \times 10^3 \text{ kg/m}^3)(9.8 \text{ m/s}^2)(15 \text{ m}) = 147 \times 10^3 \text{ N/m}^2 =$ **1.47×10^5 Pa**

Example

What is the **total pressure** on the submerged object in the preceding example? Total pressure is often called absolute pressure.

Solution

The total pressure will equal the hydrostatic pressure of the water plus the atmospheric pressure:

$$p_{TOTAL} = p_{HYDROSTATIC} + p_{ATM}$$
$$= \rho gh + p_{ATM}$$
$$= 1.47 \times 10^5 \text{ Pa} + 1.01 \times 10^5 \text{ Pa}$$
$$= \mathbf{2.48 \times 10^5 \text{ Pa}}$$

Gauge Pressure

Gauge pressure (or hydrostatic pressure) is the difference between **total pressure** and **atmospheric pressure.**

$$p_{GAUGE} = p_{TOTAL} - p_{ATMOSPHERIC}$$

Example

What is the gauge pressure on the object in the preceding example?

Solution

Using $p_{GAUGE} = p_{TOTAL} - p_{ATMOSPHERIC}$

$$= (2.48 \times 10^5 \text{ Pa}) - (1.01 \times 10^5 \text{ Pa}) = \mathbf{1.47 \times 10^5 \text{ Pa}}$$

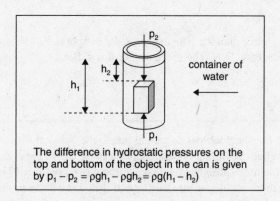

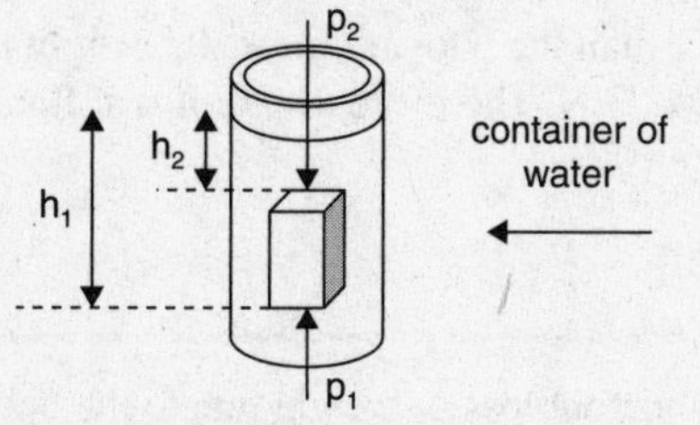

The difference in hydrostatic pressures on the top and bottom of the object in the can is given by $p_1 - p_2 = \rho gh_1 - \rho gh_2 = \rho g(h_1 - h_2)$

The preceding figure illustrates that a pressure differential between the top and bottom surfaces of a submerged object depends on their submersion depths. The higher the top surface, the lower the pressure, and the deeper the lower surface, the higher the pressure, by an inverse relationship.

Example

The rectangular object in the preceding figure is 3 cm high, the top surface is submerged 2 cm below water, and both top and bottom faces are parallel to the water's surface. What is the pressure differential between the top and bottom of the object? (1 atm = 1.0×10^5 Pa and $\rho_{WATER} = 1.00 \times 10^3$ kg/m^3)

Solution

The difference in pressure is given by $\mathbf{p_1 - p_2 = \rho g(h_1 - h_2)}$.

$p_1 - p_2 = (1.00 \times 10^3 \text{ kg/m}^3)(9.8 \text{ m/s}^2)(0.05 \text{ m} - 0.03 \text{ m}) = \mathbf{1.96 \times 10^2 \text{ Pa}}$

Buoyancy

When an object is immersed either partly or entirely in a fluid, it experiences two vertical forces: gravitational weight and buoyant force (directly opposite the force of weight). Whatever fluid the object displaces has a weight equal to the apparent loss of weight of the object (buoyant force) due to the effect of the fluid (Archimedes' Principle). This is expressed as

$$\mathbf{F_{BUOY} = V\rho g}$$

Where F_{BUOY} is the **buoyant force**, **V** is the **volume** of fluid displaced by the immersed object, ρ is the fluid density and **g** is the acceleration of **gravity** at that location. If the buoyant force is greater than the object's weight, the object will rise. If the buoyant force is less than the object's weight, the object will sink.

Example

What is the buoyant force exerted on a ball of diameter 0.2 m if immersed in seawater at or near sea level? ($\rho_{SEAWATER} = 1.03 \times 10^3$ kg/m^3)

Solution

Using $\mathbf{F_{BUOY} = V\rho g}$, $F_{BUOY} = \left(\frac{4\pi}{3}\right)(0.1 \text{ m})^3(1.03 \times 10^3 \text{ kg/m}^3)(9.8 \text{ m/s}^2) = \mathbf{42.3\ N}$

Example

If the ball in the preceding example has a mass of 0.5 kg, does it sink?

Solution

The ball will sink if its weight is greater than the buoyant force. Its weight is $\mathbf{F_{WT}} = (mg) = (0.5 \text{ kg})(9.8 \text{ m/s}^2) = 4.9$ N, which is less than the buoyant force of 42.3 N. Therefore, **the ball will float.**

Terminal Velocity

As an object falls through a fluid, the apparent loss of weight due to the buoyant upward force increases with density of the fluid and increased velocity. When the falling object reaches a point where the upward buoyant force of fluid friction plus **drag** is equal and opposite to the weight of the object, there is no net force on the object, and, by Newton's First Law of Motion, the object cannot accelerate any further and is in a state of equilibrium. This condition is called **terminal velocity**.

If an object moves through a fluid and has a velocity perpendicular to the force of gravity, **terminal velocity** is reached at the point where the propulsive force on the object is equaled and opposed by the drag force. Since there is no net force, there is no acceleration. The object is in equilibrium.

Fluid Flow

Noncompressible fluid flowing through a tube or pipe is governed by the fact that the total amount flowing through one end flows through the other. This translates into the concept that the total mass of the fluid entering one end exits the other.

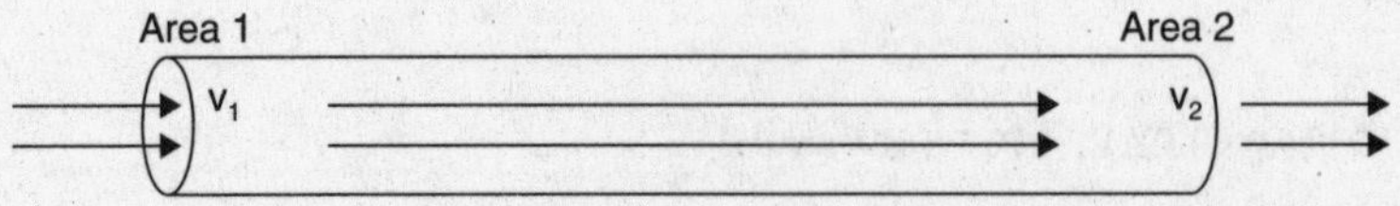

For a uniform tube or pipe, the total mass of homogeneous fluid flowing in and flowing out is simple to equate:

$$\rho A_1 v_1 \Delta t = \rho A_2 v_2 \Delta t$$

$$(\text{Density})(\text{Area}_1)(\text{Velocity}_1)(\text{Flow time}_1) = (\text{Density})(\text{Area}_2)(\text{Velocity}_2)(\text{Flow time}_2)$$

$$\downarrow \quad \downarrow \quad \downarrow \quad \downarrow \qquad \downarrow \quad \downarrow \quad \downarrow \quad \downarrow$$

$$(\text{kg/m}^3)\ (\text{m}^2)\ (\text{m/s})\ (\text{s}) = (\text{kg/m}^3)\ (\text{m}^2)\ (\text{m/s})\ (\text{s})$$

$$\text{kg} = \text{kg}$$

$$\text{mass in} = \text{mass out}$$

For a nonuniform pipe:

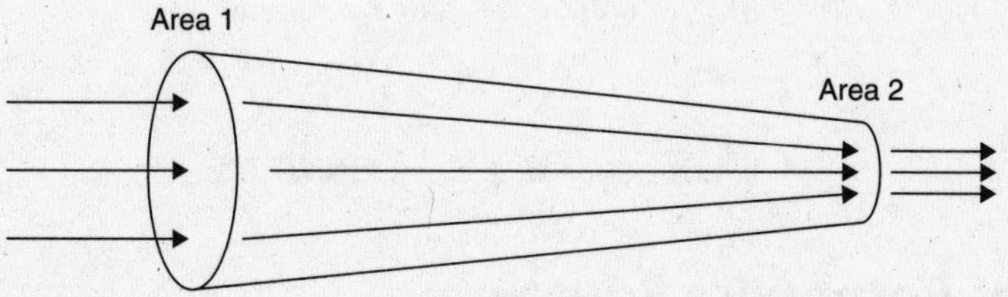

For this example of a nonuniform pipe, we will let Area 1 be twice Area 2 and again let the fluid entering and leaving be noncompressible and homogeneous. The mass entering the pipe equals the mass leaving the pipe. Use the equation:

$$\rho A_1 v_1 \Delta t_1 = \rho A_2 v_2 \Delta t_2$$

We can equate densities on both sides and substitute $2A_2$ for A_1. Taking equal flow time for both ends gives us the **continuity equation**:

$$\mathbf{A_1 v_1 = A_2 v_2}$$

This **continuity equation** dictates that the flow rate is a constant for noncompressible fluids flowing through a tube, pipe, or hose and is another way of stating that the mass flowing into a pipe is equal to the mass flowing out in equal times, regardless of differences in the pipe's cross-sectional area.

As cross-sectional area increases, velocity decreases. It is an inverse relationship. The **continuity equation** also shows that the flow rate (A)(v) is equal for both sides of a pipe with either equal or unequal cross-sectional areas.

Example

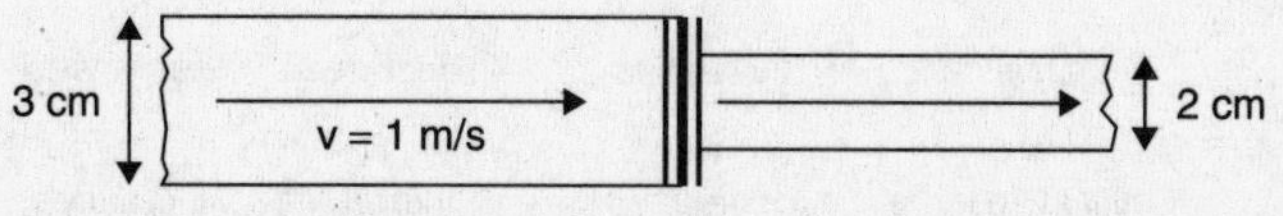

A garden hose of interior diameter 3 cm is coupled to another hose of interior diameter 2 cm. If the velocity of the water in the larger hose is 1 m/s, what is the velocity of the water in the smaller hose? What is the flow rate?

Solution

Using the **continuity equation,** $A_1v_1 = A_2v_2$, $(\pi)\left(\frac{0.03}{2}m\right)^2(1\text{ m/s}) = (\pi)\left(\frac{0.02}{2}m\right)^2(\mathbf{v_{SMALLER}})$

$$\mathbf{v_{SMALLER}} = \frac{(0.03\,\text{m})^2(1\,\text{m/s})}{(0.02\,\text{m})^2} = \mathbf{2.25\ m/s}$$

Flow rate for larger hose: $A_1\mathbf{v}_1 = (\pi)(0.03\text{m}/2)^2(1\text{ m/s}) = \mathbf{7.07 \times 10^{-4}\ m^3/s}$

Flow rate for smaller hose: $A_2\mathbf{v}_2 = (\pi)(0.02\text{m}/2)^2(2.25\text{ m/s}) = \mathbf{7.07 \times 10^{-4}\ m^3/s}$

Bernoulli's Principle

For a flowing fluid, the relationship between **pressure, velocity, and height** is stated by **Bernoulli's Equation:**

$$\mathbf{p + \rho g y + \frac{1}{2}\rho v^2 = constant}$$

In other words,

$$p_1 + \rho g y_1 + \frac{1}{2}\rho v_1^2 = p_2 + \rho g y_2 + \frac{1}{2}\rho v_2^2$$

By multiplying each part of the equation by a given constant volume of liquid:

$$Vp_1 + V\rho g y_1 + \frac{1}{2}V\rho v_1^2 = \text{constant}$$

This becomes

$$E_{PRESSURE} + E_P + E_K = \text{CONSTANT}$$

Explanation of Bernoulli's Equation

This equation is essentially a restatement of the work-energy theorem, which, *by multiplying each part of the equation by a unit volume of flowing fluid*, states that **the sum of the energy of pressure, gravitational potential energy, and the kinetic energy of a given volume of flowing fluid of constant density is equal for equal locations anywhere in a conduit.**

In other words, for a certain volume of fluid of known and constant density, when flowing through a pipe or tube whose cross-sectional area is *not* constant, the *total* amount of energy possessed by this certain volume of fluid is equal everywhere throughout the changing passageway. Note the following:

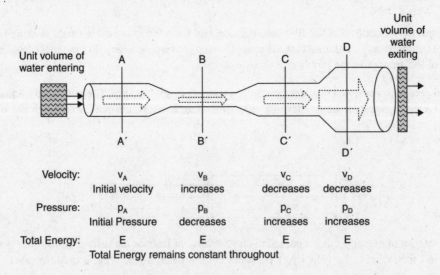

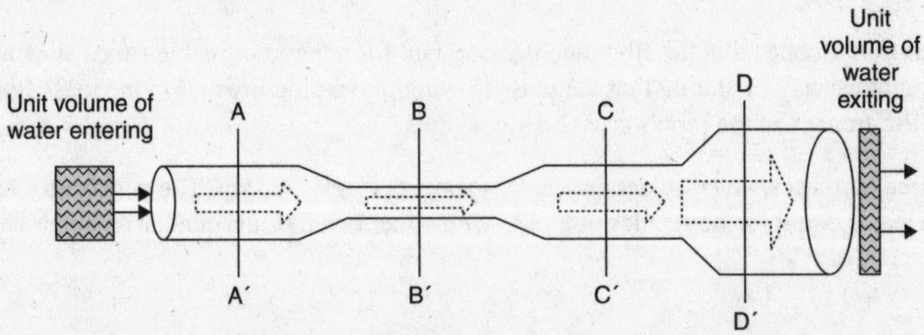

Velocity:	v_A Initial velocity	v_B increases	v_C decreases	v_D decreases
Pressure:	p_A Initial Pressure	p_B decreases	p_C increases	p_D increases
Total Energy:	E	E	E	E

Total Energy remains constant throughout

This figure depicts a unit volume of water entering and flowing through a pipe of changing cross-sectional area. Following the water through from left to right, the following points are made:

The velocity at B-B' is greater than the velocity at A-A'.

The velocity at C-C' is less than that at B-B' and is equal to that at A-A'.

The velocity at D-D' is less than that at C-C'.

The pressure at B-B' is less than that at A-A'.

The pressure at C-C' is greater than that at B-B' and is equal to that at A-A'.

The pressure at D-D' is greater than that at C-C'.

The total energy of the water *remains constant throughout.*

Example

Use the preceding figure, assigning the following measurements:

$$d_{A\text{-}A'} = 9 \text{ cm},\ d_{B\text{-}B'} = 3 \text{ cm},\ d_{C\text{-}C'} = 9 \text{ cm and } d_{D\text{-}D'} = 21 \text{ cm}$$

Initial fluid velocity: 50 cm/sec

Determine the fluid velocity in cm/sec at points A-A', B-B', C-C', D-D'.

Solution

First, determine the cross-sectional areas at those points:

$A(A\text{-}A') = 63.6 \text{ cm}^2$ $A(B\text{-}B') = 7.07 \text{ cm}^2$ $A(C\text{-}C') = 63.6 \text{ cm}^2$ $A(D\text{-}D') = 346 \text{ cm}^2$

Now, use $\mathbf{A_1v_1 = A_2v_2}$ to find flow rates:

$$A\mathbf{v} \text{ at } (A\text{-}A') = A\mathbf{v} \text{ at } (B\text{-}B') = A\mathbf{v} \text{ at } (C\text{-}C') = A\mathbf{v} \text{ at } (D\text{-}D')$$

$$(63.6 \text{ cm}^2)(50 \text{ cm/s}) = (7.07 \text{ cm}^2)(\mathbf{v_{B\text{-}B'}}) = (63.6 \text{ cm}^2)(\mathbf{v_{C\text{-}C'}}) = (346 \text{ cm}^2)(\mathbf{v_{D\text{-}D'}})$$

Solving for each yields:

$v_{A\text{-}A'} = 50 \text{ cm/s}$ $v_{B\text{-}B'} = 450 \text{ cm/s}$ $v_{C\text{-}C'} = 50 \text{ cm/s}$ $v_{D\text{-}D'} = 9.19 \text{ cm/s}$

Review Questions and Answers

B Exam Question Types

Multiple Choice

1. An object placed in a liquid rises and floats. The buoyant force on the object

(A) is less than the object's weight.
(B) is zero.
(C) is equal to the object's weight.
(D) depends on the liquid's density.
(E) depends on the object's shape.

2. The density of ice is 9.2×10^2 kg/m^3. If a chunk displaces 10^{-2} m^3, the buoyant force on the ice is most nearly

(A) 0.1 N
(B) 10 N
(C) 100 N
(D) 1000 N
(E) 10,000 N

3. Two objects with different densities

(A) may occupy equal volumes.
(B) may occupy different volumes.
(C) may have the same mass.
(D) may have different masses.
(E) may satisfy any of the above.

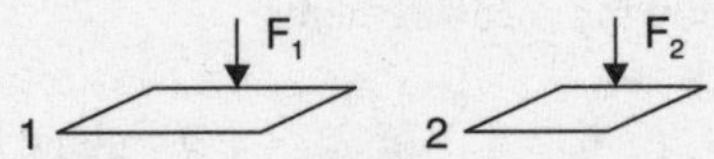

4. Surface 1 has an area of 1 m^2 and is subjected to a perpendicular force of 1 N. Surface 2 has an area of 0.5 m^2 and is subjected to a perpendicular force of 2 N. The pressure on surface 2 is

(A) $\frac{1}{4}P_1$
(B) P_1
(C) $2\,P_1$
(D) $4\,P_1$
(E) $8\,P_1$

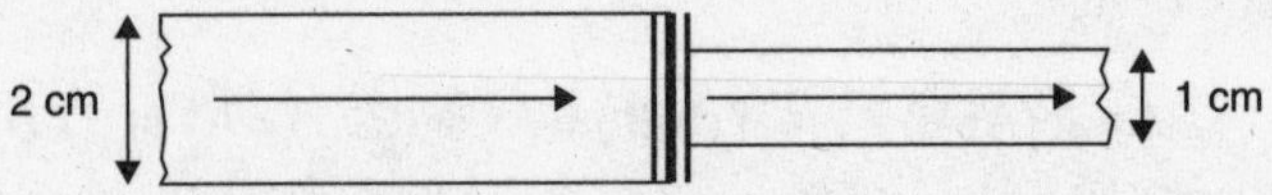

5. A pipe is 2 cm diameter and narrows to a 1 cm diameter. The velocity of the water moving from the larger to smaller side

(A) decreases by twofold.
(B) decreases by fourfold.
(C) remains constant.
(D) increases by twofold.
(E) increases by fourfold.

6.

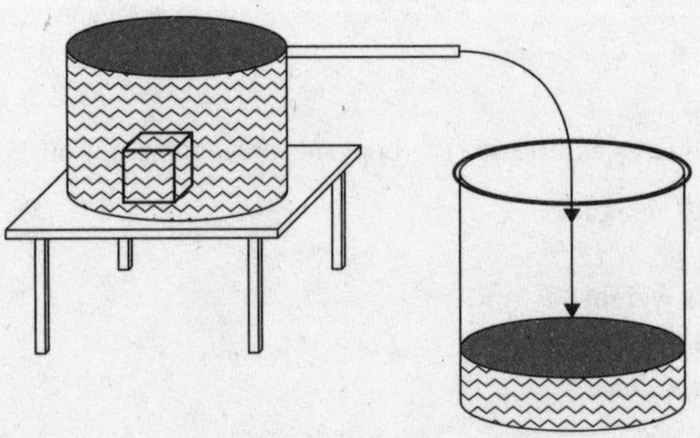

An object of mass 2.60 kg is completely immersed in a tub of water ($\rho_{WATER} = 1.00 \times 10^3$ kg/m^3). If it displaces 1.30 kg of water, the specific gravity of the object is closest to

(A) 0.02
(B) 0.20
(C) 2.00
(D) 20.0
(E) 200

7. A rock is tested to determine whether it is gold or iron sulfide (fool's gold). If the specific gravity of gold is 19.3 and the rock's mass is 0.39 kg, the mass of water it should displace is nearly

(A) 0.02 kg
(B) 0.20 kg
(C) 2.00 kg
(D) 20.0 kg
(E) 200 kg

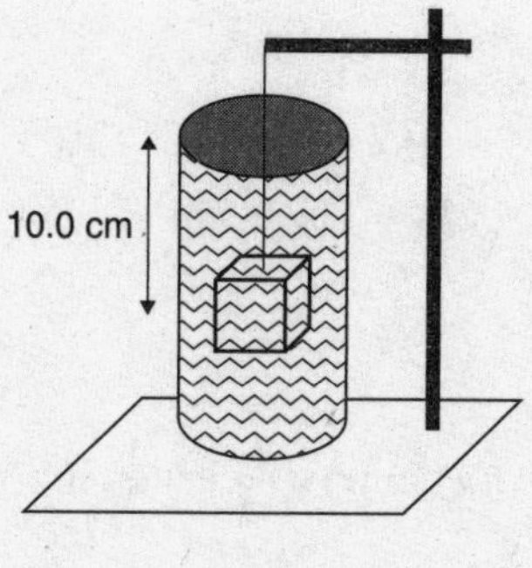

8. An aluminum cube having sides of 4.0 cm is suspended and immersed in a jar filled with water. If its center of mass is submerged 10.0 cm below the water's surface, the average hydrostatic pressure on the cube is most nearly

(A) 1 Pa
(B) 10 Pa
(C) 100 Pa
(D) 1000 Pa
(E) 10,000 Pa

9. The buoyant force on the cube in the previous problem is most nearly

(A) 0.32 N
(B) 0.64 N
(C) 3.20 N
(D) 6.40 N
(E) 32.0 N

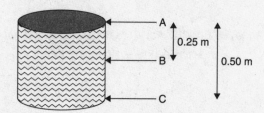

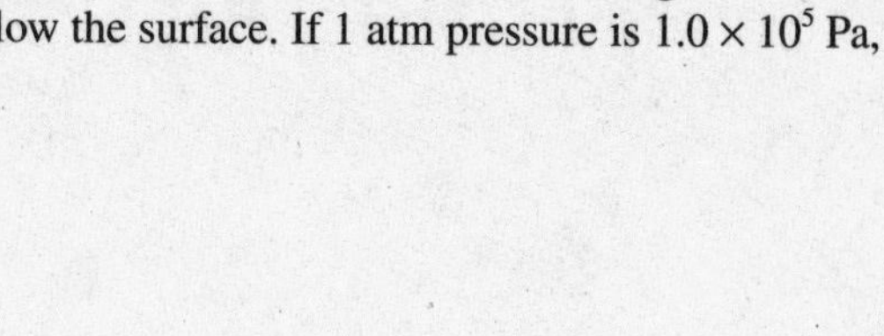

10. A pressure gauge is lowered into a water tank and shows a reading of 0 at level A and a reading of 2500 Pa at level B, a depth of 0.25 m below the surface. If 1 atm pressure is 1.0×10^5 Pa, the total pressure at level C is

(A) 1.02×10^5 Pa
(B) 1.03×10^5 Pa
(C) 1.04×10^5 Pa
(D) 1.05×10^5 Pa
(E) 1.06×10^5 Pa

Answers to Multiple-Choice Questions

1. **(D)** Because of the relationship: $\mathbf{F_{BUOY}} = \mathbf{V\rho g}$, the buoyant force depends directly on the liquid's density.

2. **(C)** Using $\mathbf{F_{BUOY}} = \mathbf{V\rho g}$, $\mathbf{F_{BUOY}} = (10^{-2}\text{m}^3)(9.2 \times 10^2\text{kg/m}^3)(10\text{ m/s}^2) = 92\text{ N}$.

   The closest answer is **100 N.**

3. **(E)** Since density equals mass divided by volume, any two objects may have different densities and occupy the same volume but have different masses or may have the same mass but occupy different volumes. **All answers are correct.**

4. **(D)** Since P = F/A,

$$P_1 = 1\text{ N}/\ 1\text{ m}^2 = 1\text{ Pa}$$

$$P_2 = 2\text{ N}/\ 0.5\text{ m}^2 = 4\text{ Pa}$$

$$P_2 = 4\ P_1$$

5. **(E)** Using $\mathbf{A_1v_1 = A_2v_2}$,

$$(\pi)(1\text{ cm})^2(\mathbf{v_1}) = (\pi)(\tfrac{1}{2}\text{ cm})^2(\mathbf{v_2})$$

$$v_2 = 4v_1$$

6. **(C)** Using sp.gr. = m of substance/m of equal vol. of water:

$$\text{sp.gr.} = 2.60\text{ kg} / 1.30\text{ kg} = \mathbf{2.00}$$

7. **(A)** Using sp.gr. = m of substance/m of equal vol. of water:

$$19.3 = 0.39\text{ kg}/x$$

$$x = 0.02\text{ kg}$$

8. **(D)** Using $\mathbf{p = \rho gh}$,

$$p = (1.00 \times 10^3\text{ kg/m}^3)(10\text{ m/s}^2)(0.10\text{ m}) = 1.00 \times 10^3\text{ N/m}^2 = \mathbf{1000\ Pa}$$

9. **(B)** Using $\mathbf{F_{BUOY}} = \mathbf{V\rho g}$,

$$\mathbf{F_{BUOY}} = (0.04\text{ m})^3(1.00 \times 10^3\text{ kg/m}^3)(10\text{ m/s}^2) = \mathbf{0.64\ N}$$

10. **(D)** Using $p_{HYD} = \mathbf{\rho gh}$,

$$p_{TOT} = p_O + \mathbf{\rho gh} = (1.0 \times 10^5\text{ Pa}) + (1.00 \times 10^3\text{ kg/m}^3)(10\text{ m/s}^2)(0.25\text{ m})$$

$$= 105{,}000\text{ Pa} = \mathbf{1.05 \times 10^5\ Pa}$$

Free Response

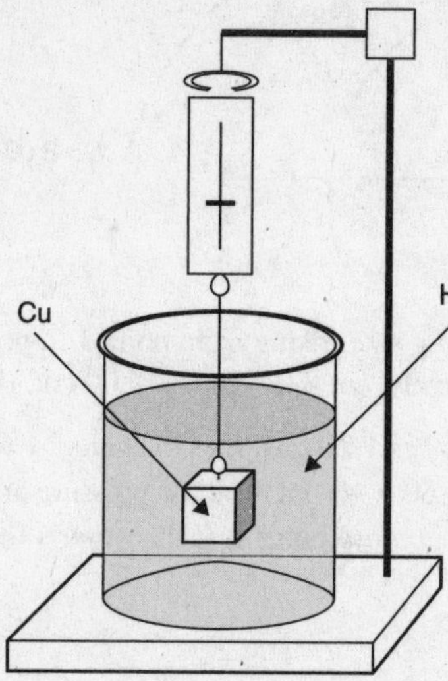

1. A copper mass of 0.75 kg and volume of 84 cm^3 is immersed in water, suspended by a spring scale, which is attached to an immovable arm. Water's density is 1.00×10^3 kg/m^3, and copper's density is 8.96×10^3 kg/m^3.

(a) What is the buoyant force on the copper mass?
(b) What apparent weight registers on the spring scale?
(c) What is copper's specific gravity?

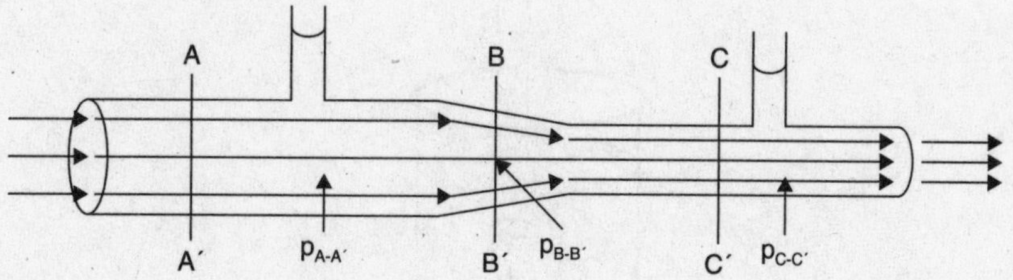

2. Water flows through a nonuniform pipe of cross-sectional areas 50 cm^2 at A-A', 40 cm^2 at B-B' and 30 cm^2 at C-C'. The initial velocity of the water as it enters the pipe is 0.5 m/s, and it is under 1.2 atm pressure.

(a) **i.** Determine the water's velocity at B-B'.
ii. Determine the water's velocity at C-C'.
(b) **i.** Determine the water's pressure at A-A'.
ii. Determine the water's pressure at B-B'.
iii. Determine the water's pressure at C-C'.

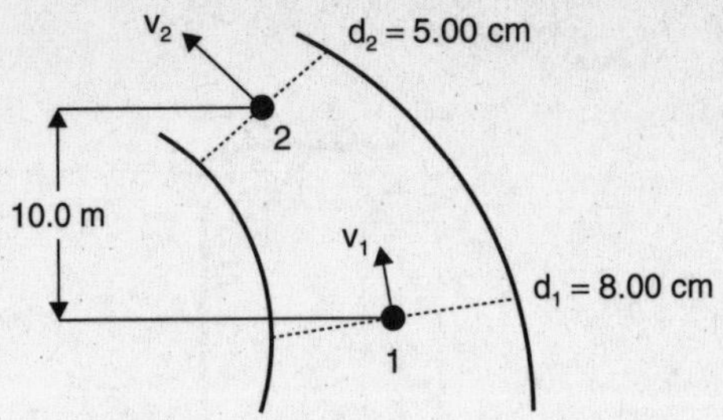

3. Water flows through a conduit of changing size, rising from point 1 to point 2 through a height of 10.0 m. The diameters at points 1 and 2, respectively, are 8.00 cm and 5.00 cm. (Water's density is 1.00×10^3 kg/m^3.)

(a) If the velocity of the water at point 1 is 1.20 m/s, find the velocity at point 2.

(b) If the water pressure at point 1 is 1.60×10^5 Pa, find the pressure at point 2.

(c) Some air enters the line somewhere before point 1. If the water's flow speed is not affected by the air, how does this affect the water's

i. density at point 2?

ii. pressure at point 2?

Answers to Free-Response Questions

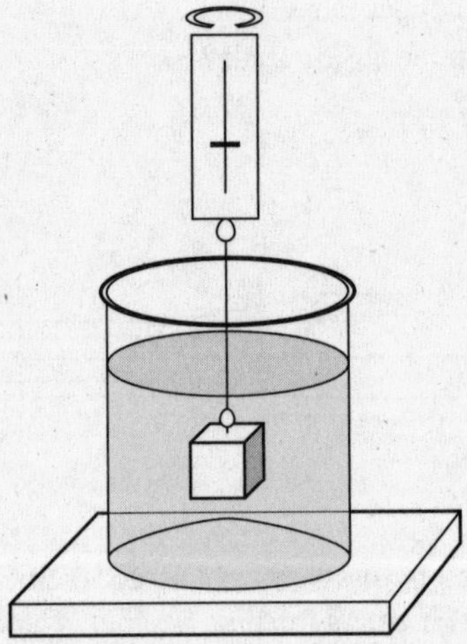

1. **(a)** Using $\mathbf{F_{BUOY}} = V\rho g$:

$$\mathbf{F_{BUOY}} = (84\text{cm}^3)(1.00 \times 10^3 \text{ kg/m}^3)(1 \text{ m}^3/10^6 \text{ cm}^3)(9.8 \text{ m/s}^2) = \mathbf{0.823\ N}$$

(b) The apparent weight would be the actual weight in air minus the buoyant force.

$$\mathbf{F_{APP\ WT}} = mg - V\rho g = (0.75 \text{ kg})(9.8 \text{ m/s}^2) - 0.823 \text{ N} = 7.35 \text{ N} - 0.823 \text{ N}$$
$$= \mathbf{6.53\ N}$$

(c) $\text{s.g.} = \rho_{CU}/\rho_{WATER} = \dfrac{(8960 \text{ kg/m}^3)}{(1.00 \times 10^3 \text{ kg/m}^3)} = \mathbf{8.96}$

2. **(a)** Using $\mathbf{A_1v_1 = A_2v_2}$, $A_{A\text{-}A'}\mathbf{v}_{A\text{-}A'} = A_{B\text{-}B'}\mathbf{v}_{B\text{-}B'} = A_{C\text{-}C'}\mathbf{v}_{C\text{-}C'}$

i. $\mathbf{v}_{B\text{-}B'} = A_{A\text{-}A'}\mathbf{v}_{A\text{-}A'} / A_{B\text{-}B'} = (50 \text{ cm}^2)(0.5 \text{ m/s}) / (40 \text{ cm}^2) = \mathbf{0.63\ m/s}$

ii. $\mathbf{v}_{C\text{-}C'} = A_{A\text{-}A'}\mathbf{v}_{A\text{-}A'} / A_{C\text{-}C'} = (50 \text{ cm}^2)(0.5 \text{ m/s}) / (30 \text{ cm}^2) = \mathbf{0.83\ m/s}$

(b) Because the flow is all horizontal, $\mathbf{y}_{A\text{-}A'} = \mathbf{y}_{B\text{-}B'} = \mathbf{y}_{C\text{-}C'}$

This reduces Bernoulli's equation to

$$p_{A\text{-}A'} + \left(\frac{1}{2}\right)\rho(\mathbf{v}_{A\text{-}A'})^2 = p_{B\text{-}B'} + \left(\frac{1}{2}\right)\rho(\mathbf{v}_{B\text{-}B'})^2 = p_{C\text{-}C'} + \left(\frac{1}{2}\right)\rho(\mathbf{v}_{C\text{-}C'})^2$$

Since ρ is equal throughout, this further reduces to

$$(p_{A\text{-}A'})(\mathbf{v}_{A\text{-}A'})^2 = (p_{B\text{-}B'})(\mathbf{v}_{B\text{-}B'})^2 = (p_{C\text{-}C'})(\mathbf{v}_{C\text{-}C'})^2$$

The water's pressure at A-A' is given: 1.2 atm

Solving for $p_{B\text{-}B'}$ and $p_{C\text{-}C'}$:

$$p_{B\text{-}B'} = \frac{(1.2\,\text{atm})(0.5\,\text{m/s})^2}{(0.63\,\text{m/s})^2} = \mathbf{0.76\ atm}$$

$$p_{C\text{-}C'} = \frac{(1.2\,\text{atm})(0.5\,\text{m/s})^2}{(0.83\,\text{m/s})^2} = \mathbf{0.44\ atm}$$

3. **(a)** Using $\mathbf{A_1v_1 = A_2v_2}$,

$$\mathbf{v}_2 = \mathbf{A}_1\mathbf{v}_1/\mathbf{A}_2 = \frac{\pi(0.04\,\text{m})^2(1.2\,\text{m/s})}{\pi(0.025\,\text{m})^2} = \mathbf{3.07\ m/s}$$

(b) Using Bernoulli's Equation: $\mathbf{p + \rho g y + \frac{1}{2}\rho v^2 = constant,}$ pressure at point 2 equals the pressure at point 1 (p.) plus the water's gravitational pressure at that point ($\mathbf{\rho g y}$) plus the water's kinetic pressure at that point $\left(\frac{1}{2}\rho\mathbf{v}^2\right)$. Solving for p_2:

$$p_2 = p_1 + \rho g y + \frac{1}{2}\rho v^2$$

$$= (1.60 \times 10^5\ \text{Pa}) + (1.00 \times 10^3\text{kg/m}^3)(9.80\ \text{m/s}^2)(10.0\ \text{m}) + \left(\frac{1}{2}\right)(1.00 \times 10^3\text{kg/m}^3)(3.07\ \text{m/s})^2$$

$$= (1.60 \times 10^5\ \text{Pa}) + (0.98 \times 10^5\ \text{Pa}) + (0.047 \times 10^5\ \text{Pa}) = \mathbf{2.63 \times 10^5\ Pa}$$

(c) i. Since the density of air is less than that of water and because air is now mixed with the water, **the density at point 2 will be reduced.**

ii. Since the density is reduced, the **pressure will also be reduced.**

Atomic and Nuclear Physics (B Exam)

Photoelectric Effect

Albert Einstein won the Nobel Prize in 1921 for work associated with the observation that light consists of tiny packets of electromagnetic energy, which are called **photons**. He built on the work of Max Planck, who noted that hot objects emit radiant energy in discrete units, called **quanta.** The higher the frequency of the light, the greater the energy contained in a **quantum** energy unit. This concept of the energy of electromagnetic radiation being directly related to its frequency is expressed by the relationship

$$E = hf$$

where E represents the energy of the quantum unit in Joules, h is **Planck's Constant** (6.63×10^{-34} J · s), and f is the frequency of the incoming electromagnetic radiation. This equation demonstrates the direct relationship between energy and frequency. In other words, *the higher the frequency of the incoming electromagnetic radiation, the greater the energy its photons possess.*

Photons of radiant energy of **higher frequency** (and lower wavelength) that travel through space and interact with atoms with which they come into contact possess **higher bursts of energy** than lower frequency photons. Einstein believed that these energy quanta, whatever their values, would be absorbed entirely when involved in interactions. Such interactions occur largely with metallic substances, which have loosely bound electrons on their outer shells or energy levels.

Philipp Lenard showed that electrons are indeed released from metallic elements when illuminated by ultraviolet light. These released or emitted electrons are called **photoelectrons**. Einstein proposed that there must be a minimum amount of energy necessary to release these metallic electrons.

The **photoelectric effect** is the phenomenon of **photoelectrons** being emitted from metallic surfaces when illuminated by electromagnetic radiation.

In conjunction with the concept that some *minimum amount of energy* is necessary to release electrons from metallic surfaces is the concept that anything less than a certain minimum incoming frequency translates into an insufficient amount of energy to release photoelectrons. This minimum frequency is sometimes called the **cutoff frequency or threshold frequency**, because anything below its value produces no photoelectron emission.

In addition, electrons not located exactly on the surface of the metal, but lying at some shallow depth, also absorb some incoming electromagnetic energy. For these *buried* electrons, some additional work must be done extracting them from that depth, owing to resistance or friction encountered while coming to the surface. The energy exacted for the purpose of freeing the electrons is called the **work function** ϕ, which understandably increases with depth.

Another aspect of the photoelectric effect is the fact that as the intensity or *brightness* of the incoming radiation is increased, the number of photoelectrons emitted shows a directly proportional increase. The energy of each photoelectron remains the same but more of them exist as more radiation is absorbed. This phenomenon is depicted by the following illustration.

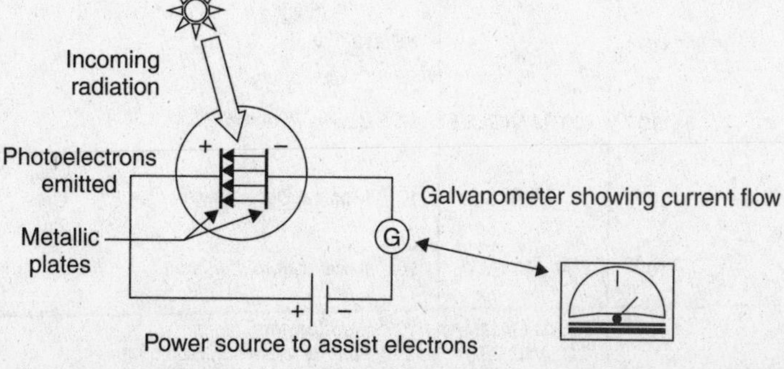

This is a pictorial representation of the photoelectric effect. A beam of incoming radiation is directed toward a pair of metallic plates that are enclosed in an evacuated glass chamber. The plates are connected to a power source to assist photoelectron emission. A galvanometer connected in series registers small amounts of current resulting in photoelectrons being emitted from the negative plate and jumping to the positive plate. If the radiant source is constant, the galvanometer shows a constant current.

Photoelectric Facts

Doubling the intensity of incoming radiation doubles the rate of photoelectron emission.

The energy of each photon at any certain frequency is equal. Higher frequencies contain higher energy photons; lower frequencies contain lower energy photons.

The kinetic energy of the emitted photoelectrons increases as the frequency of the incoming radiation increases. This means that we can equate kinetic energy and the velocity of the emitted photoelectrons from a metallic surface as follows:

$$E_K = \left(\frac{1}{2}\right) m\mathbf{v}^2 = hf$$

where m is the rest mass of an electron and **v** is its escape velocity.

There is a loss of energy for electrons emitted from a metal. Called the work function, ϕ, it results in a lessening of the kinetic energy of emitted electrons ($E = hf - \phi$). This makes the photoelectron's equation:

$$E_K = hf - \phi$$

$$\frac{1}{2}mv^2 = hf - \phi$$

If the value of the work function ϕ is greater than hf, no emission occurs, since the incoming energy is insufficient to free the electron.

ELECTROMAGNETIC SPECTRUM

λ m approximately		f (Hz) approximately
10^6	POWER	10^2 AC Electric Generators
10^2–10^3	AM RADIO	10^5–10^6 Communications
1–10	TV, FM RADIO	10^7–10^8 Communications
10^{-1}	RADAR	10^9 Communications / Sensing
10^{-1}	MICROWAVES	10^9 Communications / Cooking
10^{-4}	INFRARED	10^{12} Heat Lamps / Night Photos
7×10^{-7} – 4×10^{-7}	VISIBLE LIGHT	4.3×10^{14} Red, Orange, Yellow, Green, Blue, Indigo, 7.5×10^{14} Violet
10^{-8}	ULTRA VIOLET	10^{16} Lasers / Lamps
10^{-10}	X-RAYS	10^{18} Medical Diagnostics
10^{-12}	GAMMA RAYS	10^{20} Accelerators
10^{-16}	HARD GAMMA COSMIC RAYS	10^{24} Accelerators Natural Incoming Radiation

Example

Light of wavelength 5.8×10^{-7} m consists of photons of what energy?

Solution

Using $E_K = hf$ and $f = \mathbf{v}/\lambda$, where h is Planck's constant, **v** is c, the speed of light, and λ is given:

$$\mathbf{E_K = h\ v/\lambda} = (6.63 \times 10^{-34}\ \text{J} \cdot \text{s})(3 \times 10^{8}\ \text{m/s})/(5.8 \times 10^{-7}\ \text{m}) = \mathbf{3 \times 10^{-19}\ J}$$

Example

A certain metal has a work function of 6.4×10^{-19} J. If the visible light range is approximately 4.0×10^{-7} m (violet) to 7.0×10^{-7} m (red), will this metal emit photoelectrons if illuminated by visible light?

Solution

Using $E_K = hf - \phi$, first find the frequency range using $f = \mathbf{v}/\lambda$:

$$f_{RED} = (3 \times 10^{8}\ \text{m/s}) / (7.00 \times 10^{-7}\ \text{m}) = 4.3 \times 10^{14}\ \text{Hz}$$
$$f_{VIOLET} = (3 \times 10^{8}\ \text{m/s}) / (4.00 \times 10^{-7}\ \text{m}) = 7.5 \times 10^{14}\ \text{Hz}$$

Now, substitute these values into the equation $E_K = hf$:

$$\mathbf{E_{K\ RED}} = (6.63 \times 10^{-34}\ \text{J} \cdot \text{s})(4.3 \times 10^{14}\ \text{Hz}) = 2.85 \times 10^{-19}\ \text{J}$$
$$\mathbf{E_{K\ VIOLET}} = (6.63 \times 10^{-34}\ \text{J} \cdot \text{s})(7.5 \times 10^{14}\ \text{Hz}) = 4.97 \times 10^{-19}\ \text{J}$$

Since both of these values are less than the work function 6.4×10^{-19} J, **no electron emission occurs.**

Example

If ultraviolet light of wavelength 3.9×10^{-8}m is used, does photoelectron emission occur? If so, what is the maximum velocity of the photoelectrons?

Solution

Using $E_K = hf - \phi$, first find the frequency using $f = \mathbf{v}/\lambda$:

$$f_{UV} = (3 \times 10^{8}\ \text{m/s}) / (3.9 \times 10^{-8}\ \text{m}) = 7.7 \times 10^{15}\ \text{Hz}$$

Now, substitute this value into the equation $E_K = hf$:

$$\mathbf{E_{K\ VIOLET}} = (6.63 \times 10^{-34}\ \text{J} \cdot \text{s})(7.7 \times 10^{15}\ \text{Hz}) = 51.1 \times 10^{-19}\ \text{J}.$$

This exceeds the work function.

$$E_K = hf - \phi = 51.1 \times 10^{-19}\ \text{J} - 6.4 \times 10^{-19} = 44.7 \times 10^{-19}\ \text{J}.\ \textbf{Electron emission occurs.}$$

To determine the photoelectrons' maximum velocity, use $E_K = \left(\frac{1}{2}\right)mv^2$, where $\mathbf{E_K} = 44.7 \times 10^{-19}$ J and $m_e = 9.11 \times 10^{-31}$ kg:

$$\mathbf{v} = [(2E_K/m)]\tfrac{1}{2} = [(2)(44.7 \times 10^{-19}\ \text{J}) / (9.11 \times 10^{-31}\ \text{kg})]^{1/2} = \mathbf{3.1 \times 10^{6}\ m/s}$$

Atomic Energy Levels

A theory of the atomic model, based on quantum ideas, was proposed in 1913 by Niels Bohr. He postulated that the electrons orbiting the atomic nucleus occupy discrete energy levels. When an electron jumps from a higher energy level to a lower one, a photon is emitted. In a neon sign, for instance, electricity is passed through neon gas. As it is, electrons are *energized* and *jump* up to a higher energy level momentarily. When they jump back down to their original energy level, they radiate that extra energy back as photons of light. In the case of neon, the frequency of radiated light corresponds to a shade of the color orange. Also, because the emitted photoelectron's energy is given by **E = hf**, the radiated frequency, whether it be visible color or invisible electromagnetic radiation, is a result of the jump made between electron energy levels.

The result of *energized* electrons returning to their original stable atomic conditions is a grouping of frequencies, some visible, known as a **bright line spectrum**. The frequency or color grouping is different for different atoms.

The bright line spectrum of hydrogen is often referred to since it gives off the simplest grouping. Its frequencies extend from ultraviolet (Lyman series), through visible (Balmer series) to infrared (Paschen series) and beyond. The groupings or series are depicted in the following illustration.

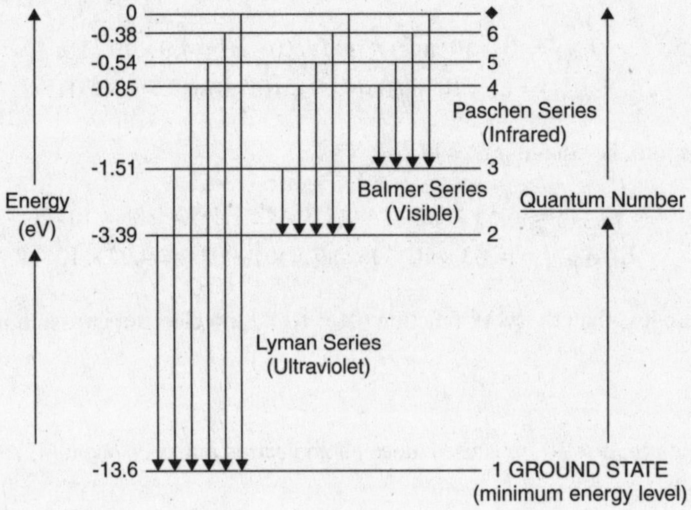

This illustration depicts electron orbits in a hydrogen atom, as permitted by the Bohr model of the atom. The **ground state** is represented by quantum number 1. Each successive quantum number level represents a higher energy level or electron shell.

An energized or excited electron that jumps from level 2 to level 1 will emit a photon of **ultraviolet** light.

An electron jumping from level 3 to level 2 will emit a photon of **visible** light and then will emit a photon of **ultraviolet** light when it continues to level 1.

An electron jumping from levels 4, 5, or 6 to level 3 will emit a photon of **infrared** light and continue to level 2, emit a photon of **visible** light, and then will emit a photon of **ultraviolet** light when it continues to level 1.

The energy equivalency for electron volts and Joules is

$$1 \text{ eV} = 1.60 \times 10^{-19} \text{ J}$$

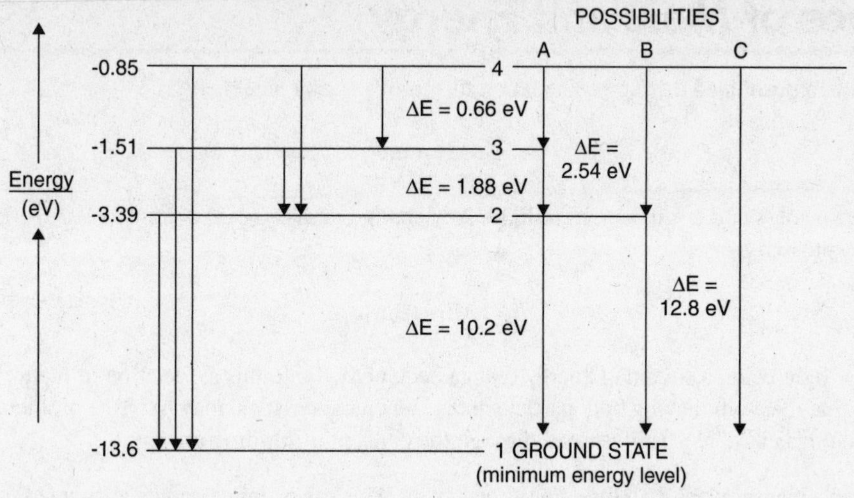

This simplified illustration shows the six possibilities for electrons jumping from energized or excited states from level 4:

A. First, an electron at level 4 with –0.85 eV jumps to level 3 at –1.51 eV of energy. In so doing, it emits a photon whose energy is the difference between –0.85 eV and –1.51 eV or **0.66 eV**.

Then, the electron jumps down from level 3 (–1.51 eV) to level 2 (–3.39 eV) and emits a photon of energy [–1.51 – (–3.39)] eV or **1.88 eV**.

Finally, the electron jumps from level 2 to level 1, emitting a photon of **10.2 eV**.

Total change in energy: 12.8 eV

B. First, an electron at level 4 with –0.85 eV jumps to level 2 at –3.39 eV of energy. In so doing, it emits a photon whose energy is the difference between –0.85 eV and –3.39 eV or **2.54 eV**.

Finally, the electron jumps from level 2 to level 1, emitting a photon of **10.2 eV**.

Total change in energy: 12.8 eV

C. The electron jumps from level 4 (–0.85 eV) to level 1 (–13.6 eV).

Total change in energy: 12.8 eV

These emission energy values may all be translated into frequencies of electromagnetic radiation by using the equations $E = hf$ and $f = \mathbf{v}/\lambda$ as done previously in this chapter.

Example

In the Balmer series, electrons jump from energy level 3 to level 2.

How much energy is radiated, and what is the frequency of electromagnetic radiation (in this case visible light)?

Solution

The electron jumps down from level 3 (–1.51 eV) to level 2 (–3.39 eV) and emits a photon of energy [–1.51–(–3.39)] eV or **1.88 eV**.

$$\text{Using } E = hf \text{ and } f = \mathbf{v}/\lambda,\ E = h\,\mathbf{v}/\lambda = h\,c/\lambda$$

$$\lambda = h\,\mathbf{c}/E = (6.63 \times 10^{-34}\ \text{J}\cdot\text{s})(3 \times 10^{8}\ \text{m/s}) / (1.88\ \text{eV})(1.60 \times 10^{-19}\ \text{J/eV}) =$$

$$6.61 \times 10^{-7}\ \text{m} \quad \text{(This corresponds to a shade of red light.)}$$

Equivalence of Mass and Energy

In 1905, Albert Einstein postulated the famous equation that unites matter and energy:

$$\mathbf{E = mc^2}$$

where E is energy, m is mass and **c** is the speed of light. Any change in mass (Δm) is immediately reflected in a change of energy (ΔE), as expressed by

$$\Delta E = (\Delta m)\mathbf{c}^2$$

Einstein's concept of light being a stream of energy quanta became tied into this association of matter and energy. One aspect was the idea that the atom, having both particle and wave characteristics, may have the problem of *missing mass* tied into an energy equivalency. This is illustrated through the concept of **binding energy**.

Nuclear masses always measure less than the sum of their parts. This **mass defect** can be accounted for by the fact that energy is always released in the formation of a nucleus. (The energy that is released is the cause of the resulting stability of the nucleus.) That energy translates into the **binding energy** of the atom's nucleus. Nuclear particles, whether protons or neutrons, often are referred to as **nucleons**. This term is used to describe the amount of binding energy per nuclear particle, or per nucleon. To calculate it, divide the amount of binding energy by the total number of nuclear particles, or **nucleons** present.

Binding energy is determined through the conversion of *missing mass* or mass defect into energy using Einstein's equation $E = mc^2$. This binding energy is associated with the strong force of nuclear cohesion.

Take, for instance, a neutral carbon-12 atom $^{12}C_6$, which consists of six protons and six neutrons in the nucleus, and six electrons encircling the nucleus. An atom of carbon-12 has a measured mass of 12.0107 u (1 u = 1.66×10^{-27} kg). The mass of a neutron is 1.67×10^{-27} kg, the proton's mass is also 1.67×10^{-27} kg and the rest mass of an electron is 9.11×10^{-31} kg (0.00055 u).

Example

Compute the **mass defect** and **binding energy** of a carbon-12 atom's nucleus.

Solution

Computing the **mass defect** and **binding energy** will be accomplished by comparing the mass of the carbon-12 atom's nucleus with the sum of the masses of its constituents:

First, subtracting the mass of the atom's six electrons will yield the mass of the nucleus:

$$12.0107\text{ u} - (6)(0.00055\text{ u}) = 12.0074\text{ u}$$

To determine the nuclear mass defect, subtract the sum of the masses of the parts of the nucleus (nucleons) from the nuclear mass:

$$(m_n = 1.00867\text{ u},\ m_P = 1.00727\text{ u})$$

- Mass of six protons = (6)(1.00727 u) = 6.0436 u
- Mass of six neutrons = (6)(1.00867 u) = **6.0520 u**
- Total nucleon mass 12.0956 u

The **mass defect** for a carbon-12 atom is (12.0956 – 12.0074)u = **0.0882 u** or $(0.0882)(1.66 \times 10^{-27}\text{ kg/ u}) = \mathbf{1.464 \times 10^{-28}}$ **kg**.

The **binding energy** for this atom is found by using Einstein's equation $E = mc^2 = (1.464 \times 10^{-28}\ kg)(3 \times 10^8\ m/s)^2 =$ $\mathbf{13.176 \times 10^{-12}\ J}$ or $(13.176 \times 10^{-12}\ J)(1\ eV/1.60 \times 10^{-19}\ J) = \mathbf{82.4\ MeV}$.

Example

How much energy per nucleon is represented in the previous example of carbon-12?

Solution

Since carbon-12 has six protons and six neutrons in its nucleus, it has 12 nucleons. Therefore, dividing the binding energy by 12 yields:

$$13.176 \times 10^{-12}\ J / 12 = 1.098 \times 10^{-12}\ J/\ nucleon$$
$$\text{or } (82.4\ MeV/12) = 6.87\ MeV/\ nucleon$$

Wave-Particle Duality

In 1923, Louis de Broglie theorized that since light exhibits qualities shared by waves and particles (propagation, reflection, and refraction) and some qualities that are not shared (interference, diffraction, travel medium), there may be similarities and/or differences for the behavior of matter. In other words, de Broglie claimed for mass something similar to that which Einstein claimed for light. If light can have particle-like behavior, could not particles have wave-like behavior?

Taking Einstein's matter-energy relationship and doing some algebraic substitutions, the following connections may be made:

$$\mathbf{E = mc^2 = mcv = mc\,(f\lambda) = hf}$$
$$\lambda = hf/mcf = h/mv = \mathbf{h/p}$$
$$\lambda = h/mv = \mathbf{h/p} \text{ (de Broglie's Equation)}$$

With this equation, de Broglie postulated the concept of matter exhibiting wave-like qualities such that matter was no longer considered just a particle but also exhibited wave-pattern characteristics. These characteristics involve effects that propagate or permeate through space. The result is that electron levels have definite and discrete energy levels associated with them. Further work done by Erwin Schrodinger mathematically depicted electron energy quantization as being a consequence of wave motion and was governed by a wave function.

In the course of scientific investigation and experimentation, it was found that the reverse of the photoelectric effect is also possible. Electrons beamed at certain substances resulted in electromagnetic emission from those substances.

In 1927, Bell Telephone Lab experimenters Davisson and Germer discovered that when electrons were beamed at a nickel surface, they were reflected almost exactly as X-rays behaved.

The mass-energy equivalency for electron volts and atomic mass units is

$$1\ u = 1.66 \times 10^{-27}\ kg = 931\ MeV/\mathbf{c}^2$$

This chapter has dealt with the equivalency of atomic mass units and mass mainly in the discussion of nuclear mass defect and nuclear binding energy.

The **electron volt** is the energy gained or lost by an electron when moving through two points that have a potential difference of one volt. A volt equals 1 Joule per Coulomb. This equivalency will be examined in more detail in the next chapter.

Nuclear Reactions

The four basic types of nuclear reactions are **nuclear fission, nuclear fusion, radioactive decay, and forced nuclear reactions.** In all nuclear reactions, charge and mass/energy are conserved.

In nuclear reactions, the subatomic particles and electromagnetic radiation involved are usually classified as follows:

$^{1}n_{0}$ **neutron—**Nucleon having a mass of 1 u and zero charge

$^{1}p^{+}{}_{1}$ or $^{1}H_{1}$ (Hydrogen nucleus) **proton—**Nucleon having a mass of 1 u and +1 charge

$^{0}e_{-1}{}^{-}$ or $^{0}B^{-}{}_{-1}$ **electron—**Has negligible atomic mass and –1 charge

$^{0}e_{+1}{}^{+}$ **positron—**Positively charged electron (electron's antimatter particle)

$^{4}\alpha_{2}$ or $^{4}He_{2}$ **alpha particle or helium nucleus—**has two protons and two neutrons, atomic mass of 4 and charge of +2

γ **gamma ray—**high-energy photon having no mass or charge

When considering any nuclear reaction, reactants and products must follow a few basic rules. Mass/energy and charge must be conserved, and the number of nucleons must remain constant. This means that the totals of mass numbers before and after the reaction must be equal, and the totals of atomic numbers must remain equal.

$$^{(\text{mass number})\ A}X_{z\ (\text{atomic number})}$$

Mass number is the *total number of nucleons* (protons + neutrons).

Atomic number is the number of *positive charges* (protons).

Isotopes are atoms that have the same atomic number but differ in their numbers of neutrons. This means that they are atoms of the same element, but some samples may be more massive if they contain more neutrons or less massive with fewer neutrons.

Example

Carbon dating is used for determining relative approximate ages of substances that contain amounts of carbon-14 ($^{14}C_6$) and its stable end product, carbon-12 ($^{12}C_6$). Why are they isotopes?

Solution

Since both carbon-14 and carbon-12 contain the same number of positive charges (6), they are the same atom. Carbon-14 has two more chargeless masses (neutrons), which make it a "heavier" variety or isotope of carbon.

Example

Uranium-238 decays into Thorium-234 and an alpha particle. This is shown by

$$^{238}U_{92} \rightarrow {}^{234}Th_{90} + {}^{4}\alpha_{2}$$

Explain how this is a balanced nuclear equation.

Solution

The equation balances because

1. A mass of 238 u decays into masses of 234 u + 4 u, thus conserving mass.
2. Total charge of +92 decays into particles whose charges are +90 and +2, thus conserving charge.
3. Uranium-238 has 92 protons and 146 neutrons. The end products have a total of 92 protons and 146 neutrons, thus conserving total nucleon number.

Nuclear fission describes the process of bombarding an atomic nucleus with high-speed subatomic particles. This has the effect of breaking the nucleus apart and releasing energy and particles. Radioactivity may be induced in this fashion.

All nuclear fission reactions obey the basic rules of conservation of mass, charge, and number of nucleons.

Example

Atoms of lithium-7 ($^7Li_3$), when bombarded by protons, produce two alpha particles and energy. Write the equation.

Solution

$^7\mathbf{Li}_3 + {}^1\mathbf{p}_1 \rightarrow 2\,{}^4\alpha_2 + \gamma$ **The energy released can be in the form of gamma or the kinetic energy of the particles.**

This illustrates that mass is conserved (total mass of 8 u remains constant); charge is conserved; (total positive charge of 4 remains constant); and number of nucleons (8) remains constant.

Nuclear fusion describes the process of atomic nuclei becoming compressed or fused into larger nuclei under conditions of extremely high pressures and temperatures. This occurs in stars. All nuclear fission reactions obey the basic rules of conservation of mass, charge, and number of nucleons.

Example

Nuclear fusion powers the sun and stars. It involves hydrogen atoms becoming compressed or fused into larger atoms of helium, releasing energy. Write the simplified equation for fusion.

Solution

Since hydrogen is normally $^1H_1$ and it combines to form $^4He_2$, conserving mass, charge, and nucleons needs to involve an isotope of hydrogen $\left({}^2H_1\right)$ called deuterium or *heavy hydrogen* in order to satisfy the conservation rules:

$$^2H_1 + {}^2H_1 \rightarrow {}^4He_2 + \text{energy}$$

Radioactive Decay

Natural radioactive decay is the process by which naturally occurring elements containing an overabundance of energy and a weak nuclear binding force emit their excess energy (decay) over a period of time by emitting alpha, beta, or gamma radiation. The period of time needed for **half** of an amount of an atomic substance to decay into a more stable end-product with a lower energy level is called the substance's **half life**.

Example

The half life of a sample of radioactive ^{76}Br is 16 hours. After how many half lives would an original 0.044 kg sample of ^{76}Br only have 0.0055 kg remaining? How long would this take?

Solution

After one half life, one half of the original sample remains, or 0.022 kg.

After two half lives, 0.011 kg remains. **After three half lives, 0.0055 kg remains.**

It would take (3) 16 hours or **48 hours.**

Exposure to electromagnetic radiant energy causes certain substances to glow or *fluoresce*. Henri Becquerel used this phenomenon to discover that radioactive substances also can cause fluorescence in some substances. The work of Marie and Pierre Curie illustrated, among other things, that radium and polonium emit such electromagnetic radiant energy. This was observably evinced by the fact that *radioactive substances' emissions leave spots on photographic film*. In addition, *radioactive emissions ionize air molecules* by bombardment and subsequent removal of electrons.

Again, as in any nuclear reactions, the basic laws or rules remain in effect. Mass, charge, and nucleon number are conserved and must remain constant.

Review Questions and Answers

B Exam Question Types

Multiple Choice

1. Photoelectron emission occurs only when incident radiant energy possesses

(A) a minimum frequency.
(B) a minimum intensity.
(C) a minimum speed.
(D) a maximum frequency.
(E) a maximum intensity.

2. Light of wavelength 6.63×10^{-7} m consists of photons of energy

(A) 3×10^{-33} J
(B) 3×10^{-19} J
(C) 3×10^{-2} J
(D) 3×10^{8} J
(E) 3×10^{19} J

3. Photoelectron emission rate is a direct function of radiation

(A) frequency.
(B) speed.
(C) intensity.
(D) energy.
(E) wavelength.

4. An electron in its ground state becomes energized and jumps up to level 4. When it returns to its original state, the number of possible photon emissions is

(A) 7
(B) 6
(C) 5
(D) 4
(E) 3

5. The de Broglie wavelength of a 3000 kg vehicle cruising at a velocity of 20 m/s is most nearly

(A) 1.1×10^{-48}m
(B) 1.1×10^{-38}m
(C) 2.2×10^{-38}m
(D) 1.1×10^{-30}m
(E) 2.2×10^{-30}m

6. The binding energy for an atom's nucleus accounts for its

(A) nucleons.
(B) atomic number.
(C) atomic mass number.
(D) electron orbits.
(E) mass defect.

7. The half life of ^{127}Xe is 36.4 days. For a 0.004 kg sample after four half-lives, the amount of ^{127}Xe remaining is

(A) 0.000125 kg
(B) 0.00025 kg
(C) 0.005 kg
(D) 0.001 kg
(E) 0.002 kg

$$^{214}Pb_{82} \rightarrow \_\_\_\_\_\_$$

8. In this radioactive decay of lead-214, the products are

(A) $^{210}Bi_{83}$ + alpha particle.
(B) $^{211}Bi_{83}$ + 2 beta particles.
(C) $^{212}Bi_{83}$ + 2 protons + gamma ray.
(D) $^{213}Bi_{83}$ + neutron.
(E) $^{214}Bi_{83}$ + beta particle + gamma ray.

$$^{10}B_5 + {}^4He_2 \rightarrow \_\_\_\_ + {}^1n_0$$

9. In the preceding nuclear equation, the missing product is

(A) $^{11}B_5$
(B) $^{12}C_6$
(C) $^{13}N_7$
(D) $^{16}O_8$
(E) $^{19}F_9$

$$^{239}Np_{93} \rightarrow {}^{239}Pu_{94} + \_\_\_\_$$

10. In the above nuclear decay, Neptunium-239 decays into Plutonium-94 plus

(A) a gamma ray.
(B) an electron.
(C) a positron.
(D) a neutron.
(E) an alpha particle.

Answers to Multiple-Choice Questions

1. **(A)** Because of the relationship **E = hf**, the **minimum frequency** of radiation that contains enough energy to produce photoelectron emission is the cutoff or threshold frequency.

2. **(B)** Using $\mathbf{E_K = hf}$ and $\mathbf{f = v/\lambda}$, where h is Planck's constant, **v** is **c**, the speed of light, and λ is given.

   $\mathbf{E_K = h\,v/\lambda} = (6.63 \times 10^{-34}\ \mathrm{J \cdot s})(3 \times 10^{8}\ \mathrm{m/s}) / (6.63 \times 10^{-7}\ \mathrm{m}) = \mathbf{3 \times 10^{-19}\ J}$

3. **(C)** Increasing intensity of incoming radiant energy directly increases the number of photoelectrons emitted.

4. **(B)** Possibilities include 4-3, 4-2, 4-1, 3-2, 3-1, 2-1. Total = **6.**

5. **(B)** Using $\boldsymbol{\lambda = h/mv}$, $\lambda = (6.63 \times 10^{-34}\ \mathrm{J \cdot s}) / (3 \times 10^{3}\ \mathrm{kg})(20\ \mathrm{m/s}) = (2.2/2) \times 10^{-38}\ \mathrm{m} = \mathbf{1.1 \times 10^{-38} m.}$

6. **(E)** Binding energy is determined through the conversion of *missing mass* or mass defect into energy using Einstein's equation E = $\mathrm{mc^2}$.

   This energy accounts for the strong force of nuclear cohesion.

7. **(B)** After one half-life, half of 0.004 kg or 0.002 kg remains. After two half-lives, 0.001 kg remains, after three half-lives, 0.0005 kg remains, and after four half-lives, **0.00025 kg** remains.

8. **(E)** Since mass, charge, and nucleon number must be conserved, only E can be correct:

   $^{214}Pb_{82} \rightarrow {}^{214}Bi_{83}$ + beta particle + gamma ray

   ↓

   $^{0}_{-1}e^-$ or $^{0}_{-1}B^-$ **electron** – no atomic mass, –1 charge

   +γ (energy but no mass or charge)

9. **(C)** Since mass, charge, and nucleon number must be conserved, only C can be correct:

   $^{10}B_5 + {}^4He_2 \rightarrow$ _____ $+ {}^1n_0$

   mass: 14 reactant mass = x + 1. The missing mass must equal 13.

   charge: +7 = y + 0: The missing charge must equal 7.

   The only answer satisfying conservation of mass, charge, and nucleons is $\mathbf{^{13}N_7}$.

10. **(B)** Since mass, charge, and nucleon number must be conserved, only C can be correct:

    $^{239}Np_{93} \rightarrow {}^{239}Pu_{94}$ + _____

    mass: 239 reactant mass = 239 + x. The missing mass must equal 0.

    charge: 93 = 94 + y. The missing charge must equal –1.

    The only answer satisfying conservation of mass, charge, and nucleons is $\mathbf{^0e_{-1}}$ **(an electron)**.

Free Response

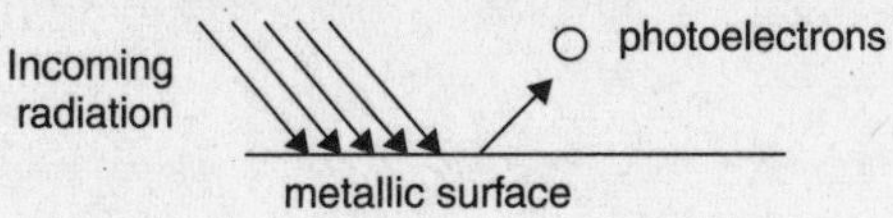

1. Incoming radiation illuminates a metallic surface which emits photoelectrons when the wavelength is 5.5×10^{-7} m or less.

   (a) Determine the cutoff frequency.
   (b) Determine the energy of the incoming radiation.
   (c) Determine the maximum velocity of the photoelectrons emitted from the surface.

2. **(a)** Calculate the de Broglie wavelength of electrons traveling at $\frac{1}{20}$ the speed of light.
   (b) Calculate the frequency of electrons traveling at that speed.

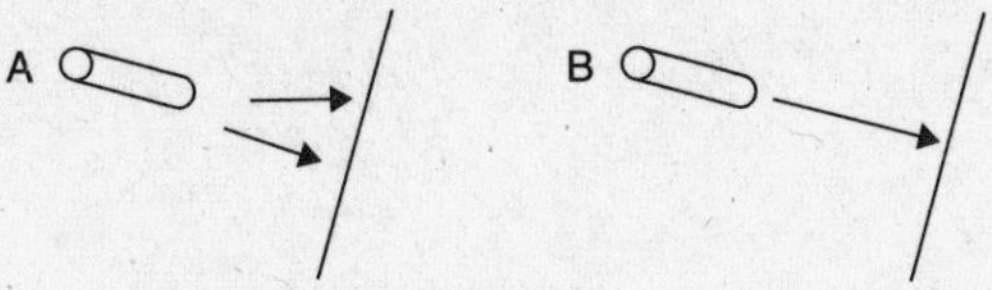

3. Light A is infrared and has a wavelength of 8000 Å. Light B is a laser with a wavelength of 633 nm. Determine the energy in Joules and eV for photons emitted by both lamps.

Answers to Free-Response Questions

1. **(a)** Using $\mathbf{f = v/\lambda}$, $f = (3 \times 10^8 \text{ m/s}) / 5.5 \times 10^{-7} \text{ m} = \mathbf{5.5 \times 10^6 \text{ Hz}}$

   (b) Using $\mathbf{E = hf}$, $E = (6.63 \times 10^{-34} \text{ J} \cdot \text{s})(5.5 \times 10^6 \text{ Hz}) = \mathbf{3.6 \times 10^{-27} \text{ J}}$

   (c) Using $\mathbf{E_K = \left(\frac{1}{2}\right) m v^2}$, $\mathbf{v} = [(2)(E_K)/(m)]^{\frac{1}{2}} = [(2)(3.6 \times 10^{-27} \text{ J}) / (9.11 \times 10^{-31} \text{ kg})]^{\frac{1}{2}} = \mathbf{89 \text{ m/ s}}$

2. **(a)** Using $\mathbf{\lambda = h/mv}$, $\lambda = (6.63 \times 10^{-34} \text{ J} \cdot \text{s}) / (9.11 \times 10^{-31} \text{ kg})(3/20 \times 10^8 \text{ m/s}) = \mathbf{4.9 \times 10^{-11} \text{ m}}$

   (b) Using $\mathbf{f = v/\lambda}$, $f = (3/20 \times 10^8 \text{ m/s}) / 4.9 \times 10^{-11} \text{ m} = \mathbf{3.0 \times 10^{17} \text{ Hz}}$

3. Light A: Using $E = hf$ and $f = \mathbf{v/\lambda}$,

   $E = hf = hv/\lambda = (6.63 \times 10^{-34} \text{ J} \cdot \text{s})(3 \times 10^8 \text{ m/s}) / (8.0 \times 10^{-7} \text{ m}) = 2.5 \times 10^{-19} \text{ J}$

   $E = (2.5 \times 10^{-19} \text{ J})(1 \text{ eV}/1.60 \times 10^{-19} \text{ J}) = 1.6 \text{ eV}$

   Light B: Using $E = hf$ and $f = \mathbf{v/\lambda}$,

   $E = hf = hv/\lambda = (6.63 \times 10^{-34} \text{ J} \cdot \text{s})(3 \times 10^8 \text{ m/s}) / (6.33 \times 10^{-7} \text{ m}) = 3.1 \times 10^{-19} \text{ J}$

   $E = (3.1 \times 10^{-19} \text{ J})(1 \text{ eV}/1.60 \times 10^{-19} \text{ J}) = 1.94 \text{ eV}$

Electric Fields and Forces

Electrostatics

Electrostatics deals with electric charge that does not flow but remains stationary, builds up, radiates an electric field, or jumps from one point to another. Electrostatic and electromagnetic forces permeate the universe, with possible implications regarding the nuclear strong force, the force of gravitation, and the modern hypotheses and debates involving string and super string theory, wormholes, dark matter, and beyond. These forces are among the most ubiquitous and powerful forces in the Cosmos, yet can be among the least visible or understood.

Electrification, the phenomenon of an object acquiring an electric charge, was described by the ancient Greeks, whose word **elektron** described hardened tree sap or amber. Rubbing amber with certain other substances caused it to acquire properties of attraction. We experience the same phenomenon whenever we brush our hair, especially on a dry day. The nylon, plastic, or rubber bristles acquire a negative charge from the electrons stripped off the surface of strands of hair. This leaves the brush **negatively charged** and the hair strands **positively charged.** As a result, the hair strands repel each other and stand apart, giving us a "bad hair day".

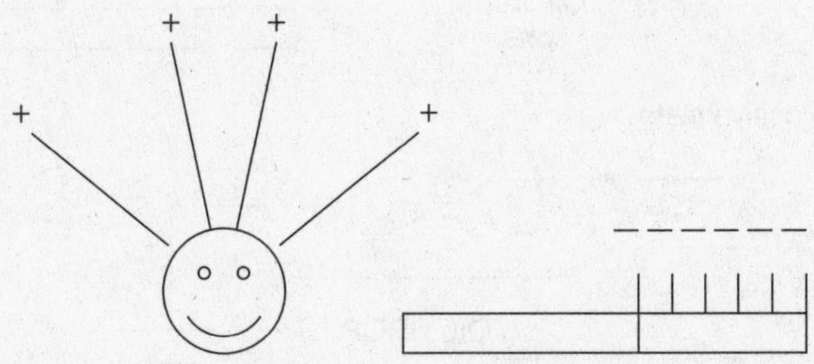

The same thing happens when a person walks through a carpeted room and touches anything that might be an electrical conductor. The carpet is made of tiny strands, which also yield electrons from their strands to the soles of the shoes. The electrons permeate through the shoes to the skin, where they spread out equally over the person's skin and the person becomes negatively charged. Touching any object that has fewer electrons will cause electrons to be transferred from the person to the object, whether it might be a door handle, any metallic substance or another person — any conductor of electrons. When this happens, the person experiences a small static electric shock. The negatively charged electrons have been transferred from the atoms in the carpet fibers, seat cover, chair, sweater, and so on, to the person's skin and finally to any electrical conductor, and become **grounded** or electrically neutralized in the process.

Benjamin Franklin is credited with having originated the terms **positive** and **negative** regarding electric charges. He described an electrified object as being positively charged when in this electrified state. He also was possibly the first person to associate the small static electric phenomena encountered in every day life with the awesome power of lightning, realizing that they are different manifestations of the same thing.

Electric Charge

When one object becomes electrically charged, either positively or negatively, by another object, the objects acquire equal and opposite charges. This **conservation of electric charge** is due to the fact that charge, like energy, cannot be created or destroyed, but changed from one *form* or location to another. When electrons leave one surface and accumulate on another, they change the electric balance of charge, which can result in the formation of electric attractive or repulsive forces, the intricacies of which are still not completely understood and have been the subject of further scientific investigation ever since the ancients rubbed amber.

Atoms are comprised of positively charged protons and uncharged neutrons in the nucleus, with the negatively charged electrons traveling around the nucleus. Interestingly, although the electron is nearly 2000 times less massive than either of the nucleons, its negative charge is exactly equal and opposite to the charge of the proton. Static electric charges usually originate with objects either gaining or losing electrons. Because electrons are the most weakly bound and the most available and mobile of subatomic particles, they are usually the easiest particles to attract, repel, or involve in reactions. Also, as discussed in the preceding chapter, because electrons exhibit some wave-like characteristics, their idiosyncrasies involve natural behaviors that go beyond point-charge restrictions. However, studies of electrostatics begin with the assumption that an electric charge is treated as a point source of electric charge.

It was Robert A. Millikan who discovered the basic charge on an electron. He sprayed oil droplets from a cologne-type atomizer. The droplets took on a negative charge when sprayed. They then were directed over a hole in the top plate of a pair of oppositely charged metallic plates and were observed as they fell between the plates, whose charge was regulated. The motion of the droplets was measured carefully and recorded. He assumed that by equating the downward gravitational force and the combination of upward buoyant and electrical forces, he could determine the amount of charge each droplet possessed. Measurements also involved size, density, and air friction on the droplets.

Millikan's observations indicated that the electrical charge on every droplet was a multiple of a certain number. That number turned out to be the individual charge on an electron, 1.60×10^{-19} Coulombs. In other words, electric charge occurs in *packages* or discrete amounts and is thus **quantized.**

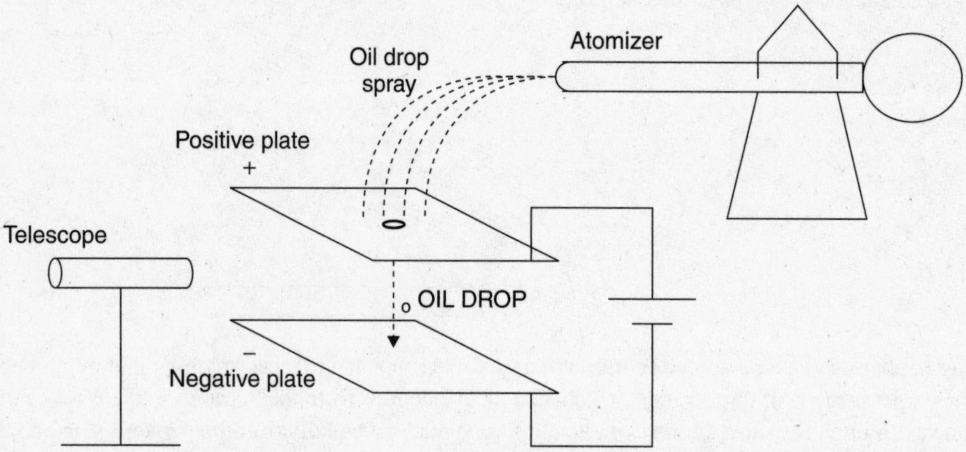

Robert A. Millikan's oil-drop apparatus was used to determine the basic charge on an electron, 1.60×10^{-19} Coulombs.

Electric Field and Electric Potential

Surrounding each electric charge is an individual electric field that radiates outward from positive charges and inward toward negative charges. Electric field lines are **perpendicular** to charge surfaces and *go out of the positive and into the negative*. The stronger or closer the charge, the stronger the resulting electric field and the closer together or denser the electric field lines are.

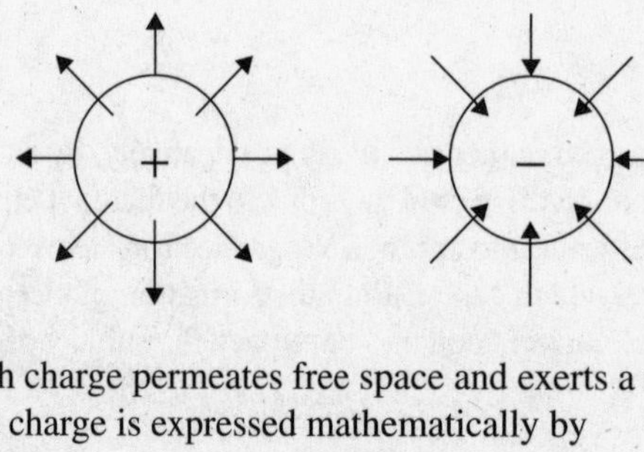

The electric field associated with each charge permeates free space and exerts a force on any other charge. The force exerted by an electric field on a point charge is expressed mathematically by

$$F = Eq$$

where **F** is the force in Newtons, **E** is the electric field strength in Newtons per Coulomb (N/C) and **q** is the charge affected by the electric field in Coulombs.

The electric field created by a point charge is then given by

$$E = \frac{1}{4\pi\varepsilon_0}\frac{q}{\mathbf{r}^2}$$

where **E** is the electric field in N/C, **q** is the charge placed in the electric field, and **r** is the distance of the charge from the origin of the electric field. ε_0 (epsilon zero) is the permittivity of free space or vacuum (how the electric field permeates through a vacuum) and ε_0 is a constant, equal to 8.85×10^{-12} C^2/Nm^2. The $\frac{1}{4\pi\mathbf{r}^2}$ is a mathematical representation of the surface of a sphere — like the bubble concept in the inverse-square relationship. For reasons of simplicity, it is noted that often the constant **k** is substituted for $\left(\frac{1}{4\pi\varepsilon_0}\right)$ and $\mathbf{k} = 9 \times 10^9$ Nm^2/C^2. This use of **k** (actually $\mathbf{k_O}$) is only valid for a vacuum or in air.

The electric field E now reduces to

$$E = \frac{kq}{\mathbf{r}^2}$$

Facts about Electric Fields

1. Electric field lines go *out of the positive* and *into the negative.*
2. Electric field lines originate and end *normal, that is perpendicular* to charge surfaces.

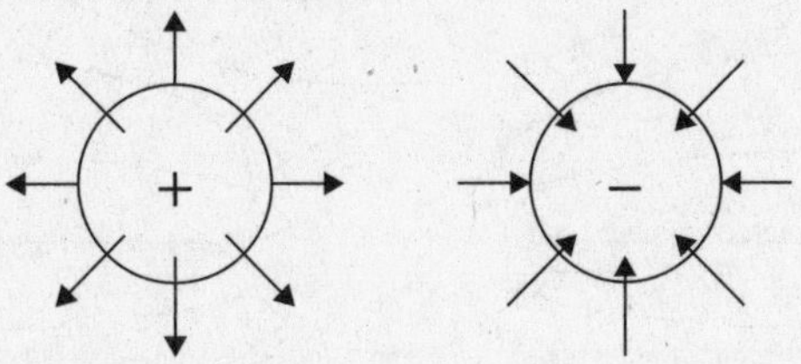

3. Electric field lines become *denser* as charge becomes *greater* or as the distance to the charge decreases.

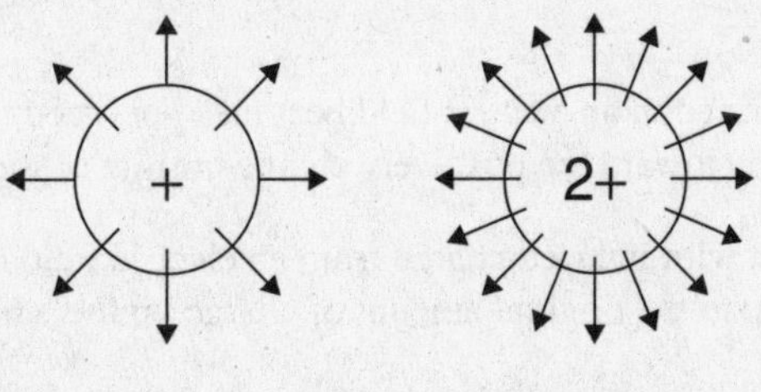

Twice as much charge means twice the electric field line density. The closer to the charge, the greater the field line density.

4. Electric field lines are closest together at *pointed* surfaces, where the charge becomes concentrated.

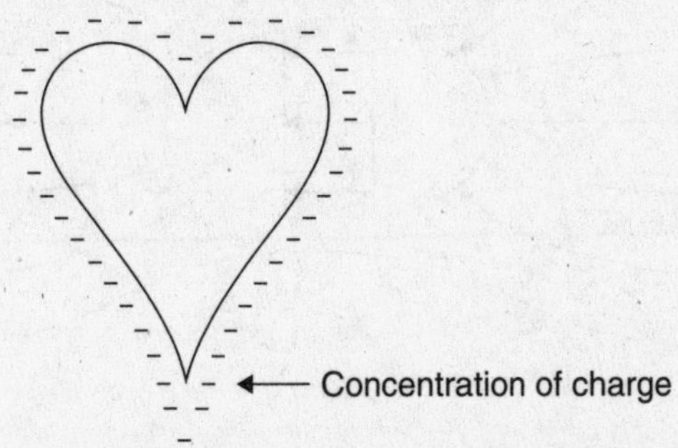

Greater charge concentration occurs at areas of greater curvature and is greatest at pointed surfaces.

5. Electric fields are *zero* inside charged objects.

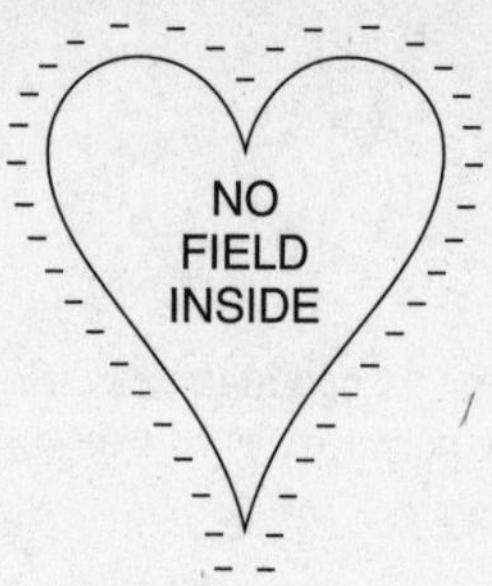

6. Electric fields between parallel charged plates are *uniform*.

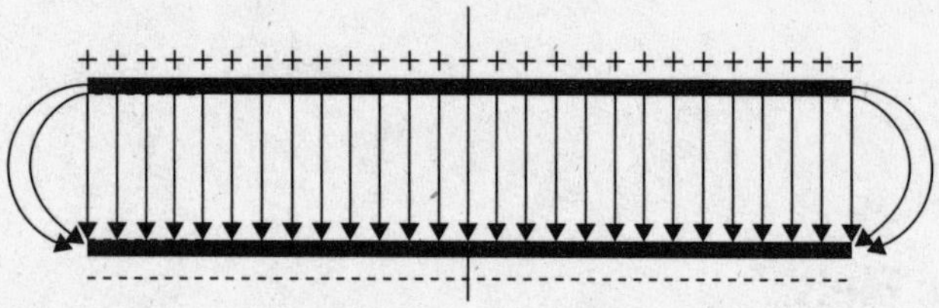

7. Objects placed in an electric field can become charged by *induction*.

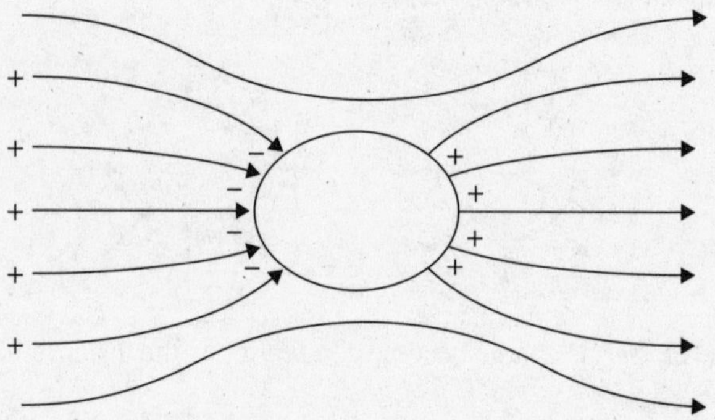

An electrically neutral object placed in an electric field becomes polarized, or charged by induction (without touching). Electrons are attracted toward the positively charged origin of the field.

8. Separating or splitting an object with induced charge from an electric field results in the separated parts possessing opposite charges that add up to the original amount of charge on the whole object before separation. In other words, charge is conserved.

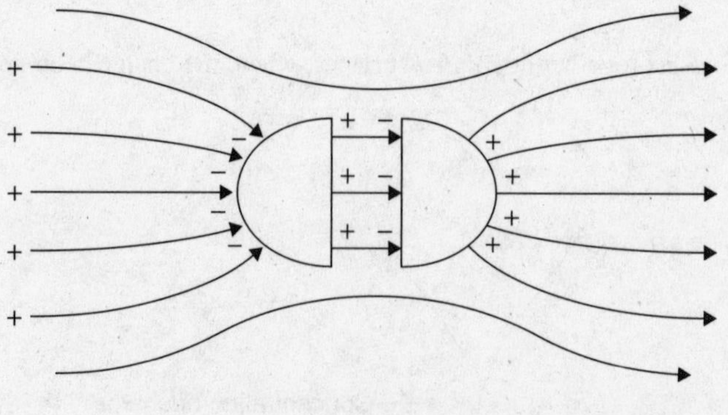

The electrically neutral object placed in the electric field in the illustration with fact number 7 developed its own polarized charge by **induction** (without touching). When the object is cut and separated, each half now is polarized and has its own induced charge. Each possesses half of the original charge as seen in this illustration.

Example

What is the electric field strength at a distance of 2 cm from a test charge of 4.8×10^{-19} C?

Solution

Using $E = \frac{kq}{r^2}$, $E = \frac{(9 \times 10^9 \text{Nm}^2)(4.8 \times 10^{-19} \text{C})}{(0.02\,\text{m})^2}$

$$= \frac{(9 \times 10^9 \text{Nm}^2)(4.8 \times 10^{-19} \text{C})}{(4.0 \times 10^{-4} \text{m}^2)} = 10.8 \times 10^{-6} \text{ N/C} = \mathbf{1.1 \times 10^{-5} \text{ N/C}}$$

Example

Two electrically neutral metallic spheres, A and B, are placed in an electric field, touched together, and separated. If sphere A has acquired a charge of +10 nC ($+10 \times 10^{-9}$C), what is the charge on sphere B?

Solution

Because **charge is conserved**, the charge on each sphere is equal and opposite the other. The charge acquired by sphere B is then **–10 nC.**

Electric Potential and Electric Potential Energy

Among similarities described by force fields are those involving the **inverse–square relationship** and associated applications of potential energy phenomena.

Although **electric potential** and **electric potential energy** are **not** the same thing, they are closely related.

Just as an object released from rest at some height above the earth's surface experiences an energy transformation from potential to kinetic, so too does a charged object released in an electric field. The gravitational field surrounding a mass exerts a gravitational force of attraction on other masses. Similarly, the electric field surrounding an electric charge exerts an electric force of attraction or repulsion on other charges, depending on whether the charges in question are positive or negative.

Because **work** is defined as **force times displacement in the same direction** and is equal to a **change in potential energy**, the concept also applies to charges moving under the influence of an electric field. The change in energy for a particular charge equals the work done on or by that charge.

Electric Potential

Electric potential V_E is analogous to altitude in a gravitational field. It is measured in Joules per Coulomb (J/C) or Volts (V), and in simple cases is the electric field strength E times the distance away from the source of the electric field, and is expressed by

$$\mathbf{V_E = Er}$$

(This may also be represented as $\mathbf{dV = -E\,dr}$ with the negative sign needed because the electric field points in the direction of decreasing voltage just as the gravitational field points in the direction of decreasing altitude), where V_E is the electric potential in J/C or V, E is the electric field strength in N/C, and **r** is the distance from the source of the electric field in m. For point charges this can be expanded to

$$\mathbf{V_E = Er} = \frac{\mathbf{kq}}{\mathbf{r}^2}(\mathbf{r}) = \frac{\mathbf{kq}}{\mathbf{r}}$$

Example

Determine the electric potential $\mathbf{V_E}$ at a point 1 Å from a proton.

Solution

Using $\mathbf{V_E}$ = kq/**r**,

$$\mathbf{V_E} = \frac{(9\times10^{9}\,\text{Nm}^2/\text{C}^2)(+1.60\times10^{-19}\,\text{C})}{10^{-10}\,\text{m}}$$

$$= \mathbf{+14.4\ J/C}$$

Example

$$\overset{+}{+} \leftarrow 1\,\text{Å} \rightarrow -$$

Determine the electric potential at a point midway between two protons and an electron that are separated by 1 Å.

Solution

Using $\mathbf{V_E}$ = kq/**r** and taking the sum for $\mathbf{V_E}$ for both charges:

$$\mathbf{V_{E\ PROTONS}} = \text{kq}/\mathbf{r} = \frac{(9\times10^{9}\,\text{Nm}^2/\text{C}^2)(2)(+1.60\times10^{-19}\,\text{C})}{(10^{-10}\,\text{m})/2} = +57.6\ \text{J/C}$$

$$\mathbf{V_{E\ ELECTRON}} = \text{kq}/\mathbf{r} = \frac{(9\times10^{9}\,\text{Nm}^2/\text{C}^2)(-1.60\times10^{-19}\,\text{C})}{(10^{-10}\,\text{m})/2} = -28.8\ \text{J/C}$$

TOTAL +28.8 J/C

Total electric potential = $\text{V}_{\text{E PROTON}} + \text{V}_{\text{E ELECTRON}}$ = **+28.8 J/C**

Potential Energy

Potential or *stored* energy (U) is measured in Newton-meters or Joules.

The **potential energy** of an electric charge is a function of the strength of the charge (q) and the **electric potential** (V_E), which is expressed by

$$\mathbf{U_E = qV_E}$$

where U_E is the potential energy in Joules, q is the charge in Coulombs, and V_E is the electric potential in Joules per Coulonb (J/C) or volts (V). For point charges this expands to

$$\mathbf{U_E = qV_E = (q_1)(kq_2/r)}$$

Example

A +20 nC charge is placed 0.25 m from a –40 nC charge. Determine the potential energy possessed by the charges.

Solution

Using $U_E = qV_E = (q_1)\left(k\frac{q_2}{r}\right)$

$$U_E = (+20\times10^{-9}\text{C})\frac{\left(9\times10^{9}\frac{\text{Nm}^2}{\text{C}^2}\right)\left(-40\times10^{-9}\text{C}\right)}{(0.25\,\text{m})} = -2.9\times10^{-5}\text{J}$$

Equipotentials

An **equipotential** is a line along which every point represents a constant amount of potential energy. For instance, a satellite moving in a circular orbit around the earth travels along a path in which all points are equidistant from the earth's center and possesses an equal amount of potential energy at any point in its orbit. The satellite's orbit describes a constant altitude or **equipotential surface or path.**

Similarly, an electric charge moving along a path of constant electric potential energy travels along an **equipotential surface or path**. In addition, since it is impossible for a mass or a charge to have more than one potential energy value at any given time, equipotential surfaces or paths can never cross or intersect.

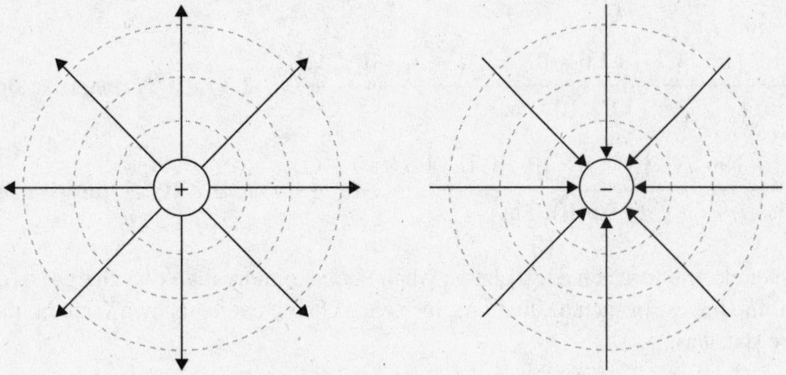

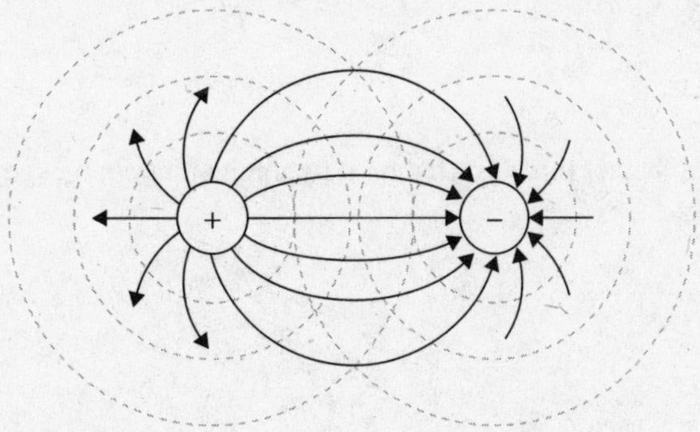

The dotted lines represent equipotentials.

Coulomb's Law

Similarities between different types of force fields involve the inverse-square relationship and relative strengths (mass, charge, sound or light intensity, magnetic pole strength) of the objects in question. Such similarities between **gravitational** and **electric** forces are as follows:

Gravitational $F_G = G\frac{m_1 m_2}{r^2}$ (where G is a constant, 6.67×10^{-11} Nm^2/kg^2, m_1 and m_2 are the masses in question, and **r** is the distance of separation.

Electric $F_E = \frac{1}{4\pi\varepsilon_0}\frac{(q_1 q_2)}{r^2} = k\frac{q_1 q_2}{r^2}$ (where k is a constant, 9×10^9 Nm^2/C^2, q_1 and q_2 are the charges in question, and **r** is the distance of separation.

This second relationship is known as **Coulomb's Law** for calculating electrostatic forces between charged particles or point charges.

Example

What is the electric force that exists between

(a) a proton and an electron separated by 1 Å (10^{-10}m)?

(b) two protons separated by 1 Å?

Solution

Using Coulomb's Law $F_E = \frac{1}{4\pi\varepsilon_0}\frac{(q_1 q_2)}{r^2} = k\frac{q_1 q_2}{r^2}$,

(a) $F_E = \frac{(9\times 10^9\,Nm^2/C^2)(+1.6\times 10^{-19}\,C)(-1.6\times 10^{-19}\,C)}{(1.0\times 10^{-10}\,m)^2} = -2.3\times 10^{-8}N$ **(negative sign signifies attraction)**

(b) $F_E = \frac{(9\times 10^9\,Nm^2/C^2)(+1.6\times 10^{-19}\,C)(+1.6\times 10^{-19}\,C)}{(1.0\times 10^{-10}\,m)^2} = +2.3\times 10^{-8}N$ **(positive sign signifies repulsion)**

When calculating the net electric force on a test charge when there are more than two charges in question, it is necessary to find the **sum** of the forces. In such a situation, since each charge exerts its own force on the test charge, Coulomb's Law can be stated as:

$$F_{E\,(NET)} = \Sigma k\frac{q_1 q_2}{r^2}$$

Example

Using q_1 as a test charge, calculate the net force exerted on it by three other charges, q_2, q_3, and q_4:

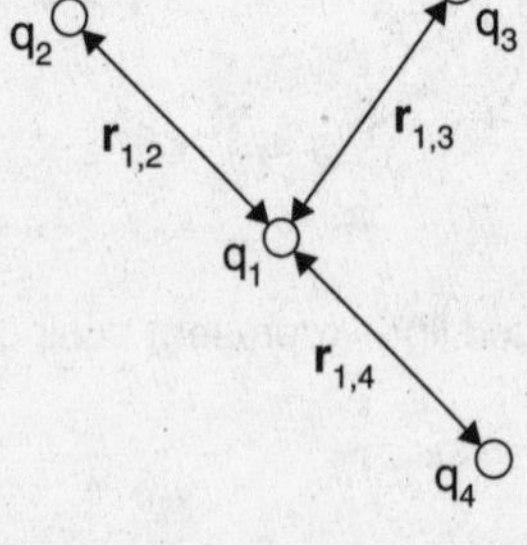

Solution

Using the vector sum application of Coulomb's Law:

$$\mathbf{F}_{E\ (NET)} = \Sigma\, k\frac{q_1 q_2}{\mathbf{r}^2}$$

$$\mathbf{F}_{E\ (NET)} = kq_1\left[\frac{q_2}{\mathbf{r}_{1-2}^{\ 2}} + \frac{q_3}{\mathbf{r}_{1-3}^{\ 2}} + \frac{q_4}{\mathbf{r}_{1-4}^{\ 2}}\right]$$

Just as in computing the electric force between any two charges, when many charges are considered, do not neglect the positive or negative signs, as they indicate the direction of the electric force vector(s):

$(+q_1)(+q_2) = +\, q_1q_2$ **REPULSION**

$(-q_1)\,(-q_2) = +\, q_1q_2$ **REPULSION**

$(-q_1)\,(+q_2) = -\, q_1q_2$ **ATTRACTION**

Geometric Charge Distribution

The following illustrations show how electric field lines curve when charges are in close proximity.

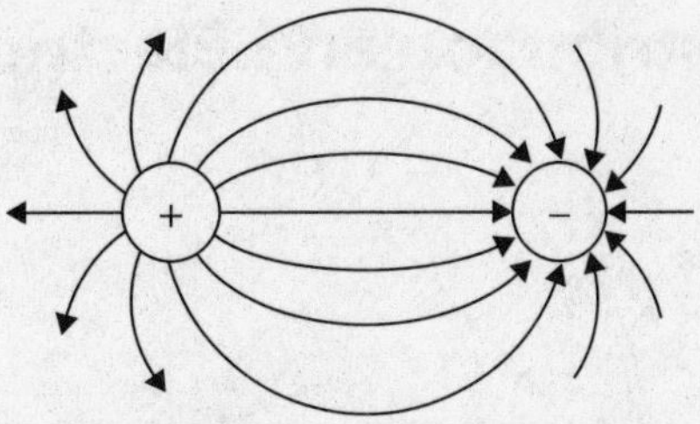

When opposite charges are in close proximity, electric field lines curve **out of** the positive and **into** the negative.

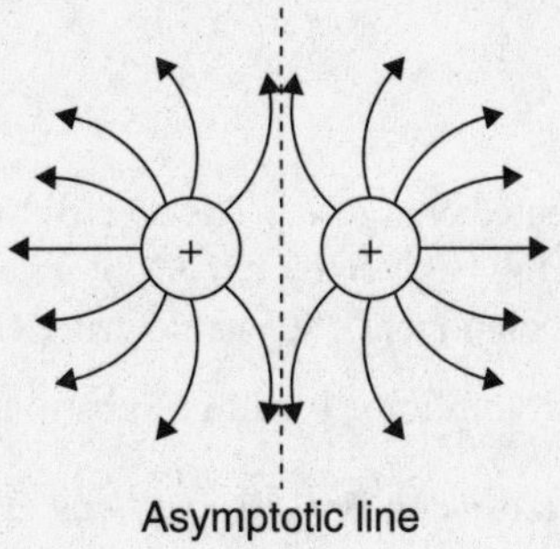

When positive charges are in close proximity, electric field lines curve **out of** of the charge and **away from** field lines from the other charge. (The dotted asymptotic line signifies that the electric field lines do not cross.)

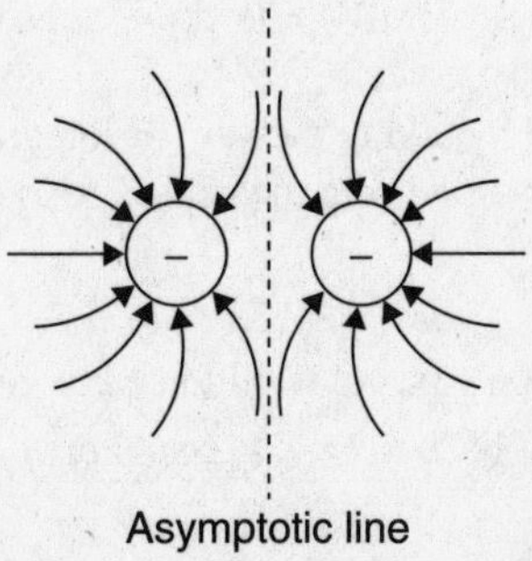

When negative charges are in close proximity, electric field lines curve **into** the charge and **away from** field lines from the other charge. (The dotted asymptotic line signifies that the electric field lines do not cross.)

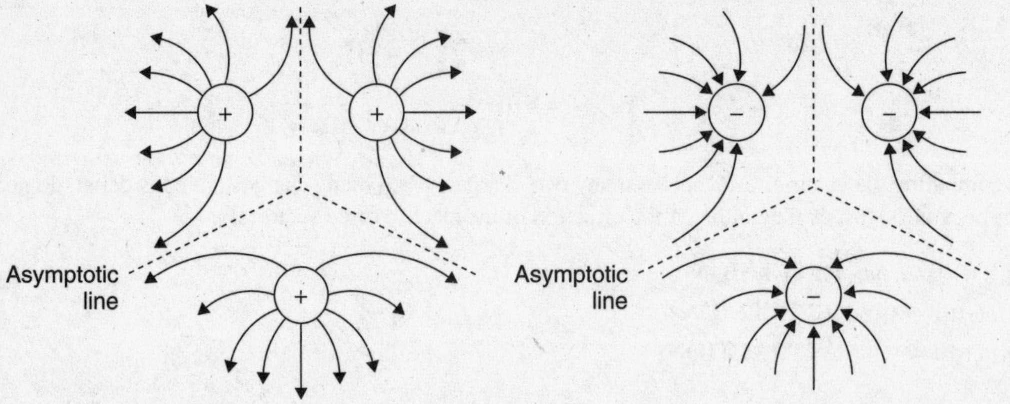

When multiple charges are in close proximity, electric field lines curve **out of** the positive charges, **into** the negative charges and **away from** field lines from the other charges. (The dotted asymptotic lines signify that the electric field lines do not cross.)

Geometric Charge Distribution and Electric Field Considerations

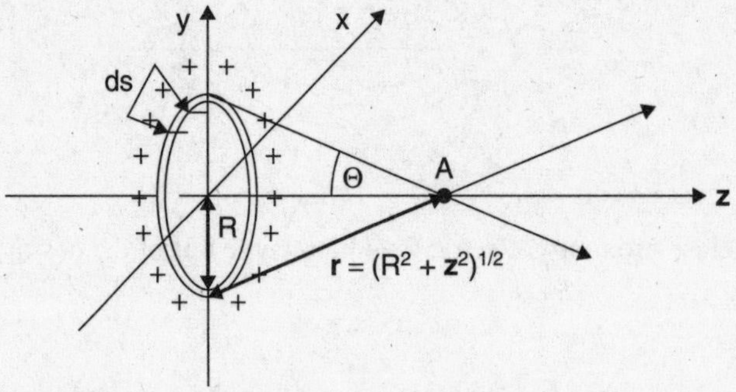

Consider a thin ring with a uniformly distributed charge +q. The electric field at point A would be determined using the fact that the distance from any point on the ring is given by $\mathbf{r} = \sqrt{R^2 + z^2}$. Since any **x** or **y** component of the electric field is equaled and negated by its **–x** or **–y** component, the net electric field is along the **z** axis and is equal to the sum of the **z** components of the electric field. For simplicity, k will be substituted for $\frac{1}{4\pi\varepsilon_0}$.

Each z component of the electric field equals r cos Θ = $\sqrt{\mathbf{R}^2 + \mathbf{z}^2}$ **cos Θ.**

This makes the total electric field at A equal to the sum of electric field components in the z direction:

$$\mathbf{E_Z} = \Sigma(\mathrm{kq}/\,\mathbf{r}^2)\cos\Theta$$

$$\text{or } \mathbf{dE_Z} = (\mathrm{k\,dq}/\,\mathbf{r}^2)\cos\Theta$$

The total electric field at point A is computed by taking the sum or integral of the **z components** (**r cos Θ or** $\sqrt{\mathbf{R}^2 + \mathbf{z}^2}$ **cos** Θ) of electric field values from the center of the ring to point P. Also, since cos Θ = **z** /**r**, this makes cos Θ = $\frac{\mathbf{z}}{\sqrt{\mathbf{R}^2 + \mathbf{z}^2}}$.

To take the sum of the electric field's **z** components, we would need to take the sum or integral of $\mathbf{dE_Z}$, which equals the sum or integral of **dE** cos Θ, which would be $\Sigma\, \mathbf{E_z} = \Sigma\, \mathrm{E}\cos\Theta$ or $\int$ dE cos Θ from 0 to 2π. The charge density λ along a length of the ring is given by $\lambda = \frac{q}{2\pi R}$.

Using $\mathbf{E} = kq/\mathbf{r}^2$ but substituting $\sqrt{R^2+z^2}$ for $\mathbf{r}$,

$$\Sigma \mathbf{E}_z = \Sigma \frac{k(q)}{(R^2+z^2)} \cos\Theta = \frac{kqz/\sqrt{R^2+z^2}}{(R^2+z^2)}$$

$$\text{which leaves } \mathbf{E}_z = \frac{kq\,z}{(R^2+z^2)^{3/2}}$$

$$\text{or using integration } \mathbf{E}_z = \frac{kq\,z/\sqrt{R^2+z^2}}{(R^2+z^2)}$$

$$= \int_0^{2\pi R} k(sq) \frac{z/\sqrt{R^2+z^2}}{R^2+z^2}\, ds$$

where **s** is a length of the ring. Removing constants and combining characters yields

$$\mathbf{E}_z = \frac{kq\,z}{(R^2+z^2)^{3/2}}$$

Now consider a line of charge, such as a wire, of length s with a charge +q uniformly distributed over its length. The charge radiates an electric field outward from the wire. At a hypothetical distance R from the wire, the electric field is cylindrical with a radius R.

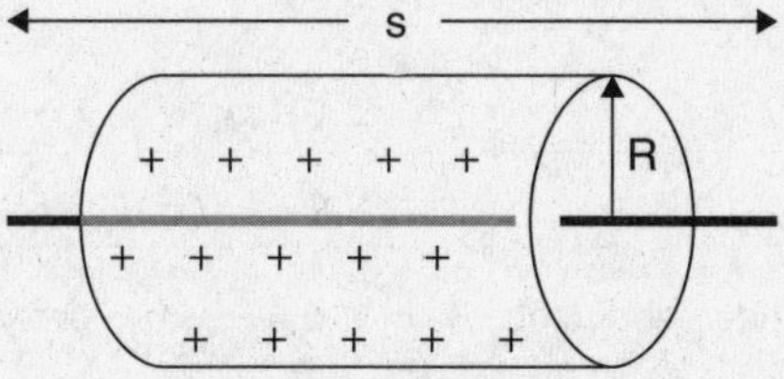

The hypothetical electric field represented by the cylinder of radius R has a surface area of $(2\pi R)(s)$ and a charge density $\lambda = q/s$. Taking the point-charge electric field equation $E = kq/\mathbf{r}^2 = (1/4\pi\varepsilon_o)(q/\mathbf{r}^2)$ and applying it to a string of point charges a distance s meters long means replacing the area of the sphere around a point charge $(4\pi\mathbf{r}^2)$ with the area of a cylinder described by the line of charge $(2\pi\mathbf{rs})$, which now yields

$$E = \frac{q}{\varepsilon_0} \frac{1}{(2\pi\mathbf{rs})}$$

and substituting the charge density relationship for q ($\lambda = q/s$) and R for $\mathbf{r}$ yields

$$E = \frac{\lambda s}{\varepsilon_0 (2\pi R s)} = \frac{\lambda}{\varepsilon_0 2\pi R}$$

The electric field radiated by a line of charge q and length s is given by

$$E = \frac{\lambda}{\varepsilon_0 2\pi R}$$

where E is the electric field in N/C, λ is the charge density in C/m, and ε_o is the permittivity or the ability of the electric field to permeate through a vacuum or free space.

In the case of two oppositely charged plates having a uniform electric field between them, the electric field is defined by the **surface charge density** σ and the **vacuum permittivity** ε_o:

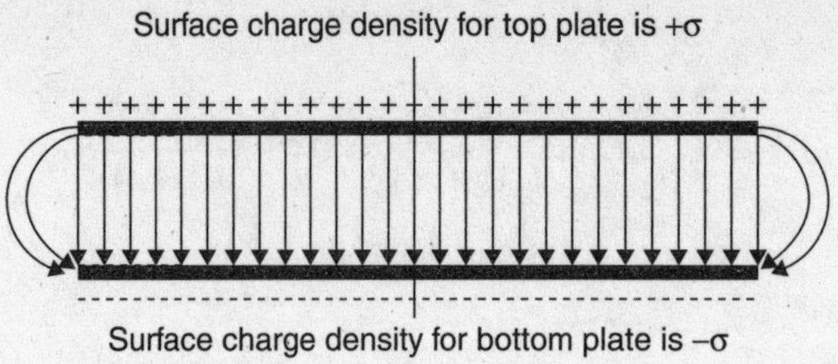

The electric field between two parallel plates, each with surface charge density σ, with air or vacuum separation, is given by

$$\mathbf{E} = \frac{\sigma}{\varepsilon_0}$$

Since capacitor plates have a uniformly distributed surface charge, the electric field anywhere between the plates is uniform.

Electric Field Summary for Common Situations

Shape or Orientation	***Electric Field***
$\left(k = \frac{1}{4\pi\varepsilon_0} = 9 \times 10^9\ \mathrm{Nm^2/C^2}\right)$	
Point charge at distance r	$E = kq/r^2$
Conductive shell of radius r and charge q	$E = kq/r^2$ (No field inside shell)
Line of charge or conducting wire or cylinder at distance R	$E = \frac{\lambda}{\varepsilon_0 2\pi R}$ (λ is charge density)
Pair of parallel, oppositely charged plates	$E = \frac{\sigma}{\varepsilon_0}$ (Assumes air or vacuum separation)

Gauss' Law (C Exam)

Gauss' Law for electrostatics is built upon the concept of a **Gaussian surface,** which is *any completely closed surface.*

Common examples include, but are not limited to, the following shapes:

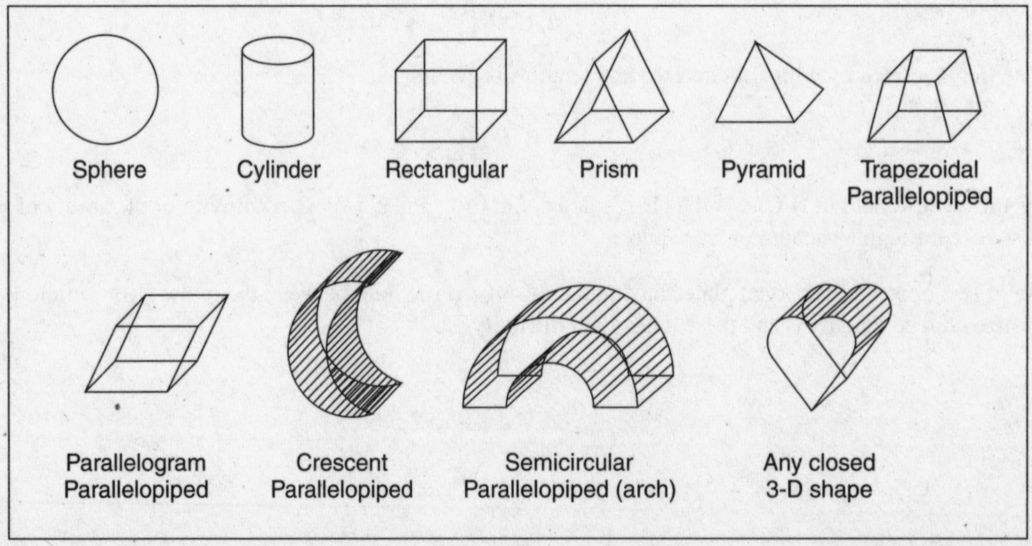

Consider a breeze flowing through an open window.

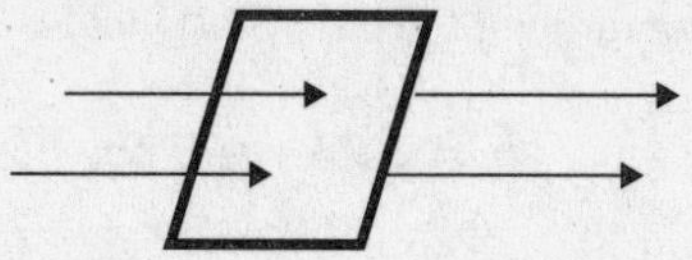

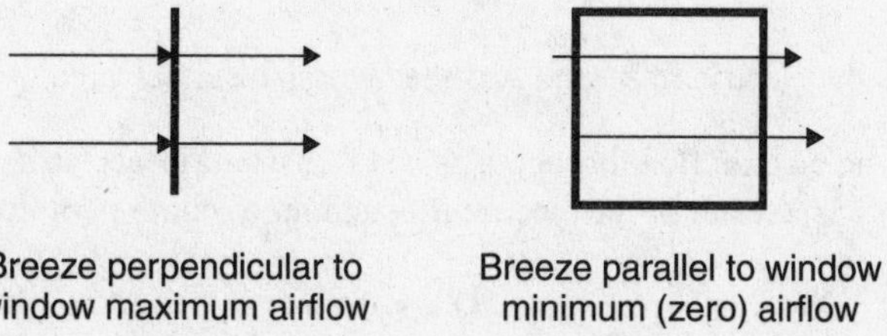

If the breeze is exactly perpendicular to the window, the maximum amount of air flows through it. If the breeze is parallel to the window, no air flows through. Also, we assume that any air entering the window exits the window. The air *flux* or *flow* through the window is a result of the amount of air entering and exiting (the speed and density of the air) and the size of the window (area). This is the basic idea behind Gauss' Law.

If the breeze enters the window at some angle other than 0° or 90°, the airflow is defined by the vector component of the flow that is perpendicular to the window.

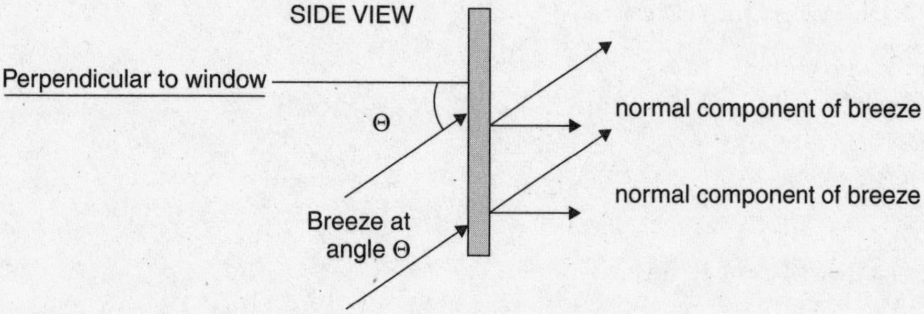

Using the part of the air flow that is perpendicular to the window, the cosine of the angle the entering air makes with the window defines the normal component of the breeze as it flows through. This is the **FLUX Φ**, or flow, and is represented as

FLUX =	**(velocity)**	**(area)**	**(cos Θ)**
Φ =	**v**	**A**	**cos Θ**

Now think of the breeze as an electric field and the window as a Gaussian surface. Whatever part of the electric field that enters the surface exits on the opposite side and *all* parts of the electric field that enter the surface also exit.

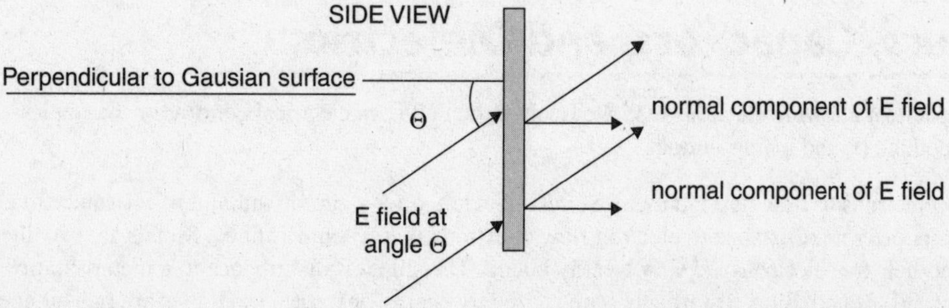

Using the same concept as for the breeze and window, the normal component of the electric field exiting the Gaussian surface equals the incoming field times the cosine of the angle. Adding up all tiny areas (integrating) to find the total normal component of the electric field passing through (electric flux) yields

$$\Phi_E = \Sigma \mathbf{EA} \cos \Theta \text{ or}$$
$$\Phi_E = \Sigma \mathbf{E} \cdot dA \text{ or}$$
$$\Phi_E = \oint \mathbf{E} \cdot dA$$

where Φ_E is the electric flux, **E** is the electric field, and A is the Gaussian surface area.

Gauss' Law is an expression of the electric flux through a closed Gaussian surface as it relates to the total charge Q enclosed by that surface. That charge is a result of the electric flux and the vacuum permittivity of free space, expressed as

$$\mathbf{Q} = \varepsilon_o \Phi$$

Combining equations expresses Gauss' Law as

$$\mathbf{Q} = \Sigma\, \varepsilon_o \mathbf{EA} \cos \Theta \text{ or}$$
$$\mathbf{Q} = \varepsilon_o \oint \mathbf{E} \cdot dA$$

Example

A conducting, spherical shell of radius r, initially uncharged, has a point charge of +q placed at its center. For any distance from its center R, determine the electric field where

(a) $R < r$
(b) $R = r$
(c) $R > r$

Solution

(a) Using Gauss' Law: $\Phi_E = \oint \mathbf{E} \cdot dA$

and $q = \varepsilon_0 \Phi_E = \varepsilon_0 \oint \mathbf{E} \cdot dA$

$+q = \varepsilon_0 \mathbf{E}A$

$$\mathbf{E} = \frac{+q}{\varepsilon_0 A} = \frac{+q}{\varepsilon_0 (4\pi R^2)} = \frac{+kq}{R^2}$$

(b) At R = r, E = 0 since there is no electric field inside a conductor.

(c) $\mathbf{E} = \frac{+q}{\varepsilon_0 A} = \frac{+kq}{\varepsilon_0 (4\pi R^2)} = \frac{+kq}{R^2}$

Placing a positive point charge at the sphere's center causes the sphere's interior surface to become negatively charged (–q) and the outer surface to attain a charge of +q.

Conductors, Capacitors, and Dielectrics

Any substance capable of allowing electrons to move freely through it is an electrical **conductor**. Examples of conductors are metals, semiconductors, and ionic solutions.

Conductors allow electrons to flow through them. Semiconductors offer a certain amount of resistance to electron flow, and superconductors offer no resistance to electron flow at extremely low temperatures. Metals are excellent conductors due to the fact that their free electrons are very weakly bound. The slightest disturbance to a metal's lattice is sufficient to push these electrons around. Such disturbances are caused by energy or forces. Examples are radiant energy, pressure, and potential difference across the metal. Heating a metal at one end causes electrons to move to the other end.

Squeezing a metal at one end does the same thing. Applying a potential difference across a metal, such as hooking it up to a battery, allows the battery's internal electrical *pressure* to push electrons through the metal.

Any conducting material allows excess charge to accumulate on its surface. There is no electric field **inside** the conductor because all the excess charge distributes itself as evenly as possible on the surface. When a person acquires a static electric charge, for example, the excess electrons build up on that person's skin. The electrons repel each other and move as far from each other as possible.

Any substance that impedes the flow of electrons is an electrical **insulator**.

Some examples of insulators are rubber, plastic, cork, cloth, glass, paper, wood, wax, paint, or shellac.

A **capacitor** is a substance, object, or component capable of storing electric charge. Some examples are rubber combs or brushes, plastic seats or seat covers, the glass surface of a television screen, metallic plates, a person, or commercially produced capacitors designed specifically for the purpose of storing charge.

The most common capacitors referred to by typical physics problems involve parallel plate capacitors. The **capacitance**, or ability of capacitors to store charge, is a function of plate area, distance of separation, and the material (if any) that separates the plates. Capacitance is expressed as

$$C = \frac{\varepsilon_0 A}{d}$$

where C is the capacitance in Farads or C/V, ε_o is the electrical permittivity of vacuum or free space (8.85×10^{-12} C^2/Nm^2), A is the plate area in m^2, and d is the distance between the plates in m.

Capacitance is also the ratio of the charge built up on a capacitor plate to the potential difference built up between those plates, expressed as

$$\mathbf{C = Q/V \text{ and } Q = VC}$$

where C is capacitance in Farads, Q is total charge on either plate in Coulombs, and V is the potential difference between the plates in volts or J/C.

In a capacitor, a potential difference builds up between the plates due to the accumulation of charge on the plates. At its maximum charge buildup, the potential difference is maximum.

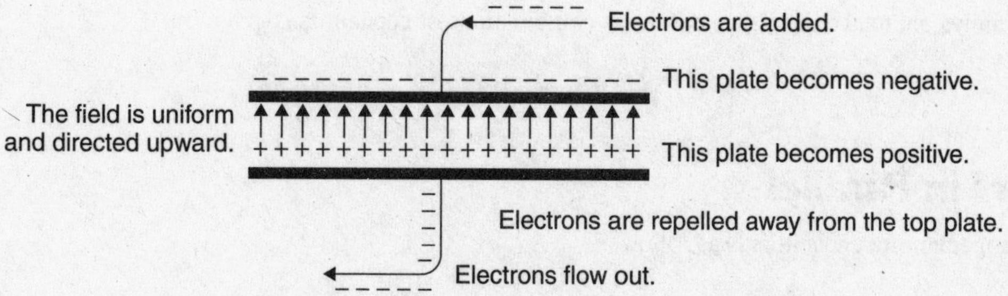

In the process of charge buildup on the top capacitor plate, the following occurs:

1. The plates begin electrically neutral, with no initial potential difference between them.
2. Electrons begin to build up on the top plate. As this happens, they repel electrons in the bottom plate. Those electrons begin to flow through the exit wire. At this point the electron flow is the maximum amount.
3. As electrons accumulate on the top plate, addition of more electrons repelling each other decreases and potential difference between the plates rises.
4. There comes a point where the addition of electrons in the top plate nears or becomes zero. This means few or no electrons are forced out the bottom wire.

   At this point, potential difference between the plates is maximum, and electron flow out the exit wire is minimum.

5. Spark discharge occurs from electron repulsion in the top plate. This neutralizes the plates, and the entire process begins again. In dry air or vacuum, spark discharge occurs at a value of about 30 kV/cm.

Example

A capacitor with a potential difference of 6.0 V has a capacitance of 3.0 pF $(1\ \text{pF} = 10^{-12}\ \text{F})$. Determine the charge on each plate.

Solution

Using **Q = VC**,

$$Q = (6.0\ \text{V})(3.0 \times 10^{-12}\ \text{F}) = \mathbf{1.8 \times 10^{-11}\ C}$$

Combining Capacitors

Capacitors may be connected in any number of combinations, but typical physics problems involve capacitors connected in **series** or **parallel** or some combination of both arrangements.z

Capacitors in Series

The series arrangement for capacitors is as follows:

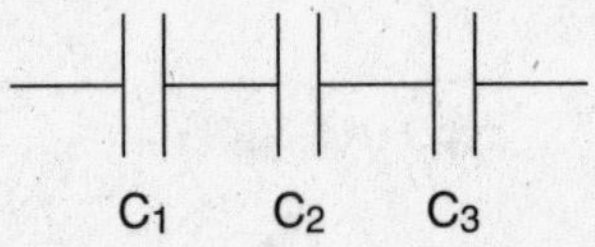

Three capacitors in series

Since charge is separated by three gaps (d in the formula for capacitance) the total capacitance is given by:

$$\frac{1}{C_T} = \frac{1}{C_1} + \frac{1}{C_2} + \frac{1}{C_3}$$

In general, the equivalent total capacitance for series combinations of capacitors is given by

$$\frac{1}{C_T} = \Sigma \frac{1}{C_n} \text{ (where n = 1, 2, 3 . . . and so on)}$$

Capacitors in Parallel

The parallel arrangement for capacitors is as follows:

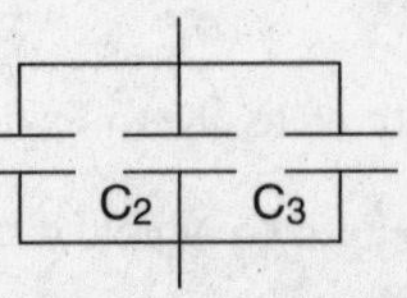

Three capacitors in parallel

Because capacitors in this arrangement are equivalent to creating more area (A in the capacitance formula) and charge accumulation occurs simultaneously, the total capacitance of this arrangement is given by

$$C_T = C_1 + C_2 + C_3$$

In general, the equivalent total capacitance for parallel combinations of capacitors is given by

$$C_T = \Sigma C_n \text{ (where n = 1, 2, 3 . . . and so on)}$$

Example

Three capacitors, 3.0 pF, 4.0 pF, and 5.0 pF are placed in a parallel arrangement.

(a) Draw the diagram.
(b) Find their total capacitance.

$$(1 \text{ pF} = 10^{-12} \text{ F})$$

Solution

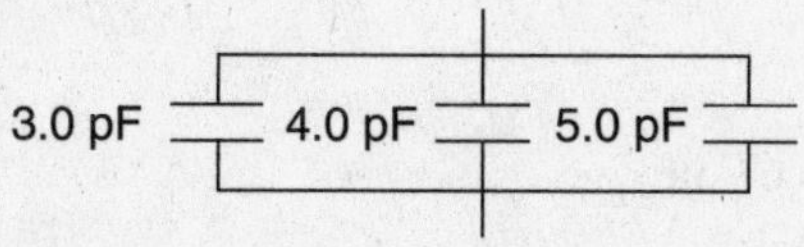

(a)

(b) Using $C_{TOT\ PARALLEL} = \Sigma C_n$,

$$C_{TOT\ PARALLEL} = 3.0 \text{ pF} + 4.0 \text{ pF} + 5.0 \text{ pF} = \mathbf{12\ pF}$$

Example

The same three capacitors from the previous example are discharged and now placed in a series arrangement and given a 6.0 V potential difference.

(a) Draw the diagram.
(b) Find the total charge Q on each plate.

Solution

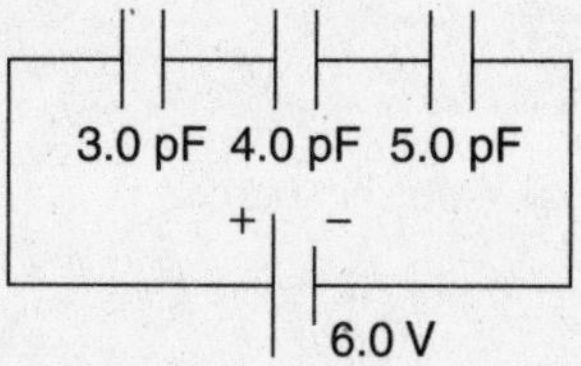

(a)

(b) First, determine the total capacitance using

$$\frac{1}{C_T} = \Sigma \frac{1}{C_n} \text{ (where n = 1, 2, 3. . . and so on)}$$

$$\frac{1}{C_T} = \frac{1}{3.0\text{ pF}} + \frac{1}{4.0\text{ pF}} + \frac{1}{5.0\text{ pF}} = \frac{47}{60.0\text{ pF}} \text{ and } C_T = 1.3 \text{ pF}$$

Now using **Q = VC,**

$$Q = (6.0 \text{ V})(1.3 \text{ pF}) = \mathbf{7.8\ pC}$$

Example

The three capacitors from the previous problem are now discharged and arranged as follows: the 3.0 pF and 4.0 pF capacitors are in series and their combination is placed in parallel with the 5.0 pF capacitor. The arrangement is reconnected to the 6.0 V potential difference.

(a) Draw the diagram.
(b) Determine the total charge Q on each plate.

Solution

(a)

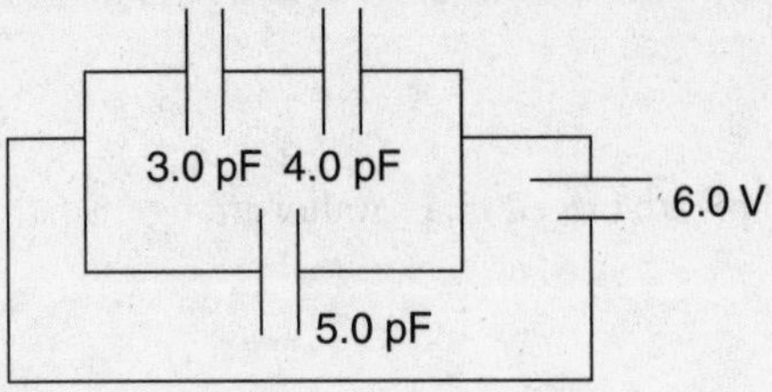

(b) First, determine the total capacitance for the 3.0 pF and 4.0 pF capacitors:

$$\frac{1}{C_{TOT}} = \frac{1}{3.0\text{ pF}} + \frac{1}{4.0\text{ pF}} = \frac{7}{12.0\text{ pF}} \text{ and } C_{TOT} = 1.7\text{ pF}$$

Then determine the total capacitance with the 5.0 pF capacitor:

$$C_{TOT} = 1.7\text{ pF} + 5.0\text{ pF} = 6.7\text{ pF}$$

Finally, using **Q = VC**,

$$Q = (6.0\text{ V})(6.7\text{ pF}) = \mathbf{40.2\text{ pC}}$$

In a capacitor, the potential energy it stores is given as the average of its potential at the time it begins to charge and the time when it is fully charged:

$$\mathbf{U_C = \left(\frac{1}{2}\right)QV}$$

where U_C is the potential or stored energy in Joules, Q is the change in Coulombs, and V is the potential difference in Volts. Combining this equation with the previous equation, Q = VC, yields

$$\mathbf{U_C = \left(\frac{1}{2}\right)V^2C}$$

Example

How many Joules of energy are stored in a 15 μC capacitor when connected to a 6.0 V power source?

Solution

Using $\mathbf{U_C = \left(\frac{1}{2}\right)QV}$,

$$U_C = \left(\frac{1}{2}\right)(15 \times 10^{-6}\text{ C})(6.0\text{ V}) = \mathbf{4.5 \times 10^{-5}\text{ J}}$$

Dielectrics

A **dielectric** is the material that separates the plates of a capacitor. In a capacitor, the dielectric material is an insulating material. The thicker or denser the dielectric material, the more difficult spark discharge becomes through that material. This results in greater charge buildup on the top plate and, therefore, greater potential difference between the plates. The ratio of capacitance of the dielectric material to that of air (or vacuum) is the **dielectric constant**, κ.

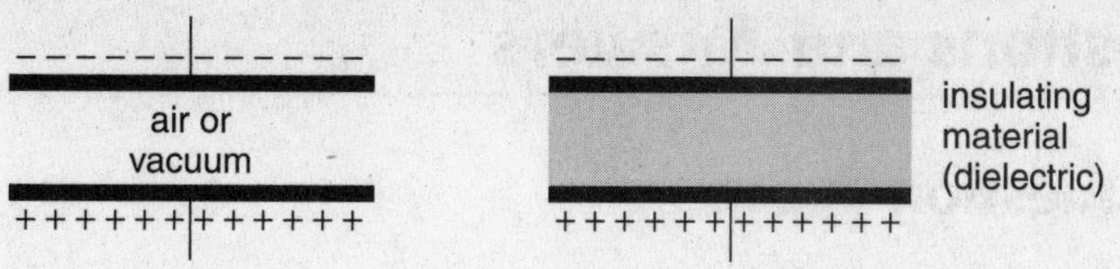

The diagram shows an air capacitor (left) and equal-sized capacitor with an insulating dielectric material separating the plates (right). Greater charge buildup occurs in the capacitor with the dielectric material by a factor κ if they are both connected to the same potential.

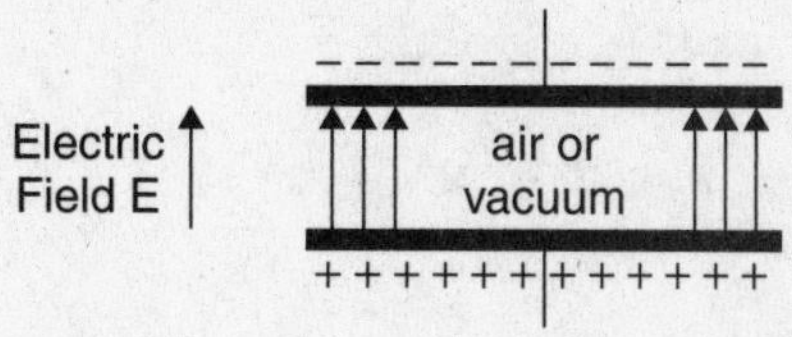

For a parallel plate capacitor with air or vacuum separating the plates, the electric field **E** is directed toward the negative plate. A negative charge released between the plates would experience a downward force and would possess potential energy that would change to kinetic.

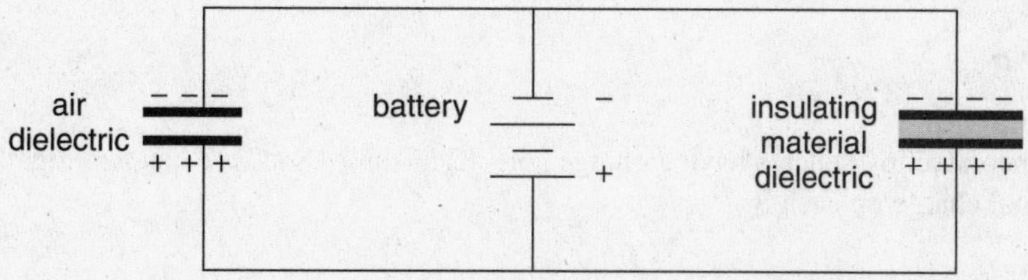

The insertion of a dielectric material other than air or vacuum in a capacitor results in a greater charge buildup. The ratio of the new charge buildup to that without the material is the dielectric constant of that material. It is a result of the density and molecular properties of the insulating material used as the dielectric.

There is also an electric field **inside** the dielectric itself. The denser the insulating dielectric material, the greater the electric field of the dielectric. However, because the direction of the electric field in the dielectric is *opposite* that of the electric field of the capacitor, the *overall* electric field is lessened:

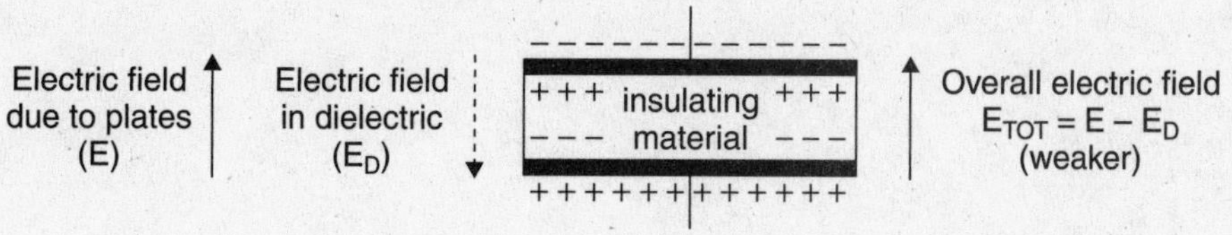

Review Questions and Answers

B & C Exam Question Types

Multiple Choice

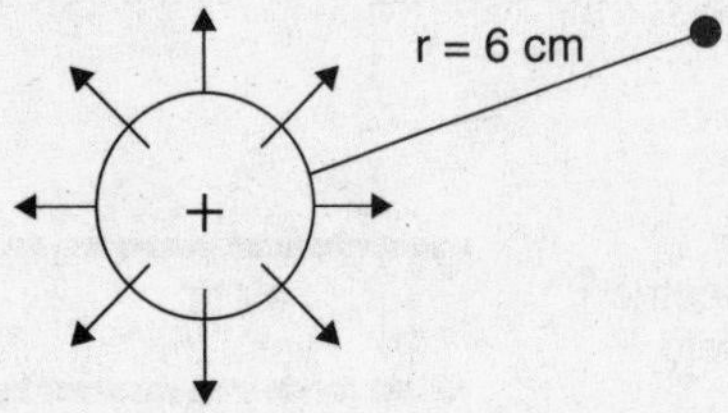

1. A certain charge radiates an electric field of $+8 \times 10^{-7}$ N/C at a distance of 6 cm. The charge is

   (A) $+12 \times 10^{-20}$ C
   (B) $+21 \times 10^{-20}$ C
   (C) $+32 \times 10^{-20}$ C
   (D) $+36 \times 10^{-20}$ C
   (E) $+48 \times 10^{-20}$ C

2. Two identical nonconducting spheres having charges of –12 nC and +8 nC are touched together and then separated. The final charge on each is

   (A) –2 nC
   (B) $-(2/3)$ nC
   (C) +2 nC
   (D) +4 nC
   (E) +48 nC

3. Tripling the distance between two electric charges changes the force between them by a factor of

   (A) $\frac{1}{9}$
   (B) $\frac{1}{6}$
   (C) $\frac{1}{3}$
   (D) 3
   (E) 9

4. The two charges in problem 2 (–12 nC and +8 nC) are placed 3.0 cm apart. The force between them is

   (A) -9.6×10^{-5} N
   (B) -9.6×10^{-4} N
   (C) $+9.6 \times 10^{-4}$ N
   (D) +9.6 N
   (E) +96 N

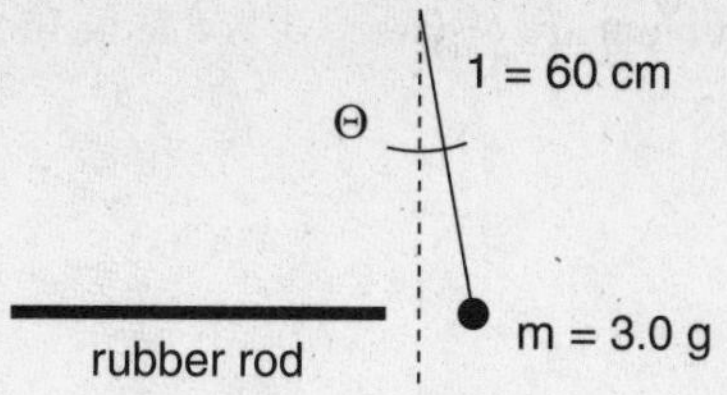

5. A 3.0 g plastic sphere hangs on the end of a 60 cm string. When a charged rubber rod is brought near it, the sphere is deflected 6.0 cm from its original position. The magnitude of the electric force on the sphere is

(A) 3.0×10^{-4} N
(B) 3.0×10^{-3} N
(C) 3.0×10^{-2} N
(D) 3.0×10^{-1} N
(E) 3.0 N

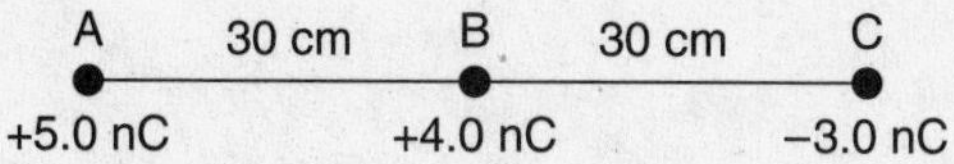

6. Three charges, +5.0 nC, +4.0 nC, and –3.0 nC, are placed at points A, B and C, respectively, in an a straight line as shown. The force exerted on the charge at point C is

(A) -2.7×10^{-9} N
(B) -9.0×10^{-9} N
(C) -16.0×10^{-7} N
(D) $+2.7 \times 10^{-7}$ N
(E) $+9.0 \times 10^{-7}$ N

7. The force exerted on the charge at point A in the previous problem is

(A) -5.0×10^{-9} N
(B) -9.0×10^{-9} N
(C) -5.0×10^{-2} N
(D) $+4.0 \times 10^{-7}$ N
(E) $+1.6 \times 10^{-6}$ N

8. Two capacitors, 0.200 μF and 0.400 μF are connected in parallel and connected to an experimental battery whose terminals have a potential difference of 90.0 V. The total charge acquired by the combination is

(A) 0.054 μC
(B) 0.54 μC
(C) 5.40 μC
(D) 54.0 μC
(E) 540 μC

9. The two capacitors in the previous problem are discharged, connected in series, and given the same potential difference. The new charge acquired is

(A) 0.135 μC
(B) 1.35 μC
(C) 11.7 μC
(D) 135 μC
(E) 1350 μC

(C Exam Type)

10. Gauss' Law is concerned with electric charge,

(A) electric potential, and gravity.
(B) charge conservation, and distance.
(C) Coulomb's Law, and the nucleus.
(D) nuclear forces, and area.
(E) electric field, and area.

Answers to Multiple-Choice Questions

1. **(C)** Using $E = kq/\mathbf{r}^2$, $q = E\,\mathbf{r}^2/k$

$$q = \frac{(+8\times10^{-7}\,\text{N/C})(.06\,\text{m})^2}{9\times10^{9}\,\text{Nm}^2/\text{C}^2} = [(+8)(36)/(9)]\times10^{-20}\,\text{C}$$

$= \mathbf{+32\times10^{-20}\,C}$

2. **(A)** –12 nC +8 nC = –4nC when touching. When separated, they share this charge equally, ending up with **–2 nC each.**

3. **(A)** Using Coulomb's Law and making a ratio of one force to the other:

$$\frac{\mathbf{F_1} = k(q_1q_2)/\mathbf{r}^2}{\mathbf{F_2} = k(q_1q_2)/(3\mathbf{r})^2}\ \text{Yields } \mathbf{F_2} = \left(\frac{1}{9}\right)\mathbf{F_1}$$

4. **(B)** Using Coulomb's Law, $\mathbf{F_1} = k(q_1q_2)/\mathbf{r}^2$

$$F = \frac{(9\times10^{9}\,\text{Nm}^2/\text{C}^2)(-12\times10^{-9}\,\text{C})(+8\times10^{-9}\,\text{C})}{(0.03\,\text{m})^2} = -96\times10^{-5}\,\text{C} = \mathbf{-9.6\times10^{-4}\,N}$$

5. **(B)** Using trigonometry and vectors:

$\tan\Theta = \mathbf{F_E/F_{WT}}$

$= \mathbf{F_E}/mg$

$= \mathbf{F_E}/(3.0\times10^{-3}\mathbf{kg})(10\,\text{m/s}^2)$

$= \mathbf{F_E}/3.0\times10^{-2}\,\text{N}$

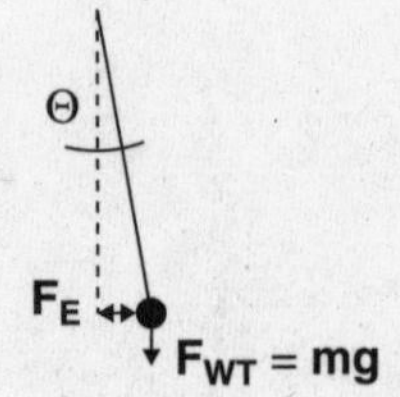

and tan Θ = 6.0 cm/60 cm

$0.1 = \mathbf{F_E} / 3.0 \times 10^{-2}$ N

and $\mathbf{F_E} = (0.1)(3.0 \times 10^{-2}\text{ N}) = \mathbf{3.0 \times 10^{-3}\ N}$

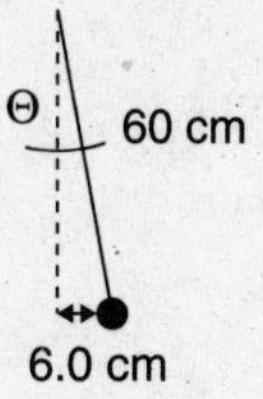

6. **(C)** The force on the charge at point C equals the sum of the forces of the other two charges on the charge at point C. Using $\mathbf{F_C} = \mathbf{F_{AC}} + \mathbf{F_{BC}}$ and Coulomb's Law, $\mathbf{F} = k\,(q_1q_2)/\mathbf{r}^2$:

$$\mathbf{F_{AC}} = \frac{(9\times10^9\,\text{Nm}^2/\text{C}^2)(+5.0\times10^{-9}\,\text{C})(-3.0\times10^{-9}\,\text{C})}{(0.6\,\text{m})^2} = \frac{(9)(5)(-3)\times10^{-9}}{36\times10^{-2}}\,\text{N} = -4.0\times10^{-7}\text{N}$$

$$\mathbf{F_{BC}} = \frac{(9\times10^9\,\text{Nm}^2/\text{C}^2)(45.0\times10^{-9}\,\text{C})(-3.0\times10^{-9}\,\text{C})}{(0.3\,\text{m})^2} = \frac{(9)(4)(-3)\times10^{-9}}{9\times10^{-2}}\,\text{N} = -12\times10^{-7}\text{N}$$

Adding both forces gives a total force at Point C of -16.0×10^{-7}N

7. **(E)** The force on the charge at point A equals the sum of the forces of the other two charges on the charge at point C. Using $\mathbf{F_A} = \mathbf{F_{AB}} + \mathbf{F_{AC}}$ and Coulomb's Law $\mathbf{F} = k\,(q_1q_2)/\mathbf{r}^2$:

$$\mathbf{F_{AB}} = \frac{(9\times10^9\,\text{Nm}^2/\text{C}^2)(+5.0\times10^{-9}\,\text{C})(+4.0\times10^{-9}\,\text{C})}{(0.3\,\text{m})^2} = \frac{(9)(5)(+4)\times10^{-9}}{9\times10^{-2}} = +20\times10^{-7}\text{N}$$

$$\mathbf{F_{AC}} = \frac{(9\times10^9\,\text{Nm}^2/\text{C}^2)(+5.0\times10^{-9}\,\text{C})(-3.0\times10^{-9}\,\text{C})}{(0.6\,\text{m})^2} = \frac{(9)(5)(-3)\times10^{-9}}{36\times10^{-2}}\,\text{N} = -3.8\times10^{-7}\text{N}$$

Adding both forces gives a total force at Point A of $+16.2 \times 10^{-7}\text{N} = \mathbf{+1.6 \times 10^{-6}\ N}$

8. **(D)** Using Q = VC, and the fact that $C_{T\,PARALLEL} = \Sigma\, C_n$

$C_{TOT} = 0.200\ \mu\text{F} + 0.400\ \mu\text{F} = 0.600\ \mu\text{F}$

$Q = (90.0\text{ V})(0.600\ \mu\text{F}) = \mathbf{54.0\ \mu C}$

9. **(C)** Using Q = VC, and the fact that $1/C_{T\,SERIES} = \Sigma\,(1 / C_n)$

$1/C_{T\,SERIES} = (1/0.200\ \mu\text{F}) + (1/0.400\ \mu\text{F}) = (3/0.400\ \mu\text{C})$

$C_T = 0.13\ \mu\text{F}$

$Q = VC = (90.0\text{ V})(0.13\ \mu\text{F}) = \mathbf{11.7\ \mu F}$

10. **(E)** Referring directly to Gauss' Law:

$q = \Sigma\varepsilon_o EA \cos\Theta$ or

$q = \varepsilon_o \oint \mathbf{E} \cdot dA$

Free Response

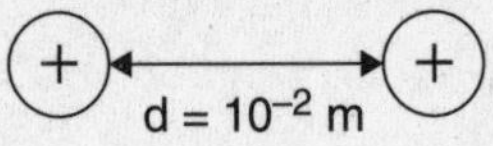

1. Two protons are separated by a distance of 10^{-2} m.

(a) Find the gravitational force $\mathbf{F_G}$ between them.
(b) Find the electric force $\mathbf{F_E}$ between them.
(c) Which is greater and by how much?
(d) What is the resultant force?
(e) Draw the electric field diagram.

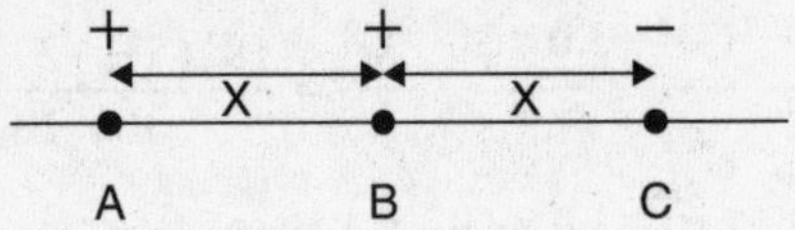

2. Charge A ($+7.0 \times 10^{-19}$ C) is separated a distance **x** from charge B ($+7.0 \times 10^{-19}$ C). Charge B is separated a distance **x** from charge C (-7.0×10^{-19} C). The charges are colinear.

(a) If $\mathbf{x} = 10^{-3}$ m, what is the total electric force on charge C?
(b) Draw the electric field diagram.
(c) Charge B is removed. Compute the total electric potential at point B.

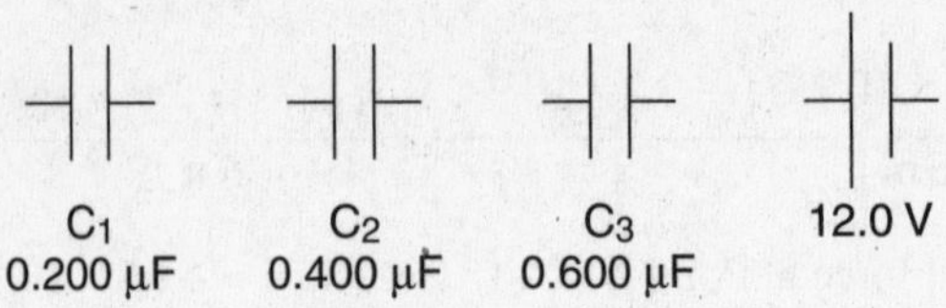

3. Given the three capacitors, C_1, C_2, C_3, and the 12.0 V power source:

(a) If capacitors C_1 and C_2 are placed in parallel with each other and their combination is connected in series with C_3 and connected to the 12.0 V source, draw the diagram.
(b) Determine the total charge acquired by the capacitors in part **(a)**.
(c) The capacitors are discharged and reconnected with C_1, now in series with the parallel combination of C_2 and C_3. The entire arrangement is now reconnected with the 12.0 V power source. Draw the diagram.
(d) Determine the total charge acquired by the capacitors in part **(c)**.

Answers to Free-Response Questions

1. **(a)** Using $\mathbf{F_G} = G\, m_1 m_2 / \mathbf{r}^2$,

$\mathbf{F_G} = (6.67 \times 10^{-11}\ Nm^2/kg^2)(1.67 \times 10^{-27}\ kg)(1.67 \times 10^{-27}\ kg)/(10^{-2}\ m)^2$

$= \mathbf{1.86 \times 10^{-60}\ N}$

(b) Using $\mathbf{F_E} = k\, q_1 q_2 / \mathbf{r}^2$,

$\mathbf{F_E} = (9 \times 10^9 Nm^2/C^2)(1.60 \times 10^{-19}\ C)(1.60 \times 10^{-19}\ C)/(10^{-2}\ m)^2$

$= \mathbf{2.30 \times 10^{-24}\ N}$

(c) The electric force is greater by a ratio of

$$\frac{\mathbf{F_E}}{\mathbf{F_G}} = \frac{2.30 \times 10^{-24}\ N}{1.86 \times 10^{-60}\ N} = \mathbf{1.24 \times 10^{36}}$$

(d) $\mathbf{F_{RES}} = \mathbf{F_E} - \mathbf{F_G} = 2.30 \times 10^{-24}\ N - 1.86 \times 10^{-60}\ N = 2.3 \times 10^{-24} N$ in the direction of the electric force, **away from each other.**

(e)

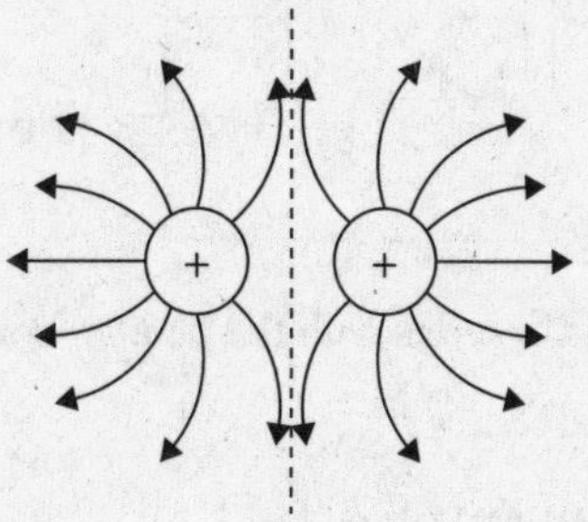

2. **(a)** Using Coulomb's Law, $\mathbf{F_E} = k\, q_1 q_2 / \mathbf{r}^2$, the total electric force on charge C is the sum of the forces on C due to charges A and B.

$\mathbf{F_C} = \mathbf{F_{AC}} + \mathbf{F_{BC}}$

$$= k\frac{q_A q_C}{(2\mathbf{x})^2} + k\frac{q_B q_C}{(\mathbf{x})^2} = \frac{k}{\mathbf{x}^2}\left(\frac{q_A q_C}{4} + q_B q_C\right)$$

$$= \frac{\left(9 \times 10^9\ {}^{Nm^2}/_{C^2}\right)}{\left(10^{-3} m\right)^2}\frac{\left[\left(+7.0 \times 10^{-19} C\right)\left(-7.0 \times 10^{-19} C\right)\right]}{4} + \left(+7.0 \times 10^{-19} C\right)\left(-7.0 \times 10^{-19} C\right)$$

$= (9 \times 10^{15}\ N/C^2)(-61.3 \times 10^{-38}\ C^2) = \mathbf{-5.52 \times 10^{-21}\ N}$ (Attracting)

(b)

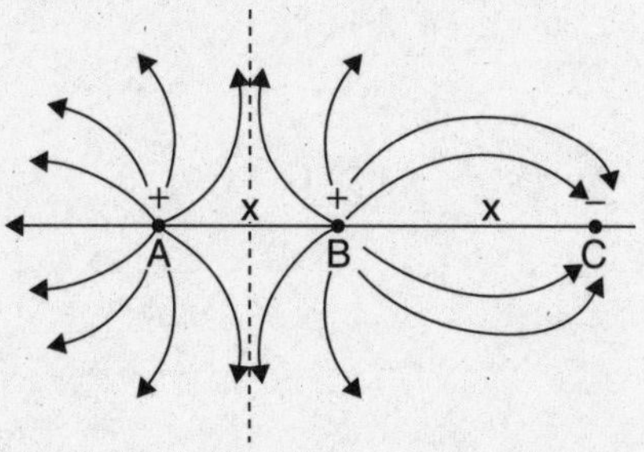

(c) Using $V_{TOT} = \Sigma\, kq/\mathbf{r} = k(\Sigma q/\mathbf{r})$

$$= (9 \times 10^9\ Nm^2/C^2)\frac{\left(-7.0 \times 10^{-19} C + 7.0 \times 10^{-19} C\right)}{\left(10^{-3} m\right)} = \mathbf{0}$$

3. (a)

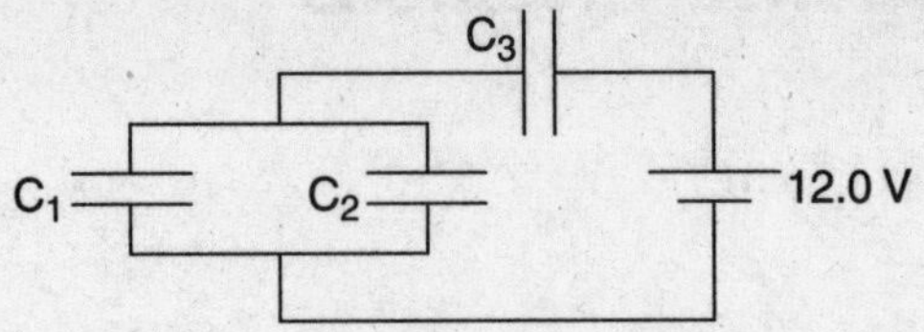

(b) Using Q = VC and the fact that C_3 is in series with the parallel combination of C_1 and C_2:

$C_{1,2} = 0.200\ \mu F + 0.400\ \mu F = 0.600\ \mu F$

$1/C_{1,2,3} = (1/C_{1,2}) + (1/C_3) = (1/0.600\ \mu F) + (1/0.600\ \mu F)$

$C_{1,2,3} = 0.300\ \mu F$

$Q = VC = (12.0\ V)(0.300\ \mu F) =$ **3.6 μC**

(c)

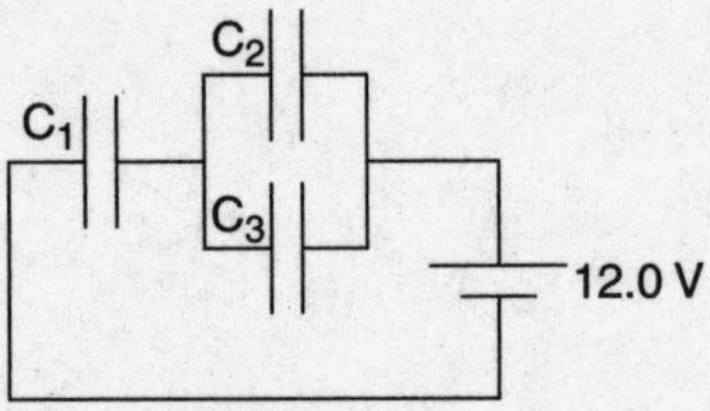

(d) Using Q = VC and the fact that C_1 is in series with the parallel combination of C_2 and C_3:

$C_{2,3} = 0.400\ \mu F + 0.600\ \mu F = 1.00\ \mu F$

$1/C_{1,2,3} = (1/C_{2,3}) + (1/C_1) = (1/1.00\ \mu F) + (1/0.200\ \mu F)$

$C_{1,2,3} = 0.167\ \mu F$

$Q = VC = (12.0\ V)(0.167\ \mu F) =$ **2.0 μC**

DC Circuits

Electricity or electric current is usually a **flow of electrons**. Electron flow occurs in a closed path, or **electric circuit.** Two types of electron flow result in two types of electric current: **Direct Current (DC)** and **Alternating Current (AC)**. In DC circuits, the electron flow is *one-way* only. In AC circuits, the current flow constantly reverses direction.

Historically, Thomas Edison was the chief proponent of direct current, and Nikola Tesla was the chief advocate for alternating current. Although a great deal of resistance developed between them, the competitive and ingenious natures of both men led them to develop their particular current to the point where we now use both forms in our everyday life. **Direct current** is most useful in small, portable appliances like flashlights; radios; portable CD, cassette; or digital music players; calculators; cell phones; and other battery-operated devices. **Alternating current** is universally used for transmitting electrical energy over long distances.

The concept is a simple one: Small appliances have relatively small amounts of wire and electrically conductive elements, which offer very little resistance to electron flow. Batteries offer relatively small, portable sources of electrical energy and yet have enough internal pressure or potential difference to maintain a fairly constant electron flow, in one direction, through the small **DC** circuits, hence the name **direct current**.

Because electrons must travel through the metallic lattice of a conductive wire, the longer the distance they travel, the more collisions they have, changing their kinetic energy into radiant energy. The wire heats up and there is energy loss.

This loss can be considerable in long or very thin wires. **AC** quickly changes the direction of electron motion back and forth over extremely small molecular distances. Tesla used this physical difference to develop dynamos capable of transmitting AC power over long distances. AC can be transformed to very high voltage, thus lowering resistance and decreasing the need for significant current to transmit power. It is this reduction in resistance that allows for efficiency. A step down transformer at the other end can return the voltage to reasonably safe levels.

This chapter deals mainly with **direct current** circuitry and associated concepts, which include current flowing from the positive terminal to the negative terminal of a power source.

Current

A useful analogy exists between electric current and flowing water. For example, when water flows through a garden hose, the rate of flow depends on the pressure forcing the water through, the diameter or cross-sectional area of the hose and the resistance inside the hose, which becomes greater with increasing length. By the same token, when electrons flow through a wire, the amount of electron flow or **current** is determined by the pressure forcing the electrons through, the diameter or cross-sectional area of the wire and the resistance inside the wire, which also becomes greater with increasing length. **Electric current** is mathematically described as

$$\mathbf{I} = \frac{\Delta Q}{\Delta t} = \frac{dQ}{dt}$$

where I is electrical current in Coulombs per second (C/s) or Amperes (A), Q is the total amount of charge passing through in Coulombs (C), and t is time in seconds. Recalling the basic unit charge, **$q = 1.60 \times 10^{-19}$ C**, the inverse gives $1\ C = 6.25 \times 10^{18}$ electrons or basic unit charges. One Ampere = 1 C/s or 6.25×10^{18} charges per second. This is the charge flow rate, which, as in a garden hose, is regulated by the pressure behind it (the **voltage** or **potential difference**), the cross-sectional **area** of the wire (the thicker the wire, the less resistance to flow), the **internal resistance** of the wire (a function of the type of metal used in the wire), and the wire's length. As a result, for a given pressure, flow area, and path resistance, just like in a garden hose, there is a certain average speed or velocity that the water flows through the hose with. For electron flow through a wire, it is called the **drift velocity**.

Example

An electronic scanner uses a 900 mA power supply. How many Coulombs of charge does this represent in 1 minute of use? How many electrons does this represent?

Solution

Since a milliAmp is 0.001 A, the current is 0.9 A or 0.9 C/s.

For one minute of operation, the scanner uses (0.9 C/s)(60 sec) or **54.0 C.**

Since $1\ C = 6.25 \times 10^{18}$ electrons, for one minute of operation the scanner uses (54.0 C)(6.25×10^{18} electrons/C) = **3.38×10^{20} electrons.**

Potential Difference and DC Circuits

Recall that energy is expended or work is done in moving a unit charge through a particular displacement in a potential difference. One volt is the potential difference present in a conductor when one Joule of work is required to move one Coulomb of charge. This becomes the basis for electrical studies involving transfer of charge in conductors that have a potential difference from one end to the other.

Resistance

Resistance to water flowing through a hose is determined by the cross-sectional area, the length, and the roughness of the inside surface of the hose. The analogy to electron flow in a wire follows the same basic concept. Electrons flow through a wire at a particular rate. If the wire becomes thinner, longer, or is replaced by a material less conducive to electron flow, this results in a retardant effect called **resistance**. If the material through which electrons flow does not allow them to move as freely as through another material, then that first material has more electrical resistance. This concept is analogous to friction acting to oppose the motion of an object. Also, if the cross-sectional area of a wire decreases and holds electrons back, the flow rate through the constriction increases, as in a garden hose. If the pressure remains the same, the electrical friction, or **resistance** increases and impedes the flow of electrons. This is represented mathematically as

$$R = \frac{\rho l}{A}$$

where R is the wire's resistance in Ohms (Ω), ρ is the **resistivity** of the particular material that the wire is made of (in units of Ω-cm or Ω-m), l is the length of the wire in meters, and A is the cross-sectional area of the wire in mm^2, cm^2 or m^2. The unit of resistance is the **Ohm (Ω)**, named after German physicist Georg Ohm.

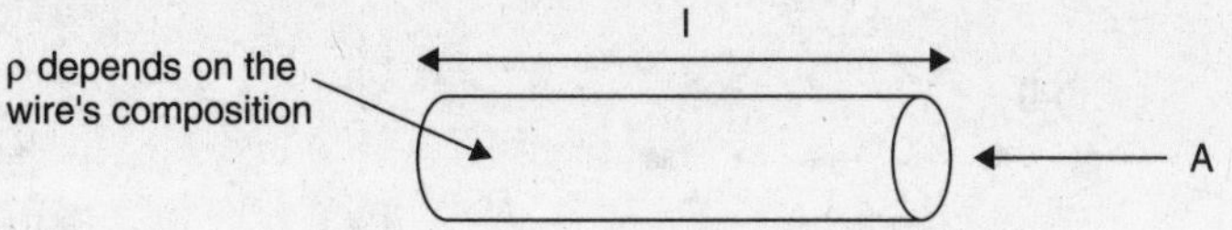

Example

A 10 m-long wire with a diameter of 0.350 mm registers a resistance of 1.8 Ω. Determine the resistivity of the wire's component metal.

Solution

Use $R = \frac{\rho l}{A}, \rho = \frac{AR}{l}$:

First, express the wire's radius as $(0.350/2) \times 10^{-3}$ m and compute the wire's cross-sectional area:

$$A = \pi r^2 = (3.1416)(0.175 \times 10^{-3}\text{ m})^2 = 9.6 \times 10^{-8}\text{ m}^2$$

Using the area in the equation yields

$$\rho = \frac{(9.6 \times 10^{-8}\text{ m}^2)(1.8\,\Omega)}{(10\text{ m})} = 17.3 \times 10^{-9}\ \Omega\text{m or } 1.73 \times 10^{-6}\ \Omega\text{cm}$$

Resistance, Resistivity, and Superconductivity

Since resistance to electrical current changes some of the electrons' **kinetic** energy into **radiant** energy, the wire heats up. If the resistivity of the wire is high, it may be used for a heater or in a toaster. A wire's resistance to electric current **increases** with increased temperature. The opposite is also true, and well-established studies into the effects of extremely low temperatures on materials have led to the fields of **cryogenics** (the study of materials' behavior at very low temperatures) and **superconductivity** (the study of electrical conductors at very low temperatures). The resistivity of metals at extremely low temperatures drops to zero. Each metal has its own such transition temperature.

DC Circuit Construction

The symbols most commonly used in discussions involving DC circuits are as follows:

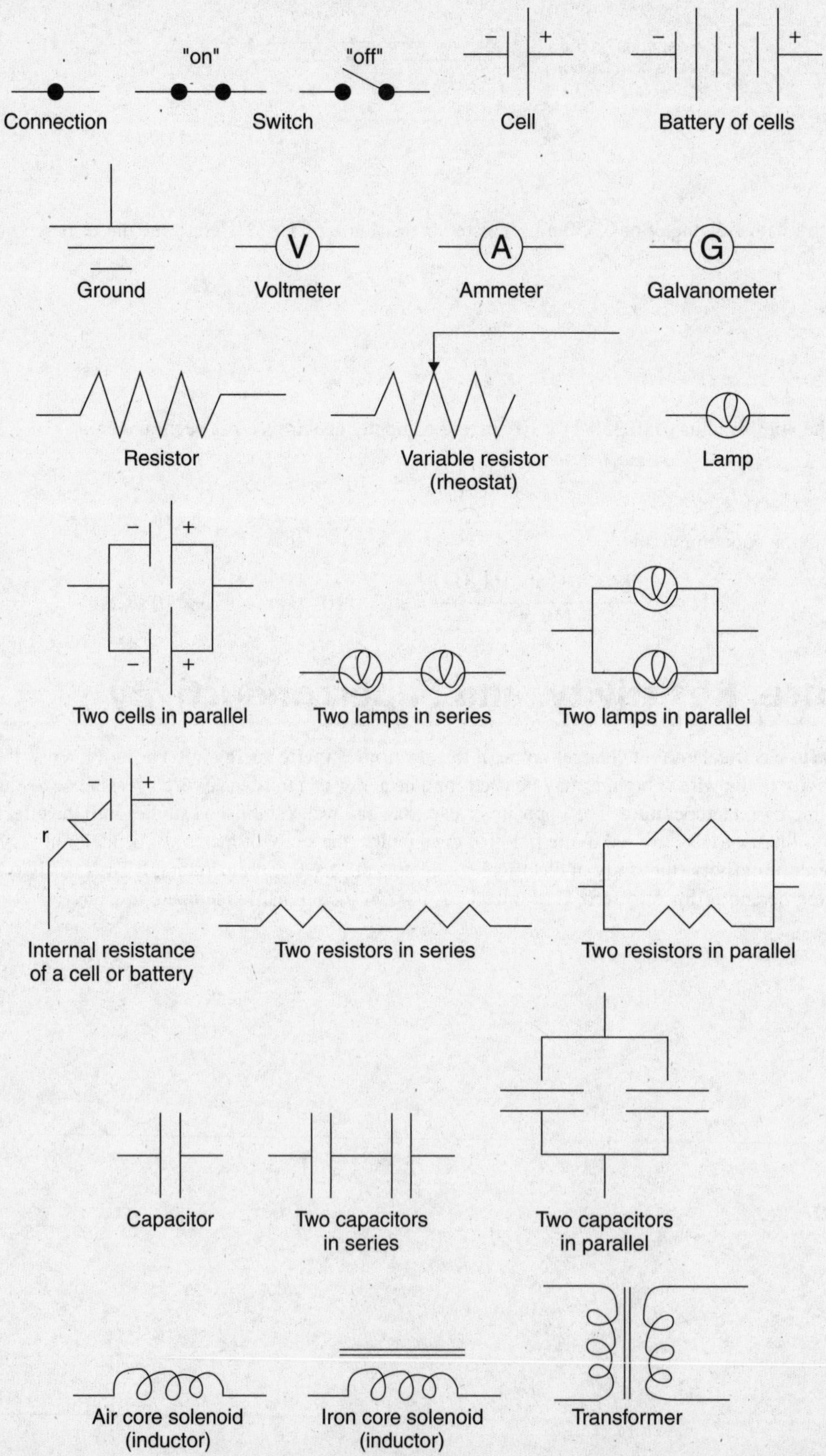

Example

The following illustration shows a simple DC circuit diagram involving a power source, a switch, a lamp, and a resistor, all in series. Since the switch is in the off position, it is an **open** circuit.

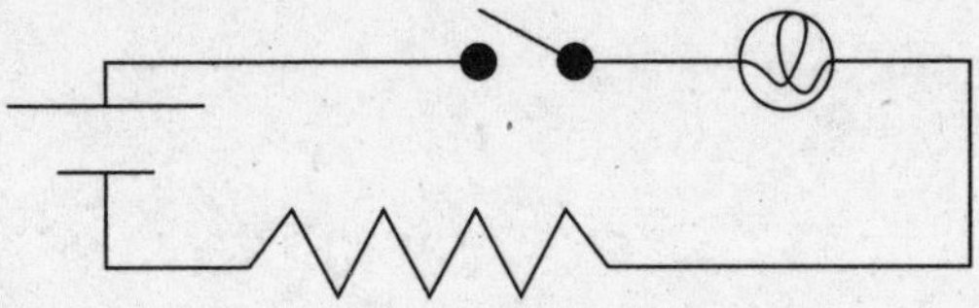

Resistors and Cells in Series and Parallel

When resistors are placed in series, the total resistance is equal to **their sum**:

$$R_{SER} = R_1 + R_2 + \ldots R_n \text{ or}$$
$$R_{SER} = \Sigma R_n$$

When resistors are placed in parallel, the total resistance is calculated as follows:

$$\frac{1}{R_{PAR}} = \frac{1}{R_1} + \frac{1}{R_2} + \ldots \frac{1}{R_n} \text{ or}$$
$$\frac{1}{R_{PAR}} = \Sigma \frac{1}{R_n}$$

For cells placed in series, their total potential difference is equal to **their sum:**

$$V_{SER} = V_1 + V_2 + \ldots V_n \text{ or}$$
$$V_{SER} = \Sigma V_n$$

For two cells of equal potential placed in parallel, the resulting potential is the potential of either cell.

$$V_{PAR} = V_1 = V_2$$

Example

Two resistors, R_1 and R_2, are 2.0 Ω and 4.0 Ω, respectively. Find their equivalent resistances when placed in (a) series and (b) parallel.

Solution

(a) Using $\mathbf{R_{SER} = \Sigma R_n}$,

$$R_{SER} = 2.0\ \Omega + 4.0\ \Omega = \mathbf{6.0\ \Omega}$$

(b) Using $\frac{1}{R_{PAR}} = \Sigma \frac{1}{R_n}$,

$$\frac{1}{R_{PAR}} = \frac{1}{2.0\Omega} + \frac{1}{4.0\Omega}, \text{ which yields } \mathbf{R_{PAR} = 1.3\ \Omega}$$

Example

Three 1.5 V cells are placed in a circuit. Find their equivalent potential difference when placed in (a) series and (b) parallel.

Solution

(a) Using $V_{SER} = \Sigma V_n$,

$$V_{SER} = 1.5\text{ V} + 1.5\text{ V} + 1.5\text{ V} = \mathbf{4.5\text{ V}}$$

(b) Using $V_{PAR} = V_1 = V_2$,

$$V_{PAR} = \mathbf{1.5\text{ V}}$$

Ohm's Law

Georg Ohm discovered that the ratio of the potential difference or electromotive force (emf, denoted by the symbol $\mathcal{E}$) to the current in a closed-loop circuit is a constant, which represents the circuit's resistance. Ohm's Law is a very important relationship in the study of electrical energy:

$$V = IR$$

where V is the potential difference in Volts, I is the current in Amperes, and R is the resistance in Ohms. One Ohm equals the resistance that will allow a current of 1 Ampere to pass between two points that have a potential difference of 1 Volt between them.

Example

If an electronic device uses 3.0 V as a power source and has a total resistance of 24 Ω, what is the current in the circuit?

Solution

Using **V = IR**, $I = V/R = 3.0\text{ V}/24\ \Omega = \mathbf{0.13\text{ A}}$

In the actual construction and operation of electronic circuitry, Ohm's Law is of great value in determining information about every part of the circuit, whether it is a component in a parallel branch or part of the external series circuit.

One important concept regards emf, electromotive force ($\mathcal{E}$). When a battery is used as a power source, the voltage value printed on the outside is the emf of the battery when it is *not connected* to a circuit. As soon as it becomes part of a circuit and the switch is turned on, its *potential energy* is less than its unconnected emf. This is due to the fact that there is some *internal resistance* in the battery. This small internal resistance is due to the energy lost by electrons in the process of leaving the battery itself and must be accounted for when calculating values for the circuit. The internal resistance of a battery or power source is often denoted by **r** near the cell symbol.

Example

A 12 V battery has an internal resistance of 2.5 Ω. If a galvanometer is used to measure the terminal current of the battery in an open (disconnected) circuit, what magnitude will it register?

Solution

Using **V = IR**, $I = V/R = 12\text{ V}/2.5\ \Omega = \mathbf{4.8\text{ A}}$

Ohm's Law and Voltmeter Ammeter Measurements

Potential difference between any two points in a circuit can be directly measured by using a voltmeter **in parallel** with the component or part of the circuit in question. The voltmeter reads the potential difference (potential drop) between the two sides of the component in question (a measure of how much voltage the component consumes). If a voltmeter is placed directly in the circuit, with no device or load between its leads, it will register zero.

Each component in a DC circuit consumes a certain number of volts.

The sum total of all the volts consumed (potential drops) adds up to the initial potential difference or emf of the source.

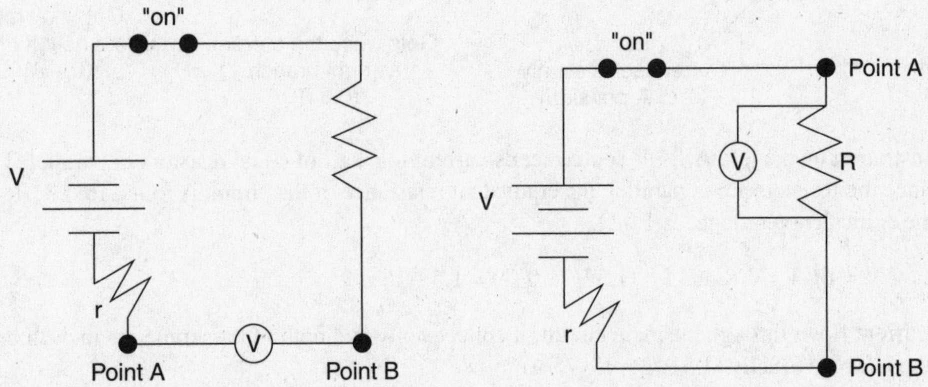

The voltmeter in the illustration on the left registers zero because there is no potential difference between Point A and Point B. The voltmeter in the illustration on the right registers some value because resistor **R** consumes some of the voltage, hence there is more electrical energy before electrons enter and less energy when electrons exit the resistor. The amount of energy consumed turns into heat.

In DC circuitry, if the components use up less than the starting emf of the power source, it can burn out one or more of the circuit's components. However, if the components require more voltage than is provided, the circuit simply won't work.

Current can be measured anywhere in a circuit by placing an ammeter (or a galvanometer for very small amounts of current) directly into (**in series with**) the part of the circuit in question, as seen in the next two illustrations.

Current Splits in a Parallel Circuit

Following the analogy with water, when electric current traveling in a DC circuit enters a junction, the **current splits** and flows through those paths in an inverse amount relative to their resistances. If the resistances are equal in a parallel circuit, an equal amount of current flows through each branch. If there are two equal resistances in parallel, half of the original current flows through each branch. For three equal resistances in parallel, the current splits evenly and one-third of the original current flows through each branch. **If the parallel resistances are unequal, the greater the resistance, the less the current flowing through.** For two parallel resistors with one resistance exactly twice that of the other, the main line current splits into two unequal parts, with the double resistance receiving half the current flowing through the single resistance.

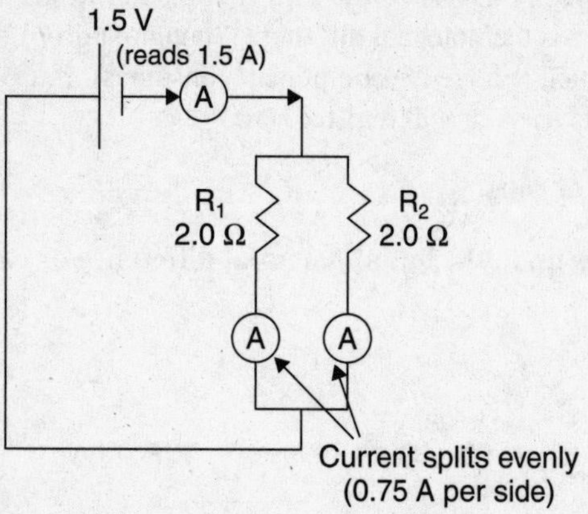

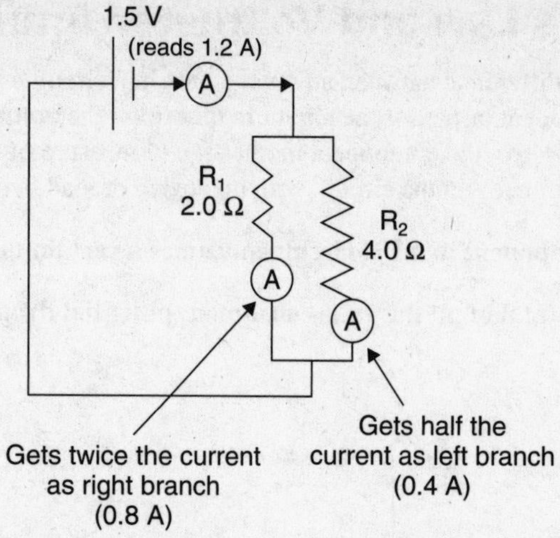

Examine the illustration on the left. A 1.5 V source sends current to a pair of equal resistors in parallel. The current splits evenly. Since the resistors are in parallel, the equivalent resistance of the circuit is found first. $(1/R_P) = (1/2.0\ \Omega) + (1/2.0\ \Omega)$ and the equivalent resistance is 1.0 Ω.

Using Ohm's Law, $V = IR$, $I = V/R$ and $I = (1.5\ V / 1.0\ \Omega) = 1.5$ A.

Since 1.5 A of current flows through the main circuit, it splits evenly and both of the ammeters in each branch in the illustration on the left read exactly (1.5 A/2) or 0.75 A.

In the illustration on the right, since the parallel resistors are unequal, the current splits unequally. The equivalent resistance of the pair is found first.

$(1/R_P) = (1/2.0\ \Omega) + (1/4.0\ \Omega)$ and the equivalent resistance R_P is 1.3 Ω.

Using Ohm's Law, $V = IR$, $I = V/R$ and $I = (1.5\ V / 1.3\ \Omega) = 1.2$ A.

Since 1.2 A of current flows through the main circuit, it splits unevenly and twice as much current flows through the 2.0 Ω side as flows through the 4.0 Ω side.

In other words, two-thirds of 1.2 A (or 0.8 A) flows through the left and one-third of 1.2 A (or 0.4 A) flows through the right side.

Note that the **ammeters** were placed directly into the circuit to measure the current for that particular circuit part or branch and **in series** with that particular part of the circuit, whereas the **voltmeters** in the previous illustrations were placed **in parallel**. The ammeter measures the flow of electrons through the point in question. The voltmeter measures the potential energy drop over that part of the circuit.

DC Circuit Problem

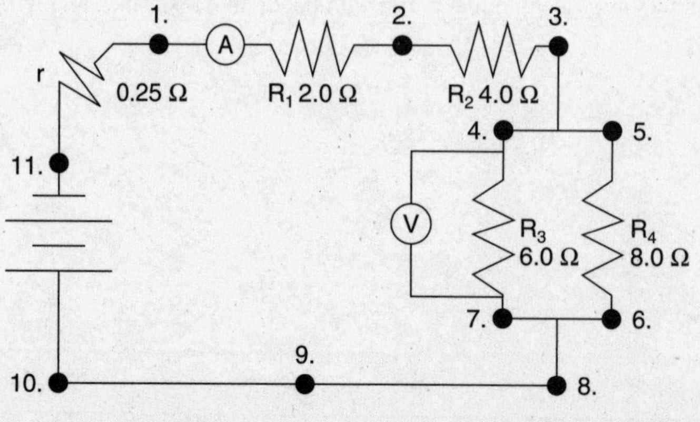

Example

In the preceding illustration, two 1.5 V cells are placed in series. The cells have a combined internal resistance **r** of 0.25 Ω. They are now placed in the circuit in series with resistor R_1 (2.0 Ω), resistor R_2 (4.0 Ω), and in series with the parallel combination of resistor R_3 (6.0 Ω), and resistor R_4 (8.0 Ω). An ammeter (A) is placed in the outside circuit. A voltmeter is placed in parallel with resistor R_3 (6.0 Ω).

(a) What is the total resistance between point 1 and point 3?

(b) What is the total resistance between point 3 and point 8?

(c) What is the total resistance between point 1 and point 10?

(d) What current does the ammeter read?

(e) What is the current at point 10?

(f) What current flows through resistor R_3?

(g) What is the potential drop across resistor R_3 (what does the voltmeter read. . . how many volts are consumed by resistor R_3)?

(h) What current flows through resistor R_4?

(i) What is the potential drop across resistor R_4 (how many volts are consumed)?

(j) What is the potential drop across the internal resistance in the battery?

(k) What is the potential drop across resistor R_1?

(l) What is the potential drop across resistor R_2?

(m) What is the sum of all potential drops in the circuit?

Solution

(a) Using $\mathbf{R_S = \Sigma R_n}$, $R_S = 2.0\ \Omega + 4.0\ \Omega = \mathbf{6.0\ \Omega}$

(b) Using $\mathbf{(1/R_P) = \Sigma (1/R_n)}$, $(1/R_P) = (1/6.0\ \Omega) + (1/8\ \Omega)$, $R_P = \mathbf{3.4\ \Omega}$.

(c) Using $\mathbf{R_S = \Sigma R_n}$, $R_S = 2.0\ \Omega + 4.0\ \Omega + 3.4\ \Omega = \mathbf{9.4\ \Omega}$

(d) Using $\mathbf{V = IR}$, $I = V/R = (1.5\ V + 1.5\ V) / (9.7\ \Omega) = \mathbf{0.31\ A}$

(e) The main line current is the same as at point 1 and the ammeter: **0.31 A**

(f) R_3 is 6/14 (43 percent of) the equivalent resistance in the parallel circuit.

R_4 is 8/14 (57 percent of) the equivalent resistance in the parallel circuit.

This means that R_3 gets 57 percent of the current: $I_3 = (0.57)(0.31\ A) = \mathbf{0.18\ A}$

(g) Using $\mathbf{V = IR}$, $V_3 = (0.18\ A)(6.0\ \Omega) = \mathbf{1.1\ V}$

(h) R_4 is 8/14 (57 percent of) the equivalent resistance in the parallel circuit.

This means that R_4 gets 43 percent of the current: $I_4 = (0.43)(0.31\ A) = \mathbf{0.13\ A}$

(Check: $I_3 + I_4 = 0.18\ A + 0.13\ A = 0.31\ A$, the main line current.)

(i) Using $\mathbf{V = IR}$, $V_4 = (I_4)(R_4) = (0.13\ A)(8.0\ \Omega) = \mathbf{1.0\ V}$

(j) Using **V = IR**, $V_B = (I_B)(R_B) = (0.31\ A)(0.25\ \Omega) = 0.078\ V$

(k) Using **V = IR**, $V_1 = (I_1)(R_1) = (0.31\ A)(2.0\ \Omega) = 0.62\ V$

(l) Using **V = IR**, $V_2 = (I_2)(R_2) = (0.31\ A)(4.0\ \Omega) = 1.24\ V$

(m) Taking the sum of all the potential drops, they add up, as they should, to **3.0 V.**

Power

In the mechanical world, power is the rate at which work is done:

$$\mathbf{P = W/t}$$

where P is in Watts (J/s), W is in N-m or Joules, and t is in seconds.

In the electrical world, **power** is the rate at which electrical energy is supplied to a circuit. The energy transfer involves both flowing current *and* heat loss in the load components. When **electrical** energy is transformed into **radiant** energy as circuit components heat up, this loss of energy over time represents a loss of **power**. This can be stated by the equation for **electric power supplied to a circuit:**

$$P = IV$$

where P is power in Watts (W), I is current in Amperes (A), and V is potential difference in Volts (V).

For circuit components through which current flows, the amount of power loss represented as heat is given as a combination of P = IV and Ohm's Law V = IR:

$$P = I^2R$$

where P again is power in Watts (W), I is current in Amperes (A), and R is resistance in Ohms (Ω).

Example

A certain lamp plugged into a 120 V wall socket in a person's home draws 0.25 A.

(a) What is the lamp's resistance?

(b) How much power does the lamp dissipate?

Solution

(a) Using **V = IR**, R = V/I = 120 V/0.25 A = **480 Ω**

(b) Using **P = IV**, P = (0.25 A)(120 V) = **30.0 W**

Alternate: Using $\mathbf{P = I^2R}$, $P = (0.25A)^2(480\ \Omega) =$ **30.0 W**

Example

A certain wire has a resistance of 6.50 Ω/m.

(a) What length of this wire would be needed for use in an electric heater designed to use 8.25 A when operated with a 120 V outlet?

(b) How much power will this wire use?

(c) **(B exam)** If an identical length of this wire were insulated and used for a water heater, how much would the temperature of 3 kg of water at an initial temperature of 10° C rise in 2.5 minutes?

Solution

(a) Using **V = IR**, R = V/I = (120 V) / (8.25 A) = 14.5 Ω are needed.

Since the wire's resistance is 6.50 Ω/m, (1 m/6.50 Ω)(14.5 Ω) = **2.23 m**

(b) Using **P = IV**, P = (8.25 A)(120 V) = **990 W**

(c) $\frac{(990\,\text{J})}{(\text{s})}(2.5\,\text{min})\frac{(60\text{s})}{(\text{min})}\frac{(1\,\text{cal})}{(4.19\,\text{J})}\frac{(1\text{g}^\circ\text{C})}{(\text{cal})}\frac{(1\,\text{kg})}{(1000\,\text{g})}\frac{(1)}{(3\,\text{kg})} = \mathbf{11.8^\circ C}$

Complex DC Circuits and Kirchoff's Laws

When studying DC circuits that involve more than simple series/parallel arrangements, it is necessary to use two important laws for calculating the currents flowing through complicated circuitry.

A complex circuit might involve two or more overlapping circuits. For such problems, we would need to refer to Kirchoff's two laws:

The Junction Law

At any junction point, or a point where currents converge or from which they diverge in a circuit, the sum of all currents entering must equal the sum of all currents exiting.

When a single current reaches a junction, it splits according to the number of paths it can take. The amount of current flowing through each path is an inverse ratio of the path's resistance. When the paths reunite, the current recombines to its original value. This is a restatement of the concept of conservation of charge.

The Potential Law (Loop Law)

The sum of all potential rises and drops in a closed circuit equals zero.

The adding of all potential energy rises of power sources and all potential drops (voltage consumed by circuit components) must equal zero. This is a restatement of the Law of Conservation of Energy.

The energy in the cell or battery is used to raise the charge to the higher voltage terminal. As the charge falls through the external circuit, it loses all of its potential and energy as it crosses the resistors it encounters. The amount of energy given to a charge by the source is completely dissipated by the circuit. That's the meaning of the *Potential Law*.

The current in a circuit flows out of the positive terminal, where it has maximum potential, through the circuit, and eventually back to the negative terminal. The battery or power source restores the potential energy of the charge and moves it over to the positive terminal, where it is pushed through the circuit again. Take that as a **positive** direction when working with complex circuits.

If there is more than one cell in a circuit, the one with the greater value pushes or overpowers the lesser one. Keep the current flow direction of the stronger cell all the way through the circuit, even through the weaker cell, whose potential will be subtracted in the process.

Assign each branch through which current flows its own designated current. Each branch current may be represented by a subscript for that particular branch.

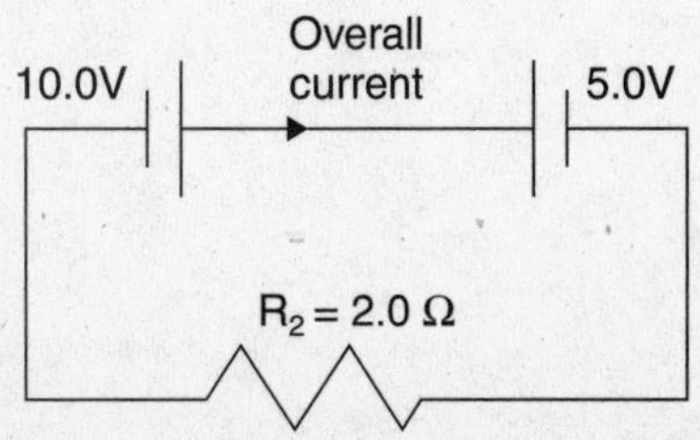

In this illustration, the 10.0 V cell overpowers the 5.0 V cell. The resultant potential difference is 5 V with the current moving clockwise through the circuit. The current in the circuit is found by using Ohm's Law, V = IR, with the current I = V/R = (10.0 V – 5.0 V)/(2.0 Ω) = 2.5 A. In this illustration, the 10.0 V battery is actually charging the 5.0 V battery.

Now we will change the configuration:

Example

Find the current I_3 through the central branch of the following complex circuit.

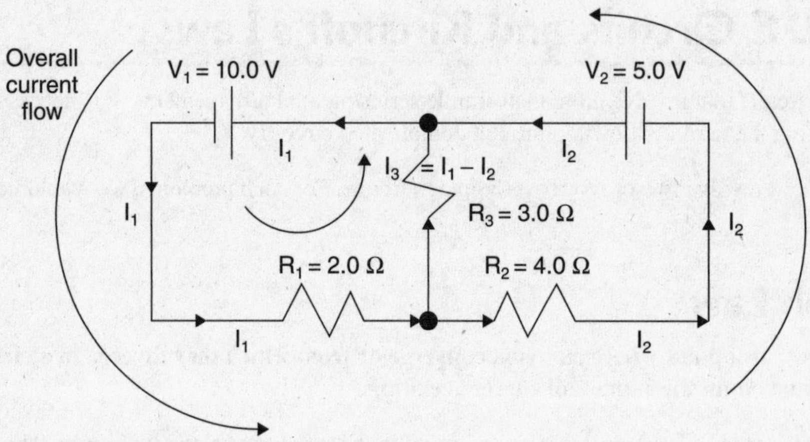

Solution

There are three currents, the current I_1 flowing through the left side, the current I_2 flowing through the right side, and the current I_3, which is the difference of I_1 and I_2, flowing through the central branch. Because the 10.0 V cell overpowers the 5.0 V cell, the result is that both currents flow in the same direction, basically counterclockwise. We want to determine the current I_3 flowing through R_3 and the central branch.

Step 1. Designate a current for every branch and assign each a direction. Assign a direction to the overall current. You may not always know the source values, so in that case assign an overall direction. If you have assigned it incorrectly, it will simply result in a *negative* current value when your computations are complete.

In this case, we know both sources, so we have chosen the counterclockwise overall current direction.

Step 2. Take each branch separately and let its cell potential equal the total of each branch's resistances times its designated current.

For the branch with I_1: $V_1 = I_1R_1 + I_3R_3$

$V_1 = I_1R_1 + (I_1 - I_2)\,R_3$

"Plugging in" numbers yields:

$10.0\text{ V} = (I_1)(2.0\ \Omega) + (I_1 - I_2)\ (3.0\ \Omega)$

Simplifying: $10.0\text{ V} = 2.0\,I_1 + 3.0\,I_1 - 3.0\,I_2$

$10.0\text{ V} = 5.0\,I_1 - 3.0\,I_2$ (BRANCH 1)

For the branch with I_2: $V_2 = -I_2R_2 + I_3R_3$

$V_2 = -I_2R_2 + (I_1 - I_2)R_3$

"Plugging in" numbers yields:

$5.0\text{ V} = (-I_2)(4.0\ \Omega) + (I_1 - I_2)(3.0\ \Omega)$

Simplifying: $5.0\text{ V} = -4.0\ I_2 + 3.0\ I_1 - 3.0\ I_2$

$5.0\text{ V} = 3.0\ I_1 - 7.0\ I_2$ (BRANCH 2)

Combining equations from BRANCH 1 and BRANCH 2:

$$\begin{array}{l} \ 10.0\text{ V} = 5.0 I_1 - 3.0 I_2 \text{ (BRANCH 1)} \\ \underline{+\ 5.0\text{ V} = 3.0 I_1 - 7.0 I_2 \text{ (BRANCH 2)}} \\ \mathbf{= 15.0\text{ V} = 8.0 I_1 - 10.0 I_2} \end{array}$$

Isolating one current (either will do):

$$I_1 = \frac{(15.0 + 10.0 I_2)}{8.0}$$

Substituting into an original equation (either will do):

$$10.0\text{ V} = 5.0\frac{(15.0 + 10.0 I_2)}{8.0} - 3.0\ I_2$$

This Yields **$I_2 = 0.192$ A**

Step 3. Find I_1 using the combined equation $15.0\text{ V} = 8.0\ I_1 - 10.0\ I_2$

This Yields **$I_1 = 2.12$ A**

Finally, $I_3 = I_1 - I_2 = 2.12\text{ A} - 0.192\text{ A} =$ **1.93 A**

R-C Circuits

An R-C circuit is one that contains one or more resistors and one or more capacitors. Capacitors temporarily store charge. Their potential difference rises as charge accumulates and current falls until there is a discharge and the process begins again.

A typical circuit containing a cell, a resistor, and a capacitor being charged by the cell is shown by the following illustration.

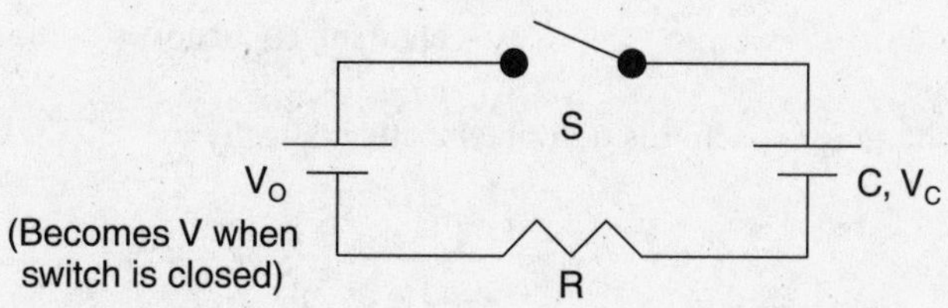

When the switch S is closed, the cell begins charging the capacitor C. Initially, when the current is maximum, both V_C and Q_C are zero. As charge and potential build up on the capacitor C, the current falls, approaching zero. Eventually, the potential differences across the capacitor and the cell become equal. The circuit is now in electrical equilibrium, and no further current flows. Potential difference across the capacitor increases directly with its accumulated charge ($Q = VC$).

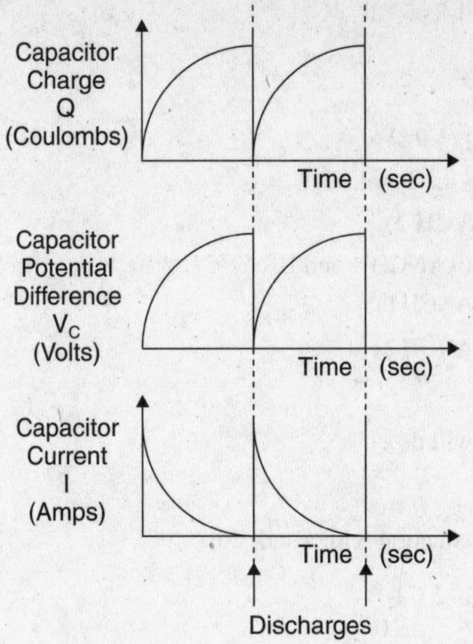

When cells, capacitors, and resistors are combined in a circuit, charge and potential difference rise as current falls in the capacitor. Upon discharge, the resistor buffers the circuit by absorbing a measured amount of the energy released by the capacitor and transforming it into a small amount of heat. Charge, potential difference, and current are all graphically represented in such a circuit by the following illustration of a charging capacitor. This might be an automobile's blinker/turn signal.

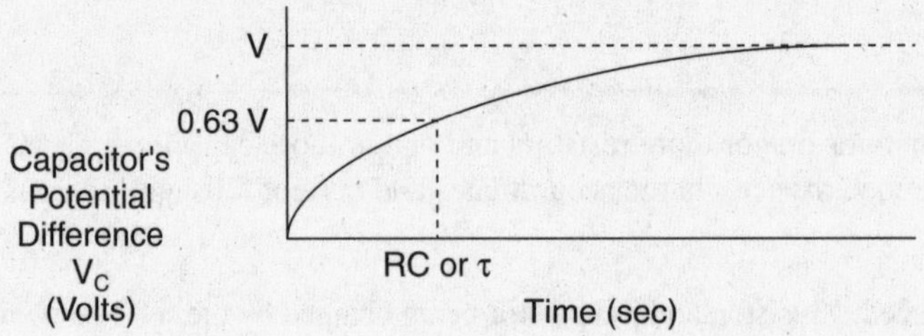

A charging capacitor

The exponential curve of this charging capacitor is described mathematically by

$$V = V_O(1 - e^{-t/RC}) \text{ or}$$
$$V = V_O(1 - e^{-t/\tau})$$

Mathematically, charge, current, and potential difference are represented as functions of time by

$$Q = Q_F(1 - e^{-t/\tau})$$
$$I = I_O e^{-t/\tau}$$
$$V_C = Q/C = V_0(1 - e^{-t/\tau})$$

where Q is the capacitor's increasing charge in Coulombs, Q_F is the capacitor's *final* or *maximum* charge, e represents the exponential expansion factor, t is time in seconds, R is resistance in Ohms, C is capacitance in Farads, I is instantaneous current, and V is potential difference in Volts.

The expression $(1 - e^{-t/\tau})$ includes the *time constant* (RC or τ), the charging time for a capacitor. When $t = \tau$, the potential difference across the capacitor is

$$V = V_O (1 - e^{-1})$$

and V is approximately 63 percent of its maximum value, when the capacitor would be fully charged. (If $t = \tau$, then $e^{-t/\tau} = e^{-1} = 0.37$, and $1 - 0.37 = 0.63$ and $V = 0.63\ V_O$.)

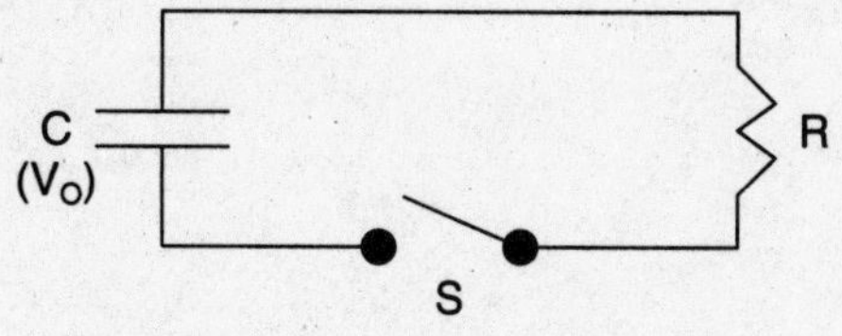

Example

A charged capacitor C, a resistor R, and a switch are connected in series as shown in the preceding illustration. It is in electrical equilibrium as shown.

(a) Determine the current through the circuit before the switch is closed.

(b) Determine the time constant for the circuit when the switch is closed.

(c) Draw the graph of instantaneous current I vs time t from t = 0 to final equilibrium.

Solution

(a) Since there is only a capacitor (whose initial voltage we will assign to have been V_O) and a resistor whose resistance we will call R, when the switch is closed, the electrical relationship at time t = o is $V_O = IR$ and at time of equilibrium is $V - IR = 0$. When the circuit is at equilibrium, there is equal potential difference across the resistor and the capacitor, so because the circuit is in equilibrium, **no current flows and I = 0.**

(b) The time constant τ represents the time from when the capacitor is fully charged to the time when its potential V has fallen by 63 percent of its maximum value.

At that time, $V = IR + Q/C$.

(c)

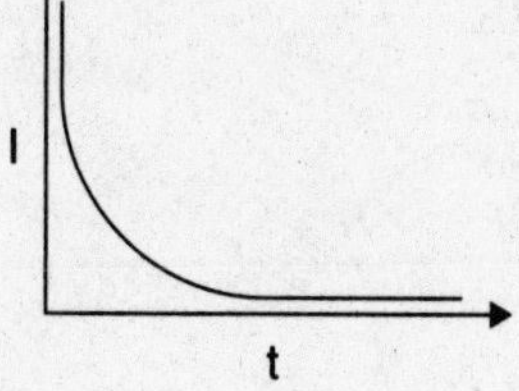

Review Questions and Answers

B & C Exam Question Types

Multiple Choice

1. Direct Current flow must have

(A) surface charge.
(B) plate separation.
(C) an equipotential surface.
(D) reversing drift velocities.
(E) a closed path.

2. A 20 m long wire with resistivity 3.0×10^{-3} Ω-cm and cross-sectional area 6.0×10^{-6} m^2 has a resistance of

(A) 10^1 Ω
(B) 10^2 Ω
(C) 10^3 Ω
(D) 10^4 Ω
(E) 10^5 Ω

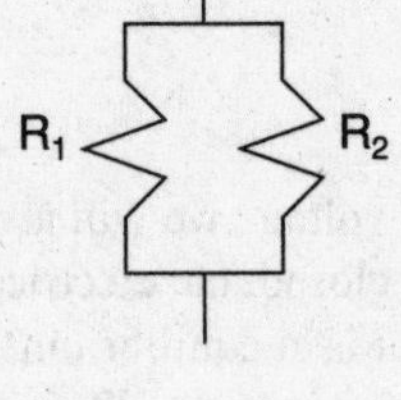

3. Resistors R_1 and R_2 are placed in parallel, as shown. If they have values of 5 Ω and 10 Ω, respectively, their combined equivalent resistance is

(A) 0.03 Ω
(B) 0.3 Ω
(C) 3.0 Ω
(D) 3.3 Ω
(E) 15 Ω

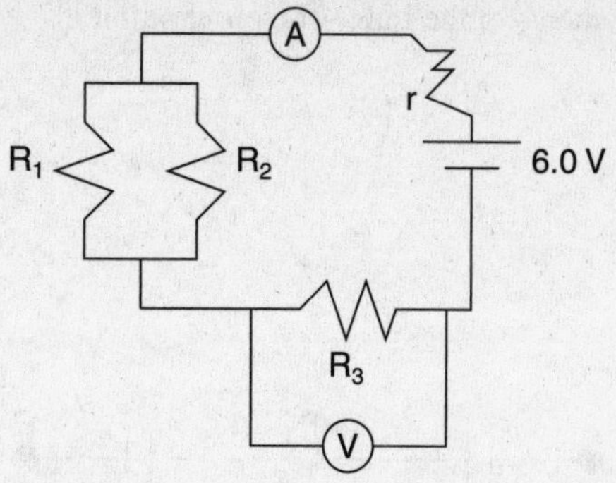

Problems 4–8 refer to the preceding illustration.

4. $R_1 = 1.0\ \Omega$, $R_2 = 1.0\ \Omega$, $R_3 = 1.0\ \Omega$, and $r = 0.10\ \Omega$. The equivalent resistance of the circuit is most nearly

(A) 1.0 Ω
(B) 1.6 Ω
(C) 3.0 Ω
(D) 9.1 Ω
(E) 16.0 Ω

5. The ammeter reads most nearly

(A) 1.8 A
(B) 2.8 A
(C) 3.8 A
(D) 4.8 A
(E) 5.8 A

6. The voltmeter reads most nearly

(A) 2.0 V
(B) 4.0 V
(C) 6.0 V
(D) 8.0 V
(E) 10.0 V

7. The voltmeter is now disconnected and reconnected across R_1. It now reads most nearly

(A) 1.0 V
(B) 2.0 V
(C) 3.0 V
(D) 4.0 V
(E) 5.0 V

8. The power used by this circuit is most nearly

(A) 0.024 W
(B) 0.24 W
(C) 2.4 W
(D) 24 W
(E) 240 W

9. Kirchoff's Laws are really restatements of the Laws of Conservation of

(A) charge and momentum.
(B) energy and momentum.
(C) energy and power.
(D) resistance and charge.
(E) charge and energy.

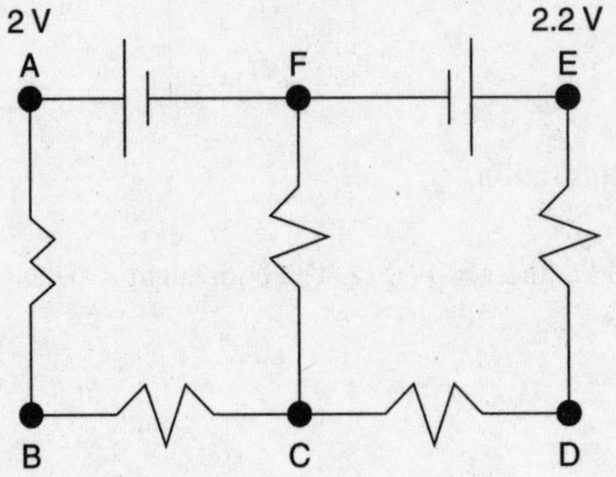

10. The current in the preceding illustration flows from

(A) A to B
(B) B to C
(C) C to D
(D) C to F
(E) D to E

Answers to Multiple-Choice Questions

1. **(E)** Current occurs on a **closed path** in one direction due to a potential difference in the circuit.

2. **(B)** Using $\mathbf{R = \frac{pl}{A}}$, $R = (3.0 \times 10^{-3}\ \Omega\text{-cm}) \frac{(1\,\text{m})}{(100\,\text{cm})} \frac{(20\,\text{m})}{(6.0 \times 10^{-6}\,\text{m}^2)} = \mathbf{1.0 \times 10^2\ \Omega}$

3. **(D)** Using $\mathbf{(1/R_P) = \Sigma\ (1/R_n)}$, $1/R_P = (1/5\ \Omega) + (1/10\ \Omega)$ and $\mathbf{R_P = 3.3\ \Omega}$

4. **(B)** Using $\mathbf{(1/R_P) = \Sigma\ (1/R_n)}$, $1/R_{1,2} = (1/1.0\ \Omega) + (1/1.0\ \Omega)$ and $R_{1,2} = 0.5\ \Omega$

Now using $\mathbf{R_S = \Sigma\ R_n}$, $R_S = 0.5\ \Omega + 1.0\ \Omega + 0.10\ \Omega = \mathbf{1.6\ \Omega}$

5. **(C)** Using $\mathbf{V = IR}$, $I = V/R = (6.0\ V / 1.6\ \Omega) = \mathbf{3.75\ A}$

6. **(B)** Using $\mathbf{V = IR}$, $V_3 = I_3 R_3 = (3.8\ A)(1.0\ \Omega) = \mathbf{3.8\ V}$

7. **(B)** Since current splits evenly between R_1 and R_2, both receive 1.9 A

Using $\mathbf{V = IR}$, $V_1 = I_1 R_1 = (1.9\ A)(1.0\ \Omega) = \mathbf{1.9V}$

8. **(D)** Using $\mathbf{P = IV}$, $P = (3.8\ A)(6.0\ V) = 22.8\ W$; the closest answer is **24 W.**

9. **(E)** The *Junction Law* is a restatement of the Law of Conservation of Charge. The *Potential Law* is a restatement of the Law of Conservation of Energy.

10. **(D)** Since the 2.2 V cell overpowers the 2.0 V cell, the current flow is clockwise, resulting in un upward flow of current through the central branch, **C to F.**

Free Response

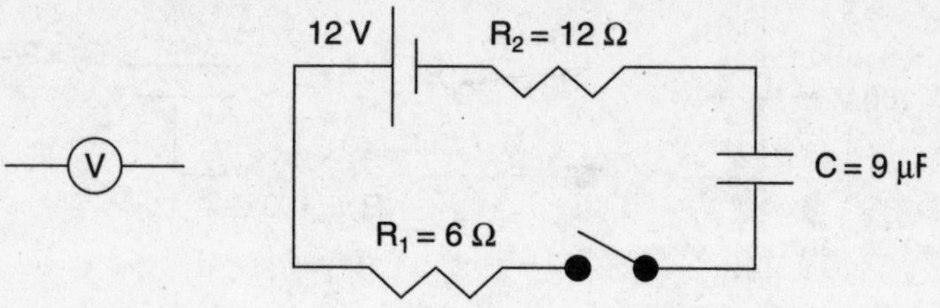

1. In the given circuit, a 12 V source is connected to resistor R_1(6 Ω), a switch, a 9 μF capacitor, and resistor R_2. The capacitor is uncharged, and the switch is in the open position. A separate voltmeter is shown near the circuit.

- **(a)** The switch is closed, and the voltmeter is used to determine the potential drop across resistor R_2. Draw the circuit with the voltmeter connected for the reading.
- **(b)** What does the voltmeter read immediately after the switch is closed?
- **(c)** After awhile the voltmeter is rechecked. What does it read at that time?
- **(d)** What is the charge on the capacitor during the reading in part **(c)**?

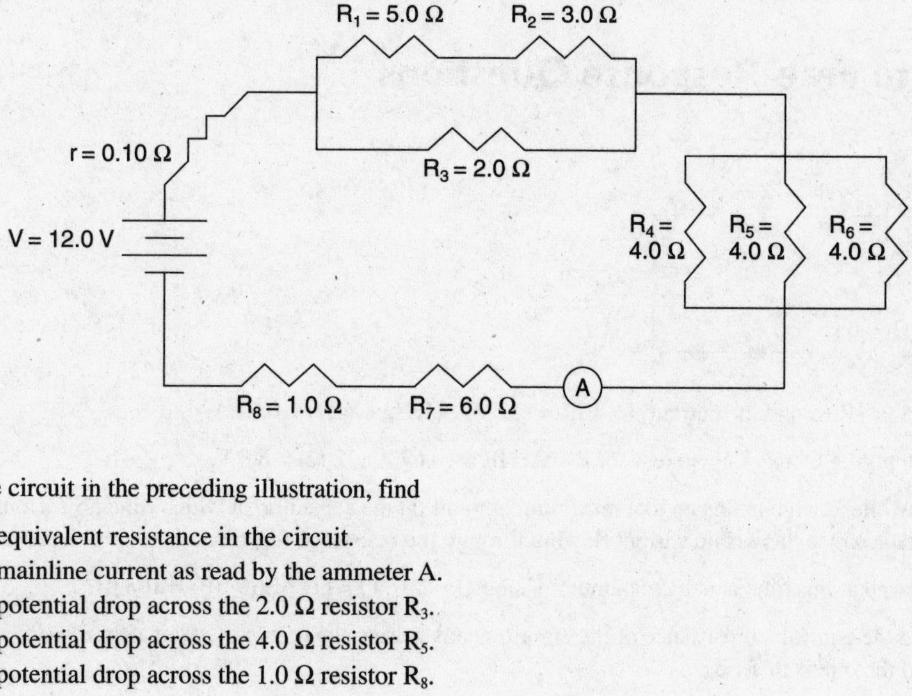

2. Given the circuit in the preceding illustration, find

- **(a)** the equivalent resistance in the circuit.
- **(b)** the mainline current as read by the ammeter A.
- **(c)** the potential drop across the 2.0 Ω resistor R_3.
- **(d)** the potential drop across the 4.0 Ω resistor R_5.
- **(e)** the potential drop across the 1.0 Ω resistor R_8.

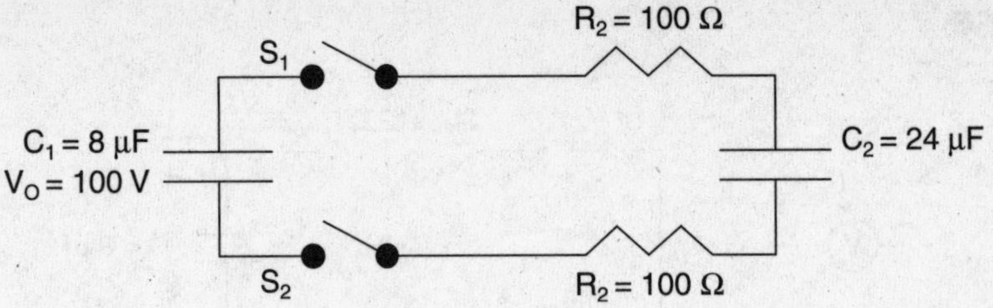

3. Capacitors C_1 (8 μF) and C_2 (24 μF) are connected in the preceding circuit, in series with two resistors R_1 and R_2 (both 100 Ω) and two switches S_1 and S_2. Initially, capacitor C_1 is charged to a voltage of 100 V and capacitor C_2 is uncharged. At a time when t = 0, both switches are closed simultaneously.

(a) After electrical equilibrium has been reached, determine the final charges on each capacitor.
(b) Is the final energy stored in C_1 and C_2 equal to the energy initially stored in C_1? If not, what is the difference and where did the energy go?
(c) Write an equation that relates the charge on capacitor C_1 with respect to time, its instantaneous current I, V_O, R_1, R_2, C_1 and C_2. (It is not necessary to solve the equation.)
(d) The current in the resistors as a function of time is given as $I = I_O e^{-t/\tau}$. Find the energy loss in the combined resistors as a function of time.

Answers to Free-Response Questions

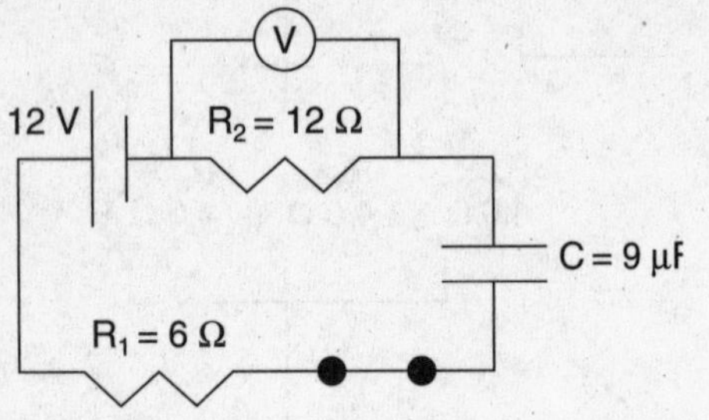

(a)

(b) Using **V = IR** to find the current, I = V/R = (12 V) / (12 Ω + 6 Ω) = 0.67 A.

Now, applying Ohm's Law to resistor R_2: V = IR = (0.67 A)(12 Ω) = **8.0 V**

(c) After awhile, charge builds up to a maximum amount on the capacitor, at which time no current flows through the circuit. Since there is no current flowing through the resistor at this time, the voltmeter reads **zero.**

(d) The charge at this time is at its maximum. Using **Q = VC**, Q = (12 V)((9 μF) = **108 μC**

2. **(a)** Finding the equivalent resistance of the circuit means getting the branches' equivalent resistances and adding them to the others in series.

For the top branches: Using $\mathbf{1/R_P = \Sigma\, 1/R_n}$ (Since R_1 and R_2 are in series, they equal 8.0 Ω.).

$$\frac{1}{R_{1,2,3}} = \frac{1}{8.0\,\Omega} + \frac{1}{2.0\,\Omega} \text{ and } R_{1,2,3} = 1.6\,\Omega$$

For the side branches: Using $\mathbf{1/R_P = \Sigma\, 1/R_n}$

$$\frac{1}{R_{4,5,6}} = \frac{1}{4.0\,\Omega} + \frac{1}{4.0\,\Omega} + \frac{1}{4.0\,\Omega} \text{ and } R_{4,5,6} = 1.3\,\Omega$$

Using $\mathbf{R_S = \Sigma\, R_n}$,

$$R_{TOT} = 0.10\,\Omega + 1.6\,\Omega + 1.3\,\Omega + 6.0\,\Omega + 1.0\,\Omega = \mathbf{10.0\,\Omega}$$

(b) The main line current as read by the ammeter is found by using Ohm's Law: **V = IR**, $I = V/R = (12.0\ V) / (10.0\ \Omega) =$ **1.20 A**

(c) For the potential drop across the 2.0 Ω resistor R_3: The main line current of 1.20 A splits inversely with resistance in the parallel circuit, which means that the 2.0 Ω resistor R_3 receives 80 percent of the main line current: $I_2 = (0.8)(1.20\ A) = 0.96\ A$.

Now using **V = IR**, $V_2 = I_2R_2 = (0.96\ A)(2.0\ \Omega) =$ **1.9 V**

(d) For the potential drop across the 4.0 Ω resistor R_5: Since the main line current of 1.20 A evenly splits three ways, $I_5 = (1.20\ A) / (3) = 0.4\ A$

Now using **V = IR**, $V_5 = I_5R_5 = (0.4\ A)(4.0\ \Omega) =$ **1.6 V**

(e) For the potential drop across the 1.0 Ω resistor R_8:

Using **V = IR**, $V_8 = I_8R_8 = (1.2\ A)(1.0\ \Omega) =$ **1.2 V**

3. (a) Using $\mathbf{Q_1 = V_1C_1}$, and since all the charge is initially on C_1, $Q_1 = Q_{TOT}$ at $t = 0$.

$$Q_{TOT} = (100\ V)(8\ \mu F) = 800\ \mu C$$

At electrical equilibrium, the potential difference or voltage across both capacitors is equal. ($V_{C1} = V_{C2}$) and $\frac{Q_1}{C_1} = \frac{Q_2}{C_2}$ and the total charge is redistributed but still adds up to the total charge that was initially on C_1, which means that $Q_{TOT} = Q_1 + Q_2$.

Since $C_2 = 3\ C_1$, C_2 has 75 percent of the redistributed total charge, which means $Q_1 = (0.25)(800\ \mu C) =$ **200 μC**

$$Q_2 = (0.75)(800\ \mu C) = \mathbf{600\ \mu C}$$

(b) Using $\mathbf{U_C = (1/2)\ CV^2}$ for potential energy stored in a capacitor, initially $C_1 = (1/2)\ (C_1)(V_1)^2 = (1/2)(8\ \mu F)(100\ V)^2 = 40{,}000\ \mu J$

$$U_{C1} = (1/2)\ C_1V_1^2 = (1/2)\ C_1\,(Q_1/C_1)^2 = (1/2)(8\ \mu F)(200\ \mu C)^2/(8\ \mu F) = 2500\ \mu J$$

$$U_{C2} = (1/2)\ C_2\,V_2^2 = (1/2)\ C_2\,(Q_2/C_2)^2 = (1/2)(24\ \mu F)(600\ \mu C)^2/(24\ \mu F) = 7500\ \mu J$$

$$\text{TOTAL} = 10{,}000\ \mu J$$

Initial – final potential = 40,000 μJ – 10,000 μJ = **30,000 μJ difference.**

It was converted into radiant energy, heating the resistors.

(c) Using Kirchoff's Laws: $\mathbf{V_{TOT} = V_{C1} + V_{C2}} + V_{R1}$ and $V_{R2} = 0$

↑ (initially charged) ↑ (initially uncharged)

So, at time $t = 0$, $V_{TOT} = V_{C1} = Q_1 / C_1$ and since charge is conserved, $Q_{TOT} = Q_1 + Q_2 = (V_{TOT})(C_1)$ at $t = 0$.

This makes $Q_2 = (V_{TOT})(C_1) - Q_1$.

Finding the potential difference for C_2 as it charges:

$$V_{C2} = Q_2 / C_2 = \frac{V_{TOT}(C_1) - Q_1}{C_2}$$

The circuit current I at any time is given by $I = \frac{-\Delta Q_1}{\Delta t}$ or $\frac{-dQ_1}{dt}$

Finding the potential difference for C_1 as it discharges:

Since $V_{TOT} = \Delta V_{C1} + \Delta V_{C2} + V_{R1}$ and V_{R2}

(discharging) (charging)

$\Delta V_{C1} = \Delta V_{C2} +$ potential converted to heat in the resistors

rewritten as $\Delta V_{C1} = \Delta V_{C2} + I(R_1 + R_2)$

Combining the preceding:

$$\frac{Q_1}{C_1} = \frac{V_{TOT}C_1 - Q_1}{C_2} + \frac{\Delta Q_1}{\Delta t}\mathbf{(R_1 + R_2)}$$

$$\textbf{or } \frac{Q_1}{C_1} = \frac{V_{TOT}C_1 - Q_1}{C_2} + \frac{dQ_1}{dt}\mathbf{(R_1 + R_2)}$$

(d) Using $\mathbf{P = IV = I^2R}$, $P = I^2R = (I_O e^{-t/\tau})^2 (R_1 + R_2)$.

Finally, $\mathbf{P = I_O^2 (R_1 + R_2)(e^{-2t/\tau})}$

Magnetic Fields and Forces

Magnetism

In nature, electromagnetic effects permeate the Universe. Humans discovered magnetism when they found magnetite-lodestone in the earth. It is magnetized iron ore whose special properties result from the magnetic field of the earth itself. The ancients found that when they placed a piece of this special stone on a balance point or hung it from a string, it would always turn and orient itself in a certain direction. The name "lead-stone" was given to it, because it would always lead or point in the same direction. The name translated into "lodestone," which is the name given to describe a piece of naturally magnetic iron ore, or magnetic magnetite. The difference is that **magnetite** is simply iron ore that is attracted by a magnet. **Lodestone** is naturally magnetic magnetite, a natural magnet. In addition, passing an electric current through a wire creates magnetic effects, which will be covered later in this chapter.

As an electric field surrounds a charged particle, a magnetic field surrounds a charged particle in **motion**. Magnetic phenomena arise from forces created by electric charges in motion. As electric charges can be **positive** or **negative**, magnetic poles can be **north** or **south**. Just as like electric charges repel and opposite charges attract, like magnetic poles repel and opposite magnetic poles attract. In addition, magnetic forces are also subject to the inverse-square law.

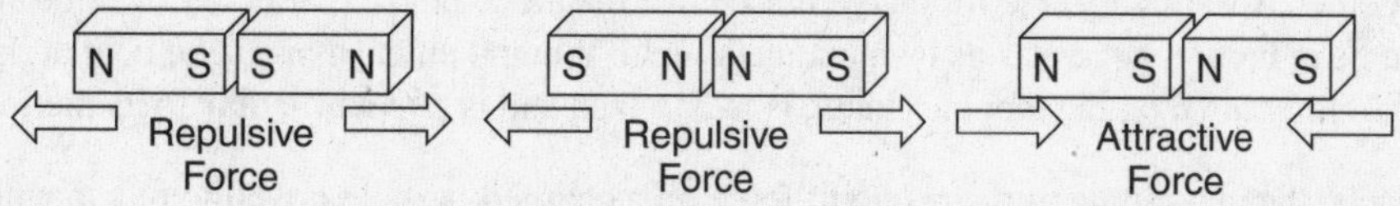

By contrast, although electric charges can be isolated or separated, the poles of a magnet seem to be impossible to isolate. Some theories suggest the existence of **magnetic monopoles** but their existence has not been conclusively proven. Cutting a bar magnet in half does not isolate the north or south poles. Instead, both smaller pieces become smaller magnets having north and south poles.

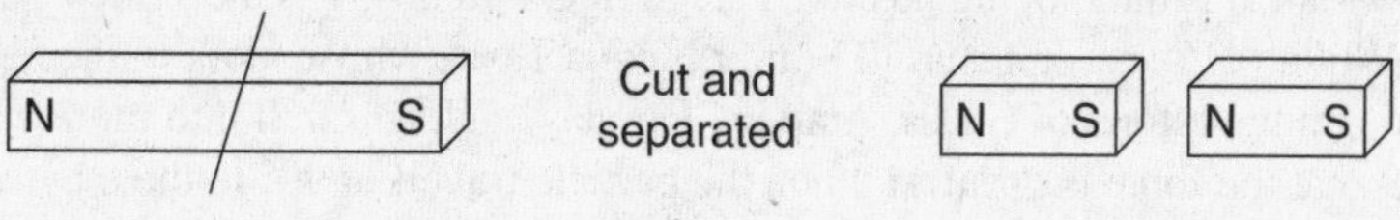

Scientists believe that the magnetic field surrounding a permanent bar magnet (comprised of iron or metallic alloys containing iron) is a result of the uniform alignment or orientation of clumps of molecules called **magnetic domains.** In a permanent magnet, the magnetic domains become oriented by cooling from a molten or semi-molten stage while in the presence of an externally applied magnetic field. The semi-mobile magnetic domains become more uniformly aligned according to the strength of the external magnetic field. A **magnetic field** may also be referred to as **magnetic induction** or **flux density**. The individual field lines are the **lines of flux**.

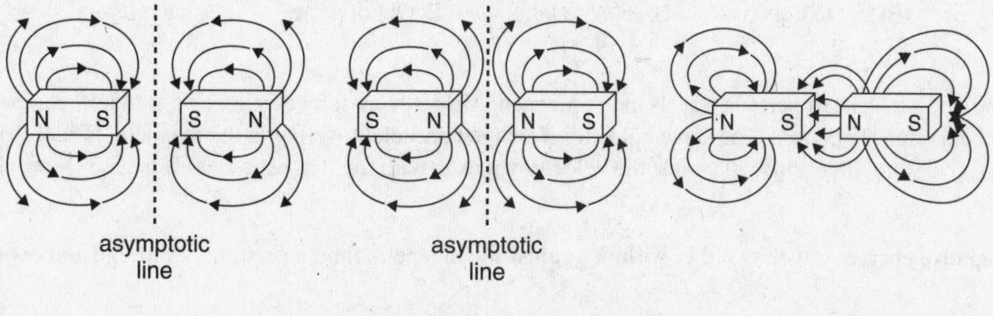

Magnetic lines of force around bar magnets

As shown in this illustration, the lines of force of a magnetic field go *out of the North Pole* and *into the South Pole*. The shapes of the repulsive and attractive force interactions are similar to those of electric charges. Since the magnetic field lines have directions, they are vector quantities. The magnetic field is denoted by **B**, which has units of Webers/m^2 or N/A·m^2 or Teslas (T).

Charges in Magnetic Fields

This book has previously made mention of the interaction between electric and magnetic phenomena. The **electromagnetic theory** developed by James Clerk Maxwell between 1860 and 1865 mathematically unified the concepts of radiant energy and electrical energy. He classified them as different electromagnetic disturbances, part of the **electromagnetic spectrum**, which travel at the speed of light in a vacuum, 3×10^8 m/s.

In short, Maxwell's mathematical unification included the results of electric and magnetic work of Gauss (Gauss' Laws for electricity and magnetism), Ampere (Ampere's Law for magnetism), and Faraday (Faraday's Law for electromagnetic induction). These mathematical laws will be listed and discussed shortly.

Just as current direction and electron flow direction are important in the study of electronic circuits, magnetic field direction is important in electromagnetic studies. It is the interaction of radiant and mechanical energy that results in most of our everyday power.

The electromagnetic spectrum is radiation that is equally divided between electric and magnetic energy components. All electromagnetic radiation has two vibratory directions, defined by the electric field and the magnetic field, which vibrate at **right angles** to each other. A stationary electric charge placed in a magnetic field experiences no force due to the magnetic field. An electric charge in motion generates its own magnetic field. When it enters a magnetic field at right angles to the magnetic field, the resulting force on the moving charge is maximized and is **perpendicular** to its motion.

Considering these facts, the magnitude of the resultant force on a charged particle traveling in a magnetic field must be a result of the **strength of the charge**, the **strength of the magnetic field**, the **velocity of the charge,** and the **angle the charge makes with the magnetic field.** Stated mathematically,

$$\mathbf{F} = q\mathbf{vB} \sin \Theta \text{ or } \mathbf{F} = q\mathbf{v} \times \mathbf{B}$$

where **F** is the force in Newtons, q is the moving positive charge in Coulombs, **v** is the velocity of the charge in m/s, and **B** is the magnetic field in Wb/m^2, N/A·m^2 or Teslas (T) (after Nikola Tesla, whose work in the fields of electromagnetism and alternating current brought the efforts of Gauss, Ampere, Faraday, and Maxwell into more widespread realization.) The cross product signifies that the force is **greatest** when the particle travels at 90° to the magnetic field and is **zero** when it travels parallel with it.

Let us examine a magnetic field. First, note the following descriptions of the direction of a magnetic field **B**:

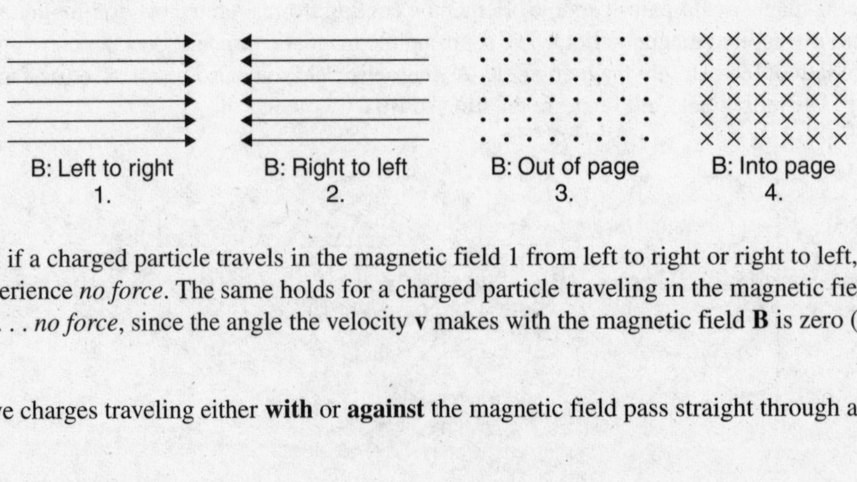

In this illustration, if a charged particle travels in the magnetic field 1 from left to right or right to left, parallel with the field, it would experience *no force*. The same holds for a charged particle traveling in the magnetic field 2 from right to left or left to right. . . *no force*, since the angle the velocity **v** makes with the magnetic field **B** is zero ($\mathbf{F} = q\mathbf{vB} \sin \Theta$ or $\mathbf{F} = q\mathbf{v} \times \mathbf{B}$).

Positive or negative charges traveling either **with** or **against** the magnetic field pass straight through and **experience no force**.

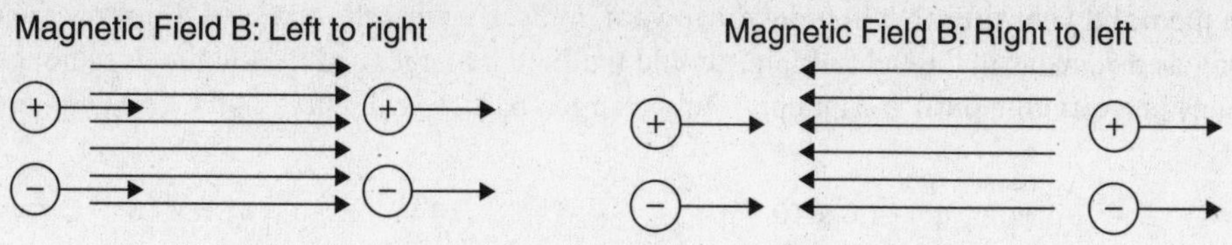

However, now consider a charged particle moving from left to right or right to left through and normal to the magnetic fields. Since the angle between the particle's velocity **v** and the magnetic field **B** is 90°, the particle experiences a maximum amount of force. As the angle decreases from 90° to 0° the force also approaches zero.

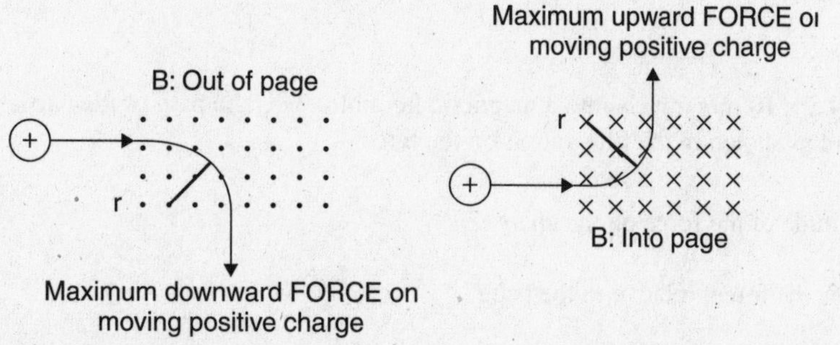

In this illustration, a **positive charge** traveling from left to right enters a magnetic field at 90°. Following **F = qvB** sin Θ or **F** = qv **X B**, point the fingers of your **right hand** (right for positive charges) in the direction of the positive charge's velocity **v**, and then turn them in the direction of the magnetic field **B.**

For the positive charge on the left, your **thumb** will point downward, indicating that the charge will follow the downward curved path as long as it remains in the field. For negative charges, **F** = (−)q**v** X **B** causes the direction of the resulting force to be reversed. The upward and downward force is only the initial direction before the particle deflects to a new direction. Should the field's area be larger and the charge does not exit the field, the charge continues in a completely **circular path**. In such a situation, the magnetic force on the charge becomes the **centripetal force** (**F** = mv^2**/r**), and various other calculations can be made regarding the path's radius, velocity, and mass. In the illustration, **r** represents the radius of curvature of the circular path. The force is centripetal even if the field is small. However, a greater part of the circle is seen if the centripetal force acts over a longer time and extent of motion of the charged particle.

For the charge on the right, your **thumb** will point upward, indicating that the charge will follow the upward curved path as long as it remains in the field. Again, should the field's area be larger and the charge does not exit the field, the charge continues in a **circular path.**

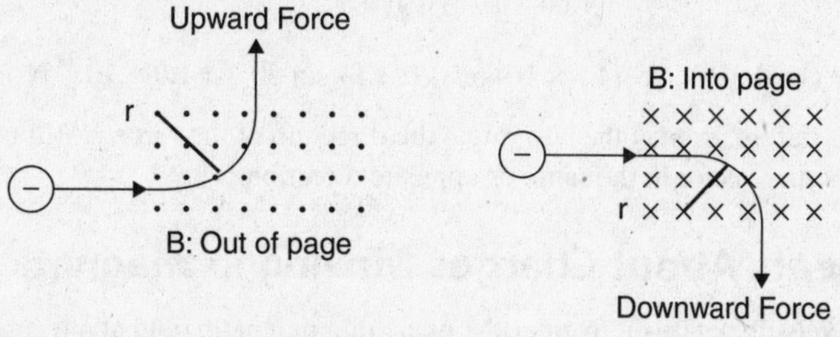

In this illustration, a **negative charge** traveling from left to right enters a magnetic field at 90°. Following **F = qvB** sin Θ or **F** = qv **X B**, point the fingers of your **left hand** in the direction of the negative charge's velocity **v**, and then turn them in the direction of the magnetic field **B.**

For the charge on the left, your **thumb** will point upward, indicating that the charge will follow the upward curved path as long as it remains in the field. Should the field be larger and the charge does not exit the field, the charge continues in a **circular path**, again involving **centripetal force**.

For the charge on the right, your **thumb** will point downward, indicating that the charge will follow the downward curved path as long as it remains in the field. Again, should the field be larger and the charge does not exit the field, the charge continues in a **circular path** and an oppositely charged particle will curve in the opposite direction.

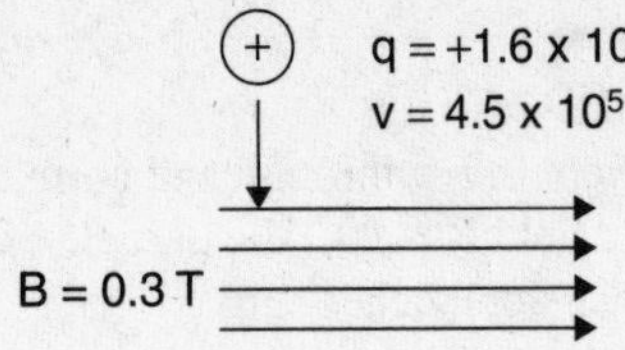

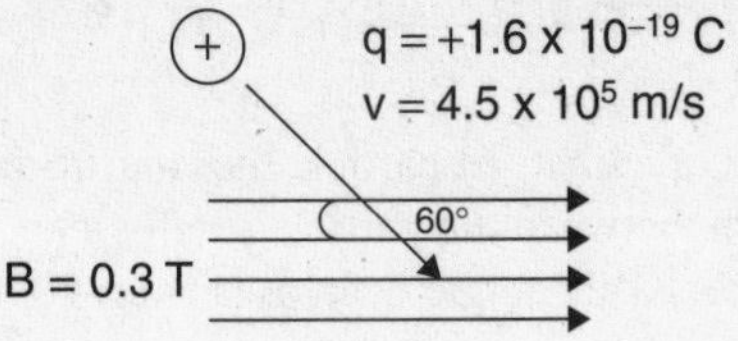

Example

A proton with velocity 4.5×10^5m/s travels into a magnetic field of 0.3 T. The path of the particle is normal to the direction of the magnetic field as shown in the illustration on the left.

(a) What is the magnitude of the force on the charge?

(b) In what direction is the force, relative to the page?

(c) What is the radius of its circular path while in the magnetic field?

(d) If the proton enters the field at an angle of 60° as shown in the illustration on the right, what is the magnitude and direction of the new force on the charge?

Solution

(a) Using **F** = qv X **B**, **F** = (q)(v)(**B**)(1) = $(+1.6 \times 10^{-19}$ C)(4.5×10^5 m/s)(0.3 T)(1) = $\mathbf{2.2 \times 10^{-14}}$ **N**

(b) Using the right hand, pointing the fingers in the direction of **v** (down), then in the direction of **B**, to the right, the thumb points in the direction of the resultant force, **out of the page**.

(c) Using **F** = qv X **B** and $\mathbf{F_C} = mv^2/\,\mathbf{r}$ and setting them equal,

$$qvB(1) = mv^2/r$$

$$\mathbf{r = mv\ /\ qB}$$

$$\mathbf{r} = \frac{(1.67 \times 10^{-27}\,\text{kg})(4.5 \times 10^5\,\text{m/s})}{(1.60 \times 10^{-19}\,\text{C})(0.3\,\text{T})} = \mathbf{1.6 \times 10^{-2}\ m\ or\ 1.6\ cm}$$

(d) Using **F** = qv X **B**, **F** = $(1.60 \times 10^{-19}$ C)(4.5×10^5 m/s)(0.3 T)(sin 60°) = $\mathbf{1.9 \times 10^{-14}}$ **N**

Using the right hand as before, even at the new angle the direction of the force is still **out of the page**. A scalar times a vector gives another vector in the same or opposite direction.

Additional Comments About Charges Moving in Magnetic Fields

1. If a charged particle's velocity causes it to enter the plane of a magnetic field at any angle between 0° and 90°, the particle takes a spiral resultant path.
2. At any instant in a particle's circular or spiral path while traveling in a magnetic field, the instantaneous velocity is directed along a path *tangent* to the circular or helical path. Because it is a circular path, the *force* and *acceleration* are directed toward the center of the circle/spiral. This concept is exactly the same as a ball swinging in a circular path on a string, insofar as the *magnetic force* acts as *centripetal force*.

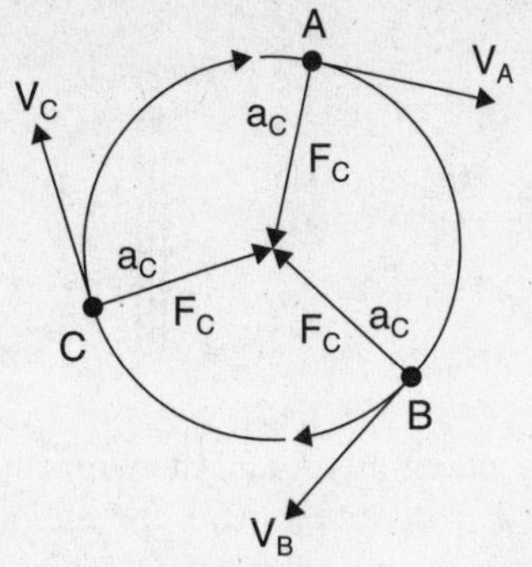

In this illustration, at any point, **A**, **B** or **C**, the acceleration of and force on the particle are toward the center (since they are both *centripetal*) and its instantaneous velocity is tangent to the circle.

Current-Carrying Wires and Magnetic Fields

It has been noted that when a charge moves, it creates its own magnetic field. For a stream of charged particles traveling in the same direction, the resultant magnetic field takes on a circular halo around the stream. This is the case when electrons flow through a wire. Each individual electron's magnetic field is compounded by those of the millions of others moving through the metallic conductor. To describe the direction of the magnetic field lines, use the right hand. Point your thumb in the direction of positive current flow. Your fingers curl around in the direction of the magnetic field.

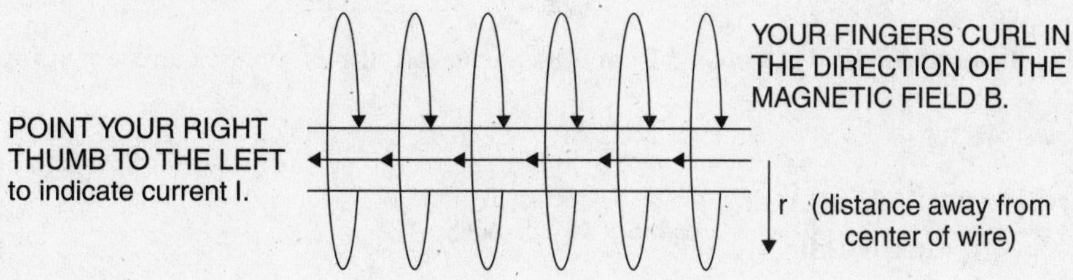

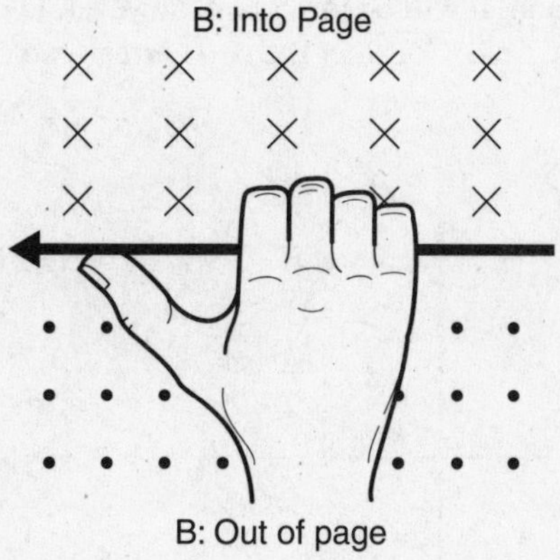

Biot-Savart's Law

When a wire carries a current, a circular magnetic field exists around the wire. The magnitude of the field at any distance from the wire will basically depend on the strength of the **current** and the **distance** from the wire. This is stated by the law of Biot and Savart, which deals with a current-carrying wire and the measure of the magnetic field at a distance **r** away from a small piece of the wire **Δs**:

$$\Delta B = \frac{\mu_o}{4\pi} \frac{I\Delta s \, X r}{r^3}$$

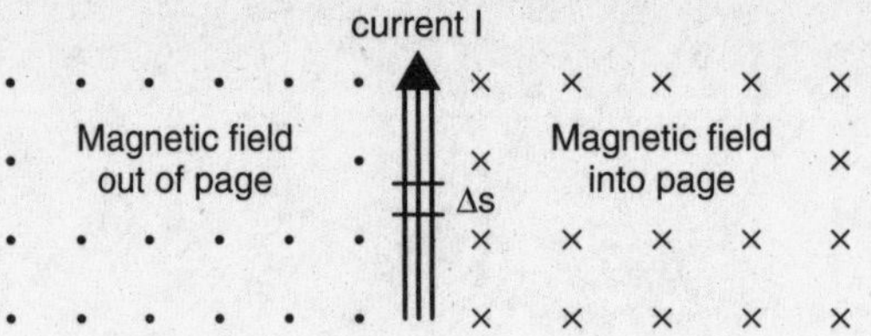

When a current flows through a wire, the magnetic field around the wire varies directly with current I and inversely with the distance **r**. (It would be a $1/r^2$ inverse-square relationship for a *point* charge radiating a bubble of force. Think of the magnetic field around a current flowing through a wire as an elongated bubble and a series of circumferences times the wire's length.) The relationship becomes

$$\mathbf{B \, \alpha \, I / r}$$

where **B** is the magnetic field (in T or N/A ·m), I is current (in A), and **r** is the radius of the magnetic field (in m). The proportionality constant is a result of the way a magnetic field permeates free space or vacuum. The **vacuum permeability (permeability of free space)** of a magnetic field is expressed as μ_0 and is a constant with a value of $4\pi \times 10^{-7}$ (T · m)/A. Its ability to cover a circumference of $2\pi r$ can be added to the previous equation to yield

$$\mathbf{B} = \frac{\mu_0 I}{2\pi r} \text{ MAGNETIC FIELD AROUND A LONG STRAIGHT WIRE}$$

Example

What is the value of the magnetic field at a point 3.5 cm from a wire carrying an electric current of 1.5 A?

Solution

Using $\mathbf{B} = \frac{\mu_0 I}{2\pi r}$, $\mathbf{B} = \frac{(4\pi \times 10^{-7} \text{T} \cdot \text{m/A})(1.5\,\text{A})}{(2\pi)(0.035\,\text{m})} = \mathbf{8.6 \times 10^{-6}\ T\ \ ANS}$

In the event that there are two or more wires, finding the magnetic field at any point nearby is a matter of adding the individual wires' magnetic fields at that point. It will be important, however, to determine the direction of each current and the ensuing direction of each magnetic field. Consider the two current-carrying wires in the following illustration (not to scale):

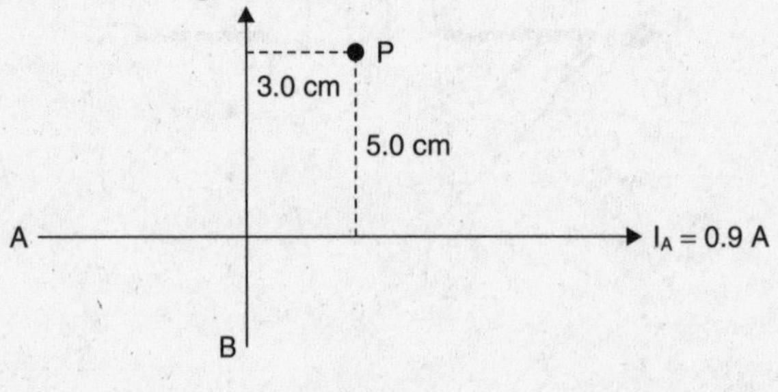

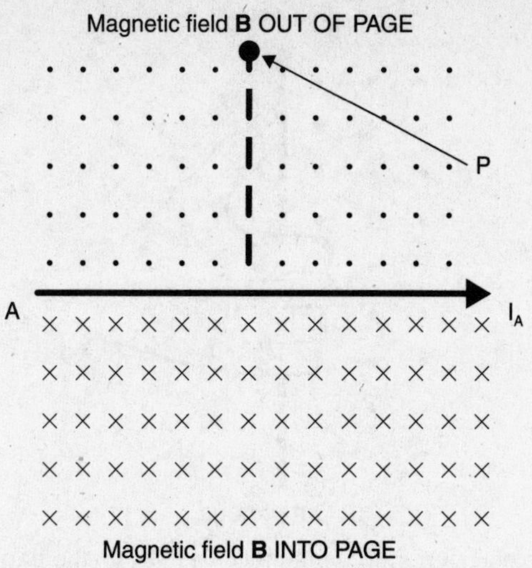

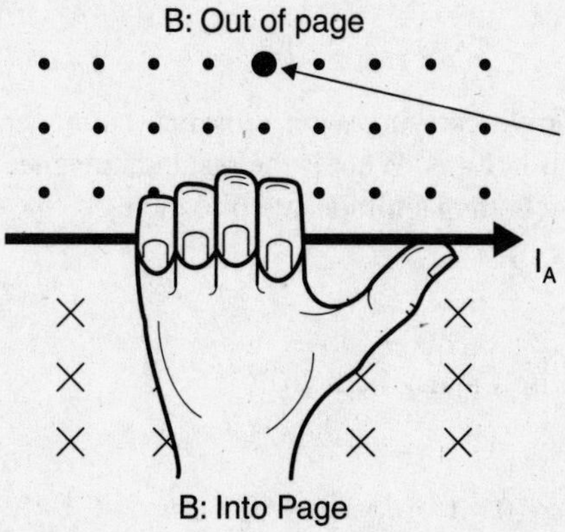

At point P, current I_A's magnetic field is out of the page.

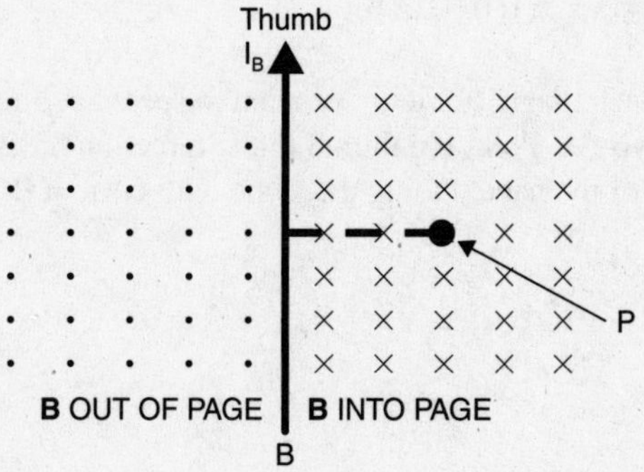

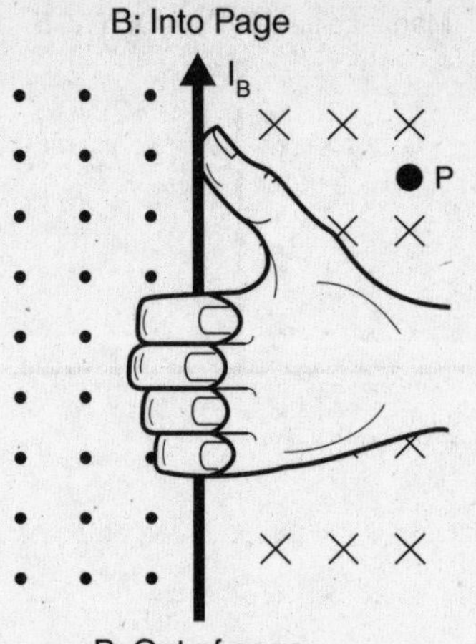

At point P, current I_B's magnetic field is into the page.

Example

The preceding illustration shows two current-carrying wires crossing (but not connected) at a 90° angle. The current in wire A is 0.9 A, and the current in wire B is 1.4 A. What is the resultant magnetic field at point P, 5.0 cm from wire A and 3.0 cm from wire B as shown? (Give both magnitude and direction.)

Solution

First determine the magnitude of each wire's magnetic field at P:

Using $\mathbf{B} = \frac{\mu_0 I}{2\pi r}$,

$$\mathbf{B}_A = \frac{\left(4\pi \times 10^{-7}\,\text{T}\cdot\text{m/A}\right)(0.9\text{A})}{(2\pi)(0.05\,\text{m})} = 3.6 \times 10^{-6}\,\text{T (OUT OF PAGE)}$$

$$\mathbf{B}_B = \frac{\left(4\pi \times 10^{-7}\,\text{T}\cdot\text{m/A}\right)(1.4\,\text{A})}{(2\pi)(0.03\,\text{m})} = 9.3 \times 10^{-6}\,\text{T (INTO PAGE)}$$

Now that we know the magnitudes of each magnetic field, we need to see what direction each is pointed in and add them. Using the right hand, start with wire A. Point the thumb in the direction of its current, and you will see that at point P, your fingers curl and point out of the page. Doing the same with current B shows that the magnetic field due to current B at point P is into the page.

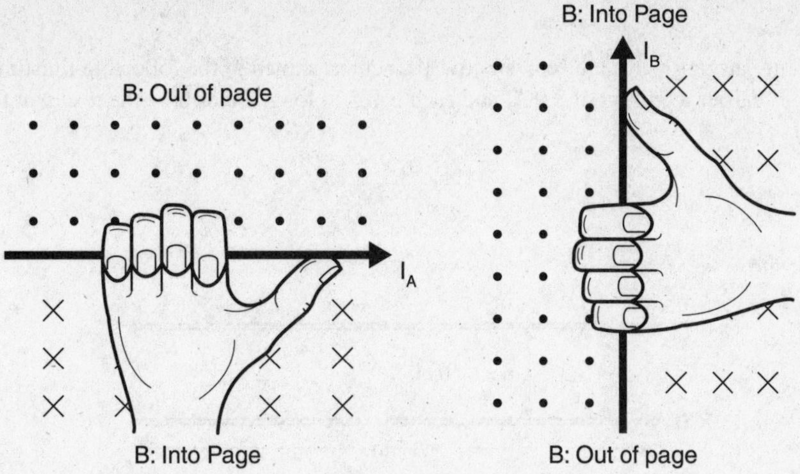

Adding fields produces the resultant field:

$$\mathbf{B_{TOT}} = 9.3 \times 10^{-6}\ \mathrm{T} - 3.6 \times 10^{-6}\ \mathrm{T} = \mathbf{5.7 \times 10\ exponent\ -6\ T\ into\ the\ page.}$$

(into page) (out of page)

Current-Carrying Wires in Magnetic Fields

In the event that two current-carrying wires are parallel and are in close proximity, the effects of the magnetic fields of each on one another are either attractive or repulsive. The current in the wire generates a magnetic field around it as indicated by the right hand model.

For two parallel wires whose currents are in the **same direction**: where the fingers curl around the first wire and intersect the fingers of a hand around the second, **parallel** current, the magnetic fields interlock and **attract** each other.

For two parallel wires whose currents are in **opposite directions:** where the fingers curl around each wire and intersect the fingers of a hand around the second, **antiparallel** current, the magnetic fields push against and **repel** one other.

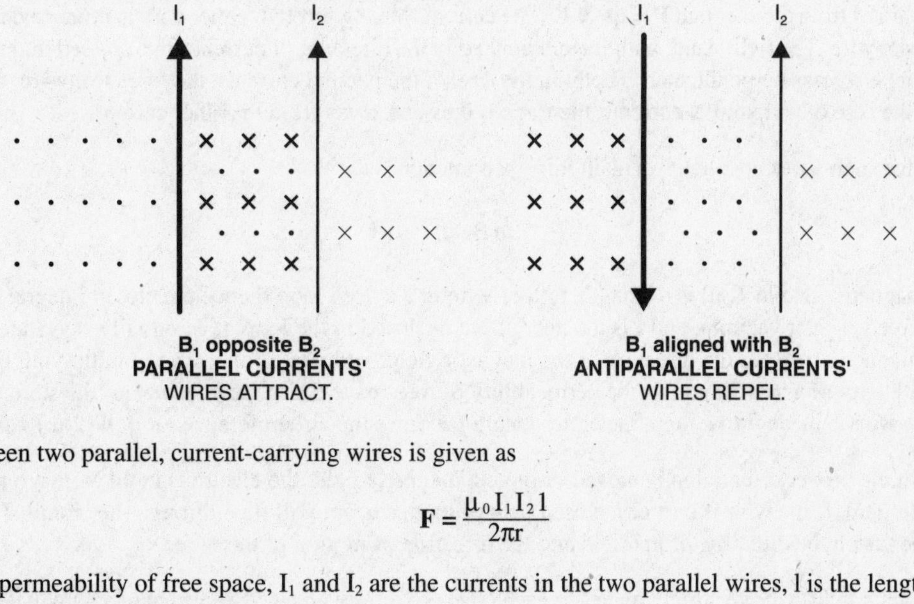

B_1 opposite B_2
PARALLEL CURRENTS'
WIRES ATTRACT.

B_1 aligned with B_2
ANTIPARALLEL CURRENTS'
WIRES REPEL.

The force between two parallel, current-carrying wires is given as

$$\mathbf{F} = \frac{\mu_0 I_1 I_2 l}{2\pi r}$$

where μ_o is the permeability of free space, I_1 and I_2 are the currents in the two parallel wires, l is the length of the wires, and r is the distance between the wires.

Example

Two long, parallel current-carrying wires are separated by 10.0 cm as shown in the following illustration. Wire A carries a current of 1.0 A, wire B carries a current of 2.0 A, and each is 0.5 m long. Determine the resultant force on each wire if the currents are

(a) parallel

(b) antiparallel (opposite)

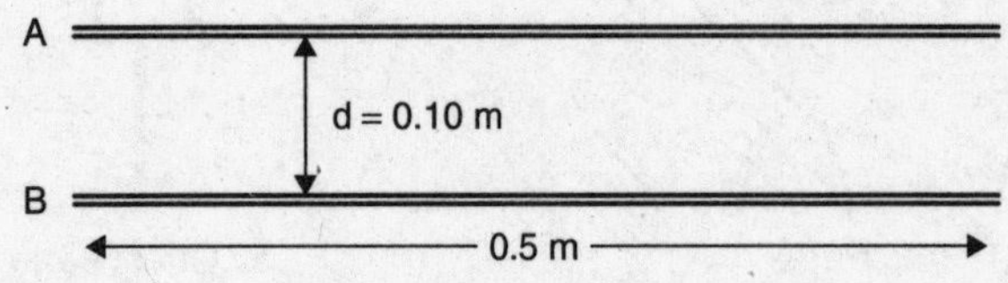

Solution

(a) Using **Fsub B** = $\frac{\mu_0 I_1 I_2 l}{2\pi r}$,

$$F_{PAR} = \frac{(4\pi \times 10^{-7}\,T \cdot m/A)(1.0\,A)(2.0\,A)(0.5\,m)}{(2\pi)(0.10\,m)}$$

$$= \mathbf{2.0 \times 10^{-6}\ N\ ATTRACTING}$$

(b) By Newton's Third Law, the force would be the same but in the opposite direction:

$$\mathbf{F_{ANTIPAR} = 2.0 \times 10^{-6}\ N\ REPELLING}$$

Ampere's Law (C Exam)

To explain the reason for the attraction or repulsion of parallel current-carrying wires, it is necessary to examine the magnetic fields outside of wires and the resultant forces that arise on and from charges moving inside those wires.

One explanation arises from the fact that **F = qv X B.** The current flowing through either wire is immersed in the magnetic field of the opposite wire. The right hand, with fingers pointed in the direction of current flow, crossed into the external magnetic field, yields a force perpendicular to both. In the case of the parallel currents, the force is inward, toward the other current. In the case of antiparallel currents, the force is outward, away from the other current.

Ampere's Law for such wires also helps explain this phenomenon:

$$\oint \mathbf{B} \cdot \mathbf{dl} = \mu_0 I$$

where **B** is the magnetic field in T, **dl** is a small length of wire in a closed loop (hence the closed integral sign), μ_0 is the permeability of free space or vacuum, and I is the net current enclosed in the loop. It essentially says that adding up all the parts of the magnetic field (in the direction of the magnetic field. . .the dot product) and multiplying those parts times each small length of wire (**dl**) equals the permeability of free space times the *net current* (the sum of all currents) in the loop. If the wire is made into a loop, the entire length becomes the circumference, or $\oint \mathbf{B}(2\pi r) = \mu_0 I$.

When a wire with the proper orientation is moved through a magnetic field, the electrons in the wire are pushed by the external magnetic field. If the wire is part of a closed path or loop, current will flow through the circuit. The direction of current flow depends on the direction of the field and the direction of motion of the wire.

Electric and magnetic fields vibrate at right angles to each other. To induce an electric current in a wire that moves through a magnetic field, a force must be applied to the wire (via mechanical energy). The wire must then cut the magnetic field lines at a right angle to give the maximum current (resulting in electrical energy).

This *interconversion* of mechanical and electrical energy is expressed mathematically as **Ampere's Law** for magnetism:

$$\Sigma \mathbf{B} \cdot \mathbf{dl} = \mu o I \text{ or } \oint \mathbf{B} \cdot \mathbf{dl} = \mu o$$

$$\text{also, } \mathbf{F} = \Sigma \mathbf{I}\, \mathbf{l}\, \mathbf{B}\, (\sin \Theta)$$

or

$$\mathbf{F} = \int \mathbf{I}\, \mathbf{dl}\, \mathbf{X}\, \mathbf{B}$$

where **F** is induced force in Newtons, **I** is the induced current in Amperes, **dl** is a small piece of wire length (**l**) in meters, and **B** is the magnetic field in T. Both equations state that the force applied to a wire moving in a magnetic field equals the sum of the added pieces of the wire times the current times the part of the magnetic field that is **perpendicular** (hence the sin or cross-product) to the wire.

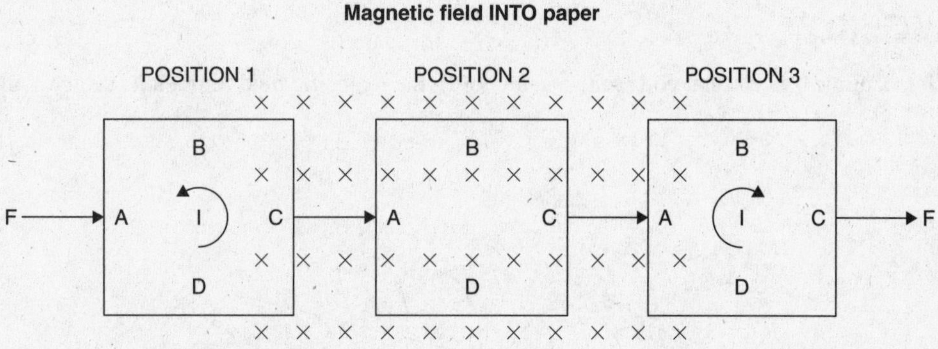

Consider a wire made into a square loop as shown in the preceding illustration. There are three positions illustrated:

1. The right side of the loop is in the magnetic field.
2. The entire loop is in the magnetic field.
3. The left side of the loop is in the magnetic field.

Of course, if the wire remains *stationary* in *any* position, *no current* will flow through any part of the wire, since there is no perpendicular mechanical interaction between the wire and the magnetic field.

If the wire is pulled to the right by the force F, in any of the three positions, no current is generated by sides B or D, since they make an angle of 0° with and do not cut across the magnetic field lines, which are into the page.

For sides A or C in motion, use $\mathbf{F} = q\mathbf{v} \times \mathbf{B}$:

In position 1, the side C will have an induced current flowing through it according to the $\mathbf{F} = q\mathbf{v} \times \mathbf{B}$ cross product rule: Using the right hand, point the fingers in the direction of motion (to the right). Cross them down, into the magnetic field (into the page). The thumb points upward, indicating that the current is counterclockwise.

In position 2, since no current is generated by loop sections on top (B) or bottom (D), and any currents through the sides (A) and (C) are equal and opposite, *there is no current.*

In position 3, the side A will have an induced current flowing through it according to $\mathbf{F} = q\mathbf{v} \times \mathbf{B}$ again: Using the right hand, point the fingers to the right, in the direction of motion, cross them down into the magnetic field (into the page). Again, the thumb points upward, in this case indicating a *clockwise* current.

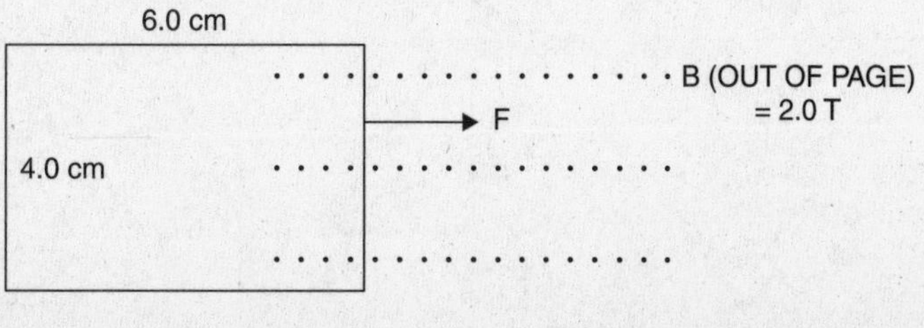

Example

A rectangle of wire moves to the right through a magnetic field of 2.0 T as a result of force **F** as shown in the preceding illustration. If the sides measure 4.0 cm, the top and bottom measure 6.0 cm, and a resulting current of 1.5 A flows through the wire as shown, find the current and its direction.

Solution

Using $\mathbf{F} = \int \mathbf{I}\,\mathbf{dl} \times \mathbf{B}$, $\mathbf{F} = (1.5\text{ A})(0.04\text{ m})(2.0\text{ T})(1) = \mathbf{0.12\ N}$

For the current direction, using $\mathbf{F} = q\mathbf{v} \times \mathbf{B}$: Point the fingers of the right hand to the right, in the direction of the force, palm up. Cross the fingers upward, out of the page, into **B**. The thumb points to the bottom of the page, indicating a *clockwise* current.

Solenoid (Coil)

For a wire bent into a circle or a circular coil (called a **solenoid**), the magnetic field through its center is illustrated by the following:

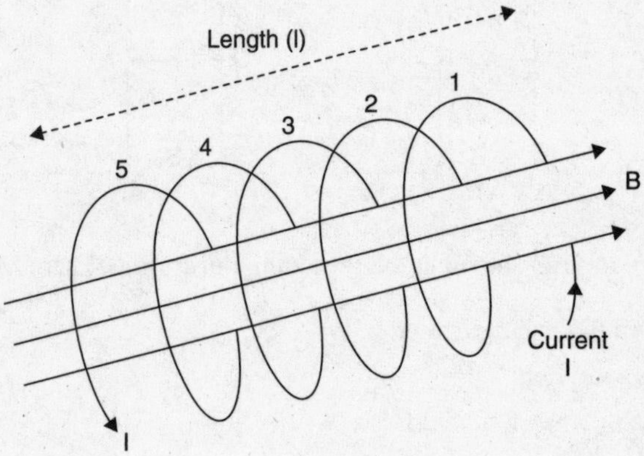

Using your right hand, wrap your fingers around the coil the way current is moving around. Your extended thumb will point in the direction of the magnetic field **B**, or using your right hand, point your thumb in the direction of the current **I**. Your fingers curl around and point in the direction of the magnetic field **B**. The magnetic field is concentrated through the center of the solenoid and is magnified by the number of turns the coil has. . . 5 in this case. It is also magnified by squeezing the number of turns into a smaller length. The magnetic field through the center of a solenoid is given as

$$\mathbf{B_S} = \mu_O\, \mathbf{n}\, \mathbf{I}$$

where $\mathbf{B_S}$ is the magnetic field strength through the center of the solenoid in Teslas (T), μ_O is the permeability of free space or vacuum in (T· m /A), n is the number of turns per unit length, and **I** is the current in Amperes (A).

Example

Using the preceding illustration, find the magnetic field strength for a 0.05 m long solenoid with 5 turns of wire and current of 900 mA.

Solution

Using $\mathbf{B_S} = \mu_O \mathbf{n\ I}$,

$$\mathbf{B_S} = (4\pi x 10^{-7}\ \mathrm{T \cdot m/A}) \frac{(5\ \mathrm{turns})}{(0.05\ \mathrm{m})} (0.9\ \mathrm{A}) = \mathbf{1.1 \times 10^{-4}\ T}$$

Example

If the solenoid in the previous example is compressed to half its original length, keeping everything else the same, what is the effect on the strength of the magnetic field?

Solution

Since length over which the turns are spread directly affects the magnetic field strength, using $\mathbf{B_S} = \mu_O \mathbf{n\ I}$,

$$\mathbf{B_S} = (4\pi \times 10^{-7}\ \mathrm{T \cdot m/A}) \frac{(5\ \mathrm{turns})}{(0.05\ \mathrm{m}/2)} (0.9\ \mathrm{A}) = \mathbf{2.3 \times 10^{-4}\ T\ (B\ DOUBLES)}$$

A solenoid consisting of a number of loops of wire with only air inside is called an **air core** solenoid. Placing a bar of iron or iron alloy, such as steel, in the center greatly enhances the magnetic field. Since the permeability is no longer that of free space, μ is now greater. Another name for an iron or steel core solenoid is an **electromagnet**, which is used in many household and automotive devices. In an electromagnet, μ_O is replaced by μ for the ferromagnetic material used in the center of the coil.

Electromagnetic Induction

It was Hans Christian Oersted who discovered the actual link between electricity and magnetism by passing current through a wire that was placed beside a magnetic compass needle. When current flowed through the wire, the needle moved to a position perpendicular to the wire.

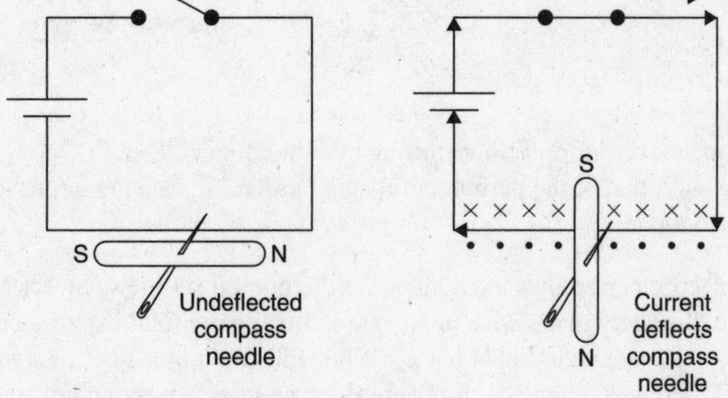

This illustration is a representation of Oersted's experiment. The circuit with no current flowing has no effect on the compass needle. When the switch is turned on, the needle moves to a position perpendicular to the current-carrying wire.

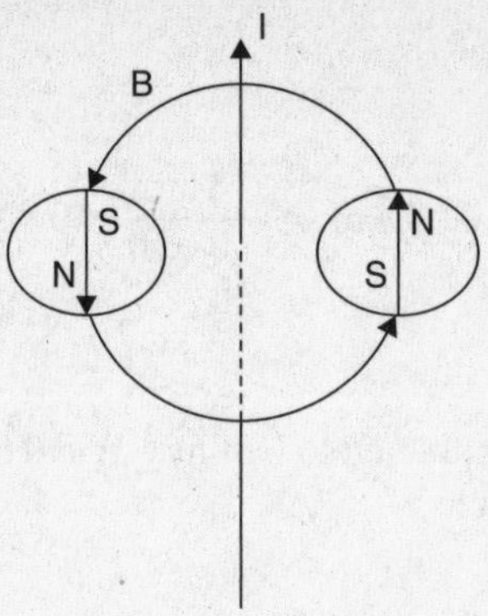

When a compass is placed normal to a current-carrying wire, the needle aligns with the circular magnetic field around the wire.

Electric meters of the analog or mechanical type such as voltmeters, galvanometers, ammeters, ohmmeters, and so on, all work on the principle of electromagnetic induction. A coil is set on a pivot with an indicating needle and a return spring attached. Around the coil is a permanent magnet. When current flows through the coil, it induces a magnetic field, which interacts with the magnetic field of the permanent magnet, producing torque on the coil, which pivots through a certain angle, depending on the strength of the current. The deflection of the indicator needle can be read on the scale.

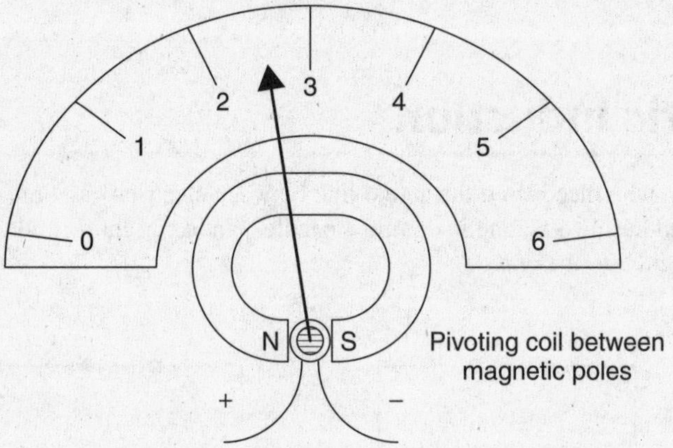

In a typical electromagnetic meter, the current flows through the pivoting coil, setting up a magnetic field around it. The coil's magnetic field interacts with that of the permanent magnet around it, causing a torque on the coil, and thus rotating it and causing the needle to move.

Oersted's discovery that an electric current has a magnetic field around it set the stage for Michael Faraday and Joseph Henry, whose work led to the discovery that a wire in an external magnetic field experiences an emf when either the position of the wire or the external magnetic field change. When either occurs, electromagnetic force is produced, and a current in the wire results if there is a complete circuit. In the case of a wire moving through a magnetic field, the emf produced is given by

$$\varepsilon = \mathbf{Blv}$$

where ε is electromotive force in volts (V), **B** is the magnetic field, sometimes called the **flux density**, in Teslas (N/A · m), l is the length of the wire, and **v** is the velocity of the wire passing through the field. The **magnetic lines of force** or **flux** are closer together nearer to a magnet and become farther apart with distance. Thus, either cutting through magnetic field lines or moving away from a magnet produces an emf in a wire.

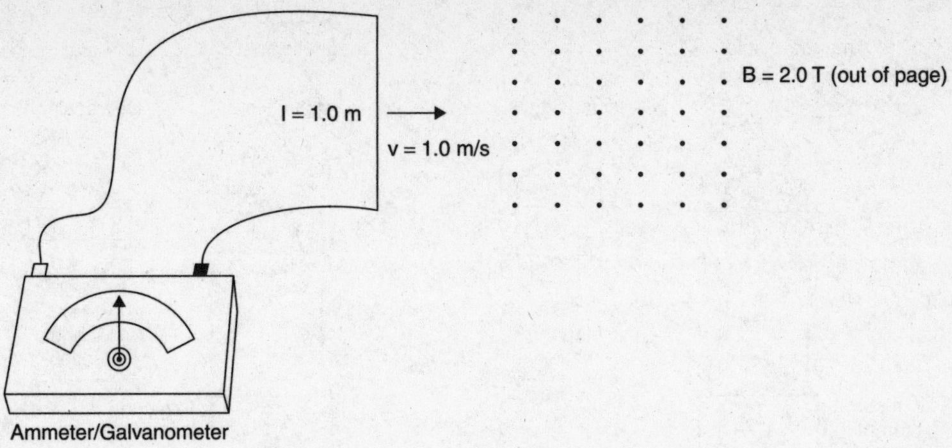

Example

A length of wire 1.0 m in length is pushed through and perpendicular to a magnetic field of 2.0 T at a speed of 1.0 m/s. Determine the emf produced in the wire (and its direction) as a result.

Solution

First, using $\mathbf{F} = q\mathbf{v} \times \mathbf{B}$, we can see that the current in the wire will flow from the **top to the bottom in the wire.**

Using $\varepsilon = \mathbf{Blv}$, we see that the emf is

$$\varepsilon = (2.0 \text{ N/A} \cdot \text{m})(1.0 \text{ m})(1.0 \text{ m/s}) = \mathbf{2.0 \text{ V}}$$

Faraday's Law of Induction

Michael Faraday recognized that an emf is produced in a wire when the magnetic flux around it changes with time. If the wire is formed into a loop, the emf is magnified by the number of turns of wire that are in the loop. Mathematically

$$\varepsilon = -N\frac{\Delta\Phi_B}{\Delta t} \text{ or } \varepsilon = -N\frac{d\Phi_B}{dt}$$

where ε is the emf in volts, N is the number of turns of wire in the loop, and $\Delta\Phi/\Delta t$ or $d\Phi/dt$ is the change of magnetic flux with time. The negative sign indicates that the emf produced is in a direction that opposes the magnetic field change. The magnetic flux Φ_B is equal to the magnetic field **B** times the area that is perpendicular to it.

Heinrich Lenz recognized that **the current produced by moving a wire through a magnetic field provides the opposition to the applied force**. This is known as **Lenz's Law**. There is a connection with Newton's Third Law here: The action is the applied force; the reaction is the opposing force due to the induced current.

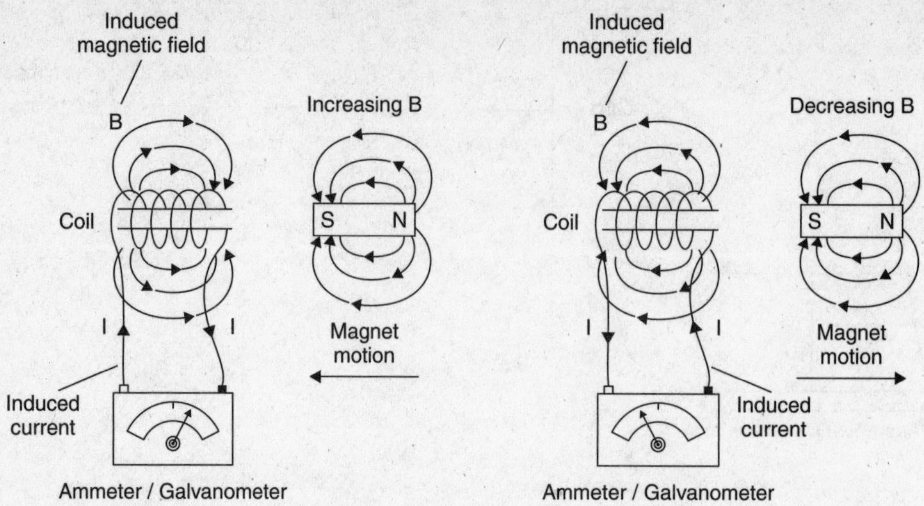

As the magnet's south pole approaches the coil, it causes current I to flow counterclockwise through the coil. A magnetic field develops in the coil, one that opposes that of the magnet. The increasing magnetic field points away from the coil, causing the counterclockwise current flow. Moving the magnet away reverses current direction.

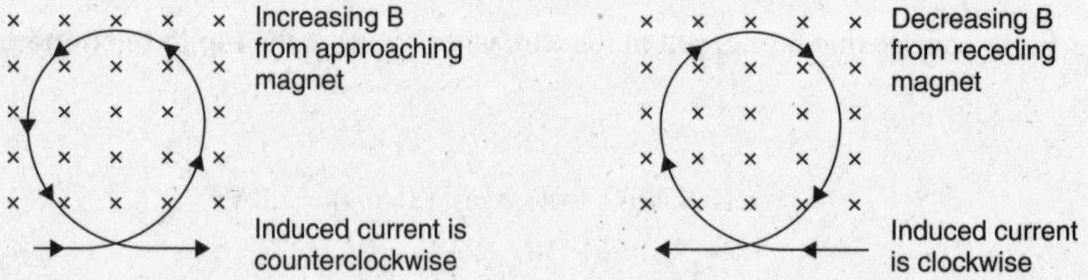

Point right thumb opposite increasing B. Fingers curl in direction of induced current in coil.

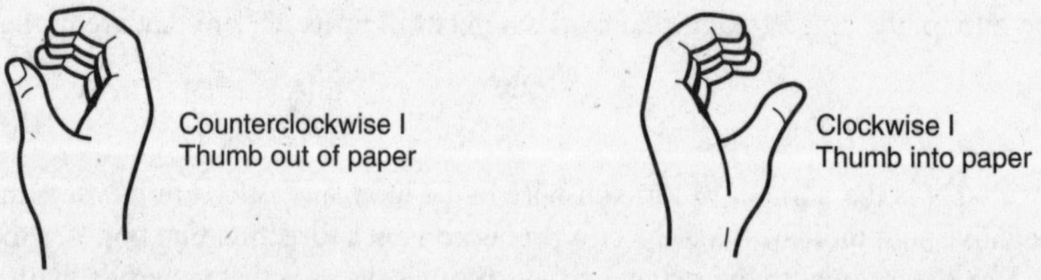

To determine the direction of the induced current in a circular wire or loop, for an increasing or approaching magnetic field, point your right thumb in the direction opposite the increasing/approaching magnetic field. The fingers curl in the direction of the induced resultant current in the loop. For a decreasing or receding magnetic field, point the thumb in the direction of B. The fingers curl in the direction of the current induced by the change in magnetic flux.

Moving the permanent magnet into the coil causes a change in magnetic flux, causing a current to flow through the coil. Moving the magnet in the opposite direction causes a current to flow in the opposite direction. The induced magnetic field in the coil causes a current to flow through the coil, which in turn produces a magnetic field in the coil. It opposes the magnet's magnetic field.

There is a direct correlation between the amount of emf produced in a wire and the **change** in an external magnetic field.

Consider a magnetic field with its lines of flux. The **flux density** is a measure of the strength of the magnetic field itself. It is an indication of the lines of flux passing perpendicularly through a given area. This is mathematically stated by **Coulomb's law of magnetic force**:

$$\mathbf{B = \Phi/A}$$

where **B** is the magnetic field (flux density) in Teslas (N/A ·m), **Φ** is the magnetic flux in Webers (Tm^2). In terms of the magnetic flux **Φ**, this is represented by

$$\Phi_B = \mathbf{B}A \cos \Theta \text{ or } \Phi_B = \int \mathbf{B} \cdot dA$$

(The magnetic field's flux is a measure of the magnetic field times the area that is perpendicular to it.)

When there is a **change in either the position of a wire in relation to a magnetic field or a change in the magnetic field around a wire**, a current will flow in the wire for as long as the change occurs. This means that a current flows **while** a wire or magnetic field at right angles move relative to each other when there is a complete circuit.

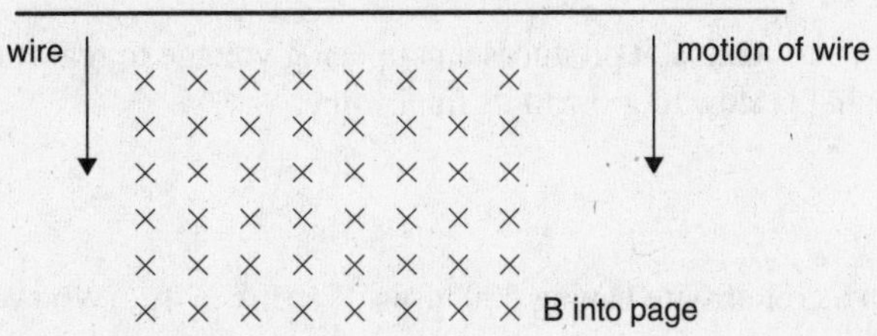

Since **F = qv × B**, current flows to the left in the wire *only* while the wire moves. Leaving the wire in the magnetic field with no current flowing through and no motion will produce no current on its own, the wire being part of a complete circuit.

If the strength of the magnetic field *changes* around a stationary wire:

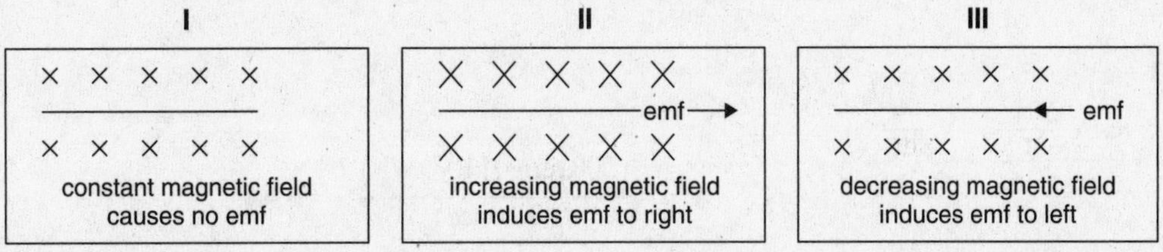

I. A constant magnetic field around a wire causes no emf because there is no change in magnetic flux with time that is "felt" by the wire.

II. Increasing the magnetic field causes a change in flux during the time of increase, causing an induced emf in the wire.

III. Decreasing the magnetic field causes a change in flux during the time of decrease, causing an oppositely directed emf in the wire.

For a conducting wire moving in and perpendicular to a magnetic field, it has been shown that the emf is a result of the field strength, the wire's length, and the wire's velocity. Now, it can be expanded to include a change in magnetic flux:

$$\varepsilon = \mathbf{Blv} = -\mathbf{N}\frac{\Delta\Phi_B}{\Delta t} = -\mathbf{N}\frac{d\Phi_B}{dt}$$

For any coil, supplying current causes the magnetic field to develop over a small period of time as the current grows. The larger the coil or the greater its resistance, the longer it takes to reach maximum magnetic flux. If the current in the coil changes (usually by turning the current on and off), so does the magnetic field and therefore so does the emf induced in the coil. This is expressed by:

$$\varepsilon = -\frac{L\Delta I}{\Delta t} \text{ or } \varepsilon = -\frac{LdI}{dt}$$

where ε is the induced emf in the coil, **L** is the **inductance** (magnetizing ability) of the coil in **henrys** H, **I** is the current in the coil in Amperes A, and t is magnetizing time in seconds s. Again, the negative sign indicates that the emf induced in the coil opposes the directional change of current flow that induced it.

Finally, there is energy gradually built up in the coil as it magnetizes. That energy is expressed by

$$U_L = \frac{1}{2}LI^2$$

where U_L is the energy stored in the inductor (or work done by the circuit in storing that energy) in Joules J, **L** is the inductance of the coil in **henrys** H, and **I** is the current in Amperes A.

Inductance acts as "electrical inertia," that is, it produces an opposing voltage to react against the increasing current. For this reason, it is the current that is slow to grow to its final value.

Example

A current of 3.0 A is supplied to a coil of wire having 500 turns. If a flux of 10^{-3} Wb (Weber) through the center of the coil develops in 0.10 sec, determine:

(a) The average emf induced in the coil.

(b) The coil's inductance.

(c) The energy stored in the coil's magnetic field.

Solution

(a) Using $\varepsilon = -\mathbf{N}\frac{\Delta\Phi_B}{\Delta t} = -\mathbf{N}\frac{d\Phi_B}{dt}$,

$$\varepsilon = -\frac{(500\text{ turns})\left(10^{-3}\text{ Wb}\right)}{(0.10\text{ sec})} = \mathbf{-5.0\ V}$$

(b) Using $\varepsilon = -\frac{L\Delta I}{\Delta t}$ or $\varepsilon = -\frac{LdI}{dt}$,

$$L = \frac{-\varepsilon}{\Delta I/\Delta t} = \frac{-\varepsilon}{dI/dt}$$

$$= \frac{-(-5.0\text{ V})}{(3.0\text{ A})/(1.0\text{ sec})} = \mathbf{0.17\ H\ ANS}$$

(c) Using $\mathbf{U_L = \frac{1}{2}LI^2}$,

$$U_L = \left(\frac{1}{2}\right)(0.17\text{ H})(3.0\text{ A})^2 = \mathbf{0.77\ J\ ANS}$$

Recap of Important Formulas Included in Maxwell's Equations

Gauss' Law for Electric Fields

"The sum of the electric field times the area of a (closed) Gaussian surface equals the total charge divided by the vacuum permittivity."

$$\oint \mathbf{E} \cdot d\mathbf{A} = \frac{q}{\varepsilon_0}$$

Ampere's Law for Magnetism

"The force on a current-carrying wire in a magnetic field equals the current times the length of wire times the perpendicular part of the magnetic field."

$$\mathbf{F} = \oint \mathbf{I}\, d\mathbf{l}\, \mathbf{X}\, \mathbf{B} \text{ and}$$

"The sum of all the parts of the magnetic field times the pieces of length of wire in a circuit equals the vacuum permeability of free space times the net current in that circuit."

$$\Sigma \mathbf{B} \cdot d\mathbf{l} = \mu_0 \mathbf{I}$$

$$\oint \mathbf{B} \cdot d\mathbf{l} = \mu_0 \mathbf{I}$$

Faraday's Law of Induction

"The induced emf is equal and opposite to the time rate of magnetic flux in a circuit."

$$\varepsilon = \oint \mathbf{E} \cdot d\mathbf{s} = \frac{d\Phi_B}{dt}$$

Coulomb's Law of Magnetic Force

"The magnetic field's flux is a measure of the magnetic field times the area that is perpendicular to it."

$$\Phi_B = BA \cos \Theta \text{ or } \Phi_B = \int \mathbf{B} \cdot d\mathbf{A}$$

Gauss' Law for Magnetism

"The magnetic flux through any (closed) Gaussian surface equals zero."

$$\oint \mathbf{E} \cdot d\mathbf{A} = 0 = \Phi_B$$

Review Questions and Answers

B & C Exam Question Types

Multiple Choice

1. A certain charge radiates an electric field. A force will be exerted on it when it

(A) travels in the direction of a magnetic field.
(B) travels against the direction of a magnetic field.
(C) remains stationary outside of a magnetic field.
(D) remains stationary inside a magnetic field.
(E) travels in a path normal to a magnetic field.

2. A positive charge enters a magnetic field and travels parallel to but opposite the field. The charge "feels" or experiences

(A) an upward force.
(B) a downward force.
(C) an accelerative force.
(D) a retardant force.
(E) no force.

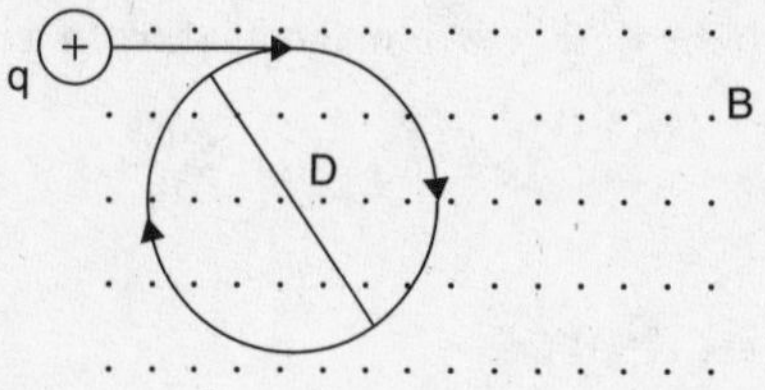

3. A proton enters a 1.0 T magnetic field as shown and attains a circular path of diameter D = 2.0 cm. The electron's speed is most nearly

(A) 10^5 m/s
(B) 10^6 m/s
(C) 10^7 m/s
(D) 10^8 m/s
(E) 10^9 m/s

4. In the previous problem, an alpha particle $^4He_2$ takes the place of the proton and the velocity and magnetic field strength are the same. The ratio of the circumference of the alpha particle's path to that of the proton is

(A) $\frac{1}{4}$
(B) $\frac{1}{2}$
(C) equal
(D) 2x
(E) 4x

5. The value of the magnetic field at a point 10 cm from a wire carrying an electric current of 2.0 A is

(A) 4.0×10^{-6}T
(B) 4.0×10^{-2}T
(C) 4.0×10^{2}T
(D) 4.0×10^{4}T
(E) 4.0×10^{6}T

6. Two long, parallel wires carry currents. If they attract each other, their currents

(A) are parallel.
(B) are antiparallel.
(C) are equal.
(D) are unequal.
(E) draw from the same source.

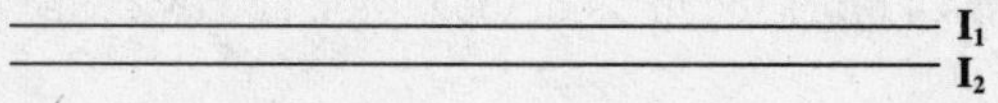

7. The force between the two parallel lengths of wire carrying I_1 and I_2 is a repulsive force. The currents in the wires are

(A) parallel.
(B) antiparallel.
(C) multiples of each other.
(D) equal.
(E) irrelevant.

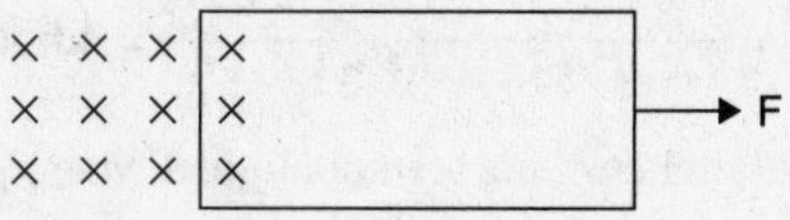

8. The current in the wire is flowing

(A) into the page.
(B) out of the page.
(C) clockwise.
(D) counterclockwise.
(E) (none of the above).

9. In the above problem, if the length of wire cutting the magnetic field is 0.5 m, its velocity is 1.0 m/s and the magnetic field has a strength of 0.75 T, the emf in the wire is most nearly

(A) 0.2 V
(B) 0.4 V
(C) 0.6 V
(D) 0.8 V
(E) 1.0 V

10. A current supplied to a coil of wire with 100 turns develops a flux of 10^{-4} Wb and grows to 1.0A in 2.0 sec. The coil's induced emf is

(A) -5.0×10^{-4} V
(B) -5.0×10^{-3} V
(C) -5.0×10^{-2} V
(D) -5.0×10^{-1} V
(E) -5.0×10^{4} V

Answers to Multiple-Choice Questions

1. (E) According to $\mathbf{F} = q\mathbf{v} \times \mathbf{B}$, a charged particle must be *moving* and have some component of its velocity be **perpendicular** to a magnetic field in order for a force to result.

2. (E) According to $\mathbf{F} = q\mathbf{v} \times \mathbf{B}$, any moving charged particle must have some component of its velocity be perpendicular to a magnetic field in order for a force to result. **The charge experiences no force.**

3. (B) Since the magnetic field exerts a force of $\mathbf{F_B = qv \times B}$ on the proton and its velocity is perpendicular to the field, the proton's circular path is defined by centripetal force ($\mathbf{F_C = mv^2/r}$). Setting the two forces equal gives us $\mathbf{F_B = F_C}$ and the following:

$$q\mathbf{v} \times \mathbf{B} = mv^2/r$$

$$\mathbf{v = rqB/m}$$

$$v = \frac{\left(1.0 \times 10^{-2}\,\text{m}\right)\left(1.6 \times 10^{-19}\,\text{C}\right)(1.0\,\text{T})}{\left(1.67 \times 10^{-27}\,\text{kg}\right)} = \mathbf{10^6\ m/s}$$

4. (D) The alpha particle $^4He_2$ has 4 times the mass of the proton and twice the proton's charge. Using and combining equations

$$\mathbf{F_B} = q\mathbf{v} \times \mathbf{B} \text{ and } \mathbf{F_C} = mv^2/r$$

For the proton: $\mathbf{r} = \frac{\mathbf{mv}}{\mathbf{qB}}$

For the alpha particle: $r_{ALPHA} = \frac{4mv}{2qB} = 2$ (twice the proton's radius)

Since the circumference of a circle is $C = 2\pi r$, the ratio remains the same as the ratio of radii, **2x.**

5. (A) Using $\mathbf{B} = \frac{\mu_0 I}{2\pi r}$,

$$B = \frac{\left(4\pi \times 10^{-7}\,\text{T}\cdot\text{m/A}\right)(2.0\,\text{A})}{(2\pi)(0.1\,\text{m})} = \mathbf{4.0 \times 10^{-6}\ T}$$

6. (A) Since the magnetic fields of parallel currents are not aligned where they intersect, the wires attract.

7. (B) Opposing or antiparallel currents in parallel wires **repel**.

8. (C) Apply $\mathbf{F} = q\mathbf{v} \times \mathbf{B}$ to the positive charges in the vertical wire still in the magnetic field. The force on the charge is upward. The charges in the other parts of the rectangular loop are not forced along the wire. Therefore, the current flows clockwise.

9. (B) Using $\varepsilon = \mathbf{Blv}$,

$$\varepsilon = (0.75\ \text{T})(0.5\ \text{m})(1.0\ \text{m/s}) = \mathbf{0.38\ V}$$

10. (B) Using $\varepsilon = -N\frac{\Delta\Phi_B}{\Delta t}$ or $\varepsilon = -N\frac{d\Phi_B}{dt}$,

$$\varepsilon = \frac{(100\,\text{turns})\left(10^{-4}\,\text{Wb}\right)}{(2.0\,\text{sec})} = -5.0 \times 10^{-3}\ \text{V}$$

Free Response

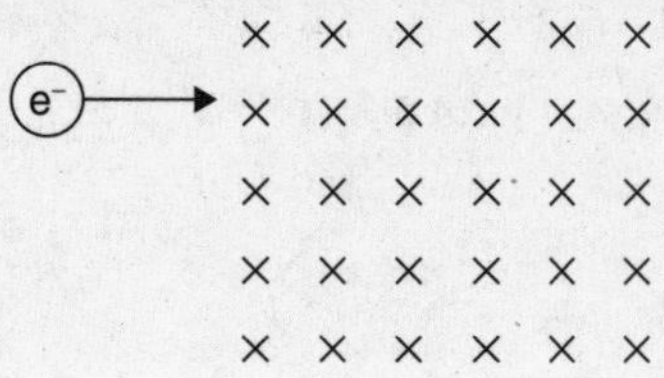

1. An electron traveling at 3.0×10^6 m/s enters a 2.0 T magnetic field directed into the page as shown.

(a) Determine the force on the electron.
(b) Draw a diagram showing the path of the electron.
(c) Determine the radius of curvature of the electron's path.
(d) Determine the electron's centripetal acceleration.
(e) Draw a diagram showing the paths of the following particles passing through the given magnetic field: i. Positron, ii. Proton, iii. Neutron, iv. Alpha particle

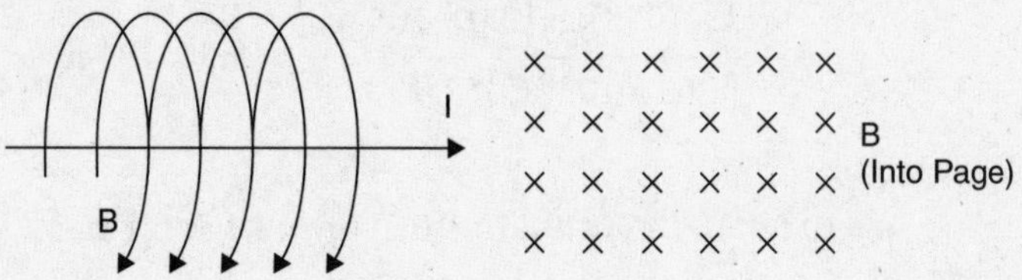

2. A wire carrying a current of 1.0 A is oriented as shown.

(a) What is the magnetic field strength at a point 5.0 cm from the wire?
(b) With the current turned off, the wire is now placed into the magnetic field on the right. When the current in the wire is restored, describe the effect on the wire.
(c) If the magnetic field strength is 5.0 T and the wire is 0.25 m long, determine the force on the wire.

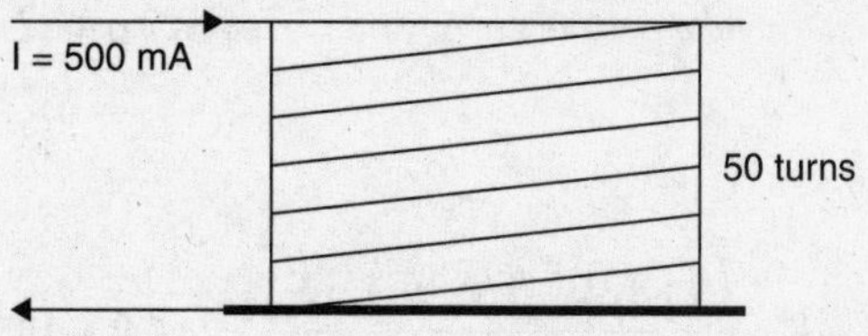

3. An increasing current of 500 mA is supplied to a coil of 50 turns and develops an emf of –9.0 V in 2.0 seconds.

(a) Draw the resulting magnetic field through the coil.
(b) Determine the change in magnetic flux over a period of 2.0 sec.
(c) Determine the inductance of the coil.
(d) How much energy is stored in the coil's magnetic field?

Answers to Free-Response Questions

1. (a) Using $\mathbf{F = qv \times B}$,

$$F = (1.60 \times 10^{-19}\text{ C})(3.0 \times 10^{6}\text{ m/s})(2.0\text{ T}) = \mathbf{9.6 \times 10^{-13}\ N}$$

(b)

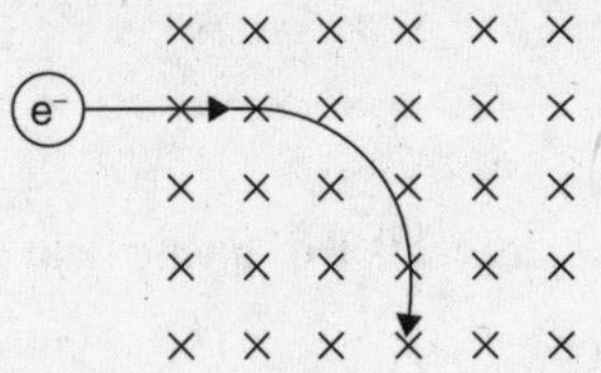

(c) Since the force on the electron is the centripetal force

$$\mathbf{F_B = F_C}$$

$$\mathbf{qv \times B = mv^2 / r}$$

$$\mathbf{r = mv / qB}$$

$$r = \frac{(9.11 \times 10^{-31}\text{ kg})(3.0 \times 10^{6}\text{ m/s})}{(1.60 \times 10^{-19}\text{ C})(2.0\text{ T})} = \mathbf{8.5 \times 10^{-6}\ m}$$

(d) Using $\mathbf{a = F / m}$,

$$\mathbf{a} = (9.6 \times 10^{-13}\text{ N}) / (9.11 \times 10^{-31}\text{ kg}) = \mathbf{1.1 \times 10^{18}\ m/s^2}$$

(e)

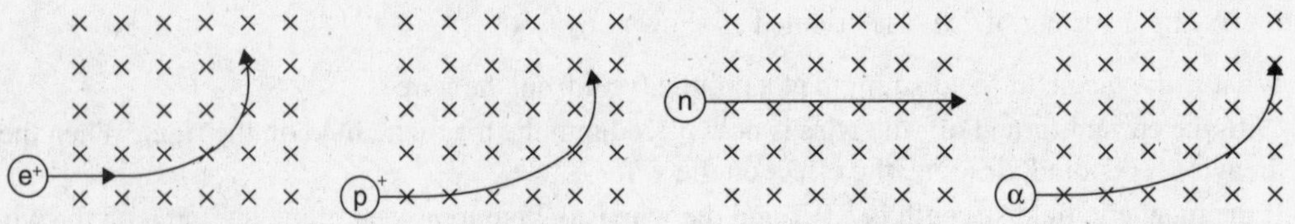

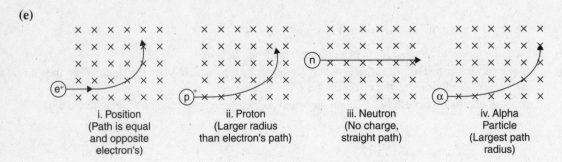

i. Position (Path is equal and opposite electron's)

ii. Proton (Larger radius than electron's path)

iii. Neutron (No charge, straight path)

iv. Alpha Particle (Largest path radius)

2. (a) Using $\mathbf{B} = \dfrac{\mu_0 I}{2\pi r}$

$$B = \frac{(4\pi \times 10^{-7}\text{ T/A}\cdot\text{m})(1.0\text{ A})}{(2\pi)(0.05\text{ m})} = \mathbf{4.0 \times 10^{-6}\ T}$$

(b) Using $\mathbf{F = qv \times B}$ ($\mathbf{F = qvB}$ sin Θ) **or** $\mathbf{F = I \times B}$ ($\mathbf{F = IB}$ sin Θ), using the right hand, pointing the fingers in the direction of the current (right), and then swinging them in the direction of the magnetic field (into the page), the thumb points **upward, toward the top of the page.**

(c) Using $\mathbf{F = I\, dl \times B}$ or $\mathbf{F = IlB}$,

$$F = (1.0\text{ A})(0.25\text{ m})(5.0\text{ T}) = \mathbf{1.3\ N}$$

3. **(a)**

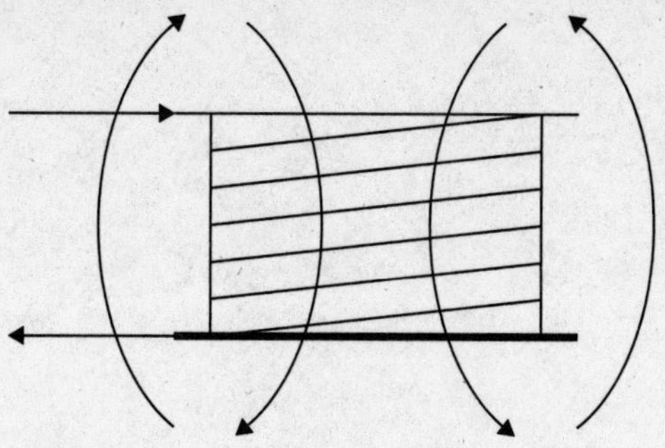

Pointing the thumb of the right hand in the direction of the current flow, the fingers curl downward, through the coil.

(b) Using $\varepsilon = -\frac{N\Delta\Phi_B}{\Delta t}$ or $\varepsilon = -\frac{Nd\Phi_B}{dt}$

$$\Delta\Phi = -\frac{\varepsilon\Delta t}{N} = -\frac{(9.0\,\text{V})(2.0\,\text{sec})}{(50\,\text{turns})} = \mathbf{0.36\ Wb}$$

(c) Using $\mathbf{L} = -\frac{\varepsilon}{\Delta I/\Delta t}$,

$$L = \frac{-(-9.0\,\text{V})}{(0.5\,\text{A})/(2.0\,\text{sec})} = \mathbf{36.0\ H\ \ ANS}$$

(d) Using $\mathbf{U_L = \frac{1}{2}LI^2}$,

$$U_L = \frac{1}{2}(36.0\,\text{H})(0.5\,\text{A})^2 = \mathbf{4.5\ J\ \ ANS}$$

PART III

PRACTICE TESTS: SAMPLE B & C EXAMS WITH ANSWERS AND COMMENTS

CliffsAP Physics B Practice Exam 1

Physics B

You must take the entire B exam as follows:

1st 90 minutes	Section I — Multiple Choice 70 Questions Percent of Total Grade — 50
2nd 90 minutes	Section II — Free Response 7 Questions Percent of Total Grade — 50

Each question in a section has equal weight. Calculators are *not permitted* on Section I of the exam but are allowed on Section II. However, calculators cannot be shared with other students and calculators with typewriter-style (QWERTY) keyboards will not be permitted. A table of information that may be helpful is found on the following page.

Use a separate piece of paper to record your answers.

Information Table

Constants and Conversions

1 Atomic mass unit	$u = 1.66 \times 10^{-27}$ kg
	$= 931$ MeV/ c^2
Proton mass	$m_p = 1.67 \times 10^{-27}$ kg
Neutron mass	$m_n = 1.67 \times 10^{-27}$ kg
Electron mass	$m_e = 9.11 \times 10^{-31}$ kg
Electron charge	$e = 1.60 \times 10^{-19}$ C
Avogadro's number	$N_0 = 6.02 \times 10^{23}$ mol^{-1}
Universal gas constant	$R = 8.31$ J / (mol · K)
Boltzmann's constant	$k_B = 1.38 \times 10^{-23}$ J / K
Speed of light	$c = 3.00 \times 10^8$ m / s
Planck's constant	$h = 6.63 \times 10^{-34}$ J · s
	$= 4.14 \times 10^{-15}$ eV · s
	$hc = 1.99 \times 10^{-25}$ J · m
	$= 1.24 \times 10^3$ eV · nm
Vacuum permittivity	$\varepsilon_0 = 8.85 \times 10^{-12}$ C^2 / N · m^2
Coulomb's law constant	$k = 1/4\pi\varepsilon_0 = 9.0 \times 10^9$ N · m^2 / C^2
Vacuum permeability	$\mu_0 = 4\pi \times 10^{-7}$ (T · m) / A
Magnetic constant	$k' = \mu_0 / 4\pi = 10^{-7}$ (T · m) / A
Universal gravitation constant	$G = 6.67 \times 10^{-11}$ Nm^2/kg^2
Acceleration due to gravity at the earth's surface	$g = 9.8$ m/s^2
1 atmosphere pressure	1 atm $= 1.0 \times 10^5$ Pa
	$= 1.0 \times 10^5$ N / m^2
1 electron volt	1 eV $= 1.60 \times 10^{-19}$ J
1 tesla	1T = 1 weber / m^2

Units

meter	m
kilogram	kg
second	s
ampere	A
kelvin	K
mole	mol
hertz	Hz
newton	N
pascal	Pa
joule	J
watt	W
coulomb	C
volt	V
ohm	Ω
henry	H
farad	F
tesla	T
degree Celsius	°C
electron volt	eV

Prefixes

Factor	Prefix	Symbol
10^9	giga	G
10^6	mega	M
10^3	kilo	k
10^{-2}	centi	c
10^{-3}	milli	m

Trigonometric Function Values for Common Angles

Θ	sin Θ	cos Θ	tan Θ
0°	0	1	0
30°	$\frac{1}{2}$	$\frac{\sqrt{3}}{2}$	$\frac{\sqrt{3}}{3}$
37°	$\frac{3}{5}$	$\frac{4}{5}$	$\frac{3}{4}$
45°	$\frac{\sqrt{2}}{2}$	$\frac{\sqrt{2}}{2}$	1
53°	$\frac{4}{5}$	$\frac{3}{5}$	$\frac{4}{3}$

Prefixes			***Trigonometric Function Values for Common Angles***			
10^{-6}	micro	μ	60°	$\frac{\sqrt{3}}{2}$	$\frac{1}{2}$	$\sqrt{3}$
10^{-9}	nano	n	90°	1	0	∞
10^{-12}	pico	p				

These conventions are used in the examination:

1. *Unless stated otherwise, inertial frames of reference are used.*
2. *The direction of electric current flow is the conventional direction of positive charge.*
3. *For an isolated electric charge, the electric potential at an infinite distance is zero.*
4. *In mechanics and thermodynamics equations, W represents work done **on** a system.*

GO ON TO THE NEXT PAGE

Physics B
Sample Exam #1
Section I

Time – 90 Minutes
70 Questions

Directions: Each of the following questions or incomplete statements is followed by five possible answers. Select the best answer and then fill in the corresponding oval on the answer sheet.

Note: You may use **g** = 10 m/s^2 to simplify calculations.

1. A 5 kg object is moving in a circle of radius 3 m at a constant speed of 2 m/s. The centripetal acceleration of the object is most nearly

(A) $\frac{1}{3}$ m/s^2
(B) 1 m/s
(C) $\frac{4}{3}$ m/s^2
(D) 2 m/s^2
(E) 9.8 m/s^2

2. Regarding the object in the previous problem, its angular acceleration with respect to an axis perpendicular to the circle's center would be

(A) 0 rad/sec^2
(B) $\frac{4}{9}$ rad/sec^2
(C) $\frac{4}{3}$ rad/sec^2
(D) 2 rad/sec^2
(E) $\frac{9}{4}$ rad/sec^2

3. Two objects, of masses M and m respectively, approach each other from opposite directions, each with speed v m/s. Upon impact, m becomes lodged in M. Their combined velocity is now

(A) Mv + mv / (M + m)
(B) Mv – mv / (M + m)
(C) Mv – mv / (M – m)
(D) Mv + mv / (M – m)
(E) Mv + mv / (M – m)^2

4. A 0.5 kg brass weight swung in a horizontal circle of radius 1.0 m at the rate of 1.0 m per sec has most nearly this centripetal force:

(A) 0.1 N
(B) 0.5 N
(C) 2.0 N
(D) 5.0 N
(E) 50 N

5. A shell fired from a cannon at a 30° angle with the ground has an initial velocity of 100 m/s. It is in the air for

(A) 2 seconds.
(B) 10 seconds.
(C) 20 seconds.
(D) 25 seconds.
(E) 30 seconds.

6. How far horizontally does the shell in the previous problem travel?

(A) $\frac{\sqrt{3}}{2}$ m
(B) $\sqrt{3}$ m
(C) $50\sqrt{3}$ m
(D) $500\frac{\sqrt{3}}{2}$ m
(E) $500\sqrt{3}$ m

GO ON TO THE NEXT PAGE

7. An object weighing 10.0 N is swung in a vertical circle of diameter 2.0 m. The object's critical velocity at sea level is most nearly

(A) 1 m/s
(B) 3 m/s
(C) 5 m/s
(D) 7 m/s
(E) 9 m/s

8. A waterfall is 100 m high. The increase in water temperature at the base is most nearly

(A) 0.2° C
(B) 0.8° C
(C) 1.2° C
(D) 2.0° C
(E) 2.8° C

9. A pendulum with a period of 2 sec at sea level is observed in a spacecraft at an altitude above the earth, which is equal to the earth's radius. The pendulum at that altitude has a period of

(A) 2 sec
(B) 4 sec
(C) 6 sec
(D) 8 sec
(E) 10 sec

10. The number of Coulombs of charge contained in an alpha particle ${}^{4}_{2}\alpha$ is

(A) 1.6×10^{-19} C
(B) 2.4×10^{-19} C
(C) 3.2×10^{-19} C
(D) 4.8×10^{-19} C
(E) 5.2×10^{-19} C

Questions 11–13 refer to the following figure.

11. Two masses, M and m, rest on a frictionless horizontal surface as shown. They are attached by a compressed spring of negligible mass, M = 2 m. The spring is released, and the masses move apart. Compared with the total momentum of the masses before release, after release, the total momentum is

(A) half as much.
(B) unchanged.
(C) twice as much.
(D) four times as much.
(E) sixteen times as much.

12. The final momentum of mass M compared with that of mass m is

(A) half as much.
(B) the same.
(C) twice as much.
(D) four times as much.
(E) sixteen times as much.

13. The final kinetic energy of mass m, as compared to that of mass M, is

(A) $\frac{1}{4}$ as great.
(B) $\frac{1}{2}$ as great.
(C) zero.
(D) twice as great.
(E) four times as great.

14. A 20 kg box is pushed along a floor with constant speed and constant force of 40 N. The coefficient of friction between the box and the floor is

(A) 0.02
(B) 0.2
(C) 0.4
(D) 0.8
(E) 2

GO ON TO THE NEXT PAGE

Questions 15–17 refer to the following figure.

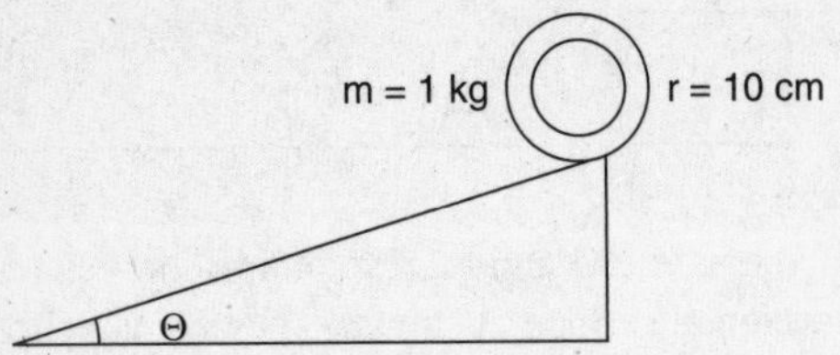

15. A 1 kg hoop of radius 10 cm rolls from rest down a ramp and makes 1.5 rev. in 1 second. Its angular acceleration is

(A) 0.2π rad/s^2
(B) 2π rad/s^2
(C) 4π rad/s^2
(D) 6π rad/s^2
(E) 20π rad/s^2

16. If I for the hoop equals mr^2, the net force acting on the hoop is closest to

(A) 2 N
(B) 4 N
(C) 6 N
(D) 8 N
(E) 10 N

17. The angle Θ of the ramp is nearest to

(A) 0°
(B) 15°
(C) 30°
(D) 45°
(E) 60°

Questions 18–19 refer to the following figure.

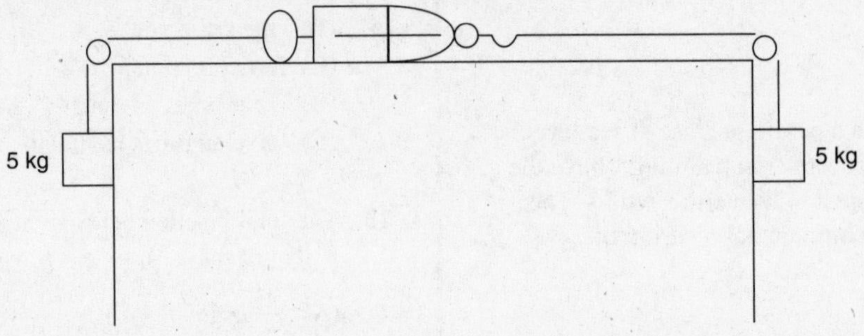

18. Two 5 kg masses are attached to a spring scale by strings that pass over frictionless pulleys at the edge of a lab table as shown here. The spring scale reads

(A) 0 N
(B) 10 N
(C) 20 N
(D) 50 N
(E) 100 N

19. A 1500 kg car is being driven over a hill of radius 40 m. What is the greatest speed that the car may attain and still maintain contact with the road?

(A) 5 m/s
(B) 10 m/s
(C) 15 m/s
(D) 20 m/s
(E) 25 m/s

GO ON TO THE NEXT PAGE

Questions 20–21 refer to the following figure.

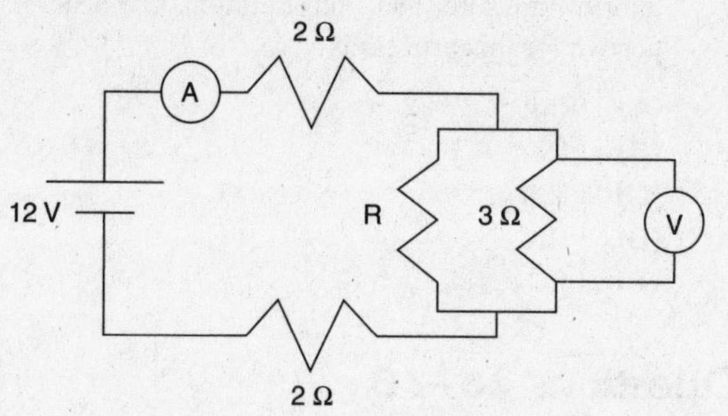

20. If the ammeter in the above illustration reads 2 A, resistor R has a value of

(A) 2 Ω
(B) 4 Ω
(C) 6 Ω
(D) 8 Ω
(E) 10 Ω

21. The voltmeter connected across the 3 Ω resistor indicates a potential drop of

(A) $\frac{1}{5}$ V
(B) $\frac{1}{3}$ V
(C) $\frac{2}{3}$ V
(D) $\frac{4}{5}$ V
(E) $\frac{4}{3}$ V

Questions 22–23 refer to the following figure.

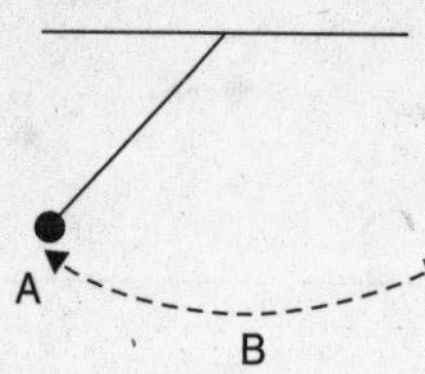

22. The above illustration shows a pendulum in simple harmonic motion. Of the following, which represent the pendulum's velocity versus time and the acceleration vs time graphs?

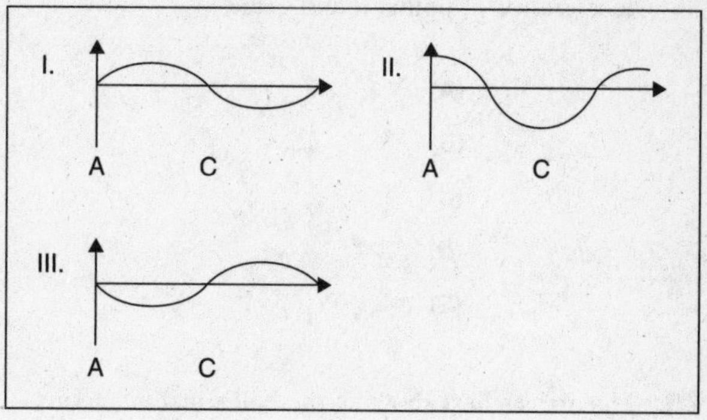

(A) I and II
(B) I and III
(C) II and III
(D) III and I
(E) III and II

23. Which are the graphs of potential energy E_P vs time and kinetic energy E_K versus time?

(A) I and II
(B) I and III
(C) II and I
(D) III and I
(E) III and II

GO ON TO THE NEXT PAGE

Questions 24–26 refer to the following figure.

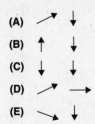

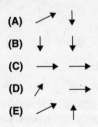

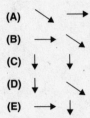

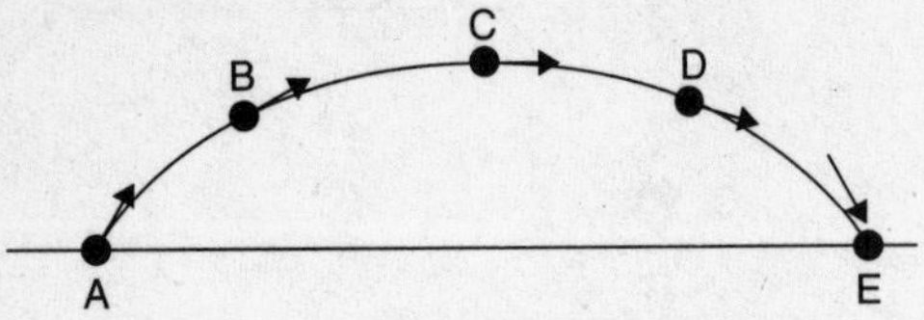

A soccer ball is kicked into the air and reaches its maximum height at point C. Points B and D are equidistant from the ground. Neglect air friction.

24. The arrows best showing the ball's instantaneous acceleration at points B and C are

(A) ↗ ↓
(B) ↑ ↓
(C) ↓ ↓
(D) ↗ →
(E) ↘ ↓

25. The arrows best showing the ball's instantaneous velocity at points A and C are

(A) ↗ ↓
(B) ↓ ↓
(C) → →
(D) ↗ →
(E) ↗ ↑

26. The horizontal and vertical components of the velocity at point D are

(A) ↘ →
(B) → ↘
(C) ↓ ↓
(D) ↓ ↘
(E) → ↓

27. A tea-heater coil of resistance R and current I is placed in a cup of water of mass m, to heat the water from room temperature to boiling at temperature T for t seconds duration. The heat needed for this process is

(A) IR/m
(B) I^2Rmt
(C) I^2Rm / t
(D) I^2Rt
(E) I^2R / t

Questions 28–30

A projectile is fired from a cannon at an angle of 30° with the horizontal and with an initial velocity of 40 m/s.

28. The time it spends in the air is

(A) 2 seconds.
(B) 4 seconds.
(C) 6 seconds.
(D) 8 seconds.
(E) 10 seconds.

29. The horizontal distance it will travel is

(A) < 100 m
(B) between 100 and 200 m
(C) between 200 and 400 m
(D) between 400 and 500 m
(E) > 500 m

30. It reaches a maximum height of

(A) 20 m
(B) 75 m
(C) 100 m
(D) 125 m
(E) 150 m

31. An X-Ray photon having a wavelength of 3 Å has an energy equivalent closest to

(A) 10^{-16} J
(B) 10^{-15} J
(C) 10^{-14} J
(D) 10^{-13} J
(E) 10^{-12} J

GO ON TO THE NEXT PAGE

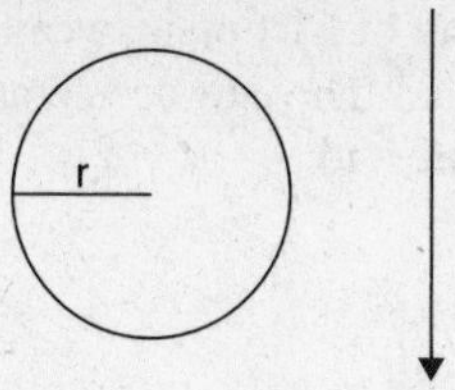

32. A circle of wire of radius r is brought near a straight wire carrying a current I, which is increasing in value. The wires are both on a flat, horizontal table, as seen from above. The false statement is

(A) The magnetic field inside the wire circle resulting from its induced current is directed into the paper.
(B) The induced current in the circle flows counterclockwise.
(C) The current in the circle depends on the changing current in the straight wire.
(D) There is a constant magnetic flux through the circle.
(E) The magnetic flux inside the circle depends on the current in the straight wire.

33. A construction worker on a 20 m high scaffolding throws an object sideways with a speed of 9 m/s. A trash container is 20 m from the base of the scaffolding. How far from the container does the object land?

(A) 2 m
(B) 3 m
(C) 4 m
(D) 5 m
(E) 6 m

34. (3 Volts) × (3 seconds) × (3 Amperes) equals

(A) 27 N
(B) 27 Ω
(C) 27 W
(D) 27 J
(E) 27 C

35. A spring with spring constant k decompresses through a distance **x** and pushes a block of mass m out along a frictionless surface with velocity **v** =

(A) **x**k/m
(B) $\mathbf{x}(m/k)^{\frac{1}{2}}$
(C) $\left(\frac{1}{2}\right)mk^2$
(D) $(k\mathbf{x})^{\frac{1}{2}}$
(E) $\mathbf{x}(k/m)^{\frac{1}{2}}$

36. A kilowatt-hour is a unit of

(A) force.
(B) current.
(C) capacitance.
(D) energy.
(E) power.

37. Each is a vector quantity except

(A) velocity.
(B) momentum.
(C) torque.
(D) energy.
(E) impulse.

38. A number of forces act on an object which is in rotational equilibrium. Which are true statements about the state of the object?

(A) There is no net force.
(B) There is no torque.
(C) There is no net torque.
(D) There is no acceleration.
(E) There is no momentum.

39. Total internal reflection occurs when

(A) sin i > sin r
(B) i < r
(C) n < 1
(D) $i > I_C$
(E) c > n

40. In order for a mass to accelerate, which is true?

(A) There must be no friction acting on it.
(B) There must be a force acting on it.
(C) There must be a net force acting on it.
(D) There must be no normal force acting on it.
(E) There must be a normal force acting on it.

GO ON TO THE NEXT PAGE

Questions 41–43 refer to the following figure.

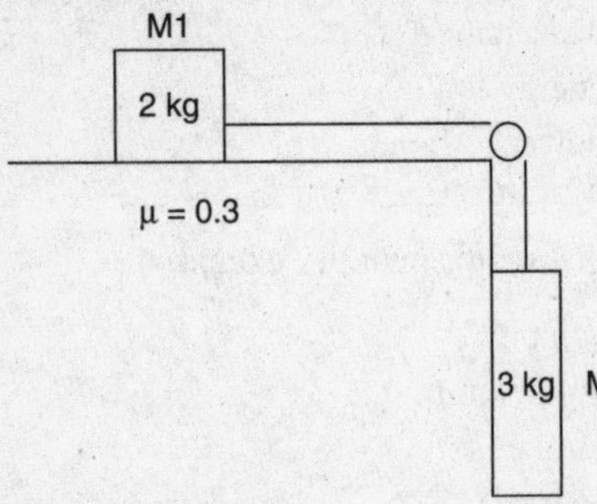

41. In the above illustration, a 2 kg mass M1 rests on a horizontal table. A second mass of 3 kg M2 hangs over the side of the table, connected to M1 by a string that passes over a frictionless pulley. The coefficient of friction between M1 and the table is 0.3. The net force on M1 is

(A) 6 N
(B) 10 N
(C) 18 N
(D) 24 N
(E) 30 N

42. The acceleration of M1 is most nearly

(A) 2 m/s^2
(B) 5 m/s^2
(C) 8 m/s^2
(D) 12 m/s^2
(E) 16 m/s^2

43. The coefficient of friction between M1 and the table is doubled. The new acceleration of the system is nearest to

(A) 2 m/s^2
(B) 3 m/s^2
(C) 4 m/s^2
(D) 5 m/s^2
(E) 6 m/s^2

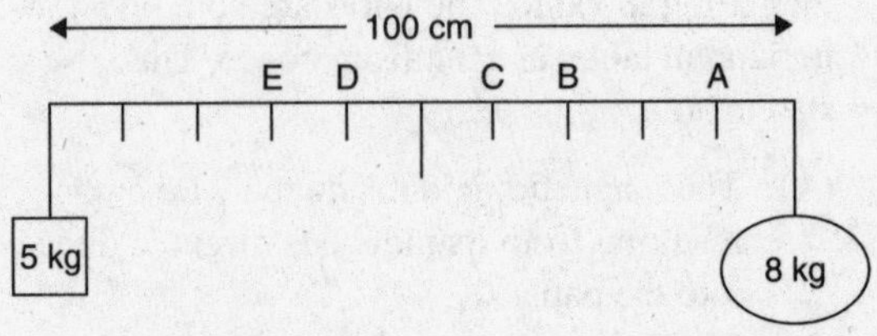

44. Two masses, 5 and 8 kg, hang from opposite ends of a meter stick of negligible mass. Closest to which point should a string be attached to suspend this system and maintain rotational equilibrium?

(A) A
(B) B
(C) C
(D) D
(E) E

Questions 45–47 refer to the following figure.

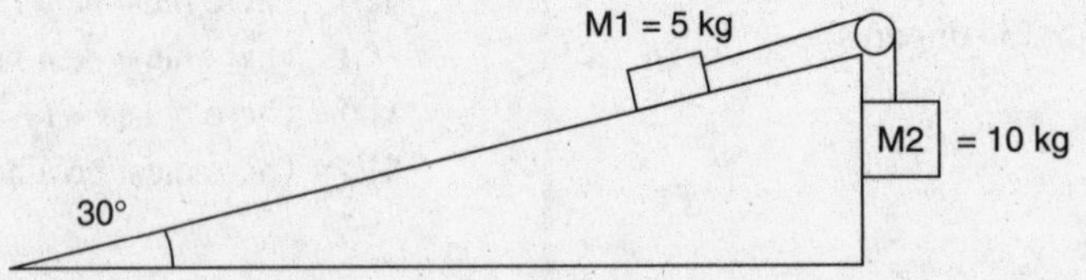

Two masses, M1 (5 kg) and M2 (10 kg) are attached by a string placed over a frictionless pulley as shown. The surface is frictionless, and the incline is 30°.

45. The net force on the system is

(A) 5 N
(B) 20 N
(C) 25 N
(D) 50 N
(E) 75 N

46. The acceleration of mass M2 is

(A) 2 m/s^2
(B) 5 m/s^2
(C) 8 m/s^2
(D) 10 m/s^2
(E) 12 m/s^2

GO ON TO THE NEXT PAGE

47. The 5 kg mass (M1) is replaced by a 10 kg mass. The new acceleration of M2 is

(A) 1.5 m/s^2
(B) 2 m/s^2
(C) 2.5 m/s^2
(D) 3 m/s^2
(E) 3.5 m/s^2

48. Isotopes contain

(A) different numbers of protons.
(B) equal numbers of neutrons.
(C) different numbers of electrons.
(D) different numbers of neutrons.
(E) equal numbers of nucleons.

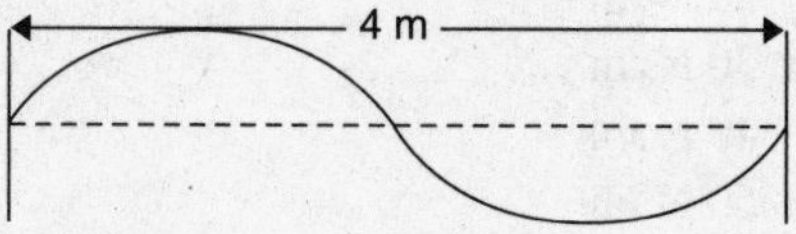

49. A standing wave of λ 4 m oscillates on a string as shown. If it oscillates at a frequency of 3 Hz, its speed is

(A) 3/4 m/s
(B) 4/3 m/s
(C) 3 m/s
(D) 6 m/s
(E) 12 m/s

50. The efficiency of a heat engine that takes in 200 J of heat and expels 140 J of heat is most nearly

(A) 10%
(B) 30%
(C) 50%
(D) 60%
(E) 70%

51. Sound waves traveling through water can best be described as

(A) torsional.
(B) transverse.
(C) undamped.
(D) longitudinal.
(E) electromagnetic.

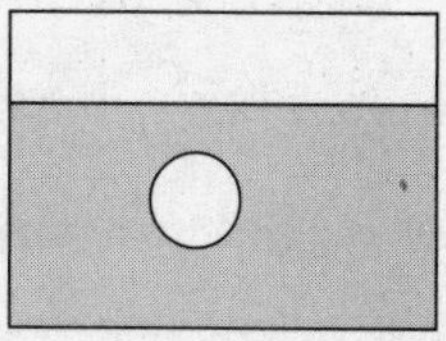

52. A sphere of mass m and diameter d is immersed in a liquid of density ρ. The buoyant force on the sphere is

(A) $\rho\pi d^2 g/6$
(B) $\rho\pi d^3 g/6$
(C) $\rho\pi d^3 g/4$
(D) $\rho\pi d^2 g/3$
(E) $\rho\pi d^3 g/3$

53. An object of mass 8 kg oscillates vertically on a spring having a spring constant of 2 N/m. Its period is

(A) π
(B) 2π
(C) 3π
(D) 4π
(E) 5π

54. A rock of mass m falls from a high cliff. If the air friction force is given by **F** = S**v**, where **v** is the velocity of the rock and S is a constant, the acceleration of the rock is

(A) (S**v** / m) – **g/2**
(B) **g** – S**v**
(C) **g** + S**v**
(D) **g** – (S**v** / m)
(E) **g** + (S**v** / m)

GO ON TO THE NEXT PAGE

Questions 55–56 refer to the following figure.

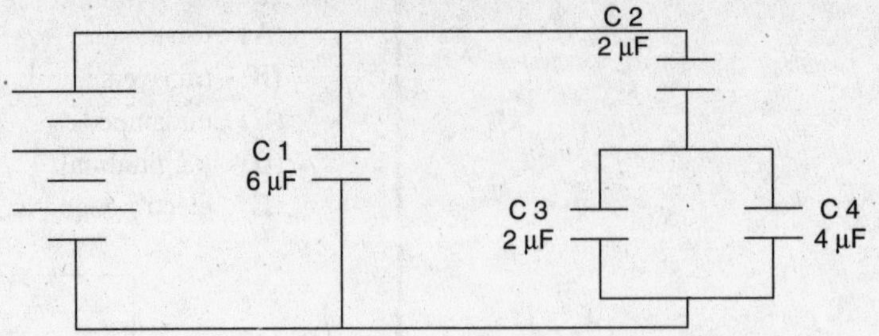

55. The equivalent capacitance of this circuit is nearest to

(A) $\frac{5}{6}$ µF
(B) $\frac{6}{5}$ µF
(C) 7 µF
(D) 7.5 µF
(E) 10.5 µF

56. The charge stored in the 6 µF capacitor C1 is nearest to

(A) 300 µC
(B) 550 µC
(C) 600 µC
(D) 750 µC
(E) 900 µC

57. A current-carrying wire can become a superconductor if

(A) it is immersed in a magnetic field.
(B) it is immersed in a sufficiently cold environment.
(C) the current is changed.
(D) the wire's diameter is sufficiently enlarged.
(E) the wire's length is sufficiently reduced.

58. A rock is dropped on an alien planet and falls a distance of 2 meters in the first second. The planet's gravitational acceleration is

(A) 1 m/s^2
(B) 2 m/s^2
(C) 4 m/s^2
(D) 6 m/s^2
(E) 8 m/s^2

59. A spring is hung from a ring stand. A 2 kg mass is now hung on the spring and stretches it a distance of 0.5 m. The spring constant is most nearly

(A) 10 N /m
(B) 20 N /m
(C) 30 N /m
(D) 40 N /m
(E) 50 N /m

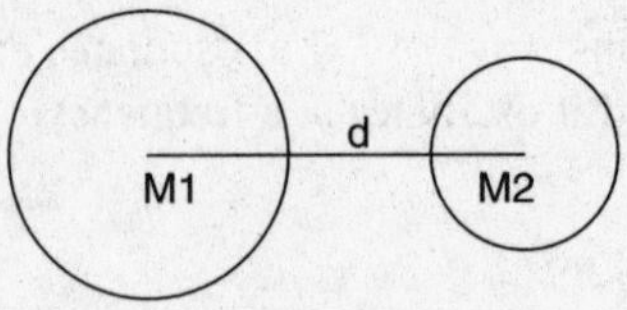

60. Two masses, M1 and M2, are simultaneously dropped in a vacuum. Their distance of separation, center to center, is d. As they begin to fall, the gravitational force they exert on each other is

(A) $G\frac{M1M2}{M1 + M2}$
(B) $G\frac{M1M2}{(M1 + M2)d^2}$
(C) $G\frac{M1M2}{(M1)d^2}$
(D) $G\frac{M1M2}{(M2)d^2/2}$
(E) $G\frac{M1M2}{d^2}$

GO ON TO THE NEXT PAGE

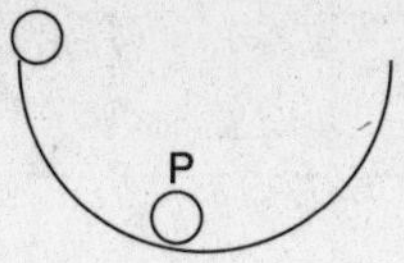

61. A marble is rolled from rest down the side of a semicircular bowl. At point P, the marble's acceleration and velocity are

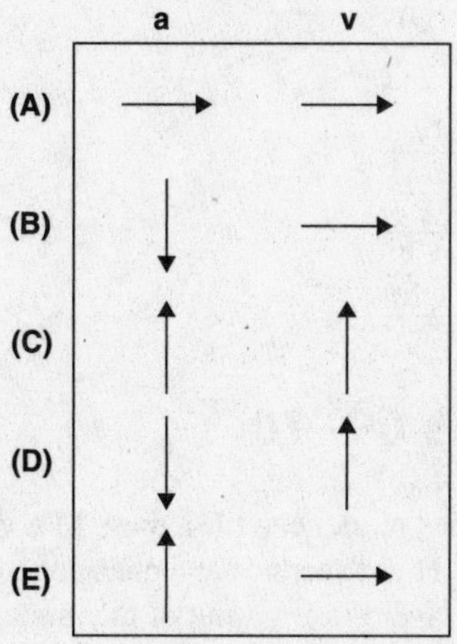

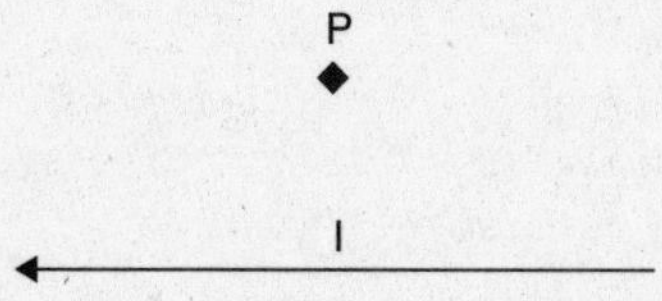

62. At point P, the direction of the magnetic field due to the wire's current I is

(A) to the left.
(B) into the page.
(C) out of the page.
(D) away from the wire.
(E) toward the wire.

Questions 63–65 refer to the following figure.

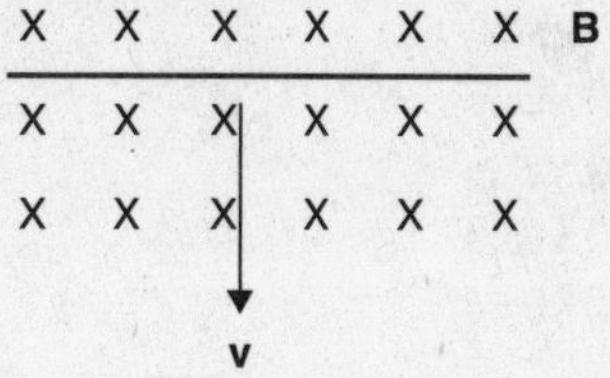

A wire of length 0.2 m moves through a constant magnetic field at a 90° angle and velocity of 0.2 m/s as shown. The 0.5 T magnetic field is directed into the paper.

63. The emf induced in the wire is most nearly

(A) 0.02 V
(B) 0.2 V
(C) 2 V
(D) 10 V
(E) 20 V

64. The current induced in the wire is directed

(A) into the page.
(B) out of the page.
(C) to the left.
(D) to the right.
(E) clockwise.

65. The graphical representation of emf and wire speed is

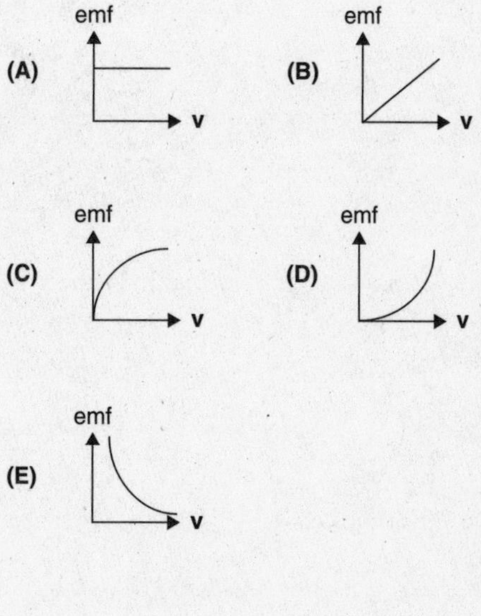

GO ON TO THE NEXT PAGE

CliffsAP Physics B Practice Exam 1

66. A sample of radioactive material has a half-life of 60 years. After how many years will 1/16 of the original substance remain?

(A) 60 years
(B) 120 years
(C) 180 years
(D) 240 years
(E) 300 years

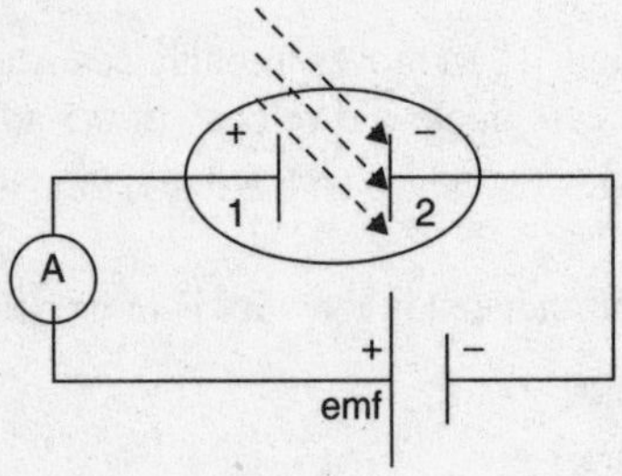

67. Radiant energy of wavelength λ is incident on a metal grid in an evacuated clear glass container as shown. A weak source of emf is included to aid electron transfer. The ammeter will register an increase in current except when

(A) radiant energy intensity is increased.
(B) emf is increased.
(C) the radiant source is constant.
(D) the separation between plates 1 and 2 is reduced.
(E) the wire in the circuit is replaced with thinner wire.

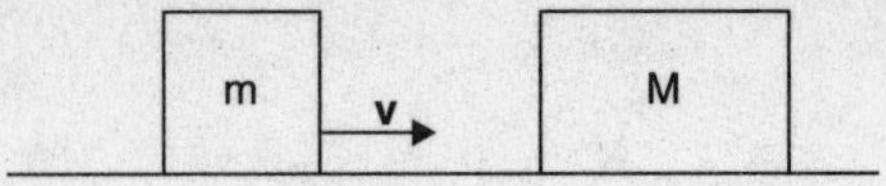

68. Mass m approaches stationary mass M with velocity **v** as shown. Upon impact, m and M become attached. They move with a final velocity **V** of

(A) $\frac{mv}{M-m}$
(B) $m-\frac{M}{mv}$
(C) $\frac{Mv}{m+M}$
(D) $\frac{mv}{M+m}$
(E) $\frac{m+M}{mv}$

Questions 69–70

69. A robotics model can lift a mass M a vertical distance H in t seconds at a constant speed. The average power output of the model is

(A) MgH
(B) MgH / t
(C) MgHt
(D) gH /mt
(E) gH / m

70. The robotics model now lifts a 3.0 kg mass through a height of 0.5 m in 1.0 seconds. The power developed is

(A) 5 W
(B) 10 W
(C) 15 W
(D) 20 W
(E) 25 W

END OF SECTION I. IF YOU FINISH BEFORE TIME IS CALLED, YOU MAY CHECK YOUR WORK ON THIS SECTION. DO NOT GO TO SECTION II UNTIL YOU ARE TOLD TO DO SO.

STOP

Physics B
Section II, Free Response Questions

Time — 90 Minutes, 7 questions of equal weight

Percent of total grade — 50

Instructions: When you are instructed to begin, carefully tear out the green insert and start working. The questions in the green insert are duplicates of the questions in this booklet. A helpful table of information and lists of equations that may be helpful to you are found on pages 1–3 of this booklet. NO CREDIT WILL BE GIVEN FOR ANYTHING WRITTEN IN THE GREEN INSERT. It is provided for reference only as you answer the free response questions.

Show all of your work. You are to write your answer to each question in the pink booklet. Additional answer pages follow each question. Credit for your answers depends on your demonstrating that you know which physical principles would be appropriate to apply in a particular situation. Credit for your answers also depends on the quality of your solutions and explanations. Be sure to write CLEARLY and LEGIBLY. If you make an error, cross it out rather than erasing it. This may save you valuable time.

Advanced Placement Physics B Equations

Newtonian Mechanics		***Electricity and Magnetism***	
$\mathbf{v} = \mathbf{v}_0 + \mathbf{a}t$	$\mathbf{a}$ = acceleration	$\mathbf{F} = \frac{1}{4\pi\varepsilon_0}\frac{q_1 q_2}{\mathbf{r}^2}$	A = area
$\mathbf{x} = \mathbf{x}_0 + \mathbf{v}_0 t + \frac{1}{2}t^2$	$\mathbf{F}$ = force	$\mathbf{E} = \frac{\mathbf{F}}{q}$	$\mathbf{B}$ = magnetic field
$\mathbf{v}^2 = \mathbf{v}_0^2 + 2\mathbf{a}(\mathbf{x}-\mathbf{x}_0)$	f = frequency	$U_E = qV = \frac{1}{4\pi\varepsilon_0}\frac{q_1 q_2}{\mathbf{r}}$	C = capacitance
$\Sigma\mathbf{F} = \mathbf{F}_{NET} = m\mathbf{a}$	h = height	$\mathbf{E}_{AVE} = -\frac{V}{\mathbf{d}}$	$\mathbf{d}$ = distance
$\mathbf{F}_{fric} \leq \mu\mathbf{N}$	$\mathbf{J}$ = impulse	$V = \frac{1}{4\pi\varepsilon_0}\Sigma(q)$	$\mathbf{E}$ = electric field
$\mathbf{a}_c = \frac{\mathbf{v}^2}{\mathbf{r}}$	K = kinetic energy	$C = \frac{Q}{V}$	ε = emf
$\tau = \mathbf{rF}\sin\Theta$	k = spring constant	$C = \frac{\varepsilon_0 A}{\mathbf{d}}$	$\mathbf{F}$ = force
$\mathbf{p} = m\mathbf{v}$	l = length	$U_c = \frac{1}{2}QV = \frac{1}{2}CV^2$	$\mathbf{I}$ = current
$\mathbf{J} = \mathbf{F}\Delta t = \Delta\mathbf{p}$	m = mass	$I_{AVE} = \frac{\Delta Q}{\Delta t}$	l = length
$K = \frac{1}{2}m\mathbf{v}^2$	$\mathbf{N}$ = normal force	$R = \rho(l/A)$	P = power
$\Delta U_g = m\mathbf{gh}$	P = power	$V = IR$	Q = charge
$\mathbf{W} = \mathbf{Fr}\cos\Theta$	$\mathbf{p}$ = momentum	$C_{PAR} = C_1 + C_2 + \ldots C_n$	q = point charge
$P_{AVE} = \frac{\mathbf{W}}{\Delta t}$	r = radius, distance	$\frac{1}{C_{SER}} = \frac{1}{C_1} + \frac{1}{C_2} = \ldots\frac{1}{C_n}$	R = resistance
$P = \mathbf{Fv}\cos\Theta$	$\mathbf{r}$ = position vector	$R_{SER} = R_1 + R_2 + \ldots R_n$	r = distance
$\mathbf{F}_s = -k\mathbf{x}$	T = period	$\frac{1}{R_{PAR}} = \frac{1}{R_1} + \frac{1}{R_2} + \ldots\frac{1}{R_n}$	t = time
$U_s = \frac{1}{2}k\mathbf{x}^2$	t = time	$\mathbf{F}_B = q\mathbf{vB}\sin\Theta$	U = potential (stored) energy
$T_P = 2\pi(l/\mathbf{g})^{\frac{1}{2}}$	U = potential energy	$\mathbf{F}_B = \mathbf{B}\,\mathbf{I}\,l\sin\Theta$	V = electric potential or potential difference
$T_S = 2\pi(m/k)^{\frac{1}{2}}$	$\mathbf{v}$ = velocity, (v) speed	$\mathbf{B} = \frac{\mu_0 I}{2\pi\mathbf{r}}$	$\mathbf{v}$ = velocity
$T = \frac{1}{f}$	$\mathbf{W}$ = work done on a system	$\phi_M = \mathbf{B}\cdot A = \mathbf{B}A\cos\Theta$	v = speed
$\mathbf{F}_G = -\frac{Gm_1m_2}{r^2}$	$\mathbf{x}$ = position	$\varepsilon_{AVE} = \frac{\Delta\phi_M}{\Delta\tau}$	ρ = resistivity
$U_G = -\frac{Gm_1m_2}{r}$	μ = coefficient of friction	$\varepsilon = \mathbf{Blv}$	ϕ_M = magnetic flux
	Θ = angle		
	τ = torque		

Fluid Mechanics and Thermal Physics

$p = p_o + \rho \mathbf{gh}$

$\mathbf{F}_{BUOY} = \rho V \mathbf{g}$

$A_1 v_1 = A_2 v_2$

$p = \rho \mathbf{gy} + 1\rho \mathbf{v}^2 = \text{const.}$

$\Delta l = \alpha l_o \Delta T$

$Q = mL$

$Q = mc\Delta T$

$p = \frac{\mathbf{F}}{A}$

$pV = nRT$

$K_{AVE} = \frac{3}{2} K_B T$

$V_{RMS} = \sqrt{\frac{3RT}{M}} = \sqrt{\frac{3K_B T}{\mu}}$

$W = -p\Delta V$

$Q = nc\Delta T$

$\Delta U = Q + W$

$\Delta U = nc_v \Delta T$

$e = \left|\frac{W}{Q_H}\right|$

$e_c = 1 - \frac{T_C}{T_H} = \frac{T_H - T_C}{T_H}$

A = area
c = specific heat
e = efficiency
F = force
K_{AVE} = average molecular kinetic energy
k_B = Bolzmann's constant
L = latent heat
l = length
M = molecular mass
m = sample mass
n = number of moles
p = pressure
Q = heat transferred to a system
T = temperature
U = internal energy
V = volume
v = velocity
v = speed
v_{RMS} = root-mean-square velocity
W = work done on a system
y = height
α = coefficient of linear expansion
μ = mass of molecule
ρ = density

Atomic and Nuclear Physics

$E = hf = \mathbf{p}c$

$K_{MAX} = hf - \phi$

$\lambda = \frac{h}{\mathbf{p}}$

$E = mc^2$

E = energy
f = frequency
h = Planck's constant
K = kinetic energy
m = mass
p = momentum
λ = wavelength
ϕ = work function

Waves and Optics

$\mathbf{v} = f\lambda$

$n = \frac{c}{v}$

$n_1 \sin\Theta_1 = n_2 \sin\Theta_2$

$\sin \Theta_c = \frac{n_2}{n_1}$

$\frac{1}{f} = \frac{1}{S_1} + \frac{1}{S_o}$

$M = \frac{-h_1}{h_o} = \frac{-S_1}{S_o}$

$f = \frac{c}{2}$

$\lambda = \frac{d \sin \Theta_m}{m}$

$x_m \cong \frac{m\lambda L}{d}$

c = radius of curvature
d = slit separation
f = frequency
f = focal length
h = height
L = distance
M = magnification
m = an integer
n = refractive index
s = distance
v = speed
x = position
λ = wavelength
Θ = angle

Geometry and Trigonometry

Rectangle
$A = bh$

Triangle
$A = \frac{1}{2} bh$

Circle
$A = \pi r^2$
$C = 2\pi r$

Paralelopiped
$V = wh$

Cylinder
$V = \pi r^2 h$
$S = 2\pi rh + 2\pi r^2$

Sphere
$V = \frac{4}{3} \pi r^3$
$S = 4\pi r^2$

Right Triangle
$a^2 + b^2 = c^2$
$\sin \Theta = a/c$
$\cos \Theta = b/c$
$\tan \Theta = a/b$

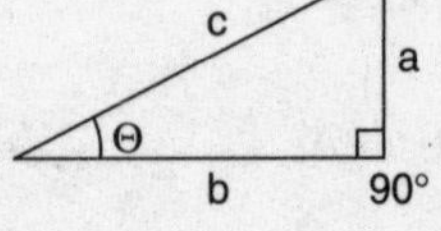

A = area
C = circumference
V = volume
S = surface area
b = base
h = height
l = length
w = width
r = radius

Physics B
Section II

Time — 90 Minutes
7 questions of equal weight

Directions: Answer all of the questions. Each question is equally weighted, but the parts within a question may not be equally weighted. Show your work. Credit for each answer depends on the quality of your explanation.

1. A truck, traveling at 10 m/s, turns a corner and a crate weighing 500 N falls out. It hits the pavement without bouncing and slides 12 m until it stops.

(a) What is the acceleration of the crate in stopping?
(b) What is the coefficient of sliding friction between the crate and the road?
(c) How much work is done stopping the crate?
(d) What is the kinetic energy of the crate just as it starts to slide?
(e) If the crate falls on a wet road having half the coefficient of friction of the dry road, how much work is done stopping the crate?
(f) If it slides 20 m to stop on a wet road, what is the coefficient of friction?

GO ON TO THE NEXT PAGE

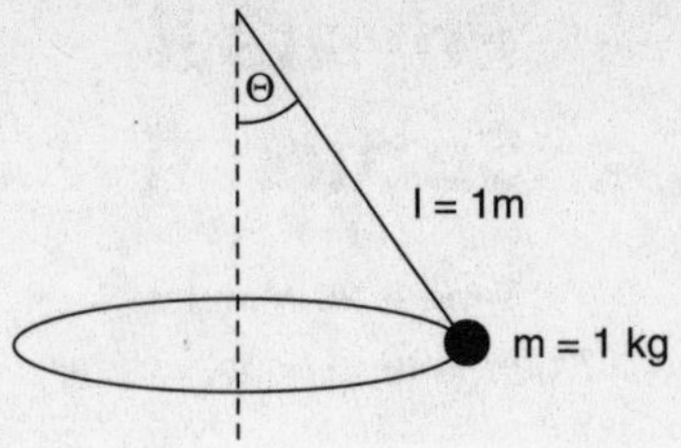

2. A 1 kg mass is swung in a conical path by a 1m long string that has a tensile breaking point at 12 N.

(a) Determine the angle with the vertical at which the string will break.

(b) What is the speed of the mass when the string breaks?

(c) The 1 kg mass is now replaced by a 1.2 kg mass.

i. Determine the new breaking angle.

ii. Determine the new speed for that angle.

(d) The string breaks when the 1.2 kg mass is spun at the top speed in part (c). If it is 1 m above the floor when this occurs, how far horizontally does the mass travel before it hits the floor?

GO ON TO THE NEXT PAGE

3. A golfer needs to "hole" a 7.8 m putt on the 18th hole to win the golf tournament. The green has a coefficient of rolling friction of 0.51 and the ball has a mass of 45g. The putter makes contact with the ball for 0.1 seconds.

(a) Determine the force of rolling friction on the ball.
(b) What minimum energy must be imparted to the ball to get it to the cup?
(c) Determine the acceleration of the ball as it rolls across the green.
(d) How fast is the ball traveling immediately after being struck by the putter if the ball just makes it to the cup?
(e) What minimum force must be applied by the putter to get the ball to the hole?

GO ON TO THE NEXT PAGE

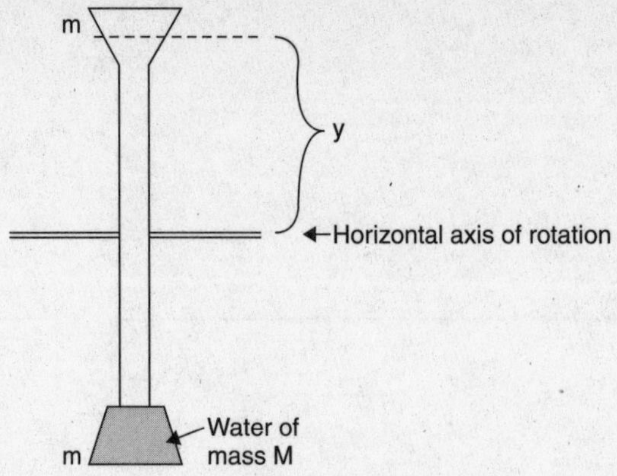

4. An experiment to determine the efficiency of a mechanical-to-thermal energy apparatus is shown. Water of mass M and specific heat of 1 cal/g$^{\circ}$C fills the lower chamber. Both upper and lower chambers are equal in size and volume, each having a mass m. The chambers are connected by a tube of negligible mass.

(a) Assuming that all the water is rotated from bottom to top, through 2y, how much work is required to rotate the apparatus through 180° about the axis?

(b) Assuming that all the work done on the system is converted to thermal energy, by how many degrees does the water's temperature increase?

(c) In the experiment, the apparatus is turned through 300 complete revolutions. If y = 0.5 m, m = 0.2 kg and M = 1 kg, calculate the theoretical increase in the water's temperature after 300 turns.

(d) The water chambers are now replaced with new ones which weigh twice as much and can hold only half as much water. How does this affect your answer to parts (a), (b), and (c)?

(e) When the experiment is performed, the actual rise in temperature is 0.41°C. What is the efficiency of the apparatus?

GO ON TO THE NEXT PAGE

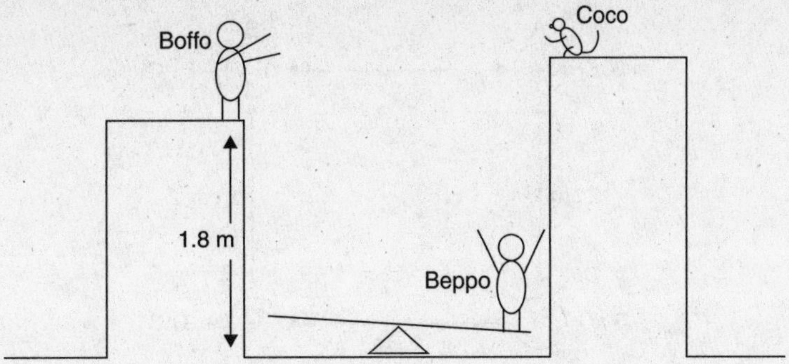

5. Boffo and Beppo, two circus acrobats, perform their famous see-saw stunt with their pet monkey, Coco. Boffo weighs 314 N and Beppo weighs 225 N. Boffo starts by jumping down 1.8 m onto his side of the see-saw, catapulting Beppo into the air. Neglect friction.

(a) How far up does Beppo go?

(b) As Beppo reaches his highest point, Coco jumps on his shoulders and they come back down together. If Coco weighs 45.5 N, how far up do they send Boffo?

(c) While Boffo is still in the air, Coco gets off Beppo's shoulders and gets a snack. When Boffo comes down again, how far up does he send Beppo without Coco?

(d) What is Boffo's speed when he lands in part C?

GO ON TO THE NEXT PAGE

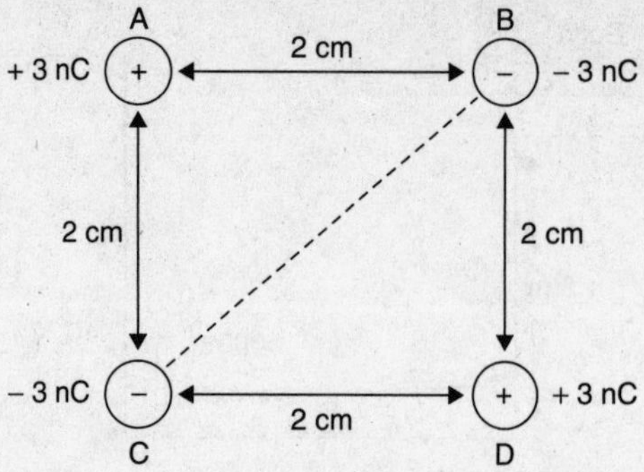

6. Four charges, A (+3.0 nC), B (–3.0 nC), C (–3.0 nC), and D (+3.0 nC) are located at the corners of a square having sides of 2 cm, as shown.

(a) Draw the electric field diagram.

(b) Calculate the electric force on charge B.

(c) Determine the electric potential at the center of the four charges. Show your calculations.

(d) What is the electric field at the center of the four charges?

GO ON TO THE NEXT PAGE

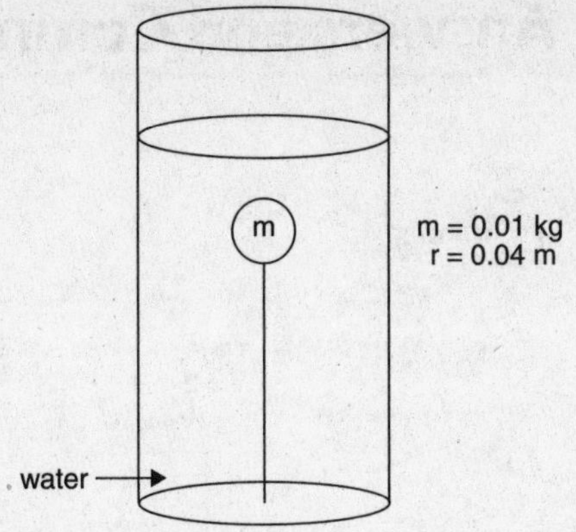

7. A plastic sphere of radius 0.04 m and mass 0.01 kg is immersed in a container of water. It is attached to a string that is attached to the bottom of the container.

(a) Determine the buoyant force on the sphere.

(b) What is the tension in the string?

(c) A larger sphere of radius 0.08 m and equal mass replaces the first one. What is the corresponding tension in the string?

END OF EXAMINATION

Sample B Exam #1 Answers and Comments

Multiple Choice

1. (C) Using $\mathbf{a_C} = v^2/\mathbf{r} = (2\ \text{m/s})^2/3\text{m} = \mathbf{4/3\ m/s^2}$

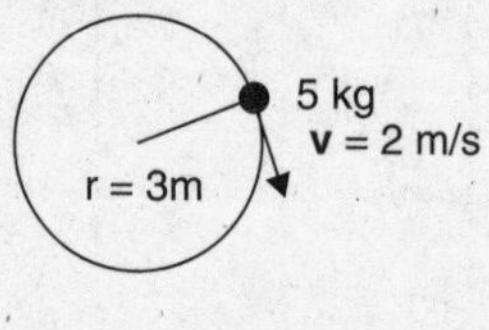

2. (A) Constant tangential speed means no linear or angular acceleration.

3. (B) **Total momentum before = total momentum after**

$$(M)(v_o) + (m)(-v_o) = (M + m)\mathbf{v_F}$$
$$\text{and } \mathbf{v_F} = (Mv - mv) / M + m$$

4. (B) Using $\mathbf{F_C} = mv^2/\mathbf{r} = (0.5\text{kg})(1.0\ \text{m/sec})^2/1.0\ \text{m} = \mathbf{0.5\ N}$

5. (B) Using $\mathbf{V_{o\ VERT}} = 100\text{m/s} \sin 30° = 50\ \text{m/s}$ and $\mathbf{d_{VERT}} = \mathbf{v_o}\,t + \frac{1}{2}\mathbf{a}t^2$

$$0 = 50\ \text{m/s}\ (t) - (5\ \text{m/s}^2)(t^2) \text{ (Total vertical displacement is zero.)}$$
$$t = 50\text{sec}/5 = \mathbf{10\ sec}$$

6. (E) Using $\mathbf{V_{o\ HORIZ}} = 100\ \text{m/s} \cos 30° = (50)\sqrt{3}\ \text{m/s}$

$$\mathbf{d_H} = (\mathbf{V_{o\ HORIZ}})(t) = (50)\sqrt{3}\ \text{m/s}\ (10\ \text{sec}) = \mathbf{(500)\sqrt{3}\ m}$$

7. (B) Using the fact that the critical velocity occurs at the top of the swing, when the object's **weight** equals the **centripetal force** on it: $\mathbf{mg = mv^2/r}$

$$\mathbf{v_{CRIT}} = (\mathbf{rg})^{\frac{1}{2}} = [(1.0\text{m})(10\ \text{m/s}^2)]^{\frac{1}{2}} = (10\ \text{m}^2/\text{s}^2)^{\frac{1}{2}} = \textbf{(closest) 3 m/s}$$

8. (A) $\mathbf{mgh} = mc\Delta T$ Pick a unit kg mass of water that falls:

$$(1\ \text{kg})(10\ \text{m/s}^2)(100\ \text{m}) = (1\ \text{kg})(1\ \text{kcal/kg°C})(4.19\ \text{kJ/kcal})(\Delta T)$$
$$1000\ \text{J°C} / 4190\ \text{J} = \mathbf{0.24°C}$$

9. (D) $\dfrac{T_{ALT} = 2\pi\,(1/\mathbf{g}_{ALT})^{\frac{1}{2}}}{T_{EARTH} = 2\pi\,(1/\mathbf{g}_{EARTH})^{\frac{1}{2}}}$ yields $T_{ALT}/T_{EARTH} = (\mathbf{g_{EARTH}})^{\frac{1}{2}}/\mathbf{g_{ALT}}^{\frac{1}{2}}$

And $T_{ALT} = \dfrac{2\sqrt{\mathbf{g}_{EARTH}}}{\sqrt{\mathbf{g}_{ALT}}}$

$$\frac{\mathbf{g}_{ALT} = Gm_E/r_{ALT}^{\ 2}}{\mathbf{g}_{EARTH} = Gm_E/r_{EARTH}^{\ 2}} \text{ yields } \mathbf{g_{ALT}/g_{EARTH} = r_{EARTH}}^2/(2\,\mathbf{r_{EARTH}})^2$$

$$\mathbf{g_{ALT}} = \left(\frac{1}{4}\right)\mathbf{g_{EARTH}} \text{ and combining with equation}$$

$$T_{ALT} = \frac{2\sqrt{g_{EARTH}}}{\sqrt{g_{ALT}}} \text{ yields } \mathbf{T_{ALT}} = \frac{2\sqrt{g_{EARTH}}}{\frac{1}{2}\sqrt{g_{EARTH}}} = \textbf{4 seconds}$$

10. **(C)** ${}^4\alpha_2$ = (+2 charge)(1.6×10^{-19}C/charge) = $\mathbf{3.2 \times 10^{-19}}$ **C**

11. **(B)** The Law of Conservation of Momentum states that the total amount of momentum **before** an interaction equals the total amount of momentum after an interaction. The final total momentum, therefore, is **unchanged.**

12. **(B)** Comparing the momentum of each after the separation: (Remember, M = 2m.)

$$MV_{FIN} = m\,\mathbf{v_{FIN}}$$
$$2m\,V_{FIN} = m\,\mathbf{v_{FIN}}$$

The final velocity of m is twice that of M, which is twice the mass of m. Momentum is conserved and the final momentum of M **equals that of m.**

13. **(D)** Comparing kinetic energies:

$$E_K \text{ (for M)} = \left(\frac{1}{2}\right)(2m)\,(v_m/2)^2 \qquad E_K\text{(for m)} = \left(\frac{1}{2}\right) m\,\mathbf{v_m}^2$$

E_K (for M) $= \left(\frac{1}{2}\right)$ (m) ($\mathbf{v_m}^2$/2) and E_K for m is **twice as great.**

14. **(B)** Using $\mu = \mathbf{F_F / F_N}$,

$$\mu = \frac{(40\,\text{N})}{(20\,\text{kg})(10\,\text{m/s}^2)} = \mathbf{0.2}$$

15. **(D)** Using $\boldsymbol{\Theta = \omega_O t + \left(\frac{1}{2}\right)\alpha t^2}$,

$$\alpha = 2\Theta / t^2$$
$$\alpha = 2\,(1.5\text{ rev})(2\pi\text{rad/rev})/(1\text{ s})^2 = \mathbf{6\pi rad/s^2}.$$

16. **(A)** Using $\boldsymbol{\tau = Fr = I\alpha}$,

$$\mathbf{F = I\alpha / r}$$
$$= (mr^2)\,\alpha / r$$
$$= (1\text{ kg})(0.1\text{ m})^2(6\pi\text{ rad/s}^2)/0.1\text{ m}) = 0.6\,\pi\text{ N} = \textbf{approx 1.86 N}$$

17. **(B)** Since the net force accelerating the hoop down the ramp is the component of the hoop's weight acting parallel to the ramp, $\mathbf{F_{NET} = F_P}$

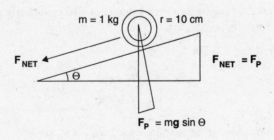

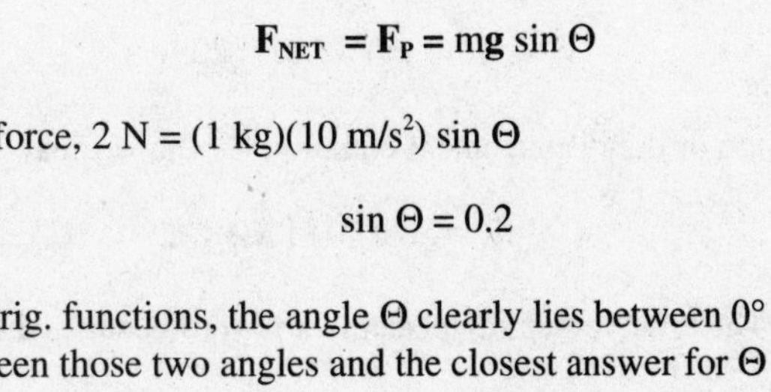

$$\mathbf{F_{NET} = F_P = mg\sin\Theta}$$

Using 2 N as the closest net force, 2 N = (1 kg)(10 m/s^2) sin Θ

$$\sin\Theta = 0.2$$

Checking the given table of trig. functions, the angle Θ clearly lies between 0° and 30°. The value for the sin is approximately halfway between those two angles and the closest answer for Θ is **15°.**

18. **(D)** The most force either mass can exert is $\mathbf{F_{WT}} = \mathbf{mg}$ = (5 kg)(10 m/s^2) = 50 N. If one of the masses is replaced by an immovable object, such as a wall attached to the string, the scale would have to read 50 N. Think of the second 5 kg mass as a wall. It is obvious that the scale can only read **50 N.**

19. **(D)** The situation where the car just maintains contact with the road is due to the weight of the car just being equal and opposite its inertia ("centrifugal force"), or equal to the magnitude of the centripetal force. This happens at the car's **critical velocity ($\mathbf{v_{CRIT}}$)**: Using $\mathbf{mg = mv^2/r}$,

Since the mass cancels, for ANY MASS: $\mathbf{v_{CRIT} = (rg)^{\frac{1}{2}}}$

$\mathbf{v_{CRIT}} = [(40\text{ m})(10\text{ m/s}^2)]^{\frac{1}{2}} = (400\text{ m}^2/\text{s}^2)^{\frac{1}{2}} = \mathbf{20\text{ m/s}}$

20. **(C)** Using **V = IR, R = V / I**

The total resistance in the circuit must be: R_{TOTAL} = 12 V / 2 A = 6 Ω

Since both of the 2 Ω resistors in the main circuit are in series with the two parallel resistors (R and 3 Ω), the parallel combination must equal 2 Ω. Thus, using $\frac{1}{R_{PAR}} = \frac{1}{R_R} + \frac{1}{R_{3\Omega}}$

$$\frac{1}{R_R} = \frac{1}{2\Omega} - \frac{1}{3\Omega} \text{ and } \mathbf{R = 6\ \Omega}$$

21. **(E)** Using V = IR, $V_{3\Omega} = (I_{3\Omega})(R_{3\Omega})$

Since the main line current of 2 A splits unevenly through the parallel resistors, such that twice as much current goes through the 3 Ω resistor as through the 6 Ω resistor R, x + 2x = 2A. Solving for x yields the current in the 6 Ω resistor R:

$$\frac{2}{3}\text{ A and } \frac{2}{3}(2\text{ A}) = \frac{4}{3}\text{ A}$$

22. **(A)** Initial velocity at point A is zero, where the acceleration is greatest.

23. **(C)** Potential energy at points A and C is maximum and is minimum at point B. Kinetic energy is zero at points A and C and is maximum at point B.

24. **(C)** After the initial applied force to launch it, any projectile is **only under the acceleration of gravity.**

25. **(D)** The initial velocity is at the greatest angle with the horizontal at point A. At point C, the velocity is completely horizontal.

26. **(E)** The horizontal velocity component is **constant and parallel with the ground** during the **entire time** any projectile is in the air. The vertical component of the velocity is also **constantly perpendicular to the ground** during the **entire time** any projectile is in the air.

27. **(D)** Using $\mathbf{Q = mc\Delta T}$, $\mathbf{P = I^2R}$ and $\mathbf{P = Q/t}$

$$\frac{mc\Delta T}{t} = I^2R$$

$$\mathbf{mc\Delta T = I^2Rt}$$

28. **(B)** First, find the vertical component of the initial velocity: $\mathbf{v_{O\,V} = v_O}$ sin 30°

$$\mathbf{v_{O\,V}} = (40\text{ m/s})(0.5) = 20\text{ m/s}$$

Now, using the position function or the displacement equation for the vertical:

$$\mathbf{d_Y = v_{O\,V}(t) + \left(\frac{1}{2}\right)at^2}$$

0 = (20 m/s)(t) – (5 m/s^2)(t^2). The total vertical displacement is zero because the projectile begins and ends its flight at the ground level.

$$\mathbf{t = 4\text{ seconds}}$$

29. (B) First, find the horizontal component of the initial velocity: $\mathbf{v_{O\,H}} = \mathbf{v_O} \cos 30°$

$$\mathbf{v_{O\,H}} = (40\ \text{m/s})\frac{(3)^{\frac{1}{2}}}{2} = (20)(3)^{\frac{1}{2}}\ \text{m/s}$$

Now, using $\mathbf{d_H} = (\mathbf{v_H})(\mathbf{t}) = (20)(3)^{\frac{1}{2}}$ m/s(4 sec) = $(80)(3)^{\frac{1}{2}}$ m or **between 100 and 200 m.**

30. (D) Using $\mathbf{d} = \mathbf{v_O t} + \left(\frac{1}{2}\right)\mathbf{a\,t^2}$,

$$\mathbf{d_V} = (\mathbf{v_{O\,V}})(\mathbf{t}) + \left(\frac{1}{2}\right)(\mathbf{g})(\mathbf{t^2})$$ It takes 2 sec to reach the highest point.

$$= (20\ \text{m/s})(2\ \text{sec}) - (5\ \text{m/s}^2)(4\text{s}^2) = 40\ \text{m} - 20\ \text{m} = \mathbf{20\ m}$$

31. (B) Using $\mathbf{f = v / \lambda}$ and $\mathbf{E = hf}$,

$$E = hv/\lambda = (6.63 \times 10^{-34}\text{J} \cdot \text{s})(3 \times 10^{8}\ \text{m/s}) / (3 \times 10^{-10}\ \text{m}) = 6.63 \times 10^{-16}\ \text{J}$$

Rounding to the nearest power of 10: $\mathbf{10^{-15}\ J}$

32. (D) The current is changing in the straight wire; therefore, the induced current in the circle is changing, as is the magnetic flux inside the circle, which is directed into the paper. The induced current flows counterclockwise in the circle.

33. (A) Using $\mathbf{d} = \mathbf{v_O t} + \left(\frac{1}{2}\right)\mathbf{at^2}$, to find the time the object is in the air:

(Position formula for vertical displacement)

$$-20\ \text{m} = -5\ \text{m/s}^2\,(t^2)$$
$$t = 2\ \text{sec in the air.}$$

For the horizontal displacement: (Position formula for horizontal displacement)

$$\mathbf{d} = \mathbf{v_O t} + \left(\frac{1}{2}\right)\mathbf{at^2}$$
$$\mathbf{d} = (9\ \text{m/s})(2\ \text{sec}) = 18\ \text{m.}$$

The object lands **2 m away from the container.**

34. (D) (3 Volts) × (3 seconds) × (3 Amperes) = (3 J / C)(3 sec)(3 C / s) = **27 J**

35. (E) Setting the energy in the compressed spring $\mathbf{E_P} = \left(\frac{1}{2}\right)\mathbf{kx^2}$ equal to the kinetic energy of the sliding mass $\mathbf{E_K} = \left(\frac{1}{2}\right)\mathbf{mv^2}$, $\left(\frac{1}{2}\right)\text{kx}^2 = \left(\frac{1}{2}\right)\text{mv}^2$

$$\mathbf{v^2} = \text{kx}^2/\ \text{m and}\ \mathbf{v} = \mathbf{x\ (k / m)^{\frac{1}{2}}}$$

36. (D) A kilowatt-hour equals 1 kW times 1 hr:

(1000 W)(3600 sec) = (1000 J / s)(3600 sec) = $\mathbf{3.6 \times 10^6}$ **J of energy.**

37. (D) Velocity is defined as speed with direction. Momentum is velocity times mass in a particular direction. Torque is a turning force and is either clockwise or counterclockwise in direction. Impulse is force over a certain amount of time in a particular direction. **Energy is not a vector.** (For instance, energy stored in a spring or a battery has a magnitude but no direction.)

38. (C) Although the object may be in rotational equilibrium, it may still be moving in a straight line, possibly accelerating, and possibly possessing momentum. The torques on the object are balanced, therefore there is **no net torque.**

39. **(D)** Total internal reflection only occurs when the angle of incidence in a refracting medium is **greater than** the critical angle for that interface.

40. **(C)** By Newton's Second Law of Motion, a **net** force acting on a mass causes it to accelerate.

41. **(B)** Tension in upper string = Tension in side string

$$\mathbf{M2g - \mu F_{NM1}} = \Sigma ma$$
$$(3\text{ kg})(10\text{ m/s}^2) - (.3)(2\text{ kg})(10\text{ m/s}^2) = (5\text{ kg})(\mathbf{a})$$
$$30\text{N} - 6\text{N} = (5\text{ kg})(\mathbf{a})$$
$$\mathbf{a} = 4.8\text{ m/s}^2 \text{ and } \mathbf{F_{NET}} = \text{ma} = (2\text{ kg})(4.8\text{ m/s}^2) = 9.6\text{N}$$

42. **(B)** Using $\mathbf{F_{NET} = ma}$,

$\mathbf{a = F_{NET} / m}$ (Here, m is the sum of the masses in the system.)

$$\mathbf{a} = (9.6\text{ N}) / (2\text{ kg}) = \mathbf{5\text{ m/s}^2}$$

43. **(C)** Doubling the friction coefficient to 0.6 means that the frictional force is doubled:

$$\mathbf{F_{NET} = F_{WT\ M2} - F_{F\ M1}} = 30\text{ N} - (0.6)(2\text{ kg})(10\text{ m/s}^2) = 30\text{ N} - 12\text{ N} = 18\text{ N}$$

Now calculating the new acceleration:

$$\mathbf{a = F_{NET} / m_{\ TOT}} = 18\text{ N} / 5\text{ kg} = \mathbf{3.6\text{ m/s}^2 \text{ (closest is } 4\text{ m/s}^2)}$$

44. **(C)** Using CW and CCW torques: (let x be the distance from the 8 kg mass)

CCW torque = CW torque

$$(5\text{kg})(10\text{ m/s}^2)(100\text{ cm} - \text{x}) = (8\text{ kg})(10\text{ m/s}^2)(\text{x})$$

x = 40 cm, position C

45. **(E)** The net force on the system is the weight of the 10 kg mass MINUS the parallel force of the 5 kg mass.

$$\mathbf{F_{NET} = (M2)(g) - F_{P\ M1}}$$
$$= (10\text{ kg})(10\text{ m/s}^2) - (5\text{ kg})(10\text{ m/s}^2)(\sin 30°)$$
$$= 100\text{ N} - 25\text{ N} = \mathbf{75\text{ N}}$$

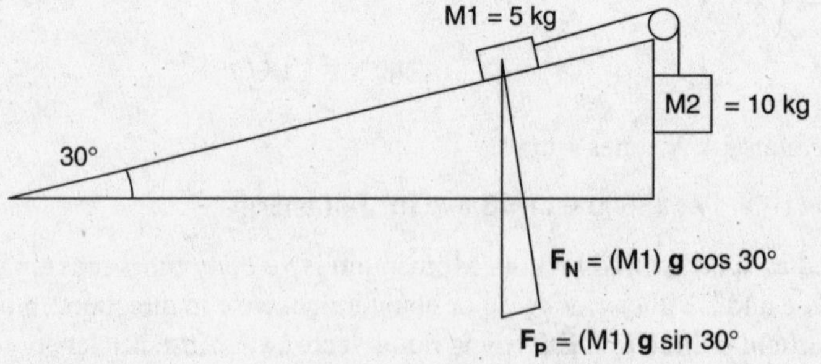

46. **(B)** Using $\mathbf{F_{NET}} = \mathbf{ma}$,

$\mathbf{a} = \dfrac{\mathrm{F_{NET}}}{\mathrm{m_{TOTAL}}}$ (Both masses will accelerate at the same rate.)

$$\mathbf{a} = (75\text{ N})/(15\text{ kg}) = \mathbf{5\text{ m / s}^2}$$

47. (C) Again, using $\mathbf{F_{NET} = ma}$,

$$\mathbf{a} = \frac{\mathbf{F_{NET}}}{\mathrm{m_{TOTAL}}} \quad \text{(Both masses will accelerate at the same rate.)}$$

$$\mathbf{a} = \frac{(10\,\text{kg})(10\,\text{m/s}^2) - (10\,\text{kg})(10\,\text{m/s}^2)(\sin 30^\circ)}{20\,\text{kg}}$$

$$= 50\text{ N} / 20\text{ N} = \mathbf{2.5\ m/s^2}$$

48. (D) Isotopes are atoms of the same element that contain different numbers of neutrons, resulting in varieties of the same element with different masses.

49. (E) Using $\mathbf{f = \frac{v}{\lambda}}$, $\mathbf{v} = f\lambda$ = (3 waves/sec)(4 m/wave) = **12 m/s**

50. (B) $\mathbf{Eff} = \frac{\Delta\text{Energy}}{\text{TOTAL Energy}} = \frac{60\text{ J}}{200\text{ J}} = \mathbf{0.3 = 30\%}$

51. (D) All sound waves are compressional (longitudinal) waves and transfer energy by pushing molecules.

52. (B) Using $\mathbf{F_{BUOY} = \rho Vg}$,

$$\mathbf{F_{BUOY}} = \rho\left(\frac{4}{3}\right)(\pi r^3)\mathbf{g} = \rho\left(\frac{4}{3}\right)(\pi)\left(\frac{d}{2}\right)^3\mathbf{g} = \mathbf{\rho\pi d^3 g/6}$$

53. (D) Using $\mathbf{T = 2\pi\,(m/k)^{\frac{1}{2}}}$,

$$T = 2\pi(8\text{ kg} / 2\text{ N/m})^{\frac{1}{2}} = \mathbf{4\pi}$$

54. (D) Using $\mathbf{F_{NET} = ma}$,

$$\mathbf{a} = \mathbf{F_{NET}} / \text{m} = \frac{\mathbf{mg - F_{FAIR}}}{\text{m}}$$

$$\mathbf{a} = \frac{\mathbf{mg - Sv}}{\text{m}} = \mathbf{g - (Sv / m)}$$

55. (D) First, since C3 and C4 are parallel, add them together. This part of the circuit is 6 μF.

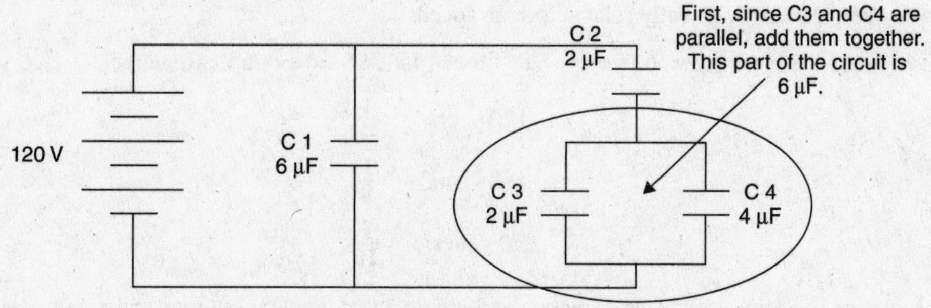

Second, C2 and the C3–C4 combination are in series, so add their inverses:

$$\frac{1}{C_{2,3-4}} = \frac{1}{2\,\mu\text{F}} + \frac{1}{6\,\mu\text{F}} \text{ and } C_{2,3\text{–}4} = 1.5\ \mu\text{F}$$

Third, since C1 is parallel with $C_{2,3\text{–}4}$, add them:

$$C_{TOTAL} = 6\ \mu\text{F} + 1.5\ \mu\text{F} = \mathbf{7.5\ \mu F}$$

56. (D) Using $\mathbf{Q = VC}$,

$$Q = (120\text{ V})(6.0\ \mu\text{F}) = \mathbf{720\ \mu C}$$

57. **(B)** Decreasing temperature extracts energy, decreasing the kinetic energy of molecular vibrations, which decreases the number of electron collisions encountered in current flow.

58. **(C)** The rock travels 2 m in the first second. Its average **speed** for the first second is 2 m/s. 2m/s is the average of (0 + 4m/s) /2. The rock picks up speed at the rate of **4 m/s/s.**

59. **(D)** Using Hooke's Law, **F = –kx,**

$$-k = \frac{(\mathbf{F/x}) = (2\,\text{kg})(10\,\text{m/s}^2)}{(0.5\,\text{m})} = \mathbf{40\ N/m}$$

60. **(E)** Using $\mathbf{F_G} = \frac{\text{G M1 M2}}{\mathbf{r}^2}$,

$$\mathbf{F_G} = \frac{\text{G M1 M2}}{\mathbf{d}^2}$$

61. **(E)** At point P, the marble's velocity is directed toward the right. At P, its acceleration is centripetal due to its circular path and is directed toward the center.

62. **(B)** Using the right hand, pointing the thumb in the direction of the current flow, the fingers curl in the direction of the magnetic field.

63. **(A)** Using ε = **Blv,**

$$\varepsilon = (0.5\ \text{T})(0.2\ \text{m})(0.2\ \text{m/s}) = \mathbf{0.02\ V}$$

64. **(D)** Using **F = qvXB,** fingers swing from **v** to **B** and the thumb points in the direction of **F**.

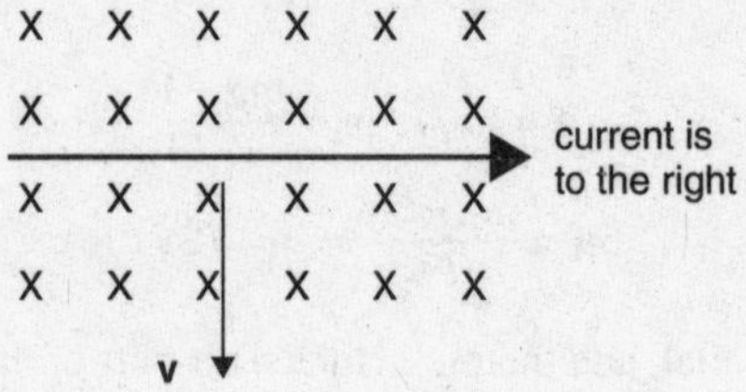

65. **(B)** Since ε = **Blv**, the emf is directly related to wire speed.

66. **(D)** Since one half-life is 60 years, 60 years / half-life × 1 half-life = $\frac{1}{2}$ sample remains

120 years	$\frac{1}{4}$
180 years	$\frac{1}{8}$
240 years	$\mathbf{\frac{1}{16}}$

67. **(A)** Increasing the intensity of the source increases the number of incident photons and results in increased photoelectron emission.

68. **(D)** Using the Law Of Conservation of Momentum:

$$\mathbf{mv} + \text{M}(0) = (\text{m} + \text{M})\ \mathbf{V}$$

$$\frac{\mathbf{mv}}{(\text{m}+\text{M})} = \mathbf{V}$$

69. **(B)** Using $\mathbf{P} = \frac{\mathbf{W}}{\text{t}} = \frac{\mathbf{Fd}}{\text{t}} = \frac{\text{MgH}}{\text{t}}$

70. **(C)** Using $\text{P} = \frac{\text{MgH}}{\text{t}}$,

$$\text{P} = (3.0\ \text{kg})(10\ \text{m/s}^2)(0.5\ \text{m})/(1.0\ \text{s}) = \mathbf{15\ W}$$

Sample B Exam #1 Answers and Comments

Free Response

1. **(a)** $\mathbf{F_{WT}} = \mathbf{F_N} = 500$ N and the mass of the crate is 500 N / 9.8 m/s^2 = 51 kg.

$\mathbf{V_O} = 10$ m/s, $\mathbf{V_F} = 0$, $\mathbf{d} = 12$m Using $\mathbf{V_F}^2 = \mathbf{V_O}^2 + 2\mathbf{ad}$, $\mathbf{a} = (-100\ \text{m}^2/\text{s}^2)/24\ \text{m} = \mathbf{-4.2\ m/s^2}$

(b) Using $\mu = \mathbf{F_F} / \mathbf{F_N}$, $\mu = \dfrac{\mathbf{ma}}{\mathbf{mg}} = \mathbf{a/g} = \dfrac{-4.2\ \text{m/s}^2}{-9.8\ \text{m/s}^2} = \mathbf{0.43}$

(c) Using $\mathbf{W} = \mathbf{Fd} = (\mathbf{ma})\mathbf{d}$, $\mathbf{W} = \dfrac{(500\ \text{N})}{(9.8\ \text{m/s}^2)}(-4.2\ \text{m/s}^2)\ (12\ \text{m}) = \mathbf{2600\ J}$

(d) The kinetic energy of the crate when it begins sliding is completely turned into heat by the work done in stopping it. Its kinetic energy was also **2600 J.**

(e) Halving the original coefficient of friction would make the crate slide farther before stopping but the **same amount of work would still be done**, just spread over a larger distance: 2600 J.

(f) Using $\mathbf{V_F}^2 = \mathbf{V_O}^2 + 2\mathbf{ad}$, $0 = (10\ \text{m/s}^2)^2 + (2)(\mathbf{a})(20\ \text{m})$

$$\mathbf{a_{WET}} = -2.5\ \text{m/s}^2$$

Using $\mu = \dfrac{\mathbf{F_F}}{\mathbf{F_N}}$, $\mu = \dfrac{(\text{ma})}{(\text{mg})} = \dfrac{\mathbf{a}}{\mathbf{g}} = \dfrac{-2.5\ \text{m/s}^2}{-9.8\ \text{m/s}^2} = \mathbf{0.26}$

2.

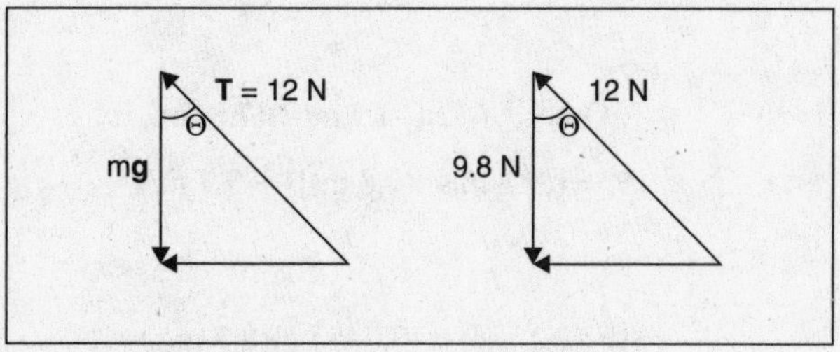

(a) $\cos\Theta = \dfrac{\mathbf{mg}}{\mathbf{T}} = \dfrac{9.8\ \text{N}}{12\ \text{N}} = 0.8166$ and $\mathbf{\Theta = 35.2°}$

(b) Since the velocity will be determined from the centripetal force,

$$\mathbf{F_C} = \frac{\mathbf{mv^2}}{\mathbf{r}}$$

$$\mathbf{F_C} = (12\ \text{N})(\sin\Theta) = (12\ \text{N})(\sin 35.2°) = 6.9\ \text{N} \qquad \frac{(1\,\text{kg})(\text{v}^2)}{(1\,\text{m})}$$

$$\mathbf{v} = \left[\frac{(6.9\,\text{N})(1\,\text{m})}{(1\,\text{kg})}\right]^{\frac{1}{2}} = \mathbf{2.6\ m/s}$$

(c) i. With the new mass: $\cos\Theta = \dfrac{\mathbf{mg}}{\mathbf{T}} = \dfrac{(1.2\,\text{kg})(9.8\,\text{m/s}^2)}{12\,\text{N}} = 0.98$ and $\Theta = \mathbf{11.5°}$

ii. Again using centripetal force:

$$\mathbf{F_C} = \frac{\mathbf{mv^2}}{\mathbf{r}}$$

$$\mathbf{F_C} = (12\ \text{N})(\sin\Theta) = (12\ \text{N})(\sin 11.5°) = 2.39\ \text{N} \qquad \frac{(1\,\text{kg})(\text{v}^2)}{(1\,\text{m})}$$

$$\mathbf{v} = \left[\frac{(2.39\,\text{N})(1\,\text{m})}{(1.2\,\text{kg})}\right]^{\frac{1}{2}} = \mathbf{1.41\ m/s}$$

(d)

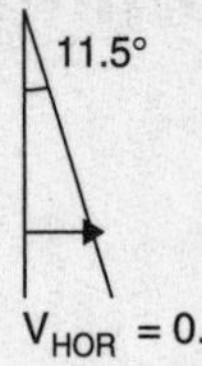

$\sin 11.5° = V_H / V_{TOTAL}$

Since $\mathbf{d_{HOR}} = (\mathbf{v_{HOR}})(t)$, find the time in the air first, using $\mathbf{d_V} = \mathbf{v_{OV}}t + \left(\frac{1}{2}\right)\mathbf{g}\,t^2$

$$-1\text{m} = (-4.9\ \text{m/s}^2)\,t^2 \text{ and } t = 0.45 \text{ sec}$$

Using the total time in the air: $\mathbf{d_H} = \mathbf{v_H t} =$ (0.63 m/s)(0.45 s) = **0.28 m**

3. **(a)** Using $\mu = \frac{\mathbf{F_F}}{\mathbf{F_N}}$,

$$F_F = \mu F_N = (0.51)(0.045\ \text{kg})(9.8\ \text{m/s}^2) = \mathbf{0.22\ N}$$

(b) Since work applied by the putter equals the work done by the grass in stopping the ball,

$$\mathbf{W = F_F d} = (0.22\ \text{N})(7.8\ \text{m}) = \mathbf{1.7\ J}$$

(c) Using **F = ma**,

$$\mathbf{a = F / m\ =\ F_F / m} = \frac{(-0.22\,\text{N})}{(0.045\,\text{kg})} = \mathbf{-4.9\ m/s^2}$$

(d) Using $\mathbf{v_f^2 = v_0^2 + 2ad}$,

$$0 = \mathbf{v_o}^2 + (2)(-4.9\ \text{m/s}^2)(7.8\ \text{m})$$

$$\mathbf{v_o} = [(9.8\ \text{m.s}^2)(7.8\ \text{m})]^{\frac{1}{2}} = \mathbf{8.7\ m/s}$$

(e) Using $\mathbf{F\Delta t = m\Delta V}$,

$$(\mathbf{F})(0.1\ \text{sec}) = (0.045\ \text{kg})(8.7\ \text{m/s})$$

$$\mathbf{F} = \frac{(0.045\,\text{kg})(8.7\,\text{m/s})}{(0.1\,\text{sec})} = \mathbf{3.9\ N}$$

4. The increase in thermal energy is derived from the potential energy of the water turning into kinetic energy as it falls, which then becomes radiant energy and heats the water.

(a) Using **W = Fd**,

W = mgh = (M)(g)(2y) since in one complete revolution, water falls through a distance of y twice.

Also, since m on the top balances m on the bottom, only the mass of water M is lifted. **W = 2Mgy**

(b) Using $\mathbf{W = E = Q = mc\Delta T}$,

$$2Mgy = Mc\Delta T \text{ and}$$

$$\Delta T = \frac{2Mgy}{Mc} = \frac{2gy}{c}$$

(c) Using $\Delta T = \frac{2gy}{c}$,

$$\Delta T = \frac{(2)(9.8\,\text{m/s}^2)(0.5\,\text{m})}{(1\,\text{cal/g}^\circ\text{C})} = \frac{9.8\,\text{m}^2\text{g}^\circ\text{C}(1\,\text{kg})}{(\text{sec}^2)(\text{cal})(1000\,\text{g})(4.19\,\text{J/cal})}$$

Look carefully. All units cancel except for °C.

$$\Delta T = 0.00234\ °\text{C for ONE REVOLUTION}$$

$$\Delta T = (0.00234\ °\text{C/rev})(300\ \text{rev}) = \mathbf{0.702\ or\ 0.7°C}$$

(d) Part (a) will now be **half** as much since only **half the mass is being lifted.**

Part (b) will be **unchanged** since **the water's mass does not affect the calculation in** $\Delta T = \frac{2gy}{c}$.

Part (c)will be **unchanged**, again since **mass does not affect the calculation in** $\Delta T = \frac{2gy}{c}$.

(e) Using **Eff** = $\frac{\text{Work Out}}{\text{Work In}} = \frac{\text{Energy Out}}{\text{Energy In}}$

or $\frac{\Delta T.\text{Observed}}{\Delta T.\text{Predicted}}$

$$\text{Efficiency} = \frac{(0.41^\circ C)}{(0.7^\circ C)} = 0.59 \text{ or } \mathbf{59\%}$$

5. **(a)** Using $\mathbf{mgh_{BOFFO} = mgh_{BEPPO}}$

$$(314\text{ N})(1.8\text{ m}) = (225\text{ N})(\mathbf{h_{BEPPO}})$$
$$\mathbf{h_{BEPPO} = 2.5\ m}$$

(b) Using $\mathbf{mgh_{BOFFO} = mgh_{BEPPO + COCO}}$

$$(314\text{ N})(\mathbf{h_{BOFFO}}) = (225\text{ N} + 45.5\text{ N})(2.5\text{ M})$$
$$\mathbf{h_{BOFFO} = 2.2\ m}$$

(c) Using $\mathbf{mgh_{BOFFO} = mgh_{BEPPO + COCO}}$

$$(314\text{ N})(2.2\text{ m}) = (225\text{ N})(\mathbf{h_{BEPPO}})$$
$$\mathbf{h_{BEPPO} = 3.1\ m}$$

(d) Using Boffo's potential energy converted to kinetic energy:

$$\mathbf{mgh_{BOFFO}} = \left(\frac{1}{2}\right)\mathbf{mv^2}$$
$$\mathbf{v_{BOFFO}} = \sqrt{2gh}$$
$$\mathbf{v_{BOFFO} = 6.6\ m/s}$$

6. **(a)**

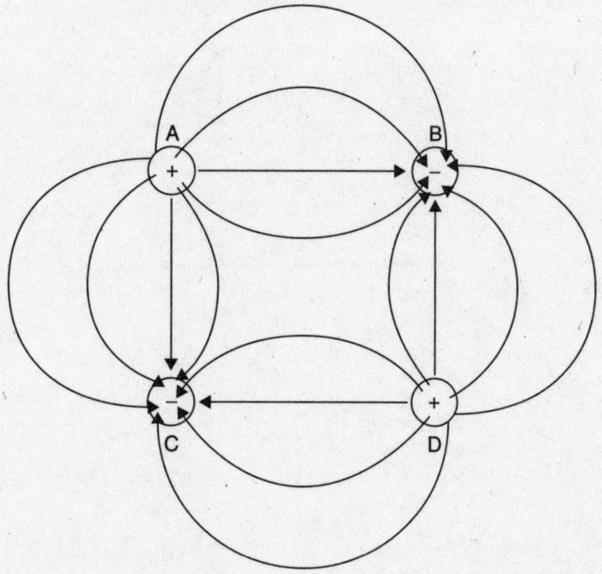

(b) The total electric force on charge B is equal to the sum of the forces of the other three charges on charge B: (k has been substituted for $\frac{1}{4\pi\varepsilon_0}$)

$$\mathbf{F_{TOTB} = F_{AB} + F_{BD} + F_{BC} = F_{BC} - F_{AB} - F_{BD}}$$

$$\mathbf{F_{BC}} = \frac{(k)(-3.0\,nC)(-3.0\,nC)}{\left[(0.02\,m)^2(0.02\,m)^2\right]} = \frac{(9.0\times10^9\,Nm^2/C^2)(9.0\times10^{-18}\,C^2)}{(8.0\times10^{-4}\,m^2)} = 1.0\times10^{-4}\,N$$

$$\mathbf{F_{AB}} = \frac{(k)(+3.0\,nC)(-3.0\,nC)}{(0.02\,m)^2} = \frac{(9.0\times10^9\,Nm/C^2)(-9.0\times10^{-18}\,C^2)}{4.0\times10^{-4}\,m^2} = -2.0\times10^{-4}\,N$$

$$\mathbf{F_{BD} = F_{AB}} = -2.0\times10^{-4}\,N$$

$$\mathbf{F_{TOTB}} = 1.0\times10^{-4}\,N - 4.0\times10^{-4}\,N = \mathbf{-3.0\times10^{-4}\,N\ directed\ toward\ charge\ C.}$$

(c) Using $\mathbf{V = \Sigma\frac{kq}{r}}$,

$$V = \frac{(9\times10^9\,Nm^2/C^2)}{(0.028/2\,m)}[+3\,nC +3\,nC -3\,nC -3\,nC] = \mathbf{0}$$

(d) Using $\mathbf{E = \Sigma\frac{F}{q} = \Sigma\frac{kq}{r}} = \frac{(9\times10^9\,Nm^2/C^2)}{(0.028\,m)^2}[+3\,nC +3\,nC -3\,nC -3\,nC] = \mathbf{0}$

There are two pairs of vectors that are anitparallel and thus add up to zero.

7. **(a)** Using $\mathbf{F_{BUOY} = V\rho g}$,

$$\mathbf{F_{BUOY}} = \left(\frac{4}{3}\right)(\pi)(0.04\,m)^3(1000\,kg/m^3)(9.8\,m/s^2) = \mathbf{2.6\,N}$$

(b) $\mathbf{F_{TENSION} = F_{BUOY} - F_{WT}} = 2.6\,N - (0.01\,kg)(9.8\,m/s^2) = \mathbf{2.5\,N}$

(c) $\mathbf{F_{TENSION} = F_{BUOY} - F_{WT}}$

$$= \left(\frac{4}{3}\right)(\pi)(0.08\,m)^3(1000\,kg/m^3)(9.8\,m/s^2) - (0.01\,kg)(9.8\,m/s^2)$$

$$= 21.0\,N - 0.098\,N \cong \mathbf{21.0\,N}$$

CliffsAP Physics B Practice Exam 2

Physics B

You must take the entire B exam as follows:

1st 90 minutes	Section I — Multiple Choice 70 Questions Percent of Total Grade — 50
2nd 90 minutes	Section II — Free Response 6 Questions Percent of Total Grade — 50

Each question in a section has equal weight. Calculators are *not permitted* on Section I of the exam but are allowed on Section II. However, calculators cannot be shared with other students and calculators with typewriter-style (QWERTY) keyboards will not be permitted. A table of information that may be helpful is found on the following page.

Use a separate piece of paper to record your answers.

Information Table

Constants and Conversions		***Units***	
1 Atomic mass unit	$u = 1.66 \times 10^{-27}$ kg	meter	m
	$= 931$ MeV/ c^2	kilogram	kg
Proton mass	$m_p = 1.67 \times 10^{-27}$ kg	second	s
Neutron mass	$m_n = 1.67 \times 10^{-27}$ kg	ampere	A
Electron mass	$m_e = 9.11 \times 10^{-31}$ kg	kelvin	K
Electron charge	$e = 1.60 \times 10^{-19}$ C	mole	mol
Avogadro's number	$N_0 = 6.02 \times 10^{23}$ mol^{-1}	hertz	Hz
Universal gas constant	$R = 8.31$ J / (mol · K)	newton	N
Boltzmann's constant	$k_B = 1.38 \times 10^{-23}$ J / K	pascal	Pa
Speed of light	$c = 3.00 \times 10^{8}$ m / s	joule	J
Planck's constant	$h = 6.63 \times 10^{-34}$ J · s	watt	W
	$= 4.14 \times 10^{-15}$ eV · s	coulomb	C
	$hc = 1.99 \times 10^{-25}$ J · m	volt	V
	$= 1.24 \times 10^{3}$ eV · nm	ohm	Ω
Vacuum permittivity	$\varepsilon_0 = 8.85 \times 10^{-12}$ C^2 / N · m^2	henry	H
Coulomb's law constant	$k = 1/4\pi\varepsilon_0 = 9.0 \times 10^{9}$ N · m^2 / C^2	farad	F
Vacuum permeability	$\mu_0 = 4\pi \times 10^{-7}$ (T · m) / A	tesla	T
Magnetic constant	$k' = \mu_0 / 4\pi = 10^{-7}$ (T · m) / A	degree Celsius	°C
Universal gravitation constant	$G = 6.67 \times 10^{-11}$ Nm^2/kg^2	electron volt	eV
Acceleration due to gravity at the earth's surface	$\mathbf{g} = 9.8$ m/s^2		
1 atmosphere pressure	1 atm $= 1.0 \times 10^{5}$ Pa $= 1.0 \times 10^{5}$ N / m^2		
1 electron volt	1 eV $= 1.60 \times 10^{-19}$ J		
1 tesla	1T = 1 weber / m^2		

Prefixes

Factor	Prefix	Symbol
10^9	giga	G
10^6	mega	M
10^3	kilo	k
10^{-2}	centi	c
10^{-3}	milli	m

Trigonometric Function Values for Common Angles

Θ	sin Θ	cos Θ	tan Θ
0°	0	1	0
30°	$\frac{1}{2}$	$\frac{\sqrt{3}}{2}$	$\frac{\sqrt{3}}{3}$
37°	$\frac{3}{5}$	$\frac{4}{5}$	$\frac{3}{4}$
45°	$\frac{\sqrt{2}}{2}$	$\frac{\sqrt{2}}{2}$	1
53°	$\frac{4}{5}$	$\frac{3}{5}$	$\frac{4}{3}$

Prefixes			*Trigonometric Function Values for Common Angles*			
10^{-6}	micro	μ	60°	$\frac{\sqrt{3}}{2}$	$\frac{1}{2}$	$\sqrt{3}$
10^{-9}	nano	n	90°	1	0	∞
10^{-12}	pico	p				

These conventions are used in the examination:

1. *Unless stated otherwise, inertial frames of reference are used.*
2. *The direction of electric current flow is the conventional direction of positive charge.*
3. *For an isolated electric charge, the electric potential at an infinite distance is zero.*
4. *In mechanics and thermodynamics equations, W represents work done* ***on*** *a system.*

GO ON TO THE NEXT PAGE

Physics B
Sample Exam #2
Section I

Time – 90 Minutes

70 Questions

Directions: Each of the following questions or incomplete statements is followed by five possible answers. Select the best answer and then fill in the corresponding oval on the answer sheet.

Note: You may use **g** = 10 m/s^2 to simplify calculations.

1. A physics experiment uses light of a certain wavelength beamed at a metal plate. As electrons are emitted from the surface, adjustments are made to produce fewer electrons per unit time but with more kinetic energy per electron. The adjustments to the beam include

(A) decreasing both the intensity and the wavelength.
(B) decreasing the intensity and increasing the wavelength.
(C) increasing the intensity and decreasing the wavelength.
(D) increasing both the intensity and the wavelength.
(E) increasing the wavelength only.

2. A charged 5 μF capacitor measures a potential difference of 120 V. The potential energy stored in the capacitor is

(A) 3.6×10^{-4} J
(B) 3.6×10^{-3} J
(C) 3.6×10^{-2} J
(D) 3.6×10^{-1} J
(E) 3.6 J

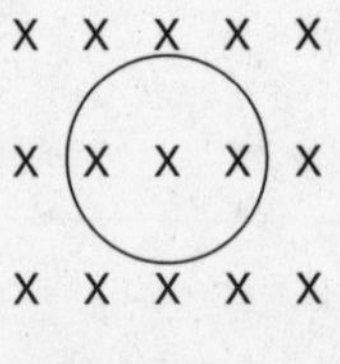

3. A circular loop of wire is placed in and perpendicular to a magnetic field directed into the page as shown. If the magnetic field's magnitude is increasing, the resulting current in the wire is

(A) zero.
(B) clockwise.
(C) counterclockwise.
(D) out of the paper.
(E) into the paper.

Questions 4–5 refer to the following figure.

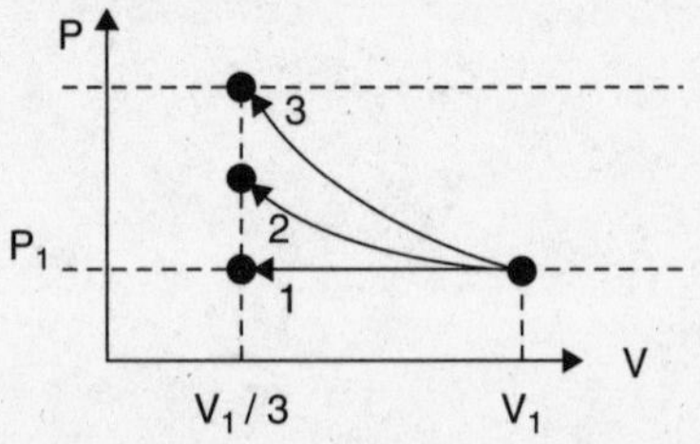

4. A tank contains an amount of an ideal gas at an initial temperature T_1, pressure P_1, and volume V_1. The gas is compressed to $\frac{1}{3}$ its original volume. As illustrated, the process may be isobaric (path 1), isothermal (path 2), or adiabatic (path 3). The mechanical work done on the gas is

(A) greatest for path 1.
(B) greatest for path 2.
(C) equal for paths 1 and 2.
(D) equal for paths 2 and 3.
(E) least for path 1.

5. The final temperature for the gas is

(A) equal for all paths.
(B) equal for paths 1 and 2.
(C) greatest for path 1.
(D) greatest for path 2.
(E) greatest for path 3.

GO ON TO THE NEXT PAGE

6. A cart of mass M moves at velocity **V** toward a second, stationary cart of mass m. Upon collision, the two carts become attached and move with final velocity

(A) MV / M-m
(B) MV / m
(C) (M + m) V / m
(D) MV / (M + m)
(E) 2 MV / (M + m)

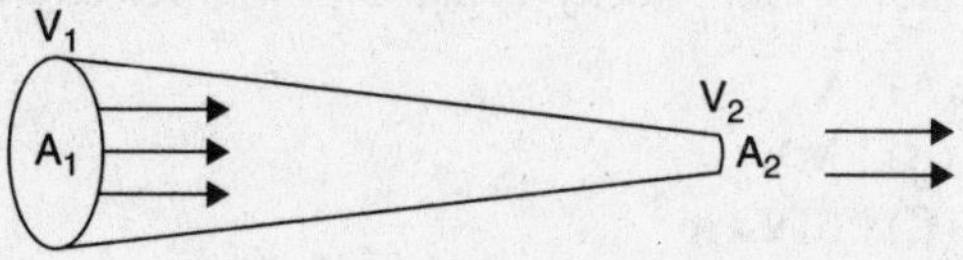

7. Liquid passing through a narrowing tube, as shown, has a speed of 0.5 m/s at A_1 (30 cm^2). At A_2 (10 cm^2) its speed is

(A) 0.5 m/s.
(B) 1.0 m/s.
(C) 1.5 m/s.
(D) 2.0 m/s.
(E) 3.0 m/s.

8. A concave mirror with radius of curvature 0.8 m is 0.6 m from an object. If the object is positioned on the mirror's principle axis, its image is

(A) virtual, inverted, and 0.8 m away.
(B) real, inverted, and 0.8 m away.
(C) virtual, the same size as the object, and 1.2 m away.
(D) real, erect, and 1.2 m away.
(E) real, inverted, and 1.2 m away.

9. The previous problem is repeated, but instead of the mirror, a convex lens of radius of curvature 1.6 m is used. It produces an image that is relative to the lens:

(A) virtual, erect, and 0.8 m away.
(B) real, erect, and 0.8 m away.
(C) virtual, equal in size with the object, and 1.2 m away.
(D) virtual, erect, and 2.4 m away.
(E) real, inverted, and 2.4 m away.

Questions 10–11 refer to the following figure.

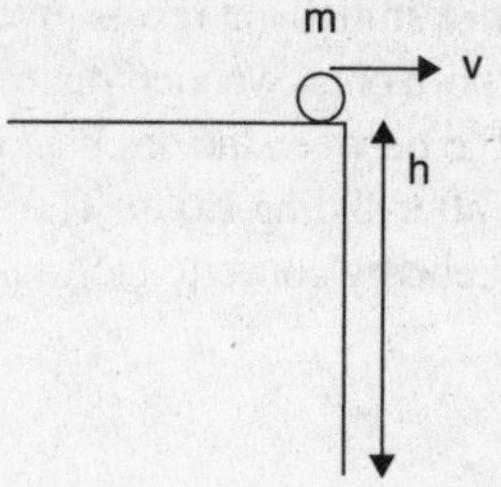

10. A ball of mass m rolls off a cliff of height **h** with a horizontal velocity **v**. It reaches the ground with vertical speed of

(A) $\sqrt{vg}$
(B) hv
(C) $\sqrt{2gh}$
(D) hgv^2
(E) $\left(\frac{1}{2}\right)hv^2$

11. The ball reaches the ground after time t =

(A) (2**h** / g)
(B) $(2h / g)^2$
(C) $\sqrt{\frac{2h}{g}}$
(D) $\sqrt{\frac{h}{2g}}$
(E) (h / 2g)

12. A charged parallel plate capacitor is connected to an emf ε with plate separation distance s. If the plate separation becomes s/2, the charge on the plates is

(A) quartered.
(B) halved.
(C) unchanged.
(D) doubled.
(E) quadrupled.

GO ON TO THE NEXT PAGE

CliffsAP Physics B Practice Exam 2

Questions 13–14 refer to the following figure.

13. A 4.0 kg brick initially at rest is given a push and slides on a horizontal wooden plank with coefficient of friction 0.6 between the brick and the plank. The brick stops after sliding 2.0 m. The amount of mechanical energy converted to radiant energy is

(A) 12 J
(B) 24 J
(C) 36 J
(D) 48 J
(E) 60 J

14. The brick's initial velocity is most nearly

(A) 1 m/s.
(B) 3 m/s.
(C) 5 m/s.
(D) 7 m/s.
(E) 9 m/s.

15. An inflated ball of mass 0.5 kg, initially at rest, is kicked into the air with a speed of 4.0 m/s at an angle of 30° with the horizontal. The impulse given to the ball is

(A) 1 Ns
(B) $\frac{2}{\sqrt{3}}$ Ns
(C) $\frac{2\sqrt{3}}{3}$ Ns
(D) $2\frac{3}{\sqrt{3}}$ Ns
(E) 2 Ns

Questions 16–17 refer to the following figure.

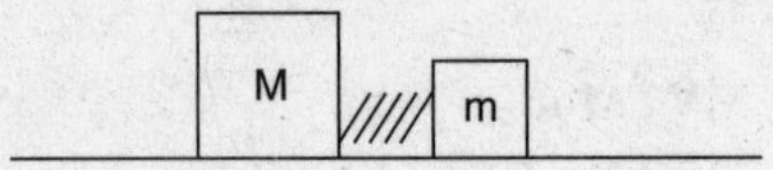

16. Two masses, M and m, are attached by a massless, compressed spring and sit motionless on a horizontal frictionless surface. The spring is released and M moves with velocity **V**. m moves with velocity **v** =

(A) **V**
(B) M**V**
(C) M**V** / m
(D) –M**V** / m
(E) –M**V**

17. Following their release, the two masses have

(A) equal kinetic energies.
(B) equal velocities.
(C) unequal speeds and equal momenta.
(D) equal nonzero acceleration.
(E) equal magnitudes of momentum.

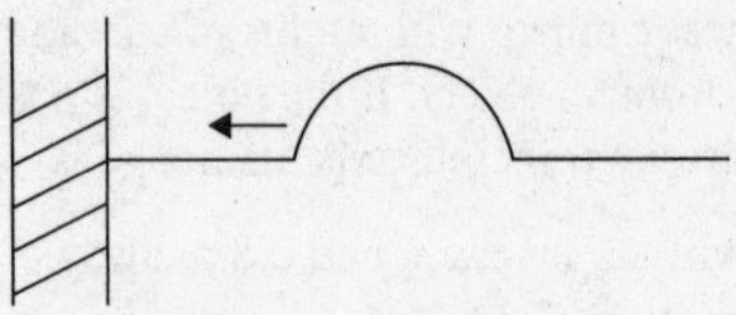

18. The transverse wave in the rope

I. will reflect on the opposite side of the rope.
II. will reflect with an equal amplitude.
III. will reflect with an equal velocity.

(A) I only
(B) I and II only
(C) II only
(D) II and III only
(E) I, II, and III

GO ON TO THE NEXT PAGE

Questions 19–20 refer to the following figure.

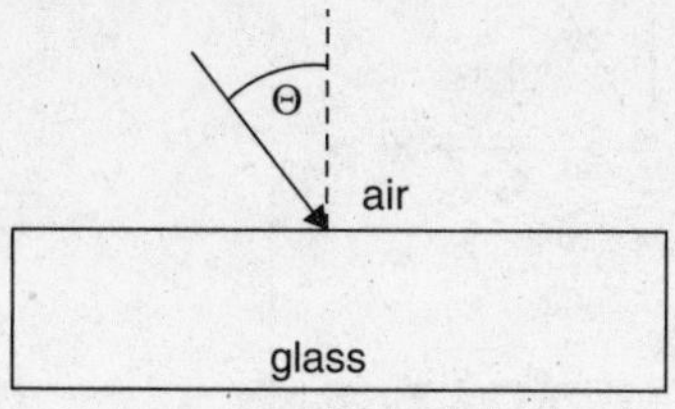

19. A beam of light is incident on a thick glass slab (n_G) an angle Θ as shown.

I. Its reflected angle is Θ.

II. Its refracted angle < Θ and $n_G > n_{AIR}$.

III. Its refracted angle < Θ and $n_G < n_{AIR}$.

(A) I only
(B) I and II only
(C) I and III only
(D) II and III only
(E) I, II, and III

20. In the previous problem, as the light passes from air into the glass, its

(A) speed and wavelength both increase.
(B) speed remains the same and its frequency increases.
(C) speed and frequency both remain the same.
(D) speed decreases and frequency remains the same.
(E) speed decreases and frequency increases.

21. In an energy exchanging experiment, 500 J of heat is added to a system while the system does 300 J of work. The change in the system's internal energy is

(A) –800 J
(B) –300 J
(C) –200 J
(D) +200 J
(E) +800 J

22. A mass M is pulled horizontally across a horizontal surface at constant velocity **V**. If the coefficient of friction between the mass and the floor is μ, at what rate is work done on the mass?

(A) Mgd
(B) μMgd
(C) μMgV
(D) μMV
(E) (μMgd) / V

23. In nuclear reactions, quantities that are conserved include

I. atomic numbers.

II. mass numbers.

III. electric charge.

(A) I only
(B) II only
(C) III only
(D) I and II only
(E) I, II, and III

24. A rock and a ping pong ball of equal volume are dropped from rest in a vacuum. At the time when each has fallen 50 cm, both objects have the same

(A) velocity.
(B) kinetic energy.
(C) momentum.
(D) potential energy.
(E) weight.

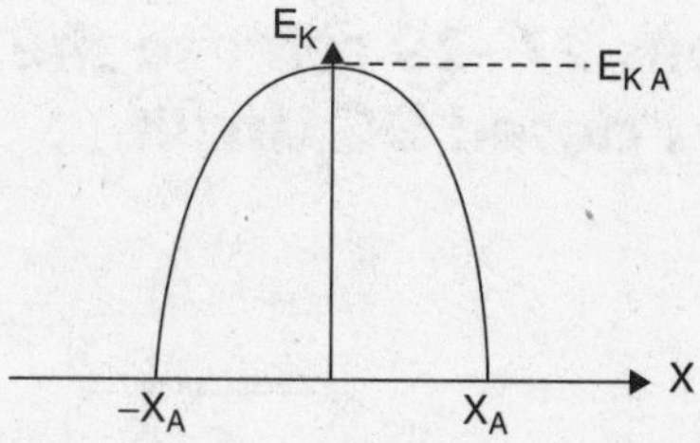

25. The illustration depicts a graph of kinetic energy E_K as a function of displacement x for an oscillating object having amplitude x_A. If the object demonstrates simple harmonic motion, the graph representing the object's potential energy E_P as a function of displacement x is

GO ON TO THE NEXT PAGE

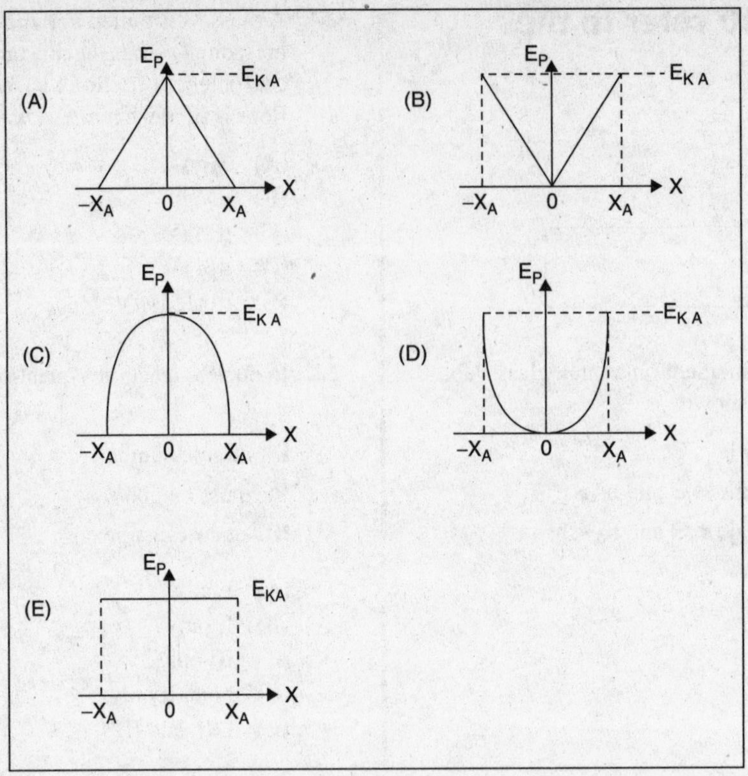

26. A 4 kg object travels in a horizontal circular path of radius 8 m at a constant speed of 6 m/s. If $I = mr^2$, the object's angular momentum, with respect to the circle's center, is

(A) 24 Nm/kg
(B) 32 m^2/s^2
(C) 64 kgm/s^2
(D) 90 Nm/kg^2
(E) 192 kgm^2/s

Questions 27–29 refer to the following diagram of part of a closed DC circuit.

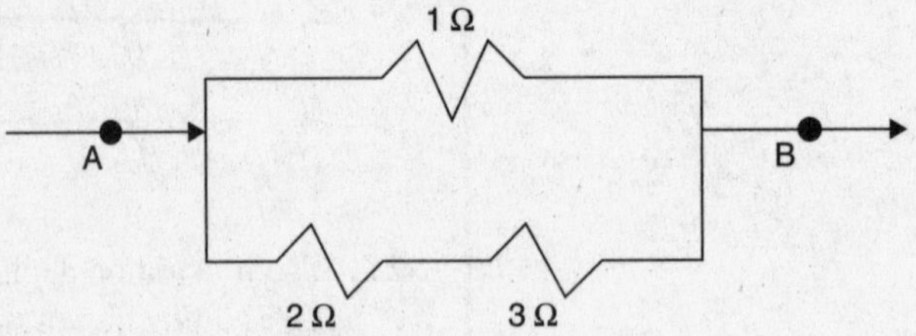

27. The equivalent resistance between points A and B is

(A) $\left(\frac{2}{3}\right)\Omega$
(B) $\left(\frac{5}{6}\right)\Omega$
(C) $1\,{}^{1}\!/_{6}\,\Omega$
(D) $1\,{}^{1}\!/_{4}\,\Omega$
(E) $2\,{}^{1}\!/_{6}\,\Omega$

GO ON TO THE NEXT PAGE

28. While a constant current flows through the circuit, the amount of charge passing a point per second is

(A) greater at point **A** than at point **B.**
(B) greater in the 2 Ω resistor than in the 3 Ω resistor.
(C) greater in the 3 Ω resistor than in the 1 Ω resistor.
(D) greater in the 1 Ω resistor than in the 2 Ω resistor.
(E) the same everywhere in the circuit.

29. A 6 A current through point A means that the current through the 3 Ω resistor is

(A) 1 A
(B) 2 A
(C) 3 A
(D) 4 A
(E) 5 A

30. Which statement describes an ideal gas?

(A) The motion of the molecules is not random.
(B) Molecular collisions are inelastic.
(C) Molecular collisions do not obey Newton's Laws.
(D) There are appreciable forces of comparable magnitude acting on the molecules other than those exerted during collisions.
(E) Molecular collisions conserve momentum.

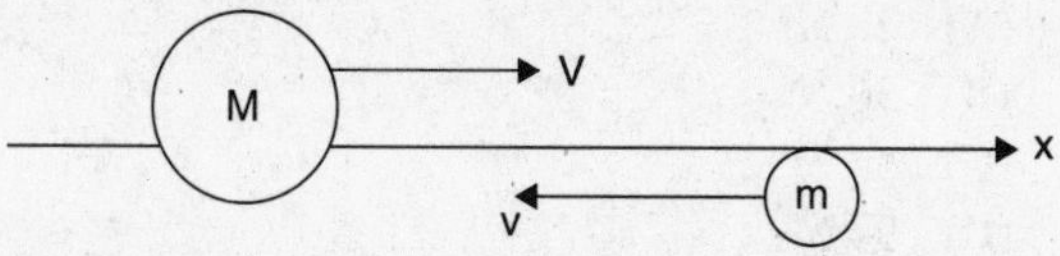

31. Two objects travel toward each other along separate paths parallel to the x-axis as shown. M = 2m. After colliding, the magnitude of the y-component of M's velocity is

(A) $\frac{1}{4}$ the y-component of m's velocity.
(B) $\frac{1}{2}$ the y-component of m's velocity.
(C) in the same direction as m's velocity.
(D) equal to the y-component of m's velocity.
(E) twice the y-component of m's velocity.

32. The buoyant force on a hollow plastic sphere is 2 N when immersed in an unknown liquid at sea level. If the sphere has a 100 cm radius, the density of the fluid is most nearly

(A) 5×10^{-6} kg/m^3
(B) 5×10^{-5} kg/m^3
(C) 5×10^{-4} kg/m^3
(D) 5×10^{-3} kg/m^3
(E) 5×10^{-2} kg/m^3

33. A convex mirror can produce an image that is

(A) real, erect, and smaller than the object.
(B) virtual, inverted, and smaller than the object.
(C) real, inverted, and smaller than the object.
(D) virtual, erect, and smaller than the object.
(E) virtual, erect, and larger than the object.

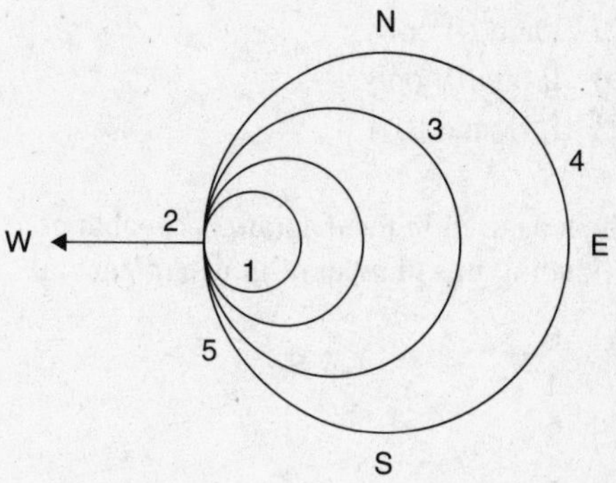

34. A sound source at position 1 moves west with velocity **v**. The apparent frequency f and wavelength λ of the sound are

(A) lowest f, lowest λ at point 2.
(B) lowest λ, highest f at point 5.
(C) lowest f, highest λ at point 3.
(D) highest λ, lowest f at point 4.
(E) highest λ, lowest f at point 1.

35. An object is placed 2 m in front of a plane mirror which produces an image that is

(A) real, erect, and located 2 m behind the mirror.
(B) real, inverted, and located 2 m behind the mirror.
(C) virtual, inverted, and located 2 m behind the mirror.
(D) virtual, erect, and located 2 m in front of the mirror.
(E) virtual, erect, and located 2 m behind the mirror.

GO ON TO THE NEXT PAGE

Questions 36–37 refer to the following figure.

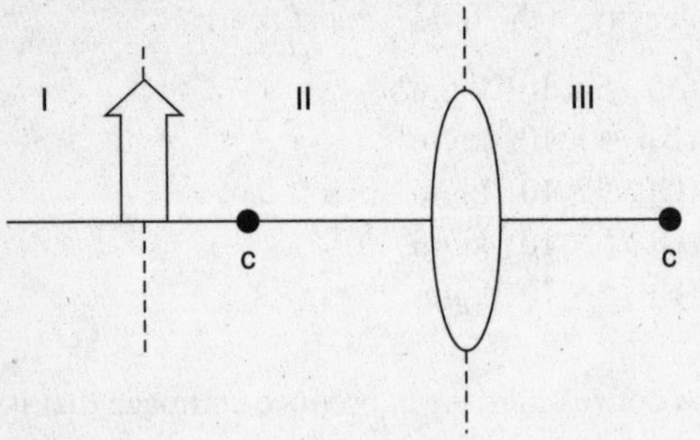

36. In the above diagram, the object in front of the convex lens can produce an image due to refraction only, in region(s)

(A) II only.
(B) III only.
(C) I and III only.
(D) II and III only.
(E) I, II, and III.

37. The lens used in the diagram is capable of producing this number of image in general:

(A) 0
(B) 1
(C) 2
(D) 3
(E) 4

38. An amount of an ideal gas is introduced into a container of unchanging volume. Thermal energy is added to the gas, which increases its temperature from 200 K to 400 K. Before heating, the average speed of the gas molecules is $\mathbf{v_O}$ and after heating, $\mathbf{v_F}$. The ratio of $\mathbf{v_F}$ to $\mathbf{v_O}$ is

(A) $\frac{1}{\sqrt{2}}$
(B) $\frac{1}{2}$
(C) 1
(D) $\sqrt{2}$
(E) 4

39. The potential energy of an object of mass m that orbits the Moon, of mass M and radius R, at an altitude R above the Moon's surface is

(A) 0
(B) GMm/4R
(C) GMm/2R
(D) $GM/2R^2$
(E) GMm/R^2

40. A 55 kg person climbs a 30° hill at a constant rate of speed of 10 m in 5 sec. What power does the person develop?

(A) 50 W
(B) 275 W
(C) 550 W
(D) 825 W
(E) 2750 W

41. A ball is launched straight up with initial velocity 15 m/s. After 2 seconds it has a total vertical displacement of

(A) 5 m
(B) 10 m
(C) 15 m
(D) 20 m
(E) 25 m

Questions 42–43 refer to the following figure.

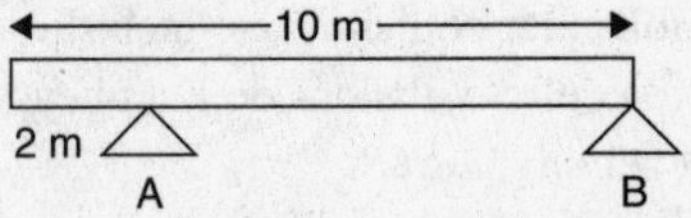

42. A uniform 10 m-long plank weighing 150 N is supported by two blocks, A and B, as shown. The upward force supplied by block A is

(A) 23 N
(B) 36 N
(C) 54 N
(D) 94 N
(E) 101 N

43. A 10 N can of paint is placed on the plank directly over block B. The upward force supplied by block B is now

(A) 66 N
(B) 82 N
(C) 104 N
(D) 160 N
(E) 244 N

GO ON TO THE NEXT PAGE

Questions 44–45 refer to the following figure.

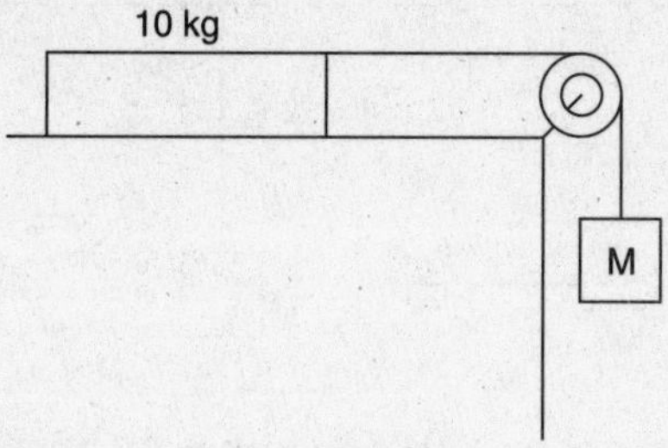

44. The 10 kg mass is pulled to the right by an unknown mass M, over a table top which has a coefficient of friction with the 10 kg block of 0.25. The pulley is frictionless. If the 10 kg mass accelerates at 2 m/s^2, the value of the unknown mass M is most nearly

(A) 2 kg
(B) 4 kg
(C) 6 kg
(D) 8 kg
(E) 10 kg

45. Another 10 kg mass is placed atop the 10 kg mass on the table. The acceleration of the system is now most nearly

(A) 0.2 m/s^2
(B) 0.4 m/s^2
(C) 0.6 m/s^2
(D) 0.8 m/s^2
(E) 1.0 m/s^2

46. The conversion of molecules to liquid from the vapor state is

(A) vaporization.
(B) melting.
(C) freezing.
(D) regelation.
(E) condensation.

Questions 47–48

47. A rock dropped into a river 80 m below will reach the water in

(A) 2 seconds.
(B) 4 seconds.
(C) 6 seconds.
(D) 8 seconds.
(E) 10 seconds.

48. When it hits the water, the rock is traveling at a velocity of

(A) 20 m/s
(B) 30 m/s
(C) 40 m/s
(D) 50 m/s
(E) 60 m/s

49. A mountain stream flows over a cliff 120 m high. The increase in water temperature directly at the base of the waterfall is most nearly

(A) 0.1° C
(B) 0.3° C
(C) 1.1° C
(D) 2.3° C
(E) 2.8° C

GO ON TO THE NEXT PAGE

Questions 50–51 refer to the following figure.

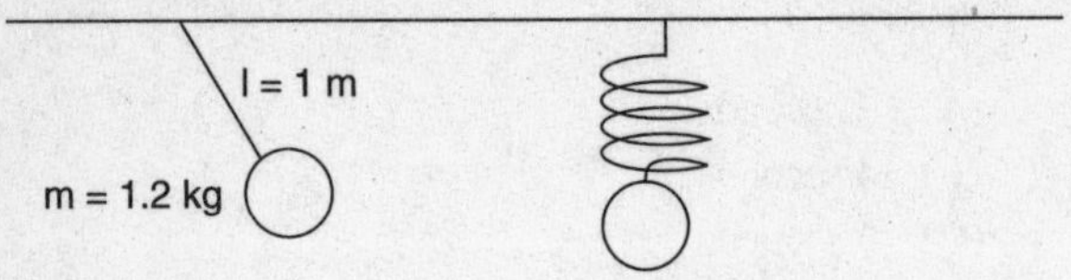

50. A 1 m long string of negligible mass has a 1.2 kg sphere attached at the bottom, swinging back and forth as shown on the left. Its period is most nearly

(A) 0.2 seconds.
(B) 1.0 second.
(C) 2.0 seconds.
(D) 3.0 seconds.
(E) 4.0 seconds.

51. The mass is detached from the string and attached to a spring of negligible mass and k of 12 N/m, as shown on the right. When set in vertical harmonic motion, its new period is most nearly

(A) 0.2 seconds.
(B) 1.0 second.
(C) 2.0 seconds.
(D) 3.0 seconds.
(E) 4.0 seconds.

52. Units of momentum may be

(A) kgm / s^2
(B) $N \cdot s^2$
(C) $kg\ m^2 / s^2$
(D) kgm / s
(E) $J / kg\ m$

53. In compressing a spring a given distance, the work required depends on the

(A) local value for **g**.
(B) spring's mass.
(C) critical velocity.
(D) applied torque.
(E) spring constant.

54. Units of power equal

(A) $J \cdot s$
(B) $N \cdot s^2$
(C) kgm^2 / s^2
(D) W / s
(E) $W \cdot s$

Questions 55–57 refer to the following figure.

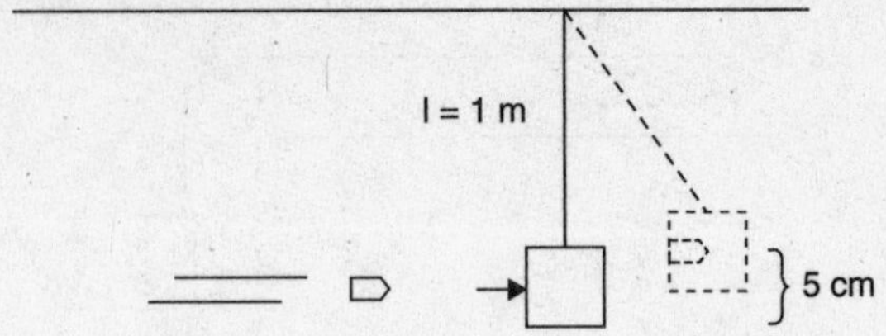

55. A bullet of mass 0.05 kg is shot at a hanging wooden block of mass 3.0 kg, which is suspended by a string of length 1 m. After striking the block, the bullet becomes embedded in it and the entire bullet-block mass rises 5 cm. The momentum of the bullet-block mass just after impact is most nearly

(A) 0.3 kg m/s
(B) 0.6 kg m/s
(C) 1.3 kg m/s
(D) 3.1 kg m/s
(E) 6.2 kg m/s

56. The kinetic energy of the system immediately after impact is

(A) 0.2 J
(B) 1.5 J
(C) 3.1 J
(D) 15 J
(E) 31 J

57. The bullet's initial velocity was most nearly

(A) 10 m/s
(B) 20 m/s
(C) 40 m/s
(D) 60 m/s
(E) 80 m/s

58. A 500 kg wagon moving at 2 m/s strikes a brick wall and stops in 0.1 sec. The average force of the wall on the wagon is

(A) 2 kN
(B) 4 kN
(C) 6 kN
(D) 8 kN
(E) 10 kN

GO ON TO THE NEXT PAGE

59. A machine's efficiency is the ratio of

(A) work in / work out.
(B) power out / power in.
(C) force out / force in.
(D) power / time.
(E) work / time.

Questions 60–61 refer to the following figure.

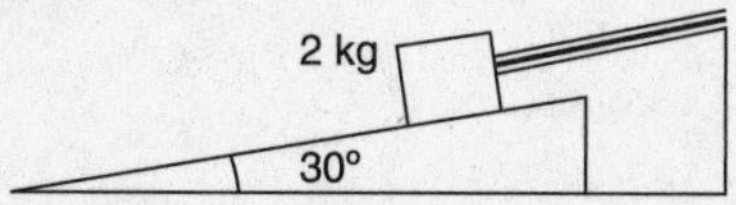

60. The 2 kg mass resting on the frictionless 30° ramp stretches the spring 10 cm. The spring constant is

(A) 50 N / m
(B) 100 N / m
(C) 150 N / m
(D) 200 N / m
(E) 250 N / m

61. If the spring lets go, after 1 second the mass will have slid a distance of

(A) 1 m
(B) 1.5 m
(C) 2 m
(D) 2.5 m
(E) 3 m

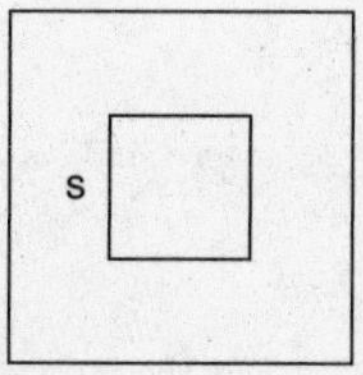

62. Two concentric squares of sides s and 2s, respectively, are made of the same type and thickness of wire. They lie in the same plane, as shown. If the resistance of the larger square is R, the resistance of the smaller square is

(A) R/4
(B) R/2
(C) R
(D) 2R
(E) 4R

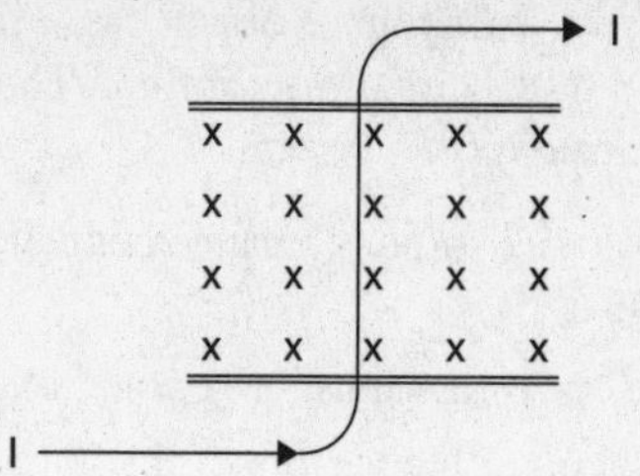

63. A current flows through a thin wire which is placed in a magnetic field as shown. It is held immobile in the field. The current flows from the bottom toward the top as shown. The resulting force on the wire is

(A) to the right.
(B) to the left.
(C) into the page.
(D) out of the page.
(E) opposite the direction of the current.

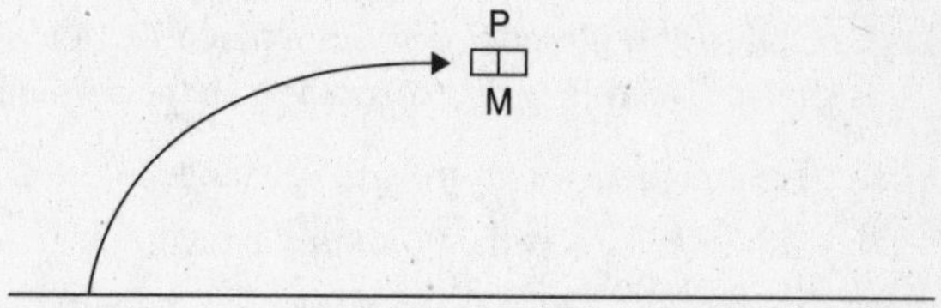

64. A fireworks package of mass M is shot at the wrong angle and fails, exploding into only two pieces of equal mass when it reaches point P. The motion of the center of mass of the remaining pieces as they continue to move is best shown by

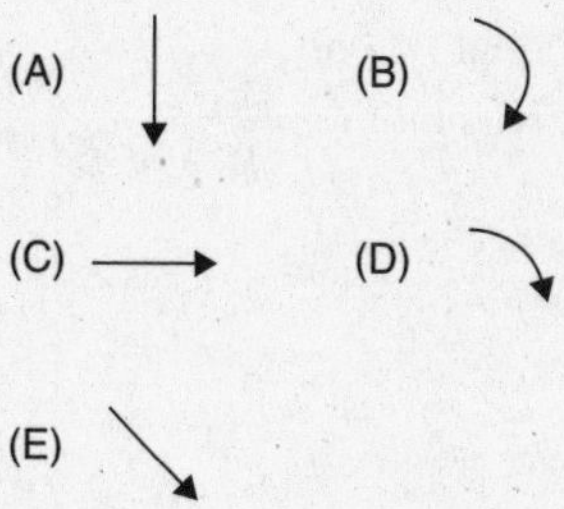

65. A 0.2 kg ball is thrown against a wall at a speed of 3 m/s. It hits the wall at a 90° angle and bounces straight back at a speed of 3 m/s. The change in the ball's momentum is

(A) zero.
(B) 0.2 kg m/s.
(C) 0.6 kg m/s.
(D) 1.2 kg m/s.
(E) indeterminate.

GO ON TO THE NEXT PAGE

66. Four forces acting on an object cause it to maintain translational equilibrium. The following must be true for the system:

I. Clockwise torques balance counterclockwise torques.

II. All four forces must have equal magnitudes.

III. The sum of the vectors of the forces equals zero.

(A) I only
(B) II only
(C) III only
(D) I and III only
(E) II and III only

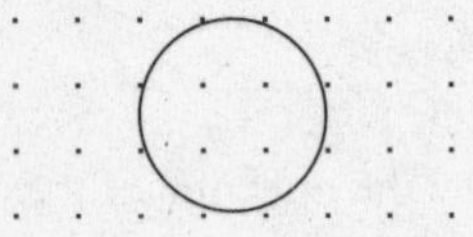

67. A metal loop is placed in a uniform magnetic field as shown. The following experiments are performed:

I. The magnetic field strength is changed.

II. The loop is moved up (out of the page).

III. The loop is pulled to the right.

IV. The loop is rotated about a diameter.

A current will flow through the loop as a result of

(A) I, II, and III only.
(B) II and III only.
(C) II, III, and IV only.
(D) I, III, and IV only.
(E) I, II, III, and IV.

68. Doing 1000 J of work on a 50 kg mass will cause it to be lifted to a shelf of height

(A) 1 m
(B) 2 m
(C) 3 m
(D) 4 m
(E) 5 m

69. If the mass in the previous problem falls, it strikes the ground in approximately

(A) 1 second.
(B) 2 seconds.
(C) 3 seconds.
(D) 4 seconds.
(E) 5 seconds.

70. A crane lifts a mass M at a constant velocity **v** and it reaches a height **h** after time t. The average power of the crane is

(A) Ma
(B) Mg
(C) Mh / t
(D) Mgh / t
(E) Mght

END OF SECTION I. IF YOU FINISH BEFORE TIME IS CALLED, YOU MAY CHECK YOUR WORK ON THIS SECTION. DO NOT GO TO SECTION II UNTIL YOU ARE TOLD TO DO SO.

STOP

Physics B
Section II, Free Response Questions

Time — 90 Minutes, 6 questions of equal weight

Percent of total grade — 50

Instructions: When you are instructed to begin, carefully tear out the green insert and start working. The questions in the green insert are duplicates of the questions in this booklet. A helpful table of information and lists of equations that may be helpful to you are found on pages 1–3 of this booklet. NO CREDIT WILL BE GIVEN FOR ANYTHING WRITTEN IN THE GREEN INSERT. It is provided for reference only as you answer the free response questions.

Show all of your work. You are to write your answer to each question in the pink booklet. Additional answer pages follow each question. Credit for your answers depends on your demonstrating that you know which physical principles would be appropriate to apply in a particular situation. Credit for your answers also depends on the quality of your solutions and explanations. Be sure to write CLEARLY and LEGIBLY. If you make an error, cross it out rather than erasing it. This may save you valuable time.

Advanced Placement Physics B Equations

Newtonian Mechanics

$\mathbf{v} = \mathbf{v}_0 + \mathbf{a}t$	$\mathbf{a}$ = acceleration
$\mathbf{x} = \mathbf{x}_0 + \mathbf{v}_0 t + \frac{1}{2}t^2$	$\mathbf{F}$ = force
$\mathbf{v}^2 = \mathbf{v}_0^{\ 2} + 2\mathbf{a}(\mathbf{x}-\mathbf{x}_0)$	f = frequency
$\Sigma\mathbf{F} = \mathbf{F}_{NET} = m\mathbf{a}$	h = height
$F_{fric} \le \mu\mathbf{N}$	$\mathbf{J}$ = impulse
$\mathbf{a}_c = \frac{\mathbf{v}^2}{\mathbf{r}}$	K = kinetic energy
$\tau = \mathbf{rF}\sin\Theta$	k = spring constant
$\mathbf{p} = m\mathbf{v}$	l = length
$\mathbf{J} = \mathbf{F}\Delta t = \Delta\mathbf{p}$	m = mass
$K = \frac{1}{2}m\mathbf{v}^2$	$\mathbf{N}$ = normal force
$\Delta U_g = m\mathbf{gh}$	P = power
$\mathbf{W} = \mathbf{Fr}\cos\Theta$	$\mathbf{p}$ = momentum
$P_{AVE} = \frac{\mathbf{W}}{\Delta t}$	r = radius, distance
$P = \mathbf{Fv}\cos\Theta$	$\mathbf{r}$ = position vector
$\mathbf{F}_s = -k\mathbf{x}$	T = period
$U_s = \frac{1}{2}k\mathbf{x}^2$	t = time
$T_P = 2\pi(l/\mathbf{g})^{\frac{1}{2}}$	U = potential energy
$T_S = 2\pi(m/k)^{\frac{1}{2}}$	$\mathbf{v}$ = velocity, (v) speed
$T = \frac{1}{f}$	$\mathbf{W}$ = work done on a system
$F_G = -\frac{Gm_1m_2}{r^2}$	$\mathbf{x}$ = position
$U_G = -\frac{Gm_1m_2}{r}$	μ = coefficient of friction
	Θ = angle
	τ = torque

Electricity and Magnetism

$\mathbf{F} = \frac{1}{4\pi\varepsilon_0}\frac{q_1q_2}{\mathbf{r}^2}$	A = area
$\mathbf{E} = \frac{\mathbf{F}}{q}$	$\mathbf{B}$ = magnetic field
$U_E = qV = \frac{1}{4\pi\varepsilon_0}\frac{q_1q_2}{\mathbf{r}}$	C = capacitance
$\mathbf{E}_{AVE} = -\frac{V}{\mathbf{d}}$	$\mathbf{d}$ = distance
$V = \frac{1}{4\pi\varepsilon_0}\Sigma\frac{q}{r}$	$\mathbf{E}$ = electric field
$C = \frac{Q}{V}$	ε = emf
$C = \frac{\varepsilon_0 A}{\mathbf{d}}$	$\mathbf{F}$ = force
$U_c = \frac{1}{2}QV = \frac{1}{2}CV^2$	$\mathbf{I}$ = current
$I_{AVE} = \frac{\Delta Q}{\Delta t}$	l = length
$R = \rho(l/A)$	P = power
$V = IR$	Q = charge
$C_{PAR} = C_1 + C_2 + \ldots C_n$	q = point charge
$\frac{1}{C_{SER}} = \frac{1}{C_1} + \frac{1}{C_2} = \ldots\frac{1}{C_n}$	R = resistance
$R_{SER} = R_1 + R_2 + \ldots R_n$	r = distance
$\frac{1}{R_{PAR}} = \frac{1}{R_1} + \frac{1}{R_2} + \ldots\frac{1}{R_n}$	t = time
$\mathbf{F}_B = q\mathbf{vB}\sin\Theta$	U = potential (stored) energy
$\mathbf{F}_B = \mathbf{B}\,\mathbf{I}\,l\sin\Theta$	V = electric potential or potential difference
$\mathbf{B} = \frac{\mu_0 I}{2\pi\mathbf{r}}$	v = velocity
$\phi_M = \mathbf{B}\cdot A = \mathbf{B}A\cos\Theta$	v = speed
$\varepsilon_{AVE} = \frac{\Delta\phi_M}{\Delta\tau}$	ρ = resistivity
$\varepsilon = \mathbf{B}l\mathbf{v}$	ϕ_M = magnetic flux

Fluid Mechanics and Thermal Physics

$p = p_o + \rho \mathbf{g} \mathbf{h}$	A = area
$\mathbf{F}_{BUOY} = \rho V \mathbf{g}$	c = specific heat
$A_1 \mathbf{v_1} = A_2 \mathbf{v_2}$	e = efficiency
$p = \rho \mathbf{g} \mathbf{y} + 1 \rho \mathbf{v}^2 = \text{const.}$	**F** = force
$\Delta l = \alpha l_o \Delta T$	K_{AVE} = average molecular kinetic energy
$Q = mL$	k_B = Bolzmann's constant
$Q = mc\Delta T$	L = latent heat
$p = \frac{\mathbf{F}}{A}$	l = length
$pV = nRT$	M = molecular mass
$K_{AVE} = \frac{3}{2} K_B T$	m = sample mass
$V_{RMS} = \sqrt{\frac{3RT}{M}}$	n = number of moles
$= \sqrt{\frac{3K_B T}{\mu}}$	p = pressure
$W = -p\Delta V$	Q = heat transferred to a system
$Q = nc\Delta T$	T = temperature
$\Delta U = Q + W$	U = internal energy
$\Delta U = nc_v \Delta T$	V = volume
$e = \left\| \frac{W}{Q_H} \right\|$	**v** = velocity
$e_c = 1 - \frac{T_C}{T_H} = \frac{T_H - T_C}{T_H}$	v = speed
	v_{RMS} = root-mean-square velocity
	W = work done on a system
	y = height
	α = coefficient of linear expansion
	μ = mass of molecule
	ρ = density

Atomic and Nuclear Physics

$E = hf = \mathbf{p}c$	E = energy
$K_{MAX} = hf - \phi$	f = frequency
$\lambda = \frac{h}{\mathbf{p}}$	h = Planck's constant
$E = mc^2$	K = kinetic energy
	m = mass
	p = momentum
	λ = wavelength
	ϕ = work function

Waves and Optics

$\mathbf{v} = f/\lambda$	c = radius of curvature
$n = \frac{c}{v}$	d = slit separation
$n_1 \sin\Theta_1 = n_2 \sin\Theta_2$	f = frequency
$\sin \Theta_c = \frac{n_2}{n_1}$	f = focal length
$\frac{1}{f} = \frac{1}{S_1} + \frac{1}{S_0}$	h = height
$M = \frac{-h_1}{h_0} = \frac{-S_1}{S_0}$	L = distance
$f = \frac{c}{2}$	M = magnification
$\lambda = \frac{d \sin \Theta_m}{m}$	m = an integer
$x_m \cong \frac{m\lambda L}{d}$	n = refractive index
	s = distance
	v = speed
	x = position
	λ = wavelength
	Θ = angle

Geometry and Trigonometry

Rectangle	
$A = bh$	A = area
Triangle	C = circumference
$A = \frac{1}{2} bh$	V = volume
Circle	S = surface area
$A = \pi r^2$	b = base
$C = 2\pi r$	h = height
Paralelopiped	l = length
$V = lwh$	w = width
Cylinder	r = radius
$V = \pi r^2 h$	
$S = 2\pi rh + 2\pi r^2$	
Sphere	
$V = \frac{4}{3} \pi r^3$	
$S = 4\pi r^2$	

Right Triangle

$a^2 + b^2 = c^2$

$\sin \Theta = a/c$

$\cos \Theta = b/c$

$\tan \Theta = a/b$

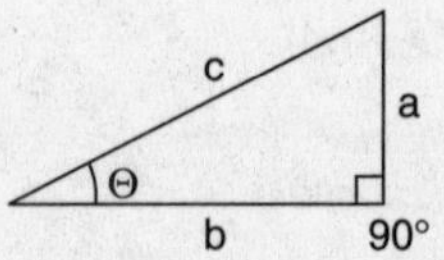

Physics B
Section II

Time – 90 Minutes
6 questions of equal weight

Directions: Answer all of the questions. Each question is equally weighted, but the parts within a question may not be equally weighted. Show your work. Credit for each answer depends on the quality of your explanation.

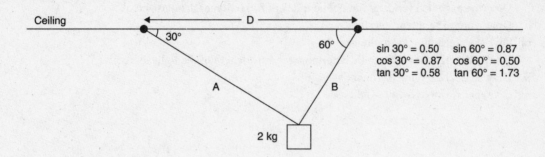

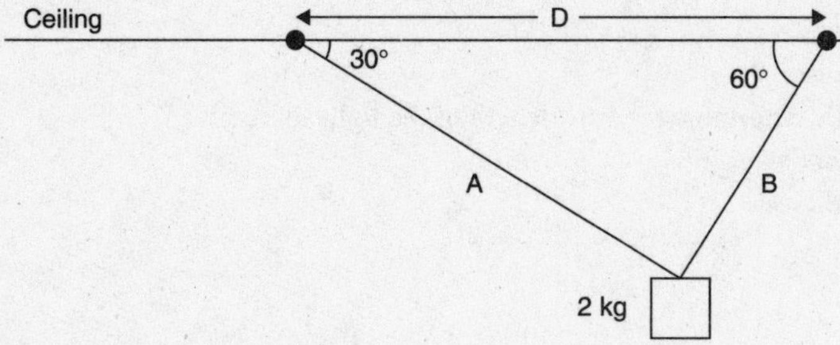

sin 30° = 0.50 sin 60° = 0.87
cos 30° = 0.87 cos 60° = 0.50
tan 30° = 0.58 tan 60° = 1.73

1. A 2 kg mass is hung from the ceiling by two strings as shown.

(a) If D = 3 m, find the lengths of strings A and B.

(b) Draw a free-body diagram on the figure below that shows all forces acting on the hanging mass. (Indicate directions but do not resolve the forces into components.)

(c) String B breaks. Compute the speed of the weight as it swings through the lowest point in its path.

(d) Determine the tension in string A as it swings through the lowest point in its path.

GO ON TO THE NEXT PAGE

2. You are given the following equipment to use in an optics experiment as described in parts (a) and (b).

A solid, rectangular glass block

An illuminator, which emits a thin beam of light

A protractor

A metric ruler

A diffraction grating of known slit spacing

A blank sheet of white paper

(a) i. Describe a method for determining the index of refraction of the glass block.
ii. Draw your experiment's design.
iii. Give the equation you would use.

(b) i. Briefly describe how you could determine the wavelength of the light source.
ii. Draw and label your experiment's design.
iii. Give the equation(s) you would use.

GO ON TO THE NEXT PAGE

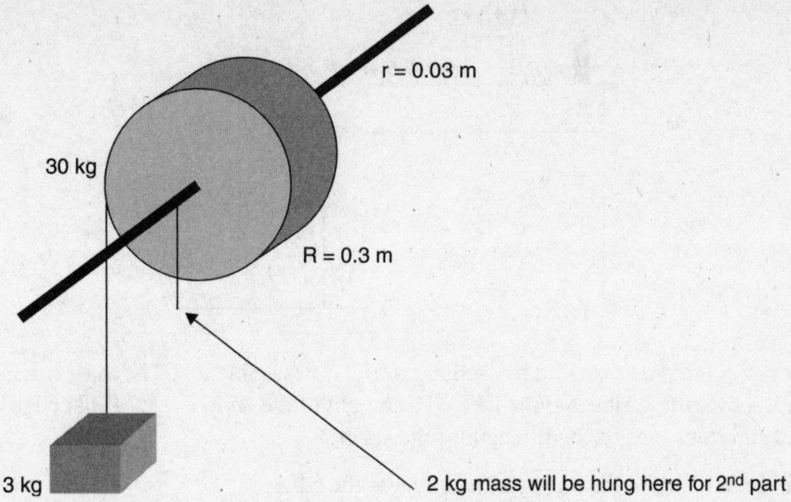

3. A 3 kg mass is hung from a solid cylinder of radius R = 0.3 m. The cylinder is fixed to an axle of negligible mass and radius r = 0.03 m as shown. The system is initially at rest and has rotational inertia $I = \left(\frac{1}{2}\right)mr^2$.

(a) Compute the angular acceleration of the cylinder.
(b) Compute the linear acceleration of the 3 kg mass.
(c) Determine the number of revolutions made by the cylinder after 5 seconds.

A second mass of 2 kg is hung on the side opposite the 3 kg mass but is hung from the axle and provides a clockwise torque.

(d) Compute the new angular acceleration of the cylinder.
(e) Compute the new linear acceleration of the 3 kg mass.
(f) Compute the new number of revolutions made by the cylinder after 5 seconds.

GO ON TO THE NEXT PAGE

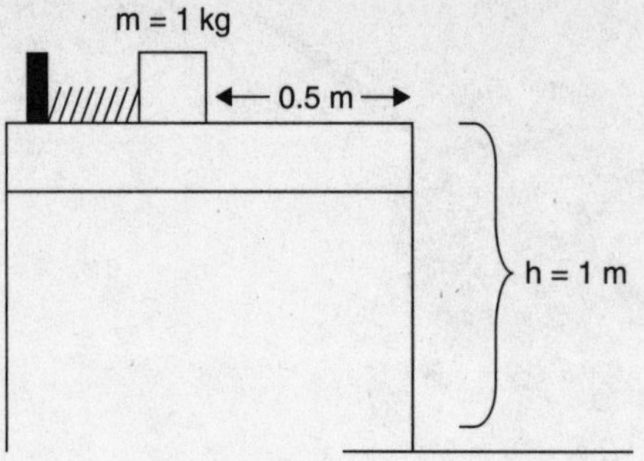

4. A spring with k = 5 N / m pushes a 1 kg mass along a frictionless table top. The mass travels a horizontal distance of 0.5 m before it leaves the spring and the table. The height of the table is 1 m. If the edge of the table also represents the equilibrium (unstretched) length of the spring:

(a) Determine the velocity of the mass just as it leaves the table.
(b) How long is the mass in the air?
(c) How far horizontally does it go after leaving the table?

The table top is now given a different surface, one that has a coefficient of friction of 0.3 with the mass. The experiment is repeated exactly, with the same spring.

(d) What is the velocity with which the mass leaves the table now?

GO ON TO THE NEXT PAGE

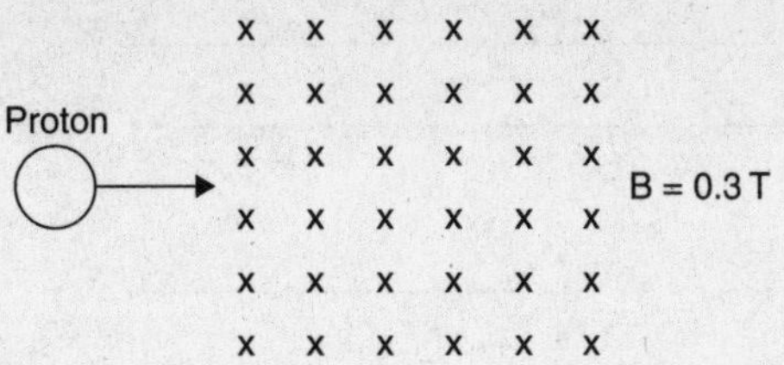

5. A proton traveling at 2.0×10^6 m/s enters a 0.3 T magnetic field as shown. Its velocity is perpendicular to the field.

(a) What is the magnitude of the resulting force on the proton?
(b) Draw the direction of the force and the direction of the path of the proton while in the field.
(c) What is the radius of curvature of the proton's path as it travels in the field?
(d) How much smaller is the radius of curvature of an electron traveling at the same velocity from the same angle?

GO ON TO THE NEXT PAGE

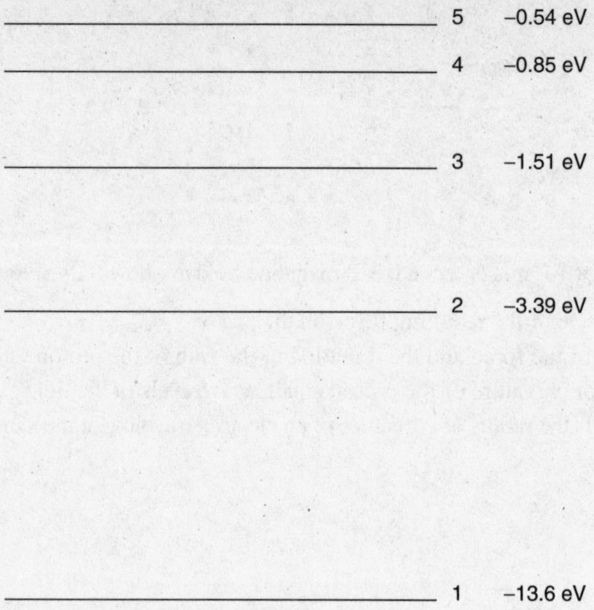

6. An electron in a hydrogen atom becomes energized and "jumps" to a higher energy level momentarily. When it "jumps" back down, it emits a photon of wavelength 4.8×10^{-7}m.

(a) What is the frequency of the emitted photon?
(b) How many Joules of energy does this photon possess?
(c) How many electron volts does the answer to (b) represent?
(d) What quantum leap has the electron made?

END OF EXAMINATION

Sample B Exam #2 Answers and Comments

Multiple Choice

1. **(A)** Decreasing intensity results in fewer emitted electrons. Using E = hf and f = **v**/λ,

   E = h**v**/λ shows that E increases with decreased wavelength.

2. **(C)** Using $U_C = \left(\frac{1}{2}\right)QV$ and Q = VC,

$$U_C = \left(\frac{1}{2}\right)CV^2$$
$$= \left(\frac{1}{2}\right)(5 \times 10^{-6}\text{ F})(120\text{ V})^2$$
$$= 3.6 \times 10^{-2}\text{ J}$$

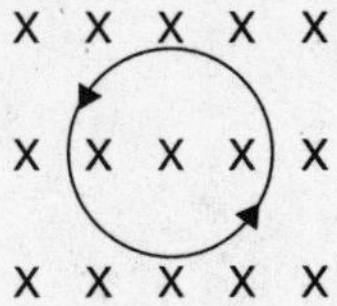

3. **(C)** Normally, the right thumb points in the direction of current flow when the fingers point in the direction of the magnetic field when it is turned on or increasing. Using $\varepsilon = -\frac{N\Delta\Phi}{\Delta t}$, the emf is oppositely directed. Reversing directions gives **counterclockwise flow.**

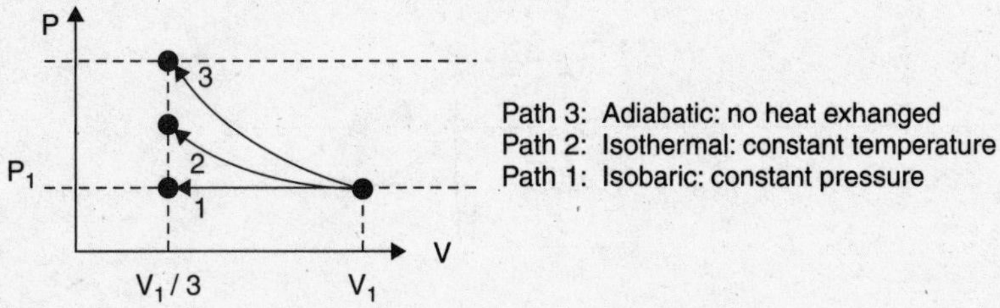

4. **(E)** Since W = PΔV, the work done on the gas is the **area under the curve**, which is greatest for path 3 and **least for path 1.**

5. **(E)** Using $\frac{P_1V_1}{T_1} = \frac{P_2V_2}{T_2}$, $T_2 = \frac{(P_2V_2)}{(P_1V_1)}T_1$

   Again, the greatest value for W = PΔV occurs for **path 3, which gives the largest final temperature.**

6. **(D)** Using the law of conservation of momentum,

$$\mathbf{p}_{\text{TOTAL}}(\text{BEFORE}) = \mathbf{p}_{\text{TOTAL}}(\text{AFTER})$$
$$M\mathbf{V} + m(0) = (M + m)\,\mathbf{v}_{\text{FINAL}}$$
$$\mathbf{v}_{\text{FINAL}} = (M\mathbf{V}) / (M + m)$$

7. **(C)** Using $\mathbf{A_1v_1 = A_2v_2}$,

$$\mathbf{v_2 = A_1v_1 / A_2}$$
$$\mathbf{v_2} = \frac{(30\text{ cm}^2)(0.5\text{ m/s})}{(10\text{ cm}^2)} = \mathbf{1.5\text{ m/s}}$$

8. (E) First, draw the diagram:

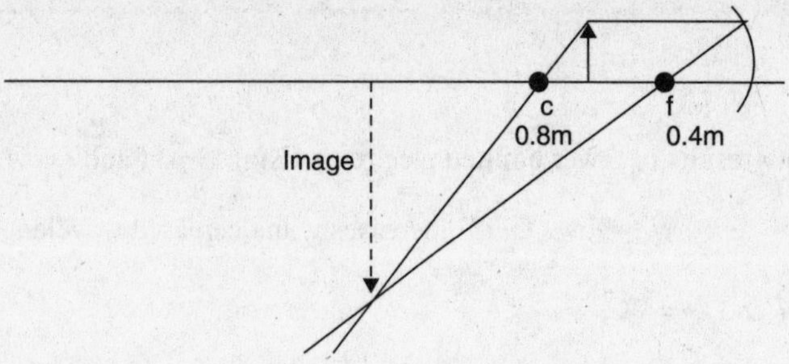

Using $1 / f = (1 / S_i) + (1 / S_O)$,

$$1/ S_i = (1 / f) - (1 / S_O)$$
$$S_i = 1.2 \text{ m}$$

The image is real, inverted, 1.2 m away.

9. (D) First, draw the diagram:

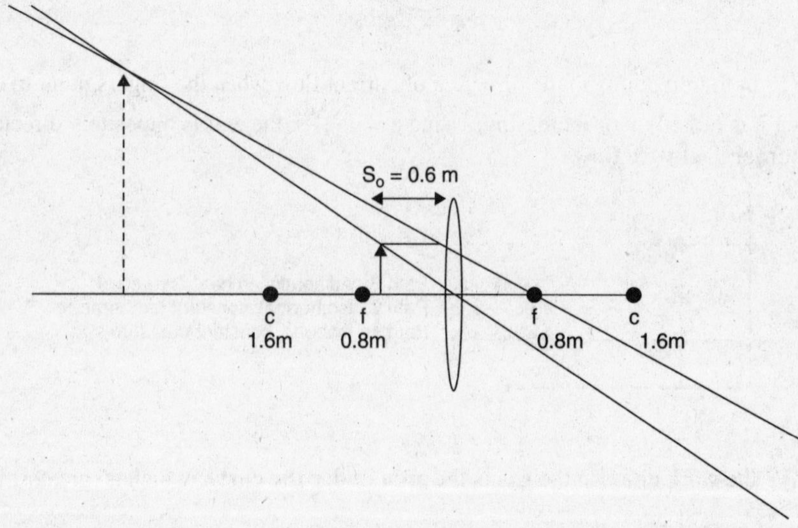

The image is virtual and erect, on the same side as the object; therefore, S_i will be negative.

Using $(1/f) = (1/S_i) + (1/S_O)$,

$$1/S_i = (1/f) - (1/S_O) = (1/0.8\text{m}) - (1/0.6\text{m})$$
$$\mathbf{S_i = -2.4 \text{ m away}}$$

10. (C) Using $\mathbf{v_{FINAL\ (VERT)}^2 = v_{INIT\ (VERT)}^2 + 2\, a_{(VERT)}\, d_{(VERT)}}$

$$\mathbf{v_{FINAL\ (VERT)} = \sqrt{2gh}}$$

11. (C) Using $\mathbf{d_{VERT} = (v_{O\ (VERT)})(t) + (1/2)gt^2}$

$$\mathbf{h = \left(\frac{1}{2}\right)gt^2}$$
$$\mathbf{t = \sqrt{\frac{2h}{g}}}$$

12. (D) Using $C = \frac{\kappa \varepsilon_0 A}{d} = \frac{Q}{V}$,

Charges increase with plate area increase and/or separation distance increase. **Therefore, $\left(\frac{1}{2}\right)$d doubles charge.**

13. (D) Using $\mu = F_F / F_N$,

$$F_F = \mu F_N,\ F_F = (0.6)(4.0\text{ kg})(10\text{ m/s}^2) = 24\text{ N}$$
$$E = W = (F_F)(d) = (24\text{ N})(2.0\text{ m}) = \mathbf{48\ J}$$

14. (C) Since all the kinetic energy is changed into radiant energy, $\left(\frac{1}{2}\right)mv_{INITIAL}^{2} = 48\text{ J}$

$$v = [(2)(48\text{J}) / (4.0\text{ kg})]^{\frac{1}{2}} = (24\text{m}^2/\text{s}^2)^{\frac{1}{2}} = \text{approx. } \mathbf{5\ m/s}$$

15. (E) Since Impulse (**J**) gives momentum (**p**),

$$J = mv = (0.5\text{ kg})(4.0\text{ m/s}) = \mathbf{2\ Ns}$$

16. (D) After the spring is released, the momentum of both sides is equal and opposite.

$$MV = mv$$
v = (MV) / m but since v is in the opposite direction,
$$v = -MV / m$$

17. (E) Momentum is conserved. The total momentum before release is zero; therefore, the total momentum after release is still zero. Although the directions of the masses are opposite, the **numerical magnitudes are equal.**

18. (B) The reflected transverse wave will be on the opposite side of the rope equal in amplitude to the original pulse, but having an opposite direction and so an opposite velocity. **(I and II only.)**

19. (B) The angle of incidence always equals the angle of reflection and n = sin i/sin r means that $n_G > n_{AIR}$. **(I and II only.)**

20. (D) When light enters a denser medium, speed decreases and frequency remains the same. (Red light stays red when it passes through glass from air.)

21. (D) 500 J of energy goes in, 300 J comes out, the increase in the system equals **+200 J**

22. (C) Power is the time rate of work or P = W/t = Fd/t = Fv

Since the force is equal and opposite to the frictional force, F_F, and $F_F = \mu F_N$,

$$P = \mu F_N v = \mu Mgv$$

23. (E) All are conserved.

24. (A) All objects fall at the same rate in a vacuum, so their velocities will be equal. Since the objects have unequal masses, their weights, momenta, E_P (or **mgh**), and E_K (or $\left(\frac{1}{2}\right)mv^2$) will all be unequal.

25. (D) As the oscillator loses potential energy, it gains kinetic energy. Since $E_K = \left(\frac{1}{2}\right)mv^2$, as it gains velocity, the energy curve is parabolic in shape.

26. (E) Using **L = Iω**,

$$L = (mr^2)(v / r) = mrv = (4\text{ kg})(8\text{ m})(6\text{ m/s}) = \mathbf{192\ kgm^2/s}$$

27. (B) Since the **series** combination of 2 Ω and 3 Ω is in parallel with the 1Ω resistor,

$$\frac{1}{R_{EQ}} = \frac{1}{1\Omega} + \frac{1}{5\Omega} = \frac{6}{5\Omega} \text{ and } R_{EQ} = (5/6)\ \Omega$$

28. (D) After passing point A, the current splits inversely to the branch resistance and recombines at point B. Thus, 5/6 of the main line current passes through the 1 Ω resistor and $\frac{1}{6}$ of the current passes through the 5 Ω branch.

29. (A) The 6 A current splits with $\frac{5}{6}$ of it going to the 1 Ω resistor and the remaining $\frac{1}{6}$ of the current going to the branch with both the 2 Ω and 3 Ω resistors.

30. **(E)** Molecular collisions are random, elastic, and are the only appreciable forces on gas molecules. Momentum is always conserved.

31. **(B)** The Law of Conservation of Momentum holds that the sum of the initial y-components equals zero. Therefore, $\mathbf{MV_Y = mv_Y}$ both before and after collision.

Substituting 2m for M, $2mV_Y = mv_Y$ and $\mathbf{V_Y = v_Y / 2}$

32. **(E)** Using $\mathbf{F_{BUOY} = \rho Vg}$,

$$\rho = \mathbf{F_{BUOY}} \;/\, \mathbf{Vg} = \frac{2\,\text{N}}{\left(\frac{4}{3}\right)(\pi)(1\,\text{m})^3\left(10\,\text{m/s}^2\right)}$$

$$= \mathbf{5 \times 10^{-2}\ kg / m^3}$$

33. **(D)** Convex mirrors can **only** produce images that are **virtual, smaller than the object, and erect.**

34. **(D)** As the sound source moves west, its waves compress in its direction of motion (where their frequency will be highest and wavelength lowest) at position 2.

The waves rarefy behind it **(where their frequency is lowest and wavelength highest) at position 4.**

35. **(E)** A plane mirror can only produce a **virtual image** and **cannot invert the image**, which appears at an **equal distance** behind the mirror.

36. **(B)** **A real image, due to refraction, will be formed in region III.**

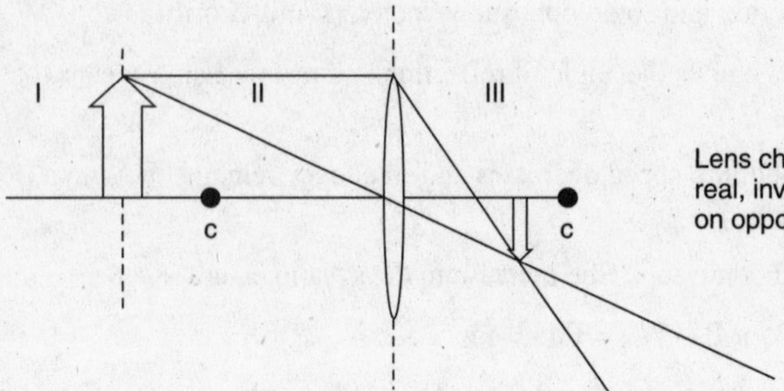

Lens characteristic:
real, inverted, between c and f,
on opposite side of lens.

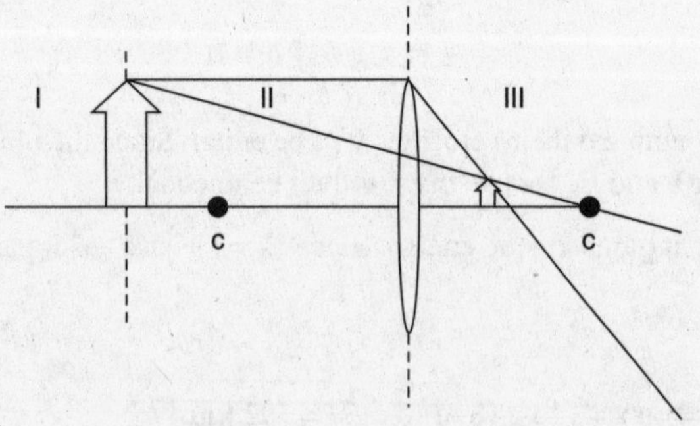

Front face-mirror characteristic:
virtual, erect, between f and lens,
on opposite side of lens.

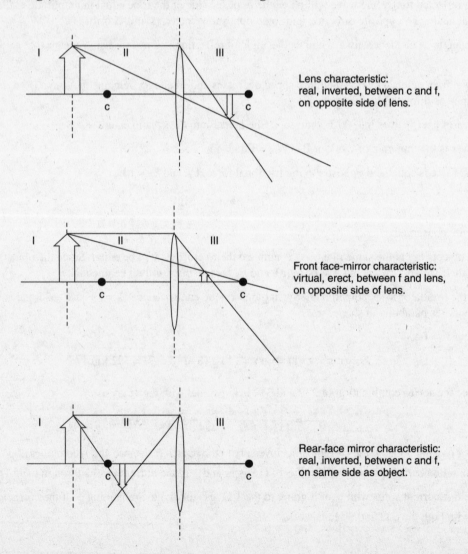

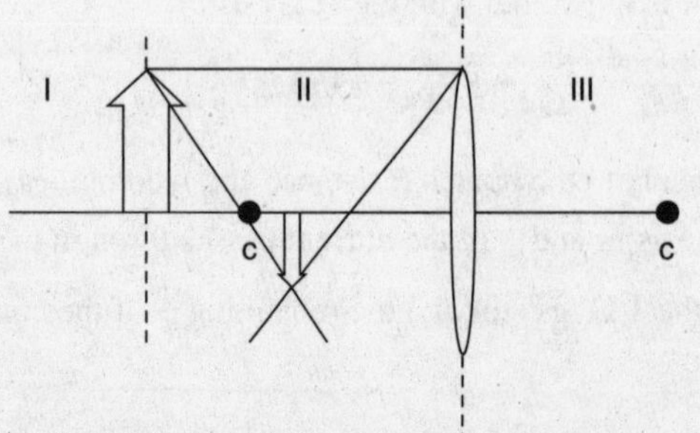

Rear-face mirror characteristic:
real, inverted, between c and f,
on same side as object.

37. (C) Convex lenses can produce real and virtual images.

38. (D) Twice the thermal energy transforms into twice the kinetic energy:

$$\mathbf{E_{K\ FINAL} = 2\ E_{K\ INITIAL}}$$

$$\left(\frac{1}{2}\right)mv_{FINAL}^{2} = (2)\left(\frac{1}{2}\right)mv_{INITIAL}^{2}$$

$$\mathbf{v_{FINAL} = \sqrt{2}\ v_{INITIAL}}$$

39. (C) Using $\mathbf{U_G = Fd = FR} = \frac{GMm}{(2R)^2}\mathbf{(2R)} = \frac{GMm}{2R}$

40. (C) The person's constant velocity is 10 m/5 s = 2 m/s up the incline, which means that the force applied is equal and opposite the person's **parallel force.**

$$\mathbf{F_P} = \text{mg(sin) } 30° = (55\text{ kg})(10\text{ m/s}^2)\left(\frac{1}{2}\right) = 275\text{ N}$$

Since P = **W/t** = **Fd/t** = **Fv** = $\mathbf{F_P v}$

$$P = (275\text{ N})(2\text{ m/s}) = \mathbf{550\ W}$$

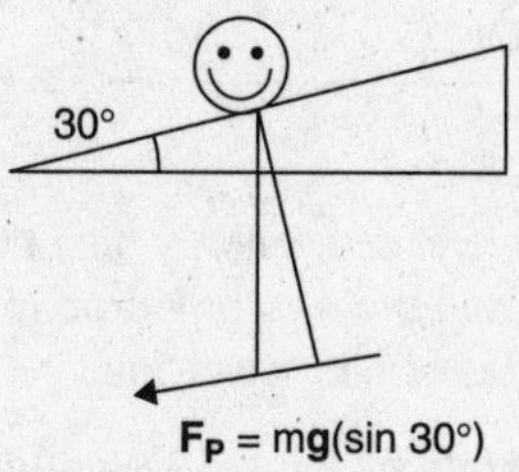

41. (B) Using $\mathbf{d} = \mathbf{v_0}t + \left(\frac{1}{2}\right)\mathbf{a}t^2$,

$$\mathbf{d} = (15\text{ m/s})(2\text{ sec}) - \left(\frac{1}{2}\right)(10\text{ m/s}^2)(2\text{ sec})^2 = \mathbf{10\ m}$$

42. (D) Solve in two steps: First, $\Sigma F_{UP} = \Sigma F_{DOWN}$

$$\mathbf{F_A + F_B = 150\ N}$$

Second: $\Sigma\tau_{CW} = \Sigma\tau_{CcW}$

Arbitrarily choosing block B as the pivot:

$$\Sigma\tau_{CW} = \Sigma\tau_{CCW}$$

$$(\mathbf{F_A})(8\text{ m}) = (150\text{ N})(5\text{ m})$$

$$\mathbf{F_A = 94\ N}$$

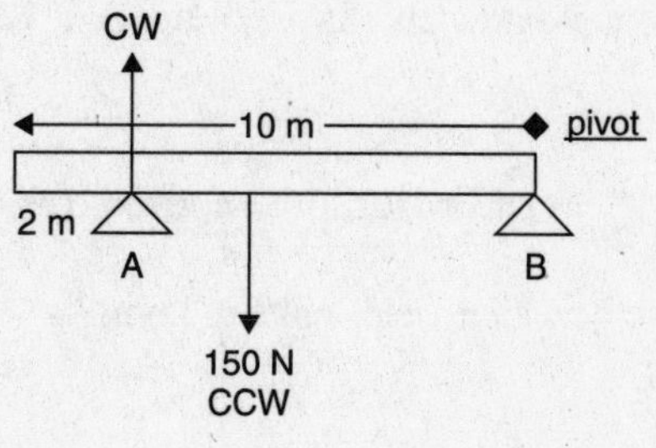

43. (A) Before the paint can was added, the upward force supplied by block B was previously $\mathbf{F_B} = 150\text{ N} - 94\text{ N} = 56\text{ N}$

Adding the can over block B supplies no additional torque if we still use B as our pivot. Therefore, now $\mathbf{F_B} = 56\text{ N} + 10\text{ N} = \mathbf{66\text{ N}}$

44. (C) The NET force causes the entire system to accelerate; that equals the weight of the unknown mass M minus the friction force between the 10 kg mass and the table.

$$\mathbf{F_{NET} = \Sigma ma}$$
$$\mathbf{Mg - F_F} = (\text{M} + 10\text{ kg})(2\text{ m/s}^2)$$
$$10\text{ M} - \mu\mathbf{F_N} = 2\text{M} + 20$$
$$8\text{ M} = 20 + (0.25)(10\text{ kg})(10\text{ m/s}^2)$$
$$\text{M} = 45/8\text{ kg} = \mathbf{5.6\text{ kg}}$$

45. (B) Once again using $\mathbf{F_{NET} = \Sigma ma}$,

$$\mathbf{a = F_{NET} / \Sigma m}$$

$$\frac{(\text{weight of hanging mass}) - (\text{friction on 20 kg masses})}{\text{total masses in the system}}$$

$$\frac{(6\text{ kg})(10\text{ m/s}^2) - (0.25)(20\text{ kg})(10\text{ m/s}^2)}{26\text{ kg}} = \mathbf{0.38\text{ m/s}^2 = approx\ 0.4\text{ m/s}^2}$$

46. (E) Molecules in the vapor state have more kinetic energy than those in the liquid state. When some of that energy is removed, the vapor is "cooled" and the resulting drop in kinetic energy of the molecules allows them to bind or stick to solid surfaces in the process of **condensation.**

47. (B) Using $\mathbf{d = v_O t + (1/2)at^2}$, $-80\text{m} = (-5\text{m/s}^2)\text{t}^2$ and $\mathbf{t = 4\ seconds}$

48. (C) Using $\mathbf{v_F{}^2 = v_O{}^2 + 2ad}$, $\mathbf{v_F} = [(2)(-10\text{ m/s}^2)(-80\text{ m})]^{1/2} = \mathbf{40\text{ m/s}}$

49. (B) The water's potential energy at the top is turned into thermal energy at the bottom. Using $\mathbf{E_P} = \text{mc}\Delta\text{t}$, $\mathbf{mgh} = \text{mc}\Delta\text{t}$

$$\Delta\text{t} = \mathbf{gh/c}$$

$$= \frac{(-10\text{ m/s}^2)(-120\text{ m})}{(1\text{ cal/g}^\circ\text{C})(4.19\text{ J/cal})} = \frac{1200\text{ m}^2/\text{s}^2}{4.19\text{ J/g}^\circ\text{C}} = \text{approx } 300\,\frac{\text{m}^2\text{g}^\circ\text{C}}{\text{J s}^2}$$

Since $1\text{ J} = 1\text{ Nm} = 1\text{ kgm}^2/\text{s}^2$:

$$\Delta\text{t} = \text{approx}\,\frac{(300\text{ m}^2\text{g}^\circ\text{C})}{(\text{kgm}^2)}\,\frac{(1\text{ kg})}{(1000\text{ g})} = \mathbf{approx\ 0.3^\circ\ C}$$

50. (C) Using $\mathbf{T = 2\pi\sqrt{\frac{l}{g}}}$,

$$\text{T} = (2)(3.1416)\sqrt{\frac{1}{10}}$$

$$= \text{approximately } (6)(0.3) = \mathbf{approx.\ 1.98\ sec}$$

51. (C) Using $\mathbf{T = 2\pi\sqrt{\frac{m}{k}}}$

$$\text{T} = (2)(3.1416)\sqrt{0.1} = \mathbf{approx.\ 1.98\ sec}$$

52. (D) Momentum is mass times velocity, which yields units of **kg m / s**

53. **(E)** Work equals applied force times distance moved in the same direction. The restoring force of a spring equals the **spring constant k** times the distance stretched x in the opposing direction.

54. **(C)** Power is work per unit of time, or J / s, which equal Nm / s or kgm^2/s^2.

55. **(D)** Using the height that the bullet-block system reaches and setting that potential energy equal to the kinetic energy of the system immediately after impact: $\mathbf{mgh} = \left(\frac{1}{2}\right)\mathbf{mv^2}$.(cancel the m, since it represents the bullet+block mass)

$$v = \left(\sqrt{2gn}\right) \text{ and momentum immediately after impact is } \mathbf{p} = (M+m)\left(\sqrt{2gn}\right)$$

$$\mathbf{p} = (3.05\text{ kg})[(2)(10\text{ m/s}^2)(0.05\text{ m})]^{\frac{1}{2}} \text{ (notice that the velocity is 1 m/s)}$$

$$= 3.05\text{ kgm/s} = \textbf{(closest) 3.1 kg m/s}$$

56. **(B)** Using $\mathbf{E_K} = \left(\frac{1}{2}\right)\mathbf{mv^2}$,

$$E_K = \left(\frac{1}{2}\right)(3.05\text{ kg})(1\text{ m/s})^2 = \mathbf{1.5\ J}$$

57. **(D)** Setting momentum before impact equal to total momentum immediately after impact,

$$(\mathbf{m_{BULLET}})(\mathbf{v_{O\,BULLET}}) = (\mathbf{M+m})(\mathbf{v_{F\,B+B}})$$

$$v_{O\,BULLET} = (3.05\text{ kg})(1\text{ m/s}) / (0.05\text{ kg}) = \mathbf{62\ m/s}$$

58. **(E)** Using **FΔ t = mΔv ("Impulse gives momentum")**,

$$F = (m\Delta v) / t = (500\text{ kg})(2\text{ m/s}) / 0.1\text{ sec} = 10,000\text{ N} = \mathbf{10\ kN}$$

59. **(B)** Efficiency = Work out / work in or **Power out / power in.**

60. **(B)** Set the parallel force equal to the spring's restoring force: $\mathbf{F_P = -kx}$

$$-k = (\text{mg sin } 30°) / x$$

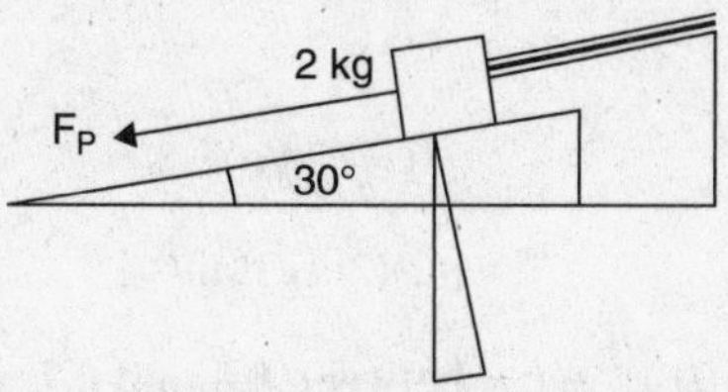

$$-k = \frac{mg\sin 30°}{x} = \frac{(2\text{ kg})(-10\text{ m/s}^2)\left(\frac{1}{2}\right)}{(0.1\text{ m})} = k = \mathbf{100\ N/m}$$

61. **(D)** Using $\mathbf{d = v_O t + (1/2)at^2}$,

$$d = \left(\frac{1}{2}\right)(10\text{ m/s}^2)(\sin 30°)(1\text{ sec})^2 = \mathbf{2.5\ m}$$

62. **(B)** Using $\mathbf{R} = \frac{\rho l}{A}$ and assigning **r** to the small square and **R** to the large one:

$$\frac{r}{R} = \frac{l}{L} \text{ and } \mathbf{r} = \frac{lR}{L} \text{ (the } \rho \text{ and A for both wires are the same)}$$

$$\mathbf{r} = 4s\,R / 8s = \left(\frac{1}{2}\right)\mathbf{R}$$

63. **(B)** Using **F = Il X B**

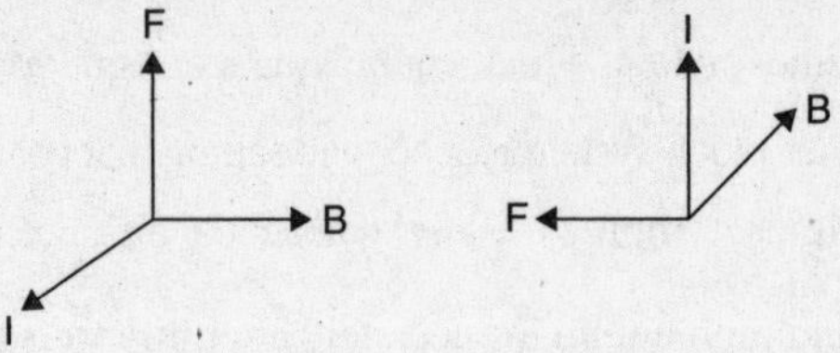

If **I** is toward the top of the page and **B** is out of the page, pointing the fingers of the right hand toward the top of the page in the direction of **I**, then swinging the fingers into the page, in the direction of **B**, the result is that the thumb points in the direction of the force **F, to the LEFT.**

64. **(D)** The center of mass continues to move in a **parabolic path**, regardless of the number of pieces it breaks up into.

65. **(D)** Using $\mathbf{\Delta p = mv_F - mv_O}$

$$\mathbf{\Delta p} = |(0.2\text{ kg})(-3\text{ m/s} - 3\text{ m/s})| = \mathbf{1.2\text{ kg m/s}}$$

66. **(C)** Only III is true. Although an object in translational equilibrium may not be in rotational equilibrium if there are unbalanced torques, the forces which keep the object in translational equilibrium must have a **vector sum of zero.**

67. **(D)** I. Changing magnetic field strength causes a current to flow ($\varepsilon = -\Delta\Phi/\Delta t$)

II. Moving the loop up and parallel to the magnetic field does not induce current.

III. Pulling the loop to the right cuts across magnetic lines of force and induces a current.

IV. Rotating the loop causes it to cut across magnetic lines of force and induces a current.

68. **(B)** Using $\mathbf{W = Fd \cos \Theta}$,

$$\mathbf{d} = \frac{\mathbf{W}}{\mathbf{F}} = \frac{1000\text{ J}}{(50\text{ kg})(10\text{ m/s}^2)} = 2\text{ m}$$

69. **(A)** Using $\mathbf{d = v_O t + \left(\frac{1}{2}\right)at^2}$,

$$t = (2d/\mathbf{a})^{\frac{1}{2}}$$

$$= [(2)(2\text{m})/10\text{m/s}^2]^{\frac{1}{2}}$$

$$= (0.4)^{\frac{1}{2}}\text{ sec} = \textbf{between 0.6 and 0.7 seconds.}$$

70. **(D)** Using $\mathbf{P = \frac{W}{t}, P = \frac{Fd}{t} = \frac{Mgh}{t}}$

Free Response

1. This is a vector problem that involves forces as well as lengths.

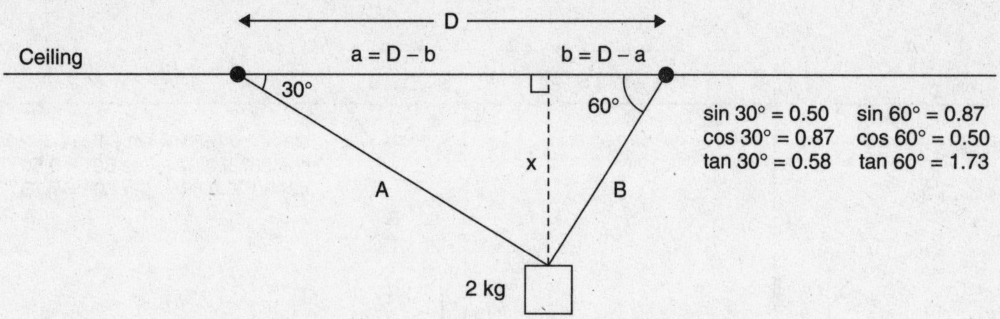

(a) Since D = 3 m, a = D – b and b = D – a and x is common to both right triangles.

x = (D – b)(tan 30°) and x = (D – a)(tan 60°). Combining equations yields a = 3b and A = 3 B.

Since a + a/3 = 3 m, a = $\frac{9}{4}$ m and b = $\frac{3}{4}$ m.

cos 30° = a/A and cos 60° = b/B. This yields: **A = 2.6 m, B = 1.5 m**

(b) mg = (2kg)(9.8 m/s^2) = 19.6 N

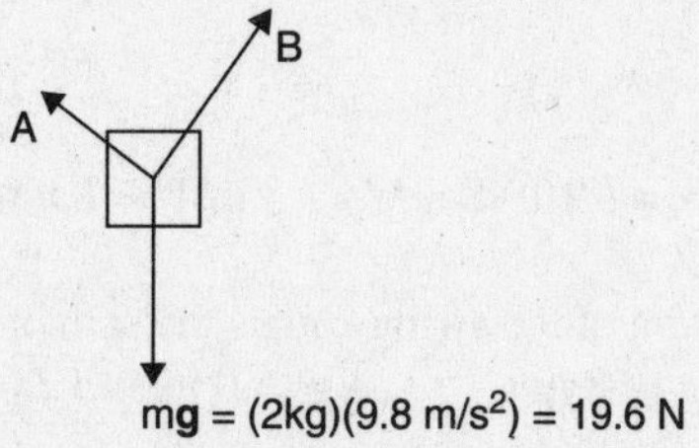

Notice the difference between the relative sizes of the FORCE VECTORS in the preceding illustration and the LENGTH VECTORS in the illustration before that. The string A may be longer than string B, but it applies less force than B, so its force vector is shorter.

Another way of viewing the force vectors is:

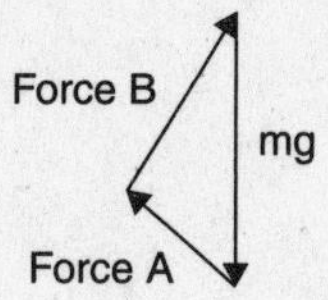

(c) If string B breaks, it will swing through height A – x. We need to determine that height and equate the **gravitational potential energy** at that point with the **kinetic energy** it will possess at the bottom of the swing (they will be equal) and compute the speed from $\mathbf{mgh} = \left(\frac{1}{2}\,\mathrm{m}\right)\mathbf{v}^2$. The mass cancels, leaving $\mathbf{v} = \sqrt{\mathbf{2gh}}$

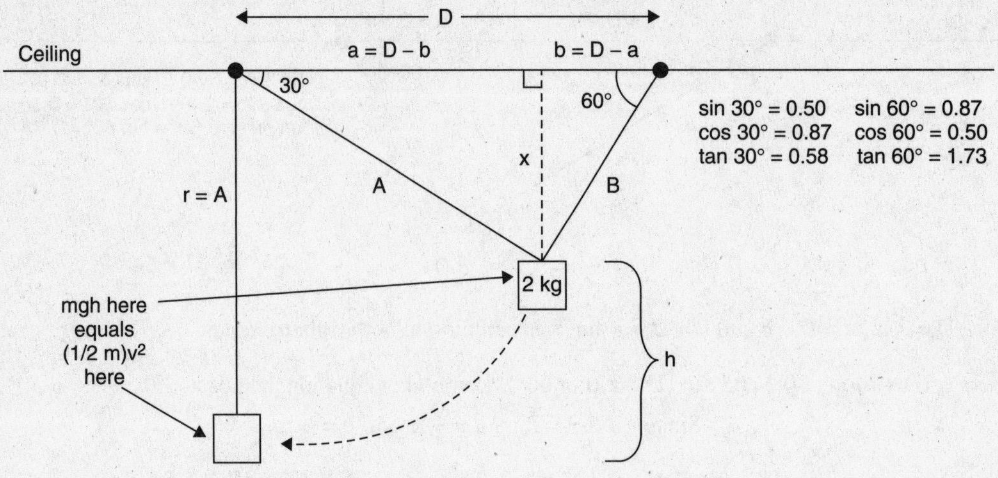

Since sin 30° = x/ A, x = (2.6 m)(0.5) = 1.3 m

$$h = A - x = 2.6\text{ m} - 1.3\text{ m} = 1.3\text{ m}$$

So, using $\mathbf{mgh} = \left(\frac{1}{2}\,\mathrm{m}\right)\mathbf{v}^2$ and $\mathbf{v} = \sqrt{\mathbf{2gh}}$

$$\mathbf{v} = [(2)(9.8\text{ m/s}^2)(1.3\text{ m})]^{\frac{1}{2}} = \mathbf{5.0\text{ m/s}}$$

(d) The tension in the string at the bottom of the swing equals the sum of the **WEIGHT** and the **CENTRIPETAL FORCE.** $(\mathbf{F_T = mg + mv^2/r})$, $\mathbf{F_T}$ = (2 kg)(9.8 m/s²) + (2 kg)(5.0 m/s)² / 2.6 m = 19.6 N + 19.2 N = **39 N**

2. **(a)** i. Put the block on the paper. Shine the light at the block at an incident angle other than 90°. Measure the incident and refracted angles.

ii.

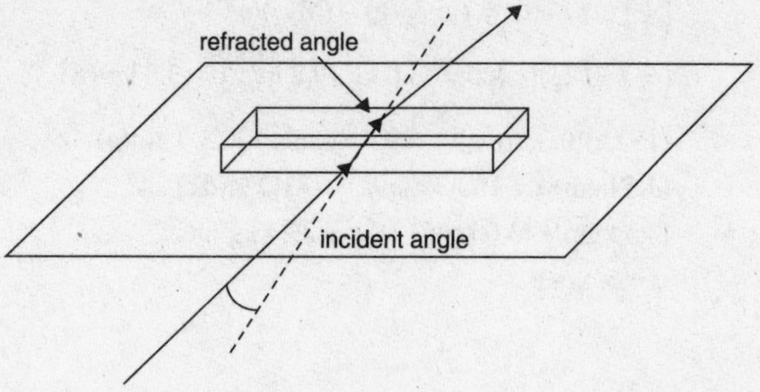

iii. $\mathbf{n} = \frac{\sin i}{\sin r}$

(b) i. Shine the light beam at the diffraction grating. Measure the angle to the first order image. Measure the distance from the principal image to the first order maximum.

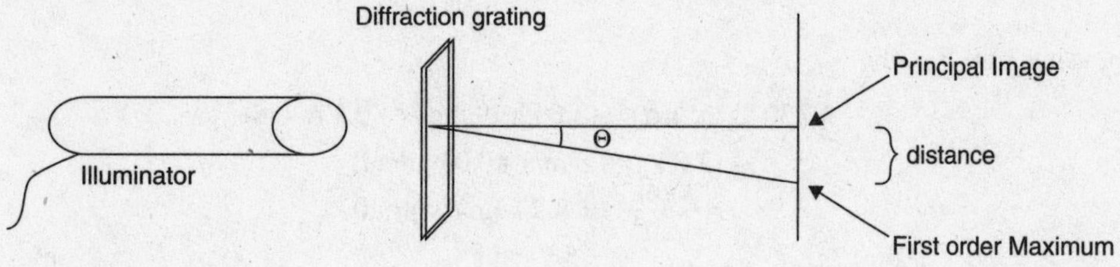

iii. $\lambda = \frac{d \sin \Theta_n}{n}$

3. **(a)** Using $\tau = Fr = I\alpha$, the net force is provided by the 3 kg mass minus the Force to accelerate it.

$$I\alpha = Fr$$
$$\left(\frac{1}{2}\right)(m)(r^2)(\alpha) = (m_{3kg})(g) - (M_{3kg})(a)$$
$$\left(\frac{1}{2}\right)(30\text{ kg})(r)(\alpha) = (3.0\text{ kg})(9.8\text{ m/s}^2) - (3.0\text{ kg})(a)$$
$$(15\text{ kg})(0.3\text{ m})(\alpha) = (29.4\text{ kg m/s}^2) - (3.0\text{ kg})(a)$$
$$(4.5\text{ kgm})\alpha = 29.4\text{ kg m/s}^2 - (3.0\text{ kg})(\alpha r)$$
$$[4.5\text{ kgm} + (3.0\text{ kg})(0.3)]\alpha = 29.4\text{ kg m/s}^2$$
$$\alpha = 5.4\text{ rad/s}^2$$

(b) Using $a = \alpha r$,

$$a = (5.4\text{ rad/s}^2)(0.3\text{ m}) = \mathbf{1.6\text{ m/s}^2}$$

(c) Using $\Theta = \omega_o t + \left(\frac{1}{2}\right)\alpha t^2$,

$$\Theta = (0) + \left(\frac{1}{2}\right)(5.4\text{ rad/s}^2)(5\text{ sec})^2(1\text{ rev}/2\pi\text{ rad})$$
$$= \mathbf{11\text{ revolutions}}$$

(d) Again, using $I\alpha = F_{NET}(r)$,

$$\left(\frac{1}{2}\right)(30\text{ kg})(r^2)(\alpha) = (3.0\text{ kg})(9.8\text{ m/s}^2)(0.3\text{ m/rad})$$
$$- (2.0\text{ kg})(9.8\text{ m/s}^2)(0.03\text{ m/rad})$$
$$1.4\frac{\text{kgm}^2}{\text{rad}^2}\alpha = 8.2\text{ kgm}^2/(\text{s}^2)(\text{rad})$$
$$\alpha = 5.9\text{ rad/s}^2$$

(e) Using $a = \alpha r$,

$$a = (5.9\text{ rad/s}^2)(0.3\text{ m/rad}) = \mathbf{1.8\text{ m/s}^2}$$

(f) Using $\Theta = \omega_o t + \left(\frac{1}{2}\right)\alpha t^2$,

$$\Theta = (0) + \left(\frac{1}{2}\right)(5.9\text{ rad/s}^2)(5\text{ sec})^2(1\text{ rev}/2\pi\text{ rad}) = \mathbf{12\text{ revolutions}}$$

4. **(a)** Equating the compressed energy of the spring to the kinetic energy of the mass as it leaves the table,

$$\frac{1}{2}kx^2 = \frac{1}{2}mv^2$$

$$v = \left(\frac{kx^2}{m}\right)^{\frac{1}{2}} = \left[\frac{(5\,N/m)(0.5\,m)^2}{(1\,kg)}\right]^{\frac{1}{2}} = 1.1\ m/s$$

(b) Using $\mathbf{d = v_0 t + \left(\frac{1}{2}\right)at^2}$ for the vertical: $-1\ m = (1/2)(-9.8 m/s^2)(t^2)$

$$t = (1/4.9)^{\frac{1}{2}}\ sec = \mathbf{0.45\ sec}$$

(c) Using $\mathbf{d = v_{AVE}t}$ for the horizontal:

$$d = (1.1\ m/s)(0.45\ sec) = 0.50\ m \text{ or } \mathbf{50\ cm}$$

(d) First, determine the frictional force against the spring using $\mu = \mathbf{F_F / F_N}$:

$$\mathbf{F_F} = \mu \mathbf{F_N} = (0.3)(1\ kg)(9.8\ m/s^2) = 2.9\ N$$

Now, to determine the NET FORCE:

$$\mathbf{F_{NET} = F_{SPRING} - F_F} = (2.5\ N - 2.9\ N) = -0.4\ N$$

Since the friction force is greater than the average force exerted by the spring, the mass will begin to move but will stop before it reaches the edge of the table.

Therefore, the mass **will not leave the table top.**

5. (a) Using $\mathbf{F_M = qv \times B}$,

$$F_M = (+1.60 \times 10^{-19}\,C)(2.0 \times 10^6\,m/s)(0.3T) = \mathbf{9.6 \times 10^{-14}\,N}$$

(b)

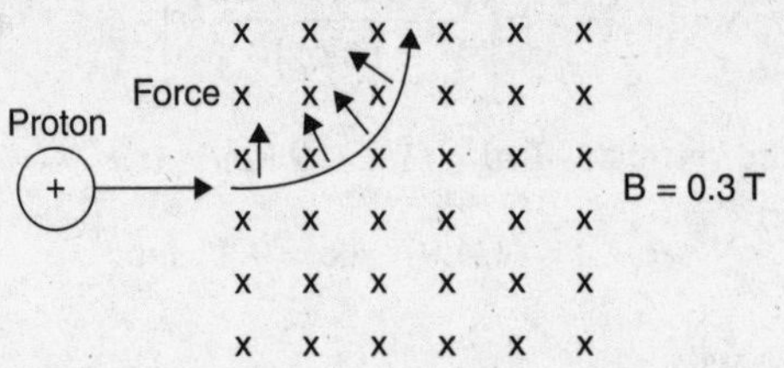

(c) Since the force that pulls the proton into the circular path is a centripetal force, equate $\mathbf{F_M}$ and $\mathbf{F_C}$:

$$\mathbf{qv \times B = \frac{mv^2}{r}}$$

$$9.6 \times 10^{-14}\,N = \frac{(1.67 \times 10^{-27}\,kg)(2.0 \times 10^6\,m/s)^2}{r} \text{ and } r = 7.0 \times 10^{-2}\,m \text{ or } 70\text{ cm}$$

(d) If the entering charge is an electron, the force due to the magnetic field is equal but in the opposite direction, since protons and electrons have equal but opposite electric charges. Therefore, the magnitude of the force would be the same as on the proton, 1.9×10^{-13} N.

Equating this force with the centripetal force for the electron:

$$\mathbf{qv \times B = \frac{mv^2}{r}}$$

$$9.6 \times 10^{-14}\,N = \frac{(9.11 \times 10^{-31}\,kg)(2.0 \times 10^6\,m/s)^2}{r} \text{ and } r = 3.8 \times 10^{-5}\,m$$

The ratio of radii is:

$$\frac{3.5 \times 10^{-2}\text{ m for the proton}}{3.8 \times 10^{-5}\text{ m for the electron}}$$

The radius of the proton's circular path is **about 920 times that of the electron.**

6. **(a)** First, determining the frequency of the emitted photon: Using $\mathbf{f = v / \lambda}$,

$$f = (3 \times 10^8 \text{ m/s}) / (4.8 \times 10^{-7} \text{ m}) = \mathbf{6.3 \times 10^{14} \text{ Hz}}$$

(b) Using $\mathbf{E = hf}$,

$$E = (6.63 \times 10^{-34} \text{ J} \cdot \text{s})(6.3 \times 10^{14} \text{ Hz}) = \mathbf{4.2 \times 10^{-19} \text{ J}}$$

(c) $(4.2 \times 10^{-19} \text{ J})\left(\dfrac{1\text{eV}}{(1.6 \times 10^{-19} \text{ J})}\right) = \mathbf{2.6 \text{ eV}}$

(d) Level 5 to 4 yields 0.31 eV

Level 5 to 3 yields 0.97 eV

Level 5 to 2 yields 2.9 eV

Level 5 to 1 yields 13.06 eV

Level 4 to 3 yields 0.66 eV

Level 4 to 2 yields 2.54 eV . . . . . . the closest to 2.6 eV

Level 4 to 1 yields 12.75 eV

Level 3 to 2 yields 1.88 eV

Level 3 to 1 yields 12.1 eV

Level 2 to 1 yields 12.2 eV

The only possible "jump" made by the electron is from **level 4 to level 2.**

CliffsAP Physics C Practice Exam 1

Physics C

You may take the entire C exam or Mechanics only or Electricity and Magnetism only as follows:

	Entire C Exam	***Mechanics Only***	***Electricity and Magnetism Only***
1st 45 min.	Sect. I—Mechanics	Sect. I—Mechanics	Sect. I—Electricity and Magnetism
	35 questions	35 questions	35 questions
	Multiple Choice	Multiple Choice	Multiple Choice
2nd 45 min.	Sect. I—Electricity and Magnetism	Sect. II—Mechanics	Sect. II—Electricity and Magnetism
	35 questions	3 questions	3 questions
	Multiple Choice	Free Response	Free Response
3rd 45 min.	Sect. II—Mechanics		
	3 questions		
	Free Response		
4th 45 min.	Sect. II—Electricity and Magnetism		
	3 questions		
	Free Response		

Separate grades are reported for Mechanics and for Electricity and Magnetism. Each section of each examination is 50 percent of the total grade.

Each question in a section has equal weight. Calculators are *not permitted* on Section I of the exam but are allowed on Section II. However, calculators cannot be shared with other students, and calculators with typewriter-style (QWERTY) keyboards will not be permitted. A table of information that may be helpful is found on the following page.

The Physics C examination contains a total of 70 multiple-choice questions. If you are taking

Mechanics only, please be careful to answer multiple choice questions 1–35.

Electricity and Magnetism only, please be careful to answer multiple choice questions 36–70.

Mechanics *and* Electricity and Magnetism (the entire examination), answer multiple-choice questions 1–70.

Information Table

Constants and Conversions

1 Atomic mass unit	$u = 1.66 \times 10^{-27}$ kg
	$= 931$ MeV/ c^2
Proton mass	$m_p = 1.67 \times 10^{-27}$ kg
Neutron mass	$m_n = 1.67 \times 10^{-27}$ kg
Electron mass	$m_e = 9.11 \times 10^{-31}$ kg
Electron charge	$e = 1.60 \times 10^{-19}$ C
Avogadro's number	$N_0 = 6.02 \times 10^{23}$ mol^{-1}
Universal gas constant	$R = 8.31$ J / (mol · K)
Boltzmann's constant	$k_B = 1.38 \times 10^{-23}$ J / K
Speed of light	$c = 3.00 \times 10^8$ m / s
Planck's constant	$h = 6.63 \times 10^{-34}$ J · s
	$= 4.14 \times 10^{-15}$ eV · s
	$hc = 1.99 \times 10^{-25}$ J · m
	$= 1.24 \times 10^3$ eV · nm
Vacuum permittivity	$\varepsilon_0 = 8.85 \times 10^{-12}$ C^2 / N · m^2
Coulomb's law constant	$k = 1/4\pi\varepsilon_0 = 9.0 \times 10^9$ N · m^2 / C^2
Vacuum permeability	$\mu_0 = 4\pi \times 10^{-7}$ (T · m) / A
Magnetic constant	$k' = \mu_0 / 4\pi = 10^{-7}$ (T · m) / A
Universal gravitation constant	$G = 6.67 \times 10^{-11}$ Nm^2/kg^2
Acceleration due to gravity at the earth's surface	$\mathbf{g} = 9.8$ m/s^2
1 atmosphere pressure	1 atm = 1.0×10^5 Pa
	$= 1.0 \times 10^5$ N / m^2
1 electron volt	1 eV = 1.60×10^{-19} J

Units

meter	m
kilogram	kg
second	s
ampere	A
kelvin	K
mole	mol
hertz	Hz
newton	N
pascal	Pa
joule	J
watt	W
coulomb	C
volt	V
ohm	Ω
henry	H
farad	F
tesla	T
degree Celsius	°C
electron Volt	eV

Prefixes

Factor	Prefix	Symbol
10^9	giga	G
10^6	mega	M
10^3	kilo	k
10^{-2}	centi	c
10^{-3}	milli	m

Trigonometric Function Values for Common Angles

Θ	sin Θ	cos Θ	tan Θ
0°	0	1	0
30°	$\frac{1}{2}$	$\frac{\sqrt{3}}{2}$	$\frac{\sqrt{3}}{3}$
37°	$\frac{3}{5}$	$\frac{4}{5}$	$\frac{3}{4}$
45°	$\frac{\sqrt{2}}{2}$	$\frac{\sqrt{2}}{2}$	1
53°	$\frac{4}{5}$	$\frac{3}{5}$	$\frac{4}{3}$

Prefixes			***Trigonometric Function Values for Common Angles***			
10^{-6}	micro	μ	60°	$\frac{\sqrt{3}}{2}$	$\frac{1}{2}$	$\sqrt{3}$
10^{-9}	nano	n	90°	1	0	∞
10^{-12}	pico	p				

These conventions are used in the examination:

1. *Unless stated otherwise, inertial frames of reference are used.*
2. *The direction of electric current flow is the conventional direction of positive charge.*
3. *For an isolated electric charge, the electric potential at an infinite distance is zero.*
4. *In mechanics and thermodynamics equations, W represents work done* ***on*** *a system.*

GO ON TO THE NEXT PAGE

Physics C
Sample Exam #1
Section I, Mechanics

Time – 45 Minutes
35 Questions

Directions: Each of the following questions or incomplete statements is followed by five possible answers. Select the best answer and then fill in the corresponding oval on the answer sheet.

Note: You may use **g** = 10 m/s^2 to simplify calculations.

Questions 1–2 refer to the following figure.

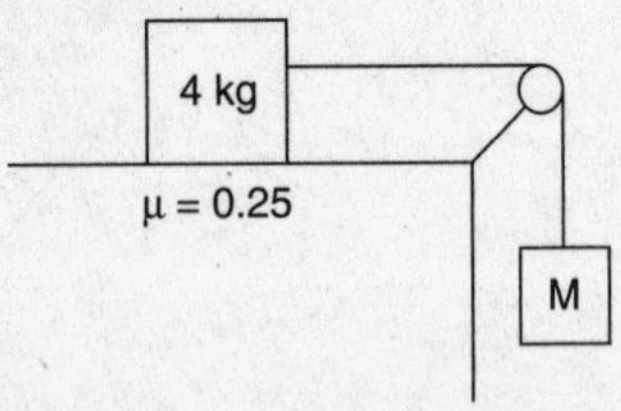

The 4 kg mass is pulled to the right by the unknown mass M, over a tabletop having a coefficient of friction of 0.25 with the 4 kg mass. The pulley is frictionless.

1. If the 4 kg mass accelerates at a rate of 2 m/s^2, the value of the unknown mass M is most nearly

(A) 2 kg
(B) 4 kg
(C) 6 kg
(D) 8 kg
(E) 10 kg

2. If both masses in the preceding question are doubled, what is the new acceleration?

(A) 2 m/s^2
(B) 4 m/s^2
(C) 6 m/s^2
(D) 8 m/s^2
(E) 10 m/s^2

3. A ball launched horizontally from a platform lands 3 m away from the base of the platform after 1 sec. From what height was it launched?

(A) 1 m
(B) 3 m
(C) 5 m
(D) 7 m
(E) 9 m

4. An object moves along the x-axis with a changing acceleration described as **a** = 7t, where **a** is in m/s^2 and t is in sec. At t = 0, its velocity and position are both zero. When t = 2 seconds, it is located closest to

(A) 1 m
(B) 3 m
(C) 5 m
(D) 7 m
(E) 9 m

GO ON TO THE NEXT PAGE

Questions 5–6 refer to the following figure.

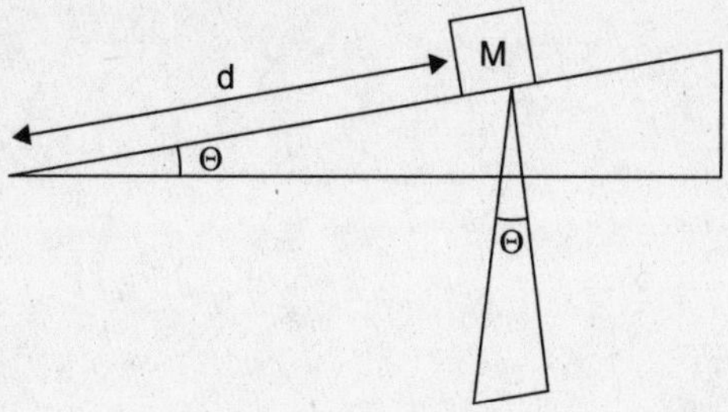

A block of mass M is released from rest at the top of a frictionless inclined plane that makes an angle Θ with the horizontal. The length of the inclined plane is d.

5. The block reaches the bottom after a time of

(A) $(g \sin \Theta\, d)^{\frac{1}{2}}$
(B) $2g \sin \Theta\, d$
(C) $2d / g \sin \Theta$
(D) $(2d / g \sin \Theta)^{\frac{1}{2}}$
(E) $(g \sin 2\Theta) / d$

6. The speed of the block at the bottom is

(A) $(g \sin \Theta\, d)^{\frac{1}{2}}$
(B) $(2g \sin \Theta\, d)^{\frac{1}{2}}$
(C) $(2g \sin \Theta) / d$
(D) $(2g \sin \Theta) / d^2$
(E) $(g \sin 2\Theta) / d$

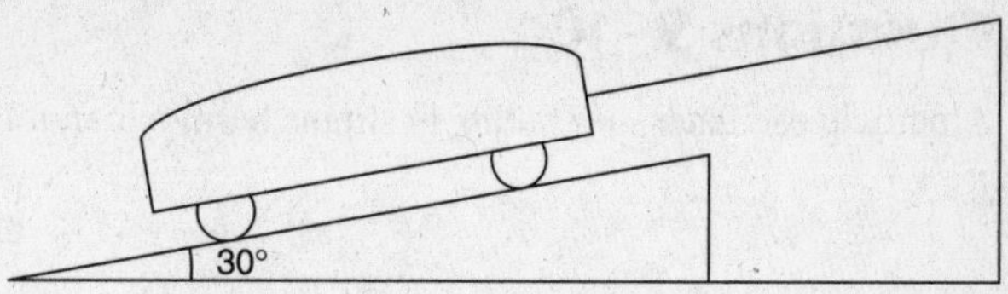

7. A 15,000 N car on a 30° incline is tied to an immovable object by a rope as shown. If the rope is just strong enough to hold the car in position, what is the minimum breaking force of the rope?

(A) 2,500 N
(B) 5,000 N
(C) 7,500 N
(D) 10,000 N
(E) 12,500 N

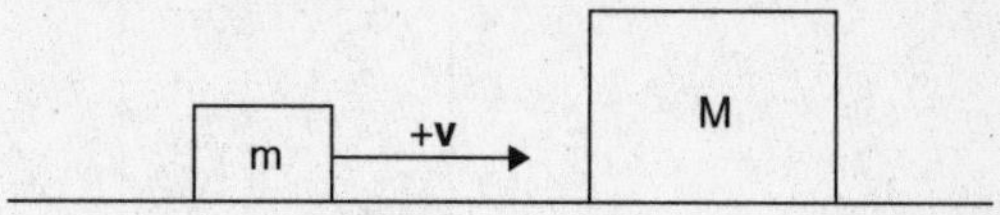

8. A mass m moves horizontally with velocity +**v** toward a stationary larger mass M on a frictionless surface. After they collide elastically

(A) Their combined momentum < their initial combined momentum.
(B) Mass m maintains a positive velocity.
(C) The final velocity of mass M is equal to the initial velocity of mass m.
(D) The total mass is M + m.
(E) Masses M and m have equal and opposite final velocities.

CliffsAP Physics C Practice Exam 1

GO ON TO THE NEXT PAGE

Questions 9–10

A particle oscillates on a spring in simple harmonic motion. Its potential energy versus displacement graph is shown here.

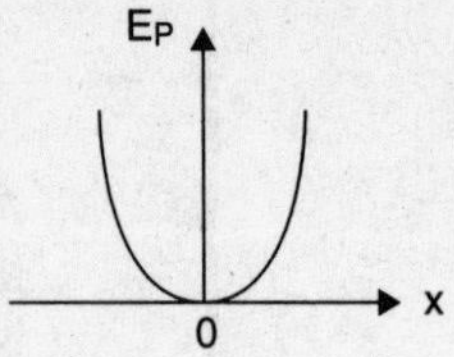

Refer to the following graphs when answering questions 9 and 10.

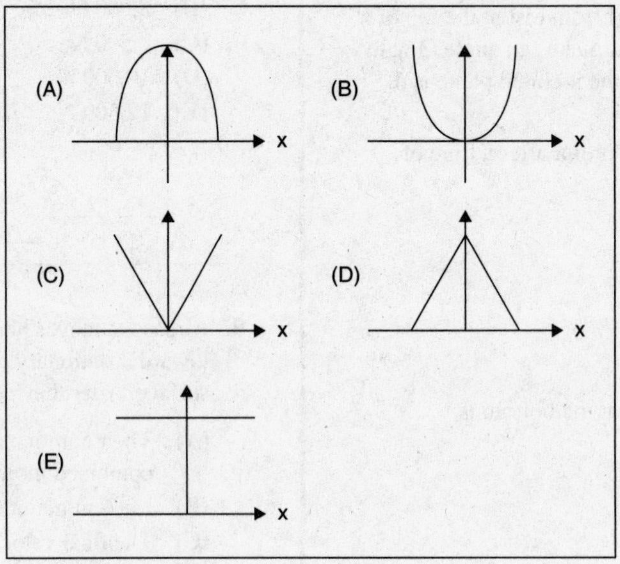

9. The graph of its total energy versus displacement is

(A) A
(B) B
(C) C
(D) D
(E) E

10. The graph of its kinetic energy versus displacement is

(A) A
(B) B
(C) C
(D) D
(E) E

GO ON TO THE NEXT PAGE

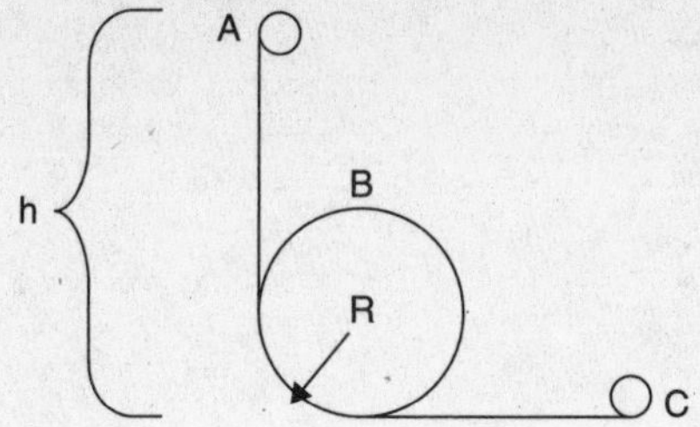

11. A ball is released from rest at point A on a frictionless track at height h above the ground. It curves into a loop-the-loop of radius R, just staying on the track at point B, and continues to point C. Height h equals

(A) 2R
(B) $\frac{2R}{\sqrt{3}}$
(C) $\frac{2R}{\sqrt{2}}$
(D) (5 / 2)R
(E) $5\sqrt{2}\,R$

Questions 12–13 refer to the following figure.

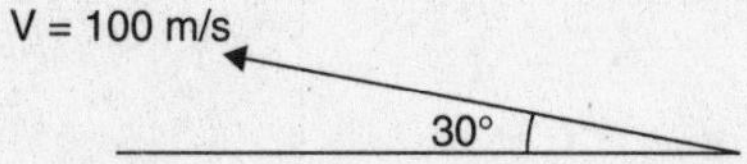

A projectile is fired from a cannon at an angle of 30° with the ground and with initial velocity of 100 m/s.

12. The total time it is in the air is

(A) 5 seconds.
(B) 10 seconds.
(C) 15 seconds.
(D) 20 seconds.
(E) 25 seconds.

13. The projectile travels a total horizontal distance of

(A) 400–500 m.
(B) 500–600 m.
(C) 600–700 m.
(D) 700–800 m.
(E) 800–900 m.

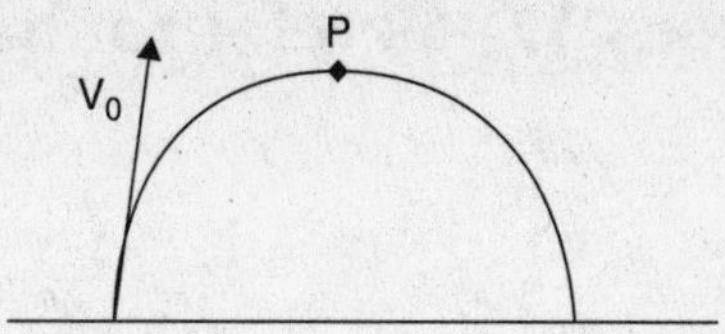

14. A projectile is launched with initial velocity $\mathbf{V_0}$. At point P it has horizontal velocity, vertical velocity, and acceleration given by

	$V_{HORIZONTAL}$	$V_{VERTICAL}$	*acceleration*
(A)	0	0	−g
(B)	$\mathbf{V_{HORIZ}}$	0	0
(C)	$\mathbf{V_{HORIZ}}$	$\mathbf{V_{VERT}}$	−g
(D)	$\mathbf{V_{HORIZ}}$	0	−g
(E)	$\mathbf{V_{HORIZ}}$	$\mathbf{V_{VERT}}$	−g

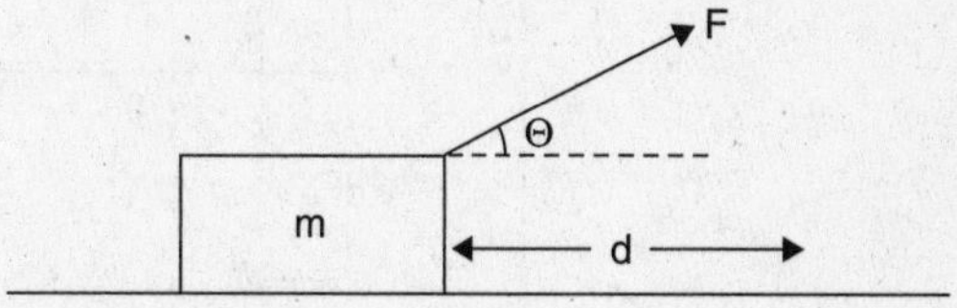

15. A force **F** is exerted on a crate of mass m at an angle Θ with the horizontal. The work done on the crate as it moves a distance **d** along the floor is

(A) **F**m cosΘ
(B) **F**m sinΘ
(C) **Fmd** cosΘ
(D) **Fd** tanΘ
(E) **Fd** cosΘ

16. A solid disk of mass m and radius r rolls along a lab table without slipping. A painted spot on the edge of the wheel moves with a linear tangential velocity **v**. Its angular momentum **L** is

(A) I **v** r
(B) I (**v** / r^2)
(C) I (**v** / r)
(D) I ($\mathbf{v}^2$ / r^2)
(E) I $\mathbf{v}^2$ r

GO ON TO THE NEXT PAGE

Questions 17–18 refer to the following figure.

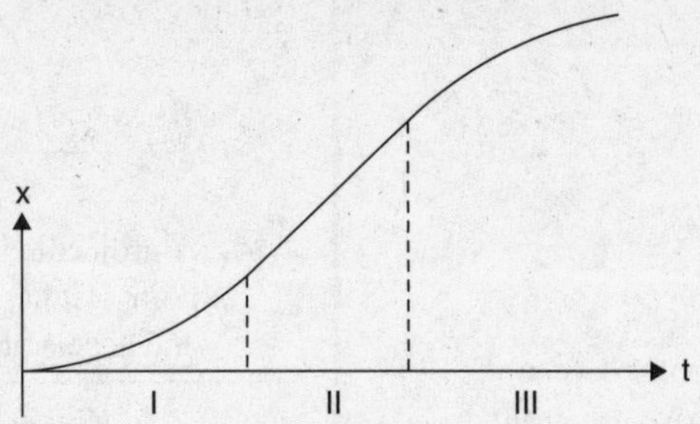

The graph here illustrates displacement **x** as a function of time t for an object moving along the **x**-axis. The five graphs that follow may indicate either **velocity versus time** or **acceleration versus time**.

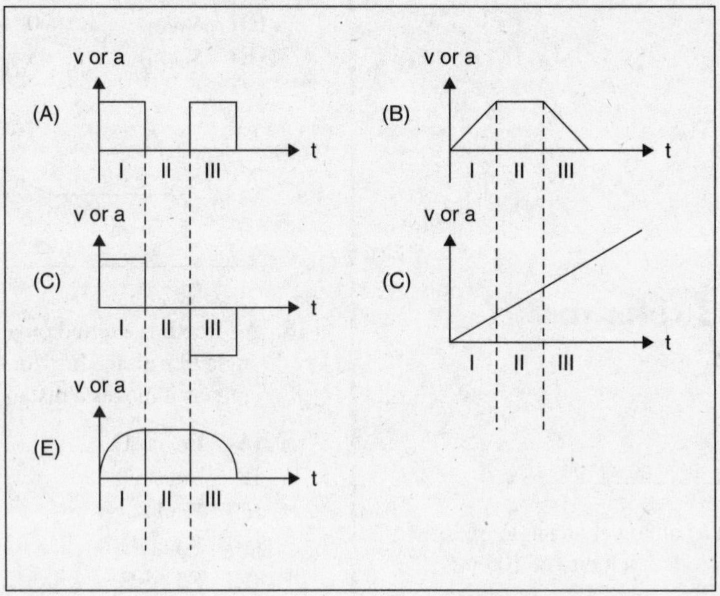

17. The graph that represents velocity **v** versus time t is

(A) A
(B) B
(C) C
(D) D
(E) E

18. The graph that represents acceleration **a** versus time t is

(A) A
(B) B
(C) C
(D) D
(E) E

19. The position of a cart moving along the **x**-axis is represented by **x** = $2t^3 + 3t^2 - 6t$, where **x** is in meters and t is in seconds. Its acceleration at time t = 2 seconds is

(A) 3 m/s^2
(B) 12 m/s^2
(C) 24 m/s^2
(D) 30 m/s^2
(E) 36 m/s^2

GO ON TO THE NEXT PAGE

Questions 20–21 refer to the following figure.

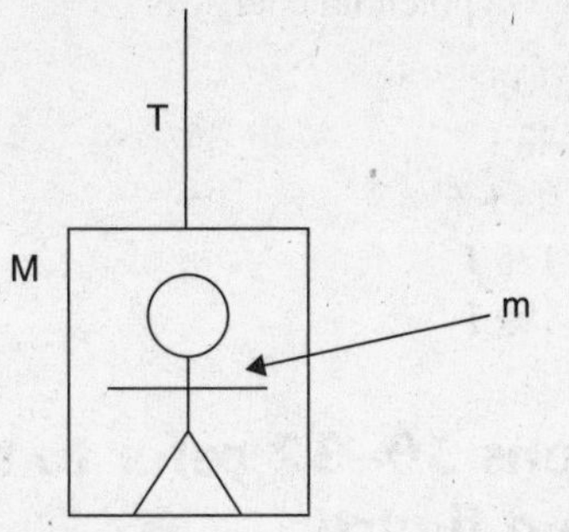

A person of mass m stands in an elevator of mass M, accelerating upward at a rate of 0.5 m/s^2.

20. The force exerted by the person on the elevator floor is

(A) M (g + 0.5 m/s^2)
(B) mg – 0.5 m/s^2
(C) mg + 0.5 m/s^2
(D) m (g + 0.5 m/s^2)
(E) M (g – 0.5 m/s^2)

21. The tension T in the cable is

(A) (M)(0.5 m/s^2 + g)
(B) (M + m)(0.5 m/s^2 + g)
(C) (M + m)(g – 0.5 m/s^2)
(D) (M – m)(g + 0.5 m/s^2)
(E) (M)(0.5 m/s^2 – g)

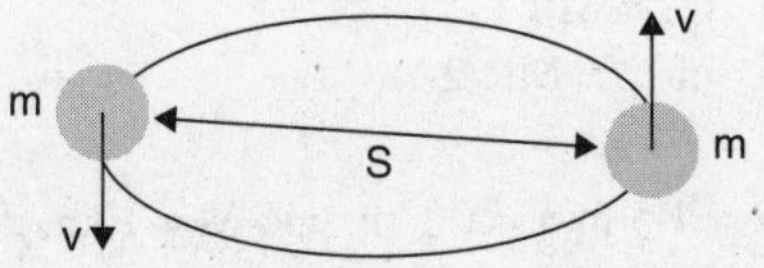

22. Two asteroids, each of mass m, orbit each other at a distance S from center to center, both with linear velocity **v**, which is represented by

(A) $(Gm / 2S)^{\frac{1}{2}}$
(B) Gm / 2S
(C) $(Gm^2 / 2S)^{\frac{1}{2}}$
(D) GmS / 2
(E) $(GmS / 2)^{\frac{1}{2}}$

23. The acceleration of either asteroid is

(A) $\mathbf{v}^2 / 2S$
(B) $\mathbf{v}^2 / S$
(C) $2\mathbf{v} / S^2$
(D) $2\mathbf{v}^2 / S$
(E) $\mathbf{v}^2 / 2S^2$

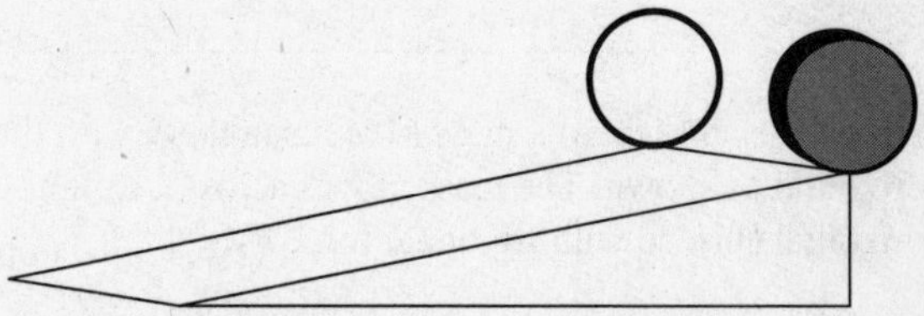

24. A hoop and a solid disk, both of mass m and radius r, roll, from rest, down the same incline simultaneously. Comparing their behavior, the disk

(A) reaches the bottom first.
(B) does not reach the bottom first.
(C) reaches the bottom at the same time as the hoop.
(D) has a greater potential energy at the top.
(E) has a smaller angular velocity at the bottom.

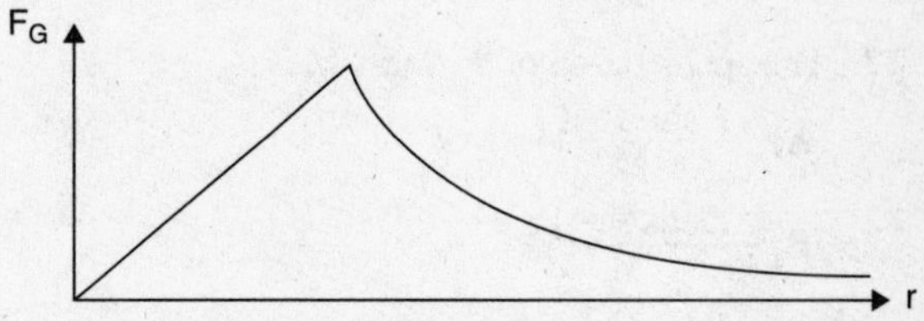

25. The graph illustrates gravitational force as a function of distance

(A) for a body in freefall.
(B) for an ideal spring.
(C) from a planet's surface.
(D) for an inclined plane.
(E) from a planet's center.

GO ON TO THE NEXT PAGE

Questions 26–27 refer to the following figure.

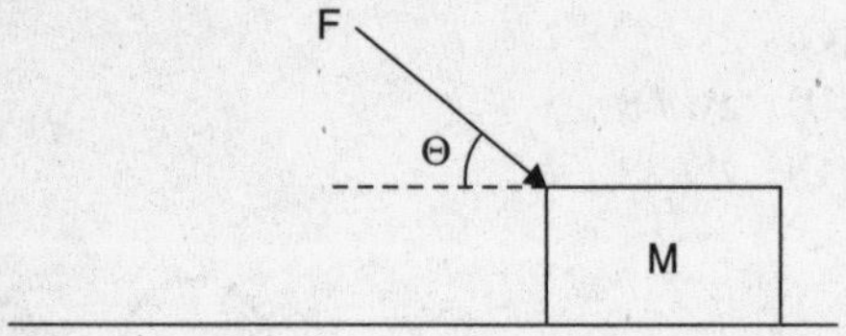

A force **F** is applied to a mass M at an angle Θ with the horizontal as shown. The mass moves across a rough horizontal surface with frictional force $\mathbf{F_F}$.

26. The coefficient of friction between the mass and the surface is

(A) $\frac{F_F}{MgF \sin\Theta}$

(B) $\frac{F_F}{Mg - F}$

(C) $\frac{F_F}{Mg + F\sin\Theta}$

(D) $\frac{F_F}{Mg - F\sin\Theta}$

(E) $\frac{F_F}{Mg\sin\Theta}$

27. The acceleration of the mass M is

(A) $\frac{F\sin\Theta - F_F}{M}$

(B) $\frac{F\cos\Theta - F_F}{M}$

(C) $\frac{F\cos\Theta + F_F}{M}$

(D) $\frac{F\cos\Theta + F_P}{M}$

(E) $\frac{F\sin\Theta + F_F}{M}$

28. Two persons on a frozen lake push each other. The larger person weighs 1000 N, and the smaller person weighs 700 N. After pushing, their center of mass moves with velocity

(A) zero.
(B) 7/10 m/s toward the 1000 N person.
(C) 1 m/s toward the 700 N person.
(D) 10/7 m/s toward the 700 N person.
(E) 1.7 m/s toward the 1000 N person.

29. The force necessary to compress a multiradial spring is given by $\mathbf{F} = 30x - 4x^2$, where **F** is in N and **x** is in m. When stretched 3 m, the spring's change in potential energy is

(A) 30 J
(B) 88 J
(C) 99 J
(D) 135 J
(E) 199 J

Questions 30–32 refer to the following figure.

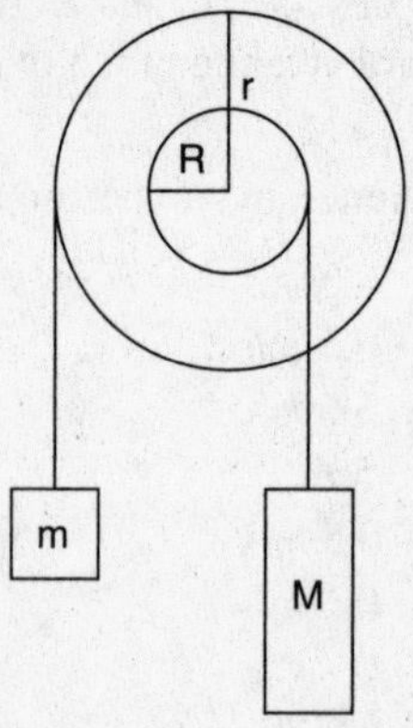

30. The two concentric attached disks of radii R and r are in rotational equilibrium. Therefore,

(A) Mr = mR
(B) $mR^2 = Mr^2$
(C) $\left(\frac{1}{2}\right)mR^2 = \left(\frac{1}{2}\right)Mr^2$
(D) mr = MR
(E) mr/2 = MR/2

31. If R = 1.5 m, r = 0.5 m, and M = 1 kg, m equals

(A) $\left(\frac{1}{3}\right)$ kg
(B) 1 kg
(C) $\left(\frac{4}{3}\right)$ kg
(D) 2 kg
(E) 3 kg

GO ON TO THE NEXT PAGE

32. If the masses are changed and M = 3 m, the angular acceleration of the system is

(A) $\frac{3mr + mRa}{I}$

(B) $\frac{3mR + mRa}{I}$

(C) $\frac{mR(r - \mathbf{a}) - mr(\mathbf{a})}{I}$

(D) $\frac{m[3(r - \mathbf{a}) - (R + \mathbf{a})]}{I}$

(E) $\frac{m[3(R + \mathbf{a}) - (r + \mathbf{a})]}{I}$

33. A crane lifts 50 kg 50 m in 10 seconds. It develops a power of

(A) 50 W
(B) 250 W
(C) 750 W
(D) 2500 W
(E) 4500 W

Questions 34–35 refer to the following figure.

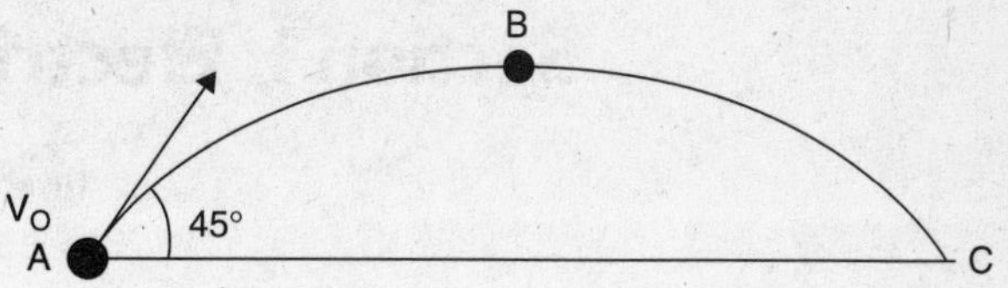

34. A projectile of mass M is launched from point A at an angle of 45° with the ground with velocity $\mathbf{v_o}$. Its speed at point B is

(A) $\mathbf{v_o}/2$

(B) $\frac{\sqrt{2}\,\mathbf{v_o}}{2}$

(C) $\frac{\sqrt{3}\,\mathbf{v_o}}{2}$

(D) $\sqrt{2}\,\mathbf{v_o}$

(E) $2\sqrt{2}\,\mathbf{v_o}$

35. If $\mathbf{v_o}$ = 200 m/s and m = 2 kg, the momentum of the projectile at point B is

(A) 100–200 kgm/s
(B) 200–300 kgm/s
(C) 300–400 kgm/s
(D) 400–500 kgm/s
(E) 500–600 kgm/s

END OF SECTION I, MECHANICS. IF YOU FINISH BEFORE TIME IS CALLED, YOU MAY CHECK YOUR WORK ON THIS SECTION ONLY. DO NOT TURN TO ANY OTHER TEST MATERIALS.

STOP

CliffsAP Physics C Practice Exam 1

Physics C
Sample Exam #1
Section I, Electricity and Magnetism

Time — 45 Minutes

35 Questions

Directions: Each of the following questions or incomplete statements is followed by five possible answers. Select the best answer and then fill in the corresponding oval on the answer sheet.

Questions 36–39

Questions 36–39 refer to the following diagram of part of a closed DC circuit.

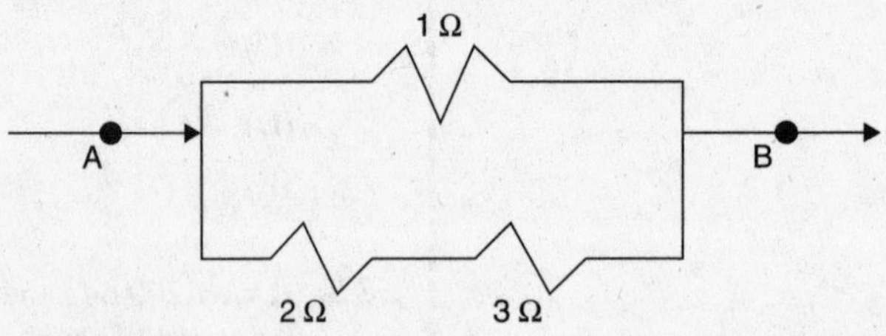

36. The equivalent resistance between points A and B is

(A) $\left(\frac{2}{3}\right)\Omega$

(B) $\left(\frac{5}{6}\right)\Omega$

(C) $1\,{}^{1}\!/_{6}\ \Omega$

(D) $1\,{}^{1}\!/_{4}\ \Omega$

(E) $2\,{}^{1}\!/_{6}\ \Omega$

37. While a constant current flows through the circuit, the amount of charge passing a point per second is

(A) greater at point A than at point B.

(B) greater in the 2 Ω resistor than in the 3 Ω resistor.

(C) greater in the 3 Ω resistor than in the 1 Ω resistor.

(D) greater in the 1 Ω resistor than in the 2 Ω resistor.

(E) the same everywhere in the circuit.

38. A 6 A current through point A means that the current through the 3 Ω resistor is

(A) 1 A

(B) 2 A

(C) 3 A

(D) 4 A

(E) 5 A

39. The voltage drop through the 3 Ω resistor is

(A) 2 V

(B) 3 V

(C) 4 V

(D) 5 V

(E) 6 V

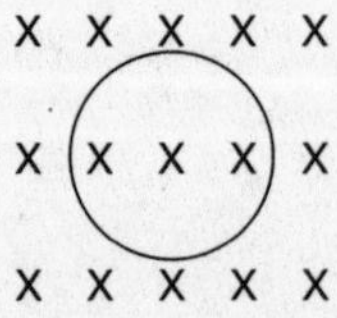

40. A circular loop of wire is placed in and perpendicular to a magnetic field directed into the page as shown. If the magnetic field's magnitude is increasing, the resulting current in the wire is

(A) zero.

(B) clockwise.

(C) counterclockwise.

(D) out of the paper.

(E) into the paper.

GO ON TO THE NEXT PAGE

Questions 41–42 refer to the following figure.

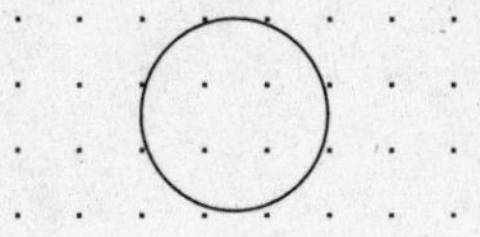

A metal loop is placed in a uniform magnetic field as shown. The following experiments are performed:

I. The magnetic field strength is changed.
II. The loop is moved up (out of the page) in the uniform field.
III. The loop is pulled to the right, and out of the field.
IV. The loop is rotated about a diameter.

41. A current will flow through the loop as a result of

(A) I, II, and III only
(B) I and III only
(C) II, III, and IV only
(D) I, III, and IV only
(E) I, II, III, and IV

42. If the magnetic field is increasing

(A) current flows clockwise.
(B) current flows into the page.
(C) current flows counterclockwise.
(D) current flows out of the page.
(E) the flux density outside the loop is greater than that inside the loop.

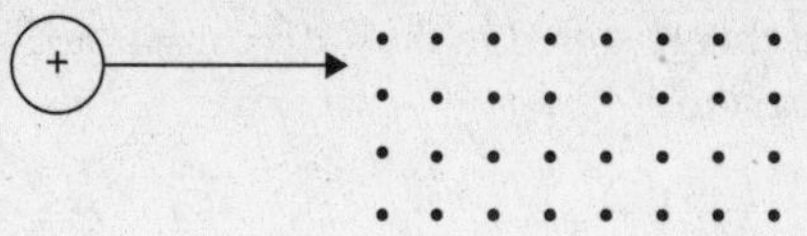

43. A positively charged particle enters a magnetic field directed out of the page. If the charge enters from the left, as shown, the resulting force is directed

(A) toward the top of the page.
(B) toward the left.
(C) out of the page.
(D) toward the bottom of the page.
(E) into the page.

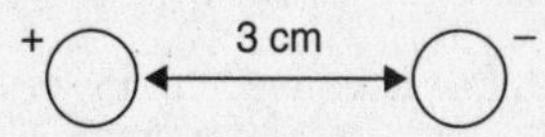

44. A proton and an electron are separated by a distance of 3 cm as shown. The electrostatic force between them is

(A) -2.56×10^{-29} N
(B) -2.56×10^{-25} N
(C) $+2.56 \times 10^{-25}$ N
(D) -2.56×10^{25} N
(E) $+2.56 \times 10^{29}$ N

Questions 45–47 refer to the following figure.

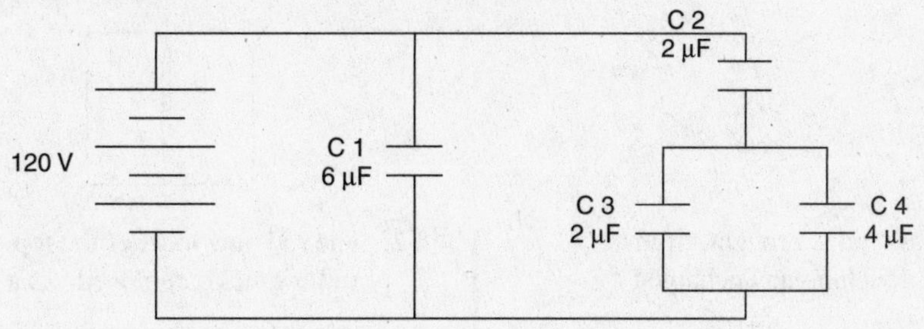

45. The equivalent capacitance of this circuit is nearest to

(A) $\frac{5}{6}$ μF
(B) $\frac{6}{5}$ μF
(C) 7 μF
(D) 7.5 μF
(E) 10.5 μF

46. The charge stored in the 6 μF capacitor C1 is nearest to

(A) 300 μC
(B) 550 μC
(C) 600 μC
(D) 750 μC
(E) 900 μC

GO ON TO THE NEXT PAGE

47. The charge stored in the 4 μF capacitor C4 is nearest to

(A) 120 μC
(B) 240 μC
(C) 360 μC
(D) 480 μC
(E) 600 μC

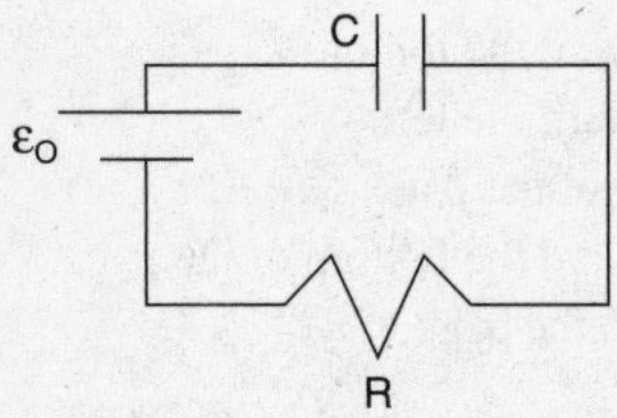

48. A cell of emf ε_O is connected, in series, to a capacitor C and a resistor R. The correct relationship regarding the circuit is

(A) $\varepsilon_0 - IR = -\frac{Q}{C}$
(B) $IR = \varepsilon_0 \frac{C}{Q}$
(C) $\varepsilon_0 = \frac{Q}{C} + IR$
(D) $C + Q + IR = \varepsilon_O$
(E) $\varepsilon_0 = \frac{Q}{C} + I^2R$

49. The number of excess electrons on a nonconducting sphere that has a charge of –9 nC is closest to

(A) 2×10^8
(B) 4×10^8
(C) 2×10^{10}
(D) 4×10^{10}
(E) 6×10^{10}

50. The number of Joules of energy delivered to a motor if 7.5 Coulombs passes through it from a 24 V battery is most nearly

(A) 100 J
(B) 200 J
(C) 300 J
(D) 400 J
(E) 500 J

Questions 51–52 refer to the following figure.

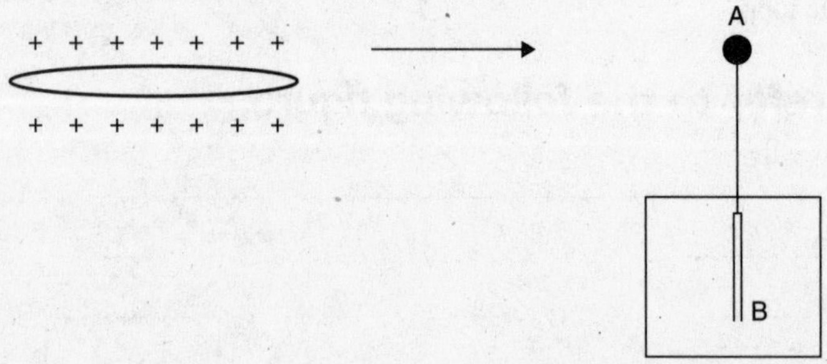

A positively charged glass rod is brought, from a distance, near to but not touching an uncharged electroscope.

51. As a result,

(A) A and B become negatively charged.
(B) A and B become positively charged.
(C) A becomes negatively charged by conduction.
(D) A becomes negatively charged by induction.
(E) B becomes positively charged by conduction.

52. The rod now touches the top of the electroscope and is quickly removed. As a result,

(A) The electroscope leaves return to their original position.
(B) The electroscope leaves become negatively charged.
(C) The top of the electroscope is grounded.
(D) The top of the electroscope becomes negatively charged.
(E) The electroscope leaves become positively charged.

GO ON TO THE NEXT PAGE

Questions 53–54 refer to the following figure.

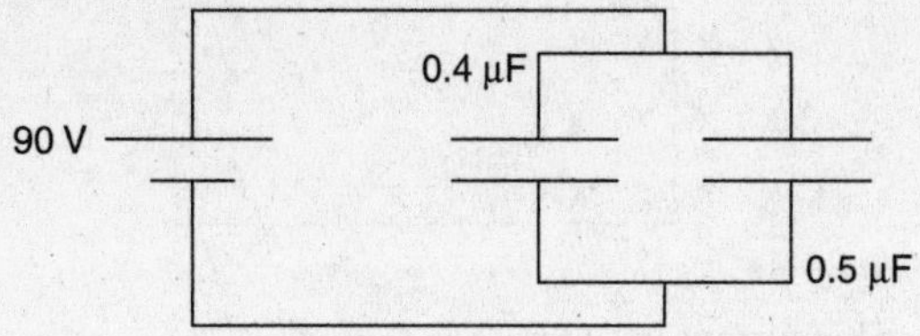

Two capacitors, 0.4 μF and 0.5 μF are connected in parallel and charged to a 90 V potential difference.

53. The total charge acquired by the capacitors is most nearly

(A) 0.08 μC
(B) 0.8 μC
(C) 8 μC
(D) 80 μC
(E) 800 μC

54. The capacitors are now discharged and reconnected in series with the same power source. The new total charge acquired is nearest to

(A) 0.2 μC
(B) 2.0 μC
(C) 20 μC
(D) 200 μC
(E) 2000 μC

Questions 55–56 refer to the following figure.

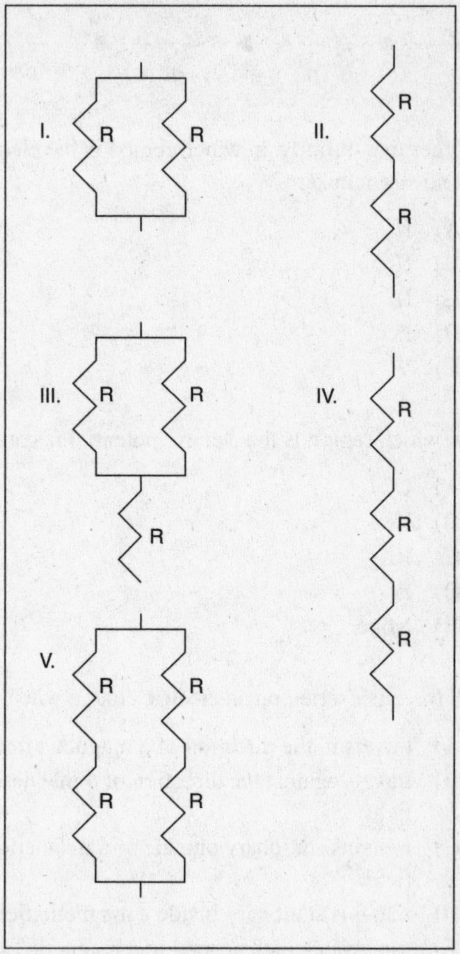

55. There are five combinations of resistors. Each resistor R has a value of 2 Ω. Each combination is connected to a 6 V battery. The combination that would dissipate 36 W of power is

(A) I
(B) II
(C) III
(D) IV
(E) V

56. The combination that would develop the least amount of heat is

(A) I
(B) II
(C) III
(D) IV
(E) V

GO ON TO THE NEXT PAGE

Questions 57–58 refer to the following figure.

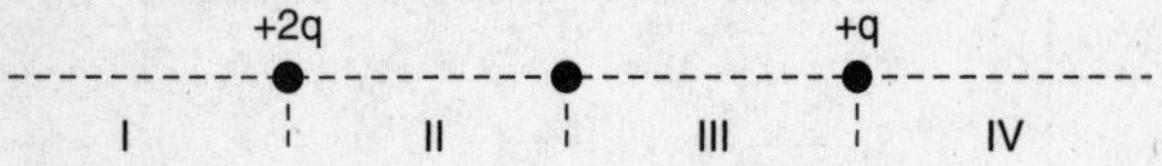

57. Other than infinity, in which region is the electric field strength zero?

(A) I
(B) II
(C) III
(D) IV
(E) V

58. In which region is the electric potential negative?

(A) I
(B) II
(C) III
(D) IV
(E) None

59. A force is exerted on an electric charge when it

(A) travels in the direction of a magnetic field.
(B) travels against the direction of a magnetic field.
(C) remains stationary outside of a magnetic field.
(D) remains stationary inside a magnetic field.
(E) travels in a path normal to a magnetic field.

60. Gauss' Law may be used to determine the electric field around each of the following *except* for a

(A) long wire.
(B) cube.
(C) sphere.
(D) hollow cylinder.
(E) heart-shaped object.

61. Two long, parallel wires carry currents. If they attract each other, their currents

(A) are parallel.
(B) are antiparallel.
(C) are equal.
(D) are unequal.
(E) draw from the same source.

Questions 62–63 refer to the following figure.

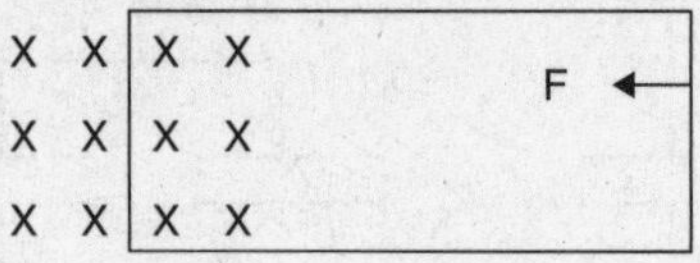

62. The current in the wire is flowing

(A) into the page.
(B) out of the page.
(C) clockwise.
(D) counterclockwise.
(E) (none of the above).

63. In the previous problem, if the length of wire cutting the magnetic field is 0.3 m, its velocity is 1.0 m/s, and the magnetic field has a strength of 0.5 T, the emf in the wire is most nearly

(A) 0.2 V
(B) 0.4 V
(C) 0.6 V
(D) 0.8 V
(E) 1.0 V

64. The work necessary to move a point charge of 3×10^{-3} C from the origin to another point 2 m away is 4.5 J. The potential difference between the two points is

(A) 4.5×10^{-4} V
(B) 1.5×10^{-3} V
(C) 3.0×10^{-2} V
(D) 10^{3} V
(E) 1.5×10^{3} V

65. A thin layer of insulating material is inserted between the plates of a parallel plate capacitor. Compared to its capacitance with air separation,

(A) it is zero.
(B) it is unchanged.
(C) it is less now.
(D) it is greater now.
(E) it cannot be determined.

GO ON TO THE NEXT PAGE

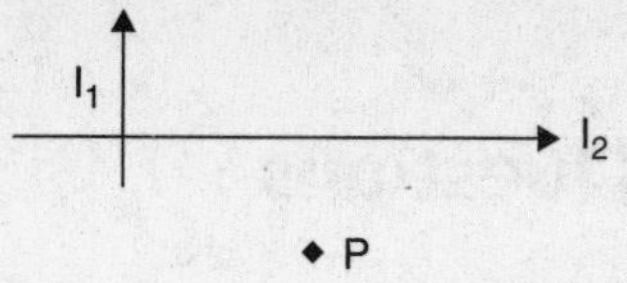

66. The magnetic field at point P is directed

(A) toward the top of the page.
(B) toward the bottom of the page.
(C) into the page.
(D) out of the page.
(E) to the right.

Questions 67–69 refer to the following figure.

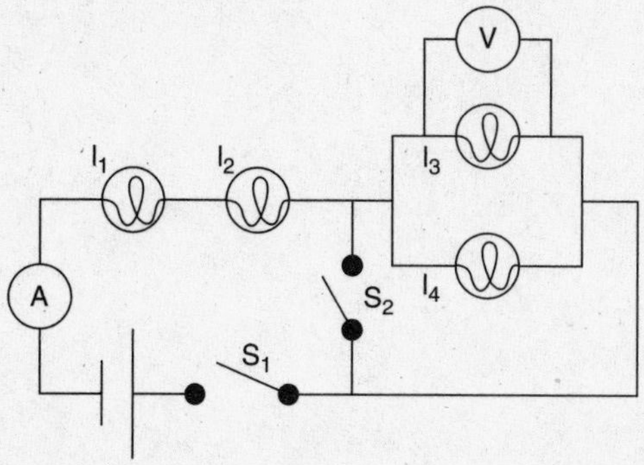

Two lamps, l_1 and l_2, are placed in series with a combination of another pair of lamps, l_3 and l_4, in parallel as shown. The circuit is powered by a battery and there are two switches, S_1 and S_2, which are both in the OFF position.

67. If each lamp has an equal resistance of 2Ω and the emf of the battery is 6 V, the current registered in the ammeter when switch S_1 is turned on is

(A) 2 / 5 A
(B) 3 / 5 A
(C) 5 / 6 A
(D) 6 / 5 A
(E) 7 / 5 A

68. With switch S_1 still on, switch S_2 is now turned on. As a result, lamps l_3 and l_4 light up and

(A) all four lamps are equally bright.
(B) lamps 1 and 2 are not as bright as before.
(C) lamps 1 and 2 are brighter than lamps 3 and 4.
(D) lamps 3 and 4 are brighter than lamps 1 and 2.
(E) all 4 lamps have different brightness.

69. When both switches are in the ON position, the voltmeter reads

(A) zero
(B) 3 / 5 V
(C) 5 / 6 V
(D) 6 / 5 V
(E) 7 / 5 V

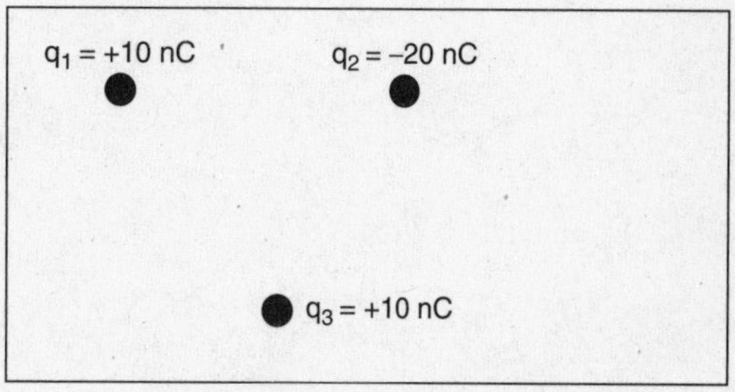

70. Three electric point charges, q_1 = +10 nC, q_2 = −20 nC, and q_3 = +10 nC, are arranged as shown. The direction of the force on charge q_3 is closest to

(A) ↙
(B) ↗
(C) →
(D) ←
(E) ↖

END OF SECTION I, ELECTRICITY AND MAGNETISM. IF YOU FINISH BEFORE TIME IS CALLED, YOU MAY CHECK YOUR WORK ON THIS SECTION ONLY. DO NOT GO ON TO SECTION II UNTIL YOU ARE TOLD TO DO SO.

STOP

CliffsAP Physics C Practice Exam 1

Physics C
Section II, Free-Response Questions

Mechanics	45 Minutes	3 required questions of equal weight
Electricity and Magnetism	45 Minutes	3 required questions of equal weight

Section II is 50 percent of the total grade for each of the two examinations.

Mark one of the three boxes to indicate which questions you will answer.

- ☐ Mechanics only
- ☐ Electricity and Magnetism only
- ☐ Both Mechanics and Electricity and Magnetism

Information Table

Constants and Conversions		*Units*	
1 Atomic mass unit	$u = 1.66 \times 10^{-27}$ kg	meter	m
	$= 931$ MeV/ c^2	kilogram	kg
Proton mass	$m_p = 1.67 \times 10^{-27}$ kg	second	s
Neutron mass	$m_n = 1.67 \times 10^{-27}$ kg	ampere	A
Electron mass	$m_e = 9.11 \times 10^{-31}$ kg	kelvin	K
Electron charge	$e = 1.60 \times 10^{-19}$ C	mole	mol
Avogadro's number	$N_0 = 6.02 \times 10^{23}$ mol^{-1}	hertz	Hz
Universal gas constant	$R = 8.31$ J / (mol · K)	newton	N
Boltzmann's constant	$k_B = 1.38 \times 10^{-23}$ J / K	pascal	Pa
Speed of light	$c = 3.00 \times 10^8$ m / s	joule	J
Planck's constant	$h = 6.63 \times 10^{-34}$ J · s	watt	W
	$= 4.14 \times 10^{-15}$ eV · s	coulomb	C
	$hc = 1.99 \times 10^{-25}$ J · m	volt	V
	$= 1.24 \times 10^3$ eV · nm	ohm	Ω
Vacuum permittivity	$\varepsilon_0 = 8.85 \times 10^{-12}$ C^2 / N · m^2	henry	H
Coulomb's law constant	$k = 1/4\pi\varepsilon_0 = 9.0 \times 10^9$ N · m^2 / C^2	farad	F
Vacuum permeability	$\mu_0 = 4\pi \times 10^{-7}$ (T · m) / A	tesla	T
Magnetic constant	$k' = \mu_0 / 4\pi = 10^{-7}$ (T · m) / A	degree Celsius	°C
Universal gravitation constant	$G = 6.67 \times 10^{-11}$ Nm^2/kg^2	electron Volt	eV
Acceleration due to gravity at the earth's surface	$\mathbf{g} = 9.8$ m/s^2		
1 atmosphere pressure	1 atm $= 1.0 \times 10^5$ Pa		
	$= 1.0 \times 10^5$ N / m^2		
1 electron volt	1 eV $= 1.60 \times 10^{-19}$ J		

Prefixes

Factor	Prefix	Symbol
10^9	giga	G
10^6	mega	M
10^3	kilo	k
10^{-2}	centi	c
10^{-3}	milli	m
10^{-6}	micro	μ
10^{-9}	nano	n
10^{-12}	pico	p

Trigonometric Function Values for Common Angles

Θ	sin Θ	cos Θ	tan Θ
0°	0	1	0
30°	$\frac{1}{2}$	$\frac{\sqrt{3}}{2}$	$\frac{\sqrt{3}}{3}$
37°	$\frac{3}{5}$	$\frac{4}{5}$	$\frac{3}{4}$
45°	$\frac{\sqrt{2}}{2}$	$\frac{\sqrt{2}}{2}$	1
53°	$\frac{4}{5}$	$\frac{3}{5}$	$\frac{4}{3}$
60°	$\frac{\sqrt{3}}{2}$	$\frac{1}{2}$	$\sqrt{3}$
90°	1	0	∞

These conventions are used in the examination:

1. *Unless stated otherwise, inertial frames of reference are used.*
2. *The direction of electric current flow is the conventional direction of positive charge.*
3. *For an isolated electric charge, the electric potential at an infinite distance is zero.*
4. *In mechanics and thermodynamics equations, W represents work done **on** a system.*

GO ON TO THE NEXT PAGE

Advanced Placement Physics C Equations

Newtonian Mechanics

$\mathbf{v} = \mathbf{v}_0 + \mathbf{a}t$	**a** = acceleration
$\mathbf{x} = \mathbf{x}_0 + \mathbf{v}_0 t + \frac{1}{2}\mathbf{a}t^2$	**F** = force
$\mathbf{v}^2 = \mathbf{v}_0^{\,2} + 2\mathbf{a}(\mathbf{x}-\mathbf{x}_0)$	f = frequency
$\Sigma\mathbf{F} = \mathbf{F}_{NET} = m\mathbf{a}$	h = height
$\mathbf{F} = \frac{d\mathbf{p}}{dt}$	I = rotational inertia
$\mathbf{J} = \int \mathbf{F}\,dt = \Delta\mathbf{p}$	**J** = impulse
$\mathbf{p} = m\mathbf{v}$	K = kinetic energy
$\mathbf{F}_{fric} \le \mu\mathbf{N}$	k = spring constant
$W = \int \mathbf{F} \cdot d\mathbf{r}$	l = length
$K = \frac{1}{2}m\mathbf{v}^2$	**L** = angular momentum
$P = \frac{dW}{dt}$	m = mass
$P = \mathbf{F} \cdot \mathbf{v}$	**N** = normal force
$\Delta U_g = m\mathbf{g}\mathbf{h}$	P = power
$\mathbf{a}_c = \frac{\mathbf{v}^2}{\mathbf{r}} = \omega^2\mathbf{r}$	**p** = momentum
$\tau = \mathbf{r} \times \mathbf{F} = \mathbf{r}\mathbf{F}\sin\Theta$	r = radius, distance
$\Sigma\tau = \tau_{NET} = I\alpha$	**r** = position vector
$I = \int \mathbf{r}^2\,dm = \Sigma m\mathbf{r}^2$	T = period
$\mathbf{r}_{CM} = \Sigma m\mathbf{r}/\Sigma m$	t = time
$\mathbf{v} = r\omega$	U = potential energy
$\mathbf{L} = \mathbf{r} \times \mathbf{p} = I\omega$	v = velocity
$K = \left(\frac{1}{2}\right)I\omega^2$	v = speed
$\omega = \omega_O + \alpha t$	W = work done on a system
$\Theta = \Theta_O + \omega_O t + \left(\frac{1}{2}\right)\alpha t^2$	**x** = position
$\mathbf{F}_s = -k\mathbf{x}$	μ = coefficient of friction
$U_s = \frac{1}{2}k\mathbf{x}^2$	Θ = angle, displacement
$T = (2\pi/\omega) = 1/f$	τ = torque
$T_P = 2\pi(l/\mathbf{g})^{\frac{1}{2}}$	ω = angular speed
$T_S = 2\pi(m/k)^{\frac{1}{2}}$	α = angular acceleration
$\mathbf{F}_G = -\frac{Gm_1 m_2}{\mathbf{r}^2}$	
$U_G = -\frac{Gm_1 m_2}{\mathbf{r}}$	

Electricity and Magnetism

$\mathbf{F} = \frac{1}{4\pi\varepsilon_0}\frac{q_1 q_2}{\mathbf{r}^2}$	A = area
$\mathbf{E} = \frac{\mathbf{F}}{q}$	**B** = magnetic field
$\oint \mathbf{E} \cdot dA = Q/\varepsilon_0$	C = capacitance
$U_E = qV = \frac{1}{4\pi\varepsilon_0}\frac{q_1 q_2}{r}$	**d** = distance
$\mathbf{E} = -\frac{dV}{d\mathbf{r}}$	**E** = electric field
$V = \frac{1}{4\pi\varepsilon_0}\Sigma\frac{(q_1)}{\mathbf{r}_1}$	ε = emf
$C = \frac{Q}{V}$	**F** = force
$C = \frac{\kappa\varepsilon_0 A}{\mathbf{d}}$	I = current
$C_{PAR} = C_1 + C_2 + \ldots C_n$	L = inductance
$\frac{1}{C_{SER}} = \frac{1}{C_1} + \frac{1}{C_2} + \ldots \frac{1}{C_n}$	l = length
$I = \frac{dQ}{dt}$	n = number of turns
$UC = \frac{1}{2}QV = \frac{1}{2}CV^2$	P = power
$R = \rho\,(l/A)$	Q = charge
$V = IR$	q = point charge
$R_{SER} = R_1 + R_2 + \ldots R_n$	R = resistance
$\frac{1}{R_{PAR}} = \frac{1}{R_1} + \frac{1}{R_2} + \ldots \frac{1}{R_n}$	**r** = distance, displacement
$P = IV$	t = time
$\mathbf{F}_B = q\mathbf{v} \times \mathbf{B} = q\mathbf{v}\mathbf{B}\sin\Theta$	U = potential (stored) energy
$\oint \mathbf{B} \cdot d\mathbf{l} = \mu_0 I$	V = electric potential difference
$F = \int I\,dl \times B$	**v** = velocity
$\mathbf{B}_s = \mu_0 nI$	v = speed
$\phi_M = \int \mathbf{B} \cdot dA$	ρ = resistivity
$\varepsilon_{AVE} = \frac{d\phi_M}{dt}$	ϕ_M = magnetic flux
$\varepsilon = -L\frac{dI}{dt}$	κ = dielectric constant
$U_L = \frac{1}{2}LI^2$	

Geometry and Trigonometry

Rectangle
$A = bh$

Triangle
$A = \frac{1}{2}bh$

Circle
$A = \pi r^2$
$C = 2\pi r$

Paralelopiped
$V = lwh$

Cylinder
$V = \pi r^2 h$
$S = 2\pi rh + 2\pi r^2$

Sphere
$V = \frac{4}{3}\pi r^3$
$S = 4\pi r^2$

A = area
C = circumference
V = volume
S = surface area
b = base
h = height
l = length
w = width
r = radius

Right Triangle
$a^2 + b^2 = c^2$
$\sin \Theta = a/c$
$\cos \Theta = b/c$
$\tan \Theta = a/b$

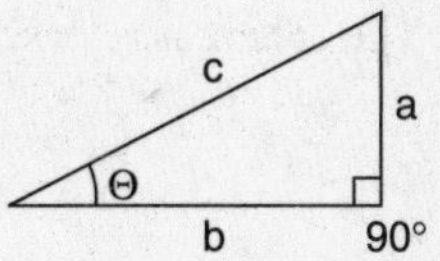

Calculus

$$\frac{df}{dx} = \frac{df}{du}\frac{du}{dx}$$

$$\frac{d}{dx}\left(x^n\right) = nx^{n-1}$$

$$\frac{d}{dx}\left(e^x\right) = e^x$$

$$\frac{d}{dx}(\ln x) = \frac{1}{x}$$

$$\frac{d}{dx}(\sin x) = \cos x$$

$$\frac{d}{dx}(\cos x) = -\sin x$$

$$\int x^n\,dx = (1/n+1)\,x^{n+1},\ n \neq -1$$

$$\int e^x\,dx = e^x$$

$$\int (dx/x) = \ln|x|$$

$$\int \cos x\,dx = \sin x$$

$$\int \sin x = -\cos x$$

GO ON TO THE NEXT PAGE

Physics C
Section II, Mechanics

Time – 45 Minutes
3 Questions

Directions: Answer all three questions. The suggested time is approximately 15 minutes for answering each question, each of which is worth 15 points. The parts within a question may *not* have equal weight.

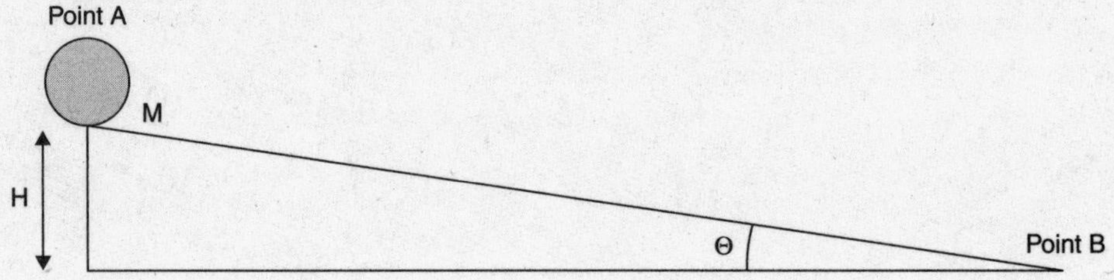

Mech. 1. A solid sphere of mass M, radius **R**, and rotational inertia 2/5 $\mathbf{MR}^2$ rolls, without slipping, down an inclined plane which makes an angle Θ with the horizontal. The sphere is released from rest at height H.

Express all solutions in terms of M, **R**, **H**, Θ, and **g**.

(a) Determine the linear velocity of the sphere at point B.

(b) On the following illustration, draw a free-body diagram of all forces acting on the sphere.

(c) What is the acceleration of the center of mass of the sphere while rolling down the plane?

(d) What is the coefficient of kinetic friction necessary to keep the sphere rolling without slipping down the incline (minimum value of μ_K).

(e) If the sphere is replaced by a different sphere having double the radius and double the mass of the original sphere, what will be the new sphere's linear velocity when it reaches point B?

GO ON TO THE NEXT PAGE

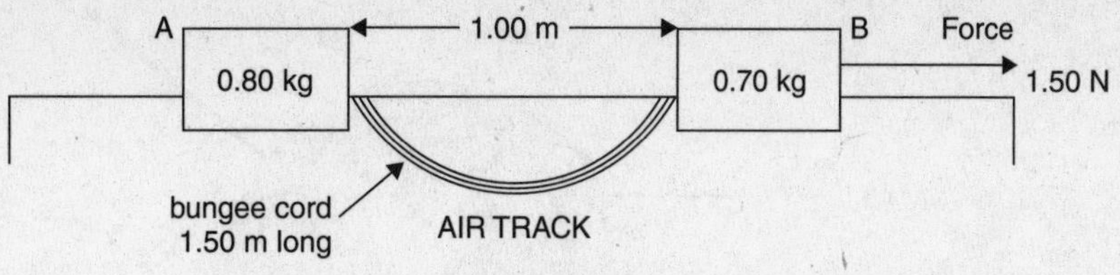

Mech. 2. Two masses, A (0.80 kg) and B (0.70 kg) are attached by a slack bungee cord of negligible mass and total unstretched length 1.50 m and k 2.00 N/m. Both masses start at rest on a frictionless air track and are initially separated by 1.00 meter.

(a) Mass B is suddenly pulled to the right by a force F of 1.50 N. Determine the acceleration of block B before the bungee cord reaches its full length and begins to stretch.

(b) When the bungee cord has stretched 0.50 m, determine the instantaneous acceleration of block A.

(c) When the bungee cord has stretched 0.50 m, determine the instantaneous acceleration of block B.

(d) Determine the kinetic energy of block B at the point where the bungee cord just begins to stretch.

(e) Determine the kinetic energy of block A at the instant when the bungee cord has stretched 0.50 m.

GO ON TO THE NEXT PAGE

CliffsAP Physics C Practice Exam 1

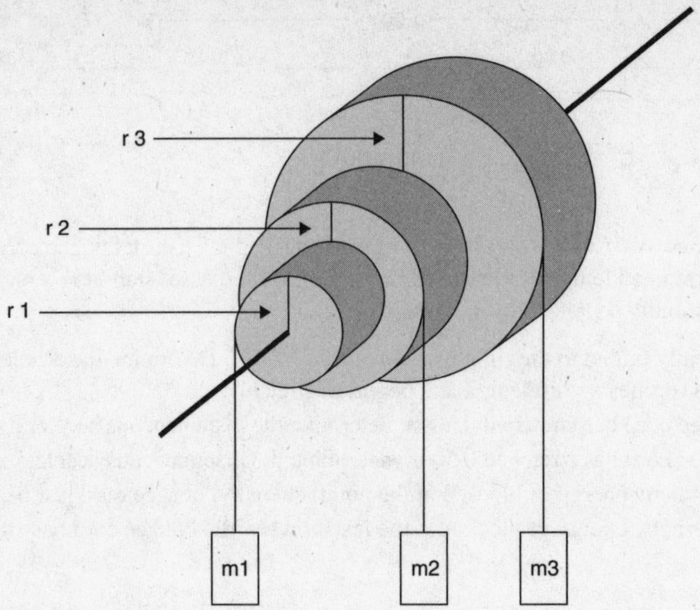

Mech. 3. Three concentric cylinders of radii r 1, r 2, and r 3, respectively, are rigidly attached to each other but rotate as a system on a frictionless axle. Each cylinder has a cord wrapped around it, each cord supporting an individual mass as shown.

Given values are: r 1 = 0.5 m, r 2 = 1.0 m, r 3 = 1.5 m, m 1 = 30 kg, m 2 = 12 kg and for the system, I = 60 kg m^2.

(a) If the entire cylindrical grouping is in rotational equilibrium, what is m3?

Mass 2 is removed and the entire grouping is allowed to rotate.

(b) Calculate the angular acceleration of the system.

(c) Calculate the tension in the rope supporting m 3.

(d) Calculate the time necessary for m 3 to rise 1.0 m.

END OF SECTION II, MECHANICS. IF YOU FINISH BEFORE TIME IS CALLED, YOU MAY CHECK YOUR WORK ON SECTION II ONLY. DO NOT TURN TO ANY OTHER TEST MATERIALS.

STOP

Physics C
Section II, Electricity and Magnetism

Time — 45 Minutes
3 Questions

Directions: Answer all three questions. The suggested time is approximately 15 minutes for answering each question, each of which is worth 15 points. The parts within a question may *not* have equal weight.

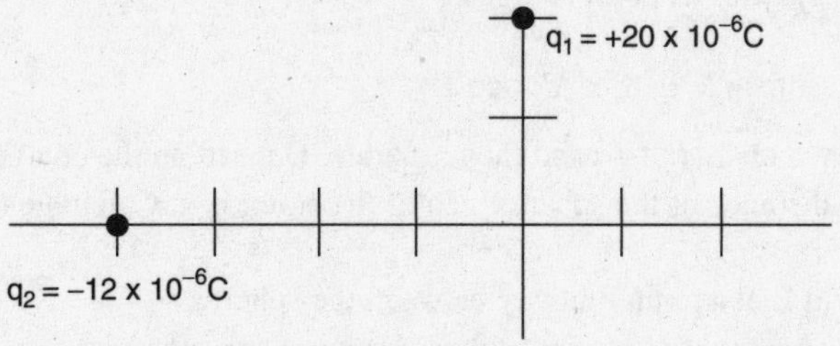

E & M. 1. A charge $q_1 = +20 \times 10^{-6}$C is located at +2 m on the y-axis. A second charge, $q_2 = -12 \times 10^{-6}$C is located at –4 m on the x-axis, as shown in the diagram.

(a) i. Calculate the magnitude of the electric field E_1 at the origin due to charge q_1.
ii. Calculate the magnitude of the electric field E_2 at the origin due to charge q_2.
iii. On the axes provided here, draw and label electric field vectors E_1 and E_2 due to each charge. In addition, draw the resultant electric field at the origin.

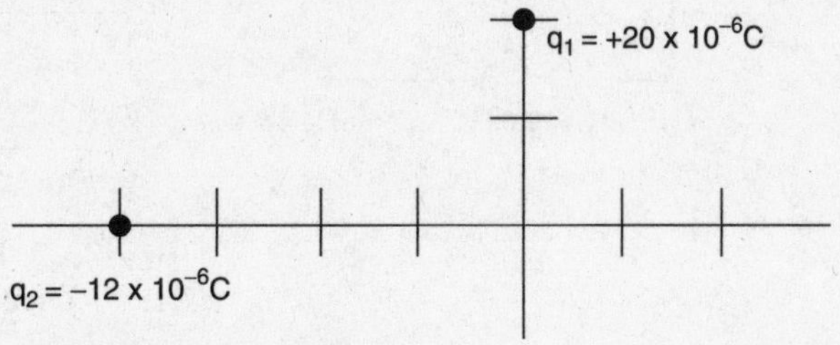

(b) What is the electric potential V at the origin?

A test charge $q_3 = +6 \times 10^{-6}$C is brought in from a considerable distance and placed at the origin.

(c) On the axes provided here, draw the direction of the resulting force on q_3 at the origin.

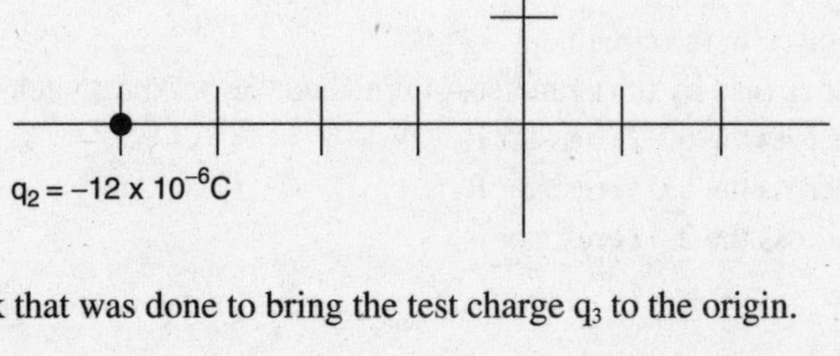

(d) Determine the work that was done to bring the test charge q_3 to the origin.

GO ON TO THE NEXT PAGE

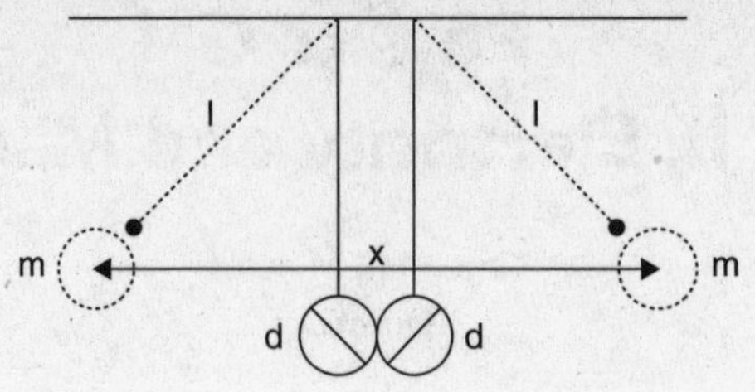

E & M. 2. Two nonconducting plastic spheres, each of mass m and initially uncharged, hang side by side and touch as illustrated. Both have diameter d and hang on strings of negligible mass and length l. Consider both spheres as point charges.

Give all your answers in terms of k, q, d, x, x', and l.

(a) The spheres are given a charge of –q and they separate. Determine the charge on each sphere.
(b) The final separation distance of the spheres, center to center, is x. Calculate the electrostatic force on each sphere.
(c) Draw the electric field E at a point midway between the spheres.
(d) Calculate the electric potential at a point midway between the spheres.
(e) The final separation angle between the spheres is Θ. Give an expression for the final separation distance in terms of l.
(f) One of the spheres is grounded, causing them to attract, touch, and repel. Repeat part (d) for this new situation, letting x' be the new distance of separation.

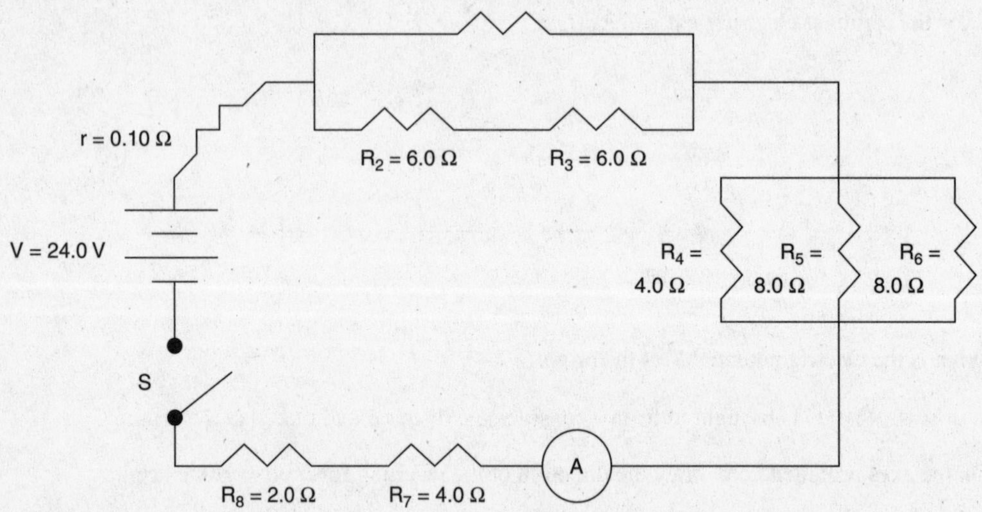

E & M. 3. Given the circuit in the above illustration, find the following:

(a) The equivalent resistance in the circuit
(b) The mainline current as read by the ammeter A immediately after the switch S is closed
(c) The potential drop across the 6.0 Ω resistor R_3
(d) The potential drop across the 8.0 Ω resistor R_5
(e) The potential drop across the 2.0 Ω resistor R_8

End of Examination

Sample C Exam #1 Answers and Comments

Mechanics

Multiple Choice

1. **(A)** Using

$$\mathbf{F_{NET} = ma}$$
$$\mathbf{Mg - F_F = (\Sigma m)(a)}$$
$$\mathbf{Mg - \mu F_N = (4\ kg + M)(2\ m/s^2)}$$
$$(M)(10\ m/s^2) - (0.25)(4\ kg)(10\ m/s^2) = 8\ N + (2\ m/s^2)(M)$$
$$\mathbf{M = 2.25\ kg}$$

2. **(A)** Again using $\mathbf{Mg - F_F = (\Sigma m)(a)}$

$$(4.5\ kg)(10\ m/s^2) - (0.25)(8\ kg)(10\ m/s^2) = (8\ kg + 4.5\ kg)(a)$$
$$\mathbf{a} = \frac{25\,N}{12.5\,kg} = \mathbf{2\ m/s^2}$$

3. **(C)** Using $\mathbf{d_{VERT} = (v_{O\,VERT})t + \left(\frac{1}{2}\right)at^2}$,

$$\mathbf{d_{VERT}} = (0) + \left(\frac{1}{2}\right)(10\ m/s^2)(1\ sec)^2$$
$$\mathbf{d_{VERT} = 5\ m}$$

4. **(E)** Since $\mathbf{a} = \frac{\mathbf{dv}}{dt} = 7t$,

$$\mathbf{v} = \int 7t\,dt = (7)\left(\frac{1}{2}\right)t^2 = \left(\frac{7}{2}\right)t^2$$
$$\mathbf{d} = \int \left(\frac{7}{2}\right)t^2\,dt = \left(\frac{28}{3}\right)m = \mathbf{9.3\ m}$$

5. **(D)** Using $\mathbf{d = v_Ot + \frac{1}{2}at^2}$,

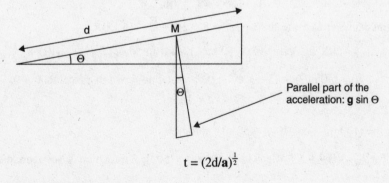

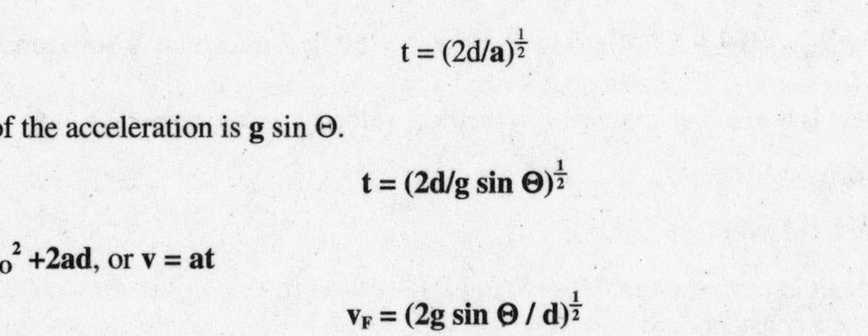

$$t = (2d/a)^{\frac{1}{2}}$$

The parallel part of the acceleration is **g** sin Θ.

$$\mathbf{t = (2d/g\ \sin\Theta)^{\frac{1}{2}}}$$

6. **(B)** Using $\mathbf{v_F^2 = v_O^2 + 2ad}$, or $\mathbf{v = at}$

$$\mathbf{v_F = (2g\ \sin\Theta / d)^{\frac{1}{2}}}$$

7. **(C)** The minimum breaking force equals the parallel part of the weight.

Using $\mathbf{F_{BREAKING\ MIN} = F_P}$,

$$= mg \sin 30°$$
$$= (15{,}000\ N)(0.5)$$
$$= \mathbf{7500\ N}$$

8. **(D)** Momentum is always conserved. Mass m has a final velocity that is **negative** and less than its initial velocity. Mass M moves with a positive final velocity that is less than both $\mathbf{v_{O\ m}}$ and $\mathbf{v_{F\ m}}$.

9. **(E)** The total amount of energy is **constant.** $\mathbf{(E_{TOT} = E_K + E_P)}$

10. **(A)** As the object loses potential energy, it gains kinetic energy by an equal amount.

11. **(D)** A height of h = 2R would just get it to the top at point B but added energy is needed to get it to point X, where it would continue to C. That added energy needed is represented by the fact that the force and acceleration at ANY point on the circle due to the curvilinear path is due to centripetal considerations.

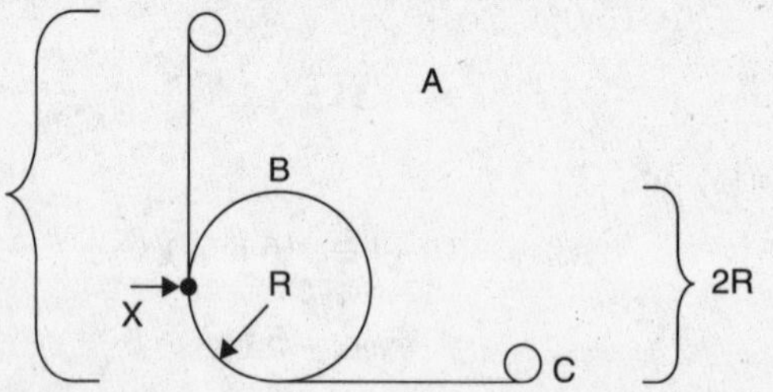

Setting the weight of the ball at point B equal to its centripetal force provides the needed velocity: Using $\mathbf{mg = mv^2 / R}$ yields needed extra speed $\mathbf{v = (gR)^{\frac{1}{2}}}$. This needed extra speed is provided by an increase in height **greater than 2R.**

We may use $\mathbf{v = (gR)^{\frac{1}{2}}}$ to determine the height needed to attain that speed: Using $\mathbf{mgh = \left(\frac{1}{2}\right)mv^2}$ to determine the transfer from potential to kinetic energy needed to attain that extra speed:

$$\mathbf{h = \frac{y^2}{2g} = \frac{gR}{2g} = \frac{R}{2}}$$

Added to the initial diameter to get to the top only: $\mathbf{h = 2R + \frac{R}{2} = (5/2)R}$

12. **(B)** First, using $\mathbf{v_{O\ VERT}} = (100\ m/s)(\sin 30°) = 50\ m/s$, then using $\mathbf{d_{VERT} = (v_{O\ VERT})(t) + \left(\frac{1}{2}\right)gt^2}$,

$$0 = (50\ m/s)t - (5\ m/s)t^2 \ldots \text{(For a total up-down displacement of 0)}$$
$$\mathbf{t = 10\ sec.}$$

13. **(E)** Using $\mathbf{v_{O\ HORIZ}} = (100\ m/s)(\cos 30°) = (50)(3)^{\frac{1}{2}}\ m/s$,

$\mathbf{d_{HORIZ} = (v_{O\ HORIZ})(t) + 0} = [(50)\sqrt{3}\ m/s](10\ sec) = (500)\sqrt{3}\ m$, which is **between 800 and 900 m.**

14. **(D)** At point P, there is horizontal velocity, zero vertical velocity, and acceleration of **g.**

15. **(E)** Using $\mathbf{W = Fd}$, $\mathbf{W = Fd \cos\Theta}$

16. **(C)** Using $\mathbf{L = I\omega}$, $\mathbf{L = I(v/r)}$

17. **(B)** Velocity increases at a constant rate for section I. In section II, velocity is constant, and in section III, the velocity decreases at a constant rate.

18. (C) The acceleration is positive during section I, zero for section II, and negative for section III.

19. (D) $v = \frac{dx}{dt} = 6t^2 + 6t - 6$

$$a = \frac{dv}{dt} = 12t + 6\ldots \text{ and when } t = 2 \text{ sec., } \mathbf{a = 30\ m/s^2}$$

20. (D) $\mathbf{F = ma = m\,(g + 0.5\ m/s^2)}$

21. (B) Using $\mathbf{F_{NET} = \Sigma ma}$, $\mathbf{F_{NET} = (M + m)(0.5\ m/s^2 + g)}$

22. (A) The gravitational force on the asteroids is also the centripetal force on them:

$$F_G = \frac{Gmm}{S^2} \text{ equals } F_C = \frac{mv^2}{2S}$$

$$\frac{Gm}{S} = 2v^2$$

$$\mathbf{v = (Gm/2S)^{\frac{1}{2}}}$$

23. (D) The acceleration of either asteroid is centripetal acceleration.

$$a = \frac{v^2}{r} = \frac{v^2}{S/2} = \frac{2v^2}{S}$$

24. (A) Since the hoop's mass is concentrated on the rim and the disk's mass is concentrated between the rim and the center, their rotational inertia values are different. ($I_{HOOP} = mr^2$ and $I_{DISK} = \left(\frac{1}{2}\right)mr^2$). Since the total gravitational potential energy (mgh) for each object becomes transformed into kinetic energy, $E_p = E_k = \frac{1}{2}mv^2 + \frac{1}{2}I\omega^2$. Since the hoop's I is greater than that of the disk, the hoop has less linear kinetic energy than the disk. As a result, the disk reaches the bottom first.

25. (E) Gravitational force varies directly with distance from planetary center to surface. Above the surface, the gravitational force follows the inverse-square law.

26. (C) Using $\mu = \frac{F_F}{F_N}$,

$$\mu = \frac{F_F}{Mg + F\sin\Theta}$$

27. (B) Using $\mathbf{F_{NET} = ma}$,

$$a = \frac{F_{NET}}{M} = \frac{F_P - F_F}{M} = \frac{F\cos\Theta - F_F}{M}$$

28. (A) The center of mass of a closed system remains stationary due to the Law of Conservation of Momentum.

29. (C) Using $\mathbf{F(x)} = -\frac{dU(x)}{dx}$,

$$\mathbf{Energy = dU(x) = |F(x)\,dx|}$$

$$= \int dU(x) = \left|\int F(x)\,dx\right|$$

$$= \int (30x - 4x^2)\,dx = \int 30x\,dx - 4\int x^2\,dx = 15x^2 - \left(\frac{4}{3}\right)x^3$$

at x = 3 m: Change in energy = **99 J**

30. (D) Since the system is in rotational equilibrium, there is no net torque. Therefore,

$$\tau_{CW} = \tau_{CCW} \text{ and therefore}$$

$$\mathbf{MR = mr}$$

31. (E) Since MR = mr,

$$m = \frac{MR}{r} = \frac{(1\,\text{kg})(1.5\,\text{m})}{(0.5\,\text{m})} = \mathbf{3\ kg}$$

32. (D) Using $\tau = I\alpha = Fr$,

$$\alpha = \frac{F_{NET}\,r}{I} = \frac{[3m(r) - (3m)\mathbf{a}] - [mR + m\mathbf{a}]}{I} = \frac{m[3(r-\mathbf{a}) - (R+\mathbf{a})]}{I}$$

33. (D) Using $\mathbf{P} = \frac{\mathbf{W}}{t} = \frac{\mathbf{Fd}}{t} = \frac{(50\,\text{kg})(10\,\text{m/s}^2)(50\,\text{m})}{(10\,\text{sec})} = \mathbf{2500\ W}$

34. (B) v at point B is $\mathbf{v_O}\cos 45° = \frac{(2)^{\frac{1}{2}}}{2}\mathbf{v_O}$

35. (B) Using $\mathbf{p = mv}$,

$$\mathbf{p} = (2\,\text{kg})(\mathbf{v_O}\cos 45°) = \frac{(2)^{\frac{1}{2}}}{2}\mathbf{v_O} = (200)\sqrt{2}\ \text{kgm/s}$$

$$= 283\ \text{kgm/s} = \mathbf{200 - 300\ kgm/s}$$

Electricity and Magnetism

Multiple Choice

36. (B) Since the **series** combination of 2 Ω and 3 Ω is in parallel with the 1 Ω resistor,

$$\frac{1}{R_{EQ}} = \frac{1}{1\Omega} + \frac{1}{5\Omega} = \frac{6}{5\Omega} \text{ and } \mathbf{R_{EQ}} = \left(\frac{5}{6}\right)\Omega$$

37. (D) After passing point A, the current splits inversely to the branch resistance and recombines at point B. Thus, $\frac{5}{6}$ of the main line current passes through the 1 Ω resistor and $\frac{1}{6}$ of the current passes through the 5 Ω branch.

38. (A) The 6 A current splits with $\frac{5}{6}$ of it going to the 1 Ω resistor and the remaining $\frac{1}{6}$ of the current going to the branch with both the 2 Ω and 3 Ω resistors.

39. (B) Using $\mathbf{V = IR}$, the voltage drop through the 3 Ω resistor is $V_{3\Omega} = I_{3\Omega} R_{3\Omega} = (1\ \text{A})(3\ \Omega) = \mathbf{3\ V}$.

40. (C) Since the increasing magnetic field is directed into the page, the resulting induced magnetic field that opposes it (according to Lenz's Law) is in the opposite direction. Point the right thumb *out of the page*, and the fingers curl in the *counterclockwise* direction of the resulting induced current in the loop.

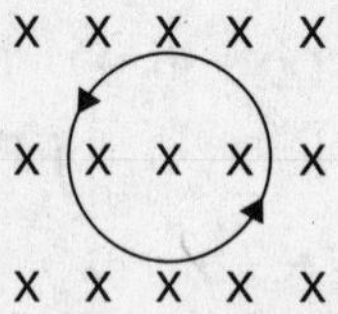

41. (B) I. Changing magnetic field strength causes a current to flow $\left(\varepsilon = -\frac{\Delta\Phi}{\Delta t}\right)$.

II. Moving the loop up and parallel to the magnetic field does not induce current.

III. Pulling the loop to the right cuts across magnetic lines of force and induces a current.

IV. Rotating the loop causes it to cut across magnetic lines of force but no net current results.

42. **(A)** Using the right hand, point the thumb in the direction opposing the increasing magnetic field (out of the page). The fingers curl in the **clockwise** direction of the current flow.

43. **(D)** Using **F = qvXB**, point the fingers of the right hand in the direction of charge motion, then swing them in the direction of the magnetic field, out of the page. The thumb points in the direction of the resulting force, **downward, toward the bottom of the page.**

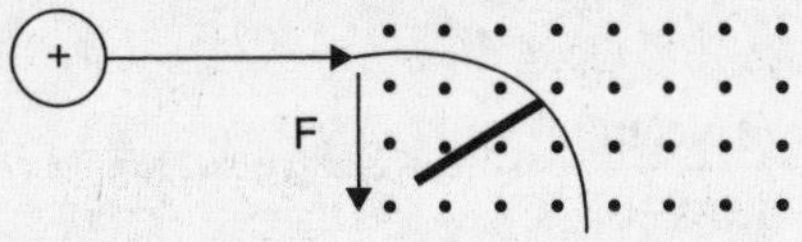

44. **(B)** Using Coulomb's Law, $\mathbf{F_E} = \frac{kq_1q_2}{\mathbf{r}^2}$ (k is substituted for $\left(\frac{1}{4}\pi\varepsilon_0\right)$ for simplicity.)

Note the two-step process in the absence of a calculator: calculate the numbers separately from the power of 10 exponents. It is a simple but effective tool.

$$\mathbf{F_E} = \frac{(9\times10^9\,\mathrm{Nm^2/C^2})(+1.6\times10^{-19}\,\mathrm{C})(-1.6\times10^{-19}\,\mathrm{C})}{(0.03\,\mathrm{m})^2} = \frac{(9)(+1.6)(-1.6)\times10^{-29}\,\mathrm{N}}{(9\times10^{-4})}$$

$$= -(1.6)^2\times10^{-29+4} = \mathbf{-2.56\times10^{-25}\ N}$$

45. **(D)** First, since C3 and C4 are parallel, add them together. This part of the circuit is 6 μF.

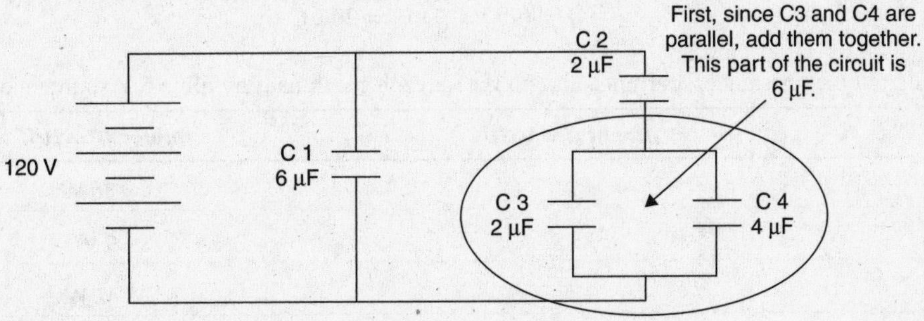

Second, C2 and the C3–C4 combination are in series, so add their inverses:

$$\frac{1}{C_{2,3-4}} = \frac{1}{2\,\mu\mathrm{F}} + \frac{1}{6\,\mu\mathrm{F}} \text{ and } C_{2,3\text{-}4} = 1.5\ \mu\mathrm{F}$$

Third, since C1 is parallel with $C_{2,3\text{-}4}$, add them:

$$C_{TOTAL} = 6\ \mu\mathrm{F} + 1.5\ \mu\mathrm{F} = \mathbf{7.5\ \mu F}$$

46. **(D)** Using **Q = VC**,

$$Q = (120\ \mathrm{V})(6\ \mu\mathrm{F}) = \mathbf{720\ \mu C}$$

47. **(A)** Since the parallel equivalence of C3 and C4 equals 6 μF, the equivalent capacitance of the entire right branch is 1.5 μF. Using Q = VC,

$Q_{1.5} = (120\ \mathrm{V})(1.5\ \mu\mathrm{F}) = 180\ \mu\mathrm{C}$ in the right branch, which is also Q in the 6 μF capacitor.

Finding V for C1: $V_{C1} = \frac{Q}{C} = \frac{180\,\mu\mathrm{C}}{6\,\mu\mathrm{F}} = 30\mathrm{V}$, this is also the potential in the right branch:

$$Q_{C4} = VC = (30\ \mathrm{V})(4\ \mu\mathrm{F}) = 120\ \mu\mathrm{C}$$

48. (C) Since **Q = VC, V = Q/C**. (This represents the voltage across the capacitor.)

The voltage drop across the resistor is **V = IR**.

The sum of all voltage drops in the circuit must equal the initial emf:

$$\varepsilon_0 = \frac{Q}{C} + IR$$

49. (E) Since one electron equals -1.6×10^{-19} C,

$$(-9 \times 10^{-9}\text{ C})\frac{(1\text{ electron})}{\left(-1.6 \times 10^{-19}\text{ C}\right)} = \mathbf{5.6 \times 10^{10}\ (closest\ to\ 6 \times 10^{10}\ electrons)}$$

50. (B) Since $1\text{V} = \frac{1\text{J}}{1\text{C}}$, J = VC = (24 V)(7.5 C) = **180 J (closest to 200 J)**.

51. (D) Electrons, attracted by the positive rod, move from the leaves up through the metal axis toward the top.

52. (E) The rod takes some electrons away with it. This leaves the entire electroscope electron-poor and positive. The leaves at the bottom remain apart.

53. (D) Using **Q = VC**, plus the fact that the capacitances, being in parallel, are added: Q = (90 V)(0.9 μC) = 8.1×10^{-5} C = **81 μC.**

54. (C) Using **Q = VC**, plus the fact that the capacitances, being in series, are added inversely: $\frac{1}{C_{TOT}} = \frac{1}{0.4\,\mu\text{F}} + \frac{1}{0.5\,\mu\text{F}}$ and $C_{TOT} = \frac{2}{9}\,\mu\text{F}$.

$$Q = (90\text{ V})\left(\frac{2}{9}\,\mu\text{F}\right) = \mathbf{20\ \mu C}$$

55. (A) Working out the resistances, currents, and power for each combination with a 6 V source yields:

Resistance	***Current (I = V/R)***	***Power (P = IV)***
I. 1 Ω	6 A	36 W
II. 4 Ω	1.5 A	9 W
III. 3 Ω	2 A	12 W
IV. 6 Ω	1 A	6 W
V. 2 Ω	3 A	18 W

56. (D) See above table.

57. (C)

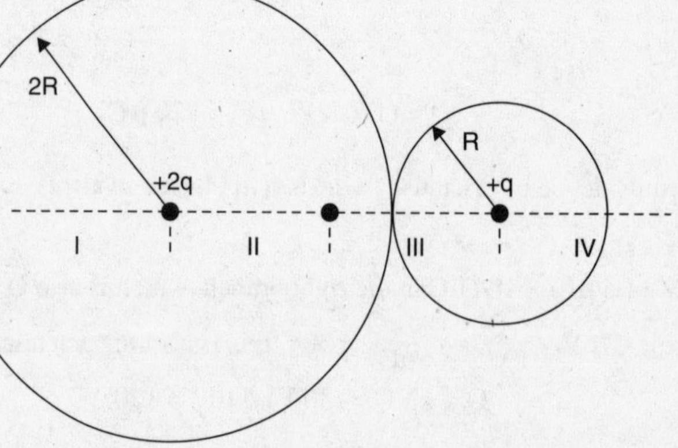

58. (E) The electric potential is only positive for these two charges.

59. (E) According to **F** = q**v** X **B**, a charged particle must be MOVING and have some component of its velocity be **perpendicular** to a magnetic field in order for a force to result.

60. (B) All Gaussian surfaces must satisfy E · dA = EdA, equal electric field magnitude at all points, and have the summary of surface area representable by $4\pi r^2$. A cube's pointed corners represent an **intersection of surfaces and electric fields that is not smooth and cannot be added easily as Gauss' Law requires.**

61. (A) Parallel currents attract each other due to their intersecting magnetic fields.

62. (D) Using **q = V X B**,

V X B = I counterclockwise

63. (A) Using **ε = Blv**,

$$\varepsilon = (0.5\text{ T})(0.3\text{ m})(1.0\text{ m/s}) = \mathbf{0.15\text{ V (closest is 0.2 V)}}$$

64. (E) Using **U = qV**,

$$V = \frac{U}{q} = \frac{4.5\text{ J}}{3 \times 10^{-3}\text{ C}} = \mathbf{1.5 \times 10^3\text{ V}}$$

65. (D) Capacitance is governed by the amount of charge that can be stored on parallel plates before electrons jump across and discharge the capacitor. Thicker insulating material between the plates allows more charge to build up before passing through to discharge.

66. (C) Using the right hand, **pointing** the thumb in the direction of the current, the fingers point **into the page**, indicating the direction of the magnetic field at point P for both currents I_1 and I_2.

67. (D) The two lamps in series have a total resistance of 4 Ω. The two lamps in parallel have a total resistance of 1 Ω. Using **V = IR**, I = V/R = 6V/5 Ω = **6 / 5 A**.

68. (C) Since lamps 1 and 2 are in the main line, more current (twice as much) passes through them than through lamps 3 and 4. Since P = IV, more current means more power and greater brightness.

69. (A) The current will take the path of least resistance, in this case, the short-circuited path directly through S_2. No current gets to the parallel bulbs.

70. (B)

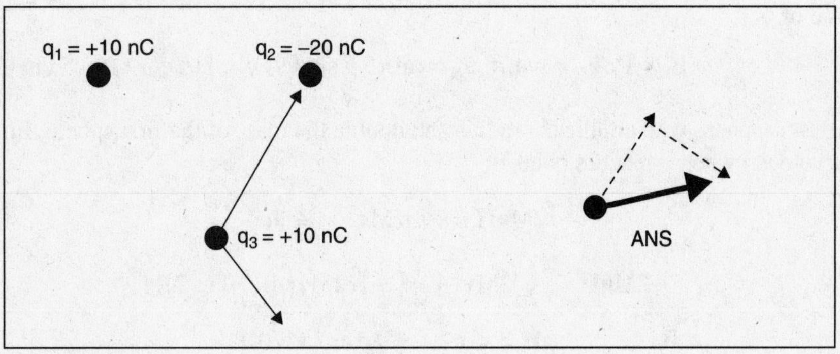

Section II, Mechanics

Free Response

Mech. 1. **(a)** To determine the linear velocity of the sphere at point B:

$$\mathbf{MgH} = \left(\frac{1}{2}\right)\mathbf{Mv}^2 + \left(\frac{1}{2}\right)\mathbf{I\omega}^2$$

$$\mathbf{MgH} = \left(\frac{1}{2}\right)\mathbf{Mv}^2 + \left(\frac{1}{2}\right)\left(\frac{2}{5}\mathbf{MR}^2\right)(\mathbf{v/R})^2$$

$$\mathbf{gH} = \mathbf{v}^2/2 + \mathbf{v}^2/5 = 7\,\mathbf{v}^2/10$$

$$\mathbf{v} = (10\,\mathbf{gH}/7)^{\frac{1}{2}}$$

(b) Free-body diagram of all forces acting on the sphere.

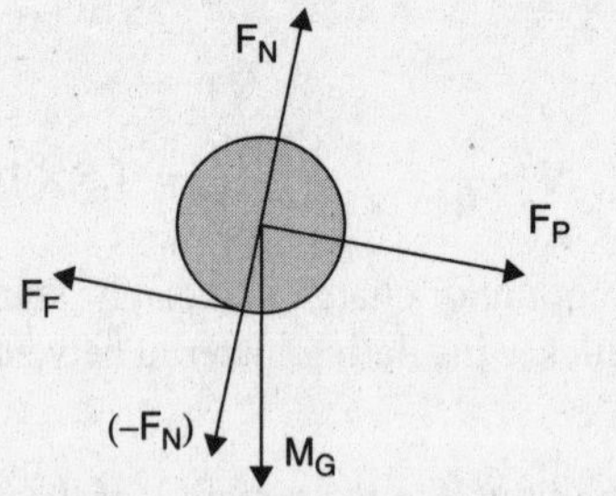

(c) For the acceleration of the center of mass of the sphere while rolling down the plane:

$$\mathbf{gH} = 7\,\mathbf{v}^2/10$$

$$\mathbf{v_F}^2 = 10\,\mathbf{gH}/7 = \mathbf{v_O}^2 + \mathbf{2ad}$$

$$10\,\mathbf{gH}/7 = \mathbf{2aH}$$

($\mathbf{a} = 5\,\mathbf{g}/7$ if sliding without friction) To add frictional effect:

$\Theta = \mathbf{a}_{CM}/5\,\mathbf{g}/7$ . . . and the acceleration is . . . $\mathbf{a}_{CM} = 5\,\mathbf{g}\sin\Theta/7$

(d) For the coefficient of kinetic friction necessary to keep the sphere rolling without slipping down the incline (minimum value of μ_K):

$$\mu_K = \mathbf{F_F}/\mathbf{F_N} = \mathbf{ma_F}/\mathbf{ma_N} = (\text{m})(5\,\mathbf{g}\sin\Theta/7)/(\text{m})\,(\mathbf{g}\cos\Theta) = 5\tan\Theta/7$$

(e) For the different sphere with double the radius and double the mass of the first sphere, finding the new sphere's linear velocity when it reaches point B:

$$2\mathbf{MgH} = \left(\frac{1}{2}\right)2\text{Mv}^2 + \left(\frac{1}{2}\right)\mathbf{I\omega}^2$$

$$2\text{M}\mathbf{gH} = \left(\frac{1}{2}\right)2\text{Mv}^2 + \frac{1}{2}\left(\frac{2}{5}\right)(2\text{M})\left(4\text{R}^2\right)(\text{v}/2\text{R})^2$$

$$\mathbf{gH} = \mathbf{v}^2/2 + \mathbf{v}^2/5 = 7\,\mathbf{v}^2/10$$

$$\mathbf{v} = (10\,\mathbf{gH}/7)^{\frac{1}{2}}$$

Mech. 2. **(a)** The acceleration of block B equals the net force divided by the mass:

$$\mathbf{a} = \frac{\mathbf{F_{NET}}}{\mathrm{m}} = \frac{1.50\,\mathrm{N}}{0.70\,\mathrm{kg}} = 2.14\ \mathrm{m/s^2}$$

(b) Since block A only begins to move when the bungee cord begins to stretch, its acceleration is due only to the force exerted by the cord through 0.50 m:

$$\mathbf{a} = \frac{\mathbf{F_{NET}}}{\mathrm{m}} = \frac{\mathbf{kx}}{\mathrm{m}} = \frac{(2.00\,\mathrm{N/m})(0.50\,\mathrm{m})}{(0.80\,\mathrm{kg})} = 1.25\ \mathrm{m/s^2}$$

(c) At the point where block B is moving and has stretched the bungee cord 0.50 m, it is no longer slack and is just ready to begin to pull back on block B. (Block A is only beginning to feel the force of the stretched bungee cord and has not yet moved.) The acceleration of block B is due to the net force on it, which is 1.5 N to the right minus the force exerted by the cord to the left plus the force needed to accelerate block A:

$$\mathbf{a} = \frac{\mathbf{F_{NET}}}{\mathrm{m}} = \frac{\mathbf{F_{RIGHT}} - \mathbf{F_{LEFT}} + \mathbf{F_{ON\ A}}}{\mathrm{m}}$$

$$= \frac{(1.50\,\mathrm{N}) - (2.0\,\mathrm{N/m})(0.50\,\mathrm{m}) + (0.8\,\mathrm{kg})(1.25\,\mathrm{m/s^2})}{(0.70\,\mathrm{kg})} = \mathbf{0.71\ m/s^2}$$

(d) To determine the kinetic energy of block B at this point, you first need to determine its velocity:

Using $\mathbf{V_{FIN}^2 = v_O^2 + 2ad}$,

$$\mathbf{V_{FIN}} = [0 + (2)(2.14\ \mathrm{m/s^2})(0.50\ \mathrm{m})]^{\frac{1}{2}} = 1.46\ \mathrm{m/s}$$

Now, using $E_K = \left(\frac{1}{2}\right)\mathbf{mv^2}$,

$$E_K = \left(\frac{1}{2}\right)(0.70\ \mathrm{kg})(1.46\ \mathrm{m/s})^2 = \mathbf{0.75\ J}$$

(e) The kinetic energy of block A after the cord has stretched 0.50 m is completely due to the force exerted on block A by the cord. First, determine the velocity of the block A at that point:

Using $\mathbf{V_{FIN}^2 = V_O^2 + 2ad}$, the average acceleration is $\frac{1.25}{2}\ \mathrm{m/s^2}$

$$V_{FIN} = [(0 + (2)(1.25/2\ \mathrm{m/s^2}))(0.50\ \mathrm{m})]^{\frac{1}{2}} = 0.79\ \mathrm{m/s}$$

Using $E_K = \left(\frac{1}{2}\right)\mathbf{mv^2}$, $E_K = \left(\frac{1}{2}\right)(0.80\ \mathrm{kg})(0.79\ \mathrm{m/s})^2 = 0.25\ \mathrm{J}$

Mech. 3. **(a)** Rotational equilibrium means that there is no net torque:

$$\boldsymbol{\tau_{CW} = \tau_{CCW}}$$
$$(m2)(r2) + (m3r2) = (m1)(r1)$$
$$(12\text{ kg})(1\text{ m}) + (m3)(1.5\text{ m}) = (30\text{ kg})(0.5\text{ m})$$
$$\mathbf{m3 = 2\ kg}$$

(b) Using $\boldsymbol{\tau = I\alpha = (F_{NET})(R) = \tau_{CCW} - \tau_{CW}}$,

$$(60\text{ kg m}^2)(\boldsymbol{\alpha}) = (m_3\,g)(r_3) - (m_3)(\mathbf{a}) - (m_1)(\mathbf{g})(r_1) + (m_1)(\mathbf{a})$$
$$(60\text{ kgm}^2)\boldsymbol{\alpha} = (m_3\mathbf{g} - m_3\mathbf{a})R_3 - (m_1\mathbf{g} + m_1\mathbf{a})R_1$$
$$\left[(2\text{ kg})(9.8\text{ m/s}^2) - (2\text{ kg})(\boldsymbol{\alpha}R_3)\right]1.5\text{ m} - \left[(30\text{ kg})(9.8\text{ m/s}^2) + (30\text{ kg})(\boldsymbol{\alpha}R_1)\right].5\text{ m}$$
$$\boldsymbol{\alpha} = (19.6\text{N} - 3\boldsymbol{\alpha})1.5\text{m} - 147\text{ J} - 7.5\boldsymbol{\alpha}\text{ m}^2$$
$$\boldsymbol{\alpha} = 29.4\text{ J} - 4.5\text{m}^2\boldsymbol{\alpha} - 147\text{ J} - 7.5\text{ m}^2\boldsymbol{\alpha}$$
$$60\boldsymbol{\alpha} + 4.5\boldsymbol{\alpha} + 7.5\boldsymbol{\alpha} = (29.4 - 147)\text{ J}$$
$$72\boldsymbol{\alpha} = -118\text{ J}$$
$$\boldsymbol{\alpha} = \mathbf{-1.64\ rad/s^2\ (CCW)}$$

(c) $T = (m_3\mathbf{g}) + (m_3\mathbf{a}) = m_3(\mathbf{g+a}) = (2.0\text{ kg})\left[\left(9.8\text{ m/s}^2\right) + \left(\frac{1.64\text{ rad}}{\text{s}^2}\right)\left(\frac{1.5\text{ m}}{\text{rad}}\right)\right]$

$$= 2.0\text{ kg }[(9.8\text{ m/s}^2) + (2.46\text{ m/s}^2)] = \mathbf{24.5\ N}$$

(d) Using $\mathbf{d = v_O t + \frac{1}{2}at^2}$,

$$1.0\text{ m} = \frac{1}{2}\left(1.64\,\frac{\text{rad}}{\text{s}^2}\right)\left(\frac{1.5\text{ m}}{\text{rad}}\right)t^2$$
$$\frac{1.0\text{ m}}{1.23\text{ m/s}^2} = t^2 = 0.81\text{ s}^2$$
$$t = \sqrt{0.81\text{ s}^2} = \mathbf{0.90\ sec}$$

Section II, Electricity and Magnetism

Free Response

E & M. 1. **(a)** i. Using $\mathbf{E} = \frac{\mathbf{kq}}{\mathbf{r}^2}$, (k is substituted for $(1/4\pi\varepsilon_0$ for brevity.)

$$\mathbf{E_1} = \frac{(9 \times 10^9\,\text{Nm}^2/\text{C}^2)(+20 \times 10^{-6}\,\text{C})}{(2\,\text{m})^2} = \mathbf{+4.5 \times 10^4\ N/C}$$

ii. $$\mathbf{E_2} = \frac{(9 \times 10^9\,\text{Nm}^2/\text{C}^2)(-20 \times 10^{-6}\,\text{C})}{(-4\,\text{m})^2} = \mathbf{-6.8 \times 10^3\ N/C}$$

iii.

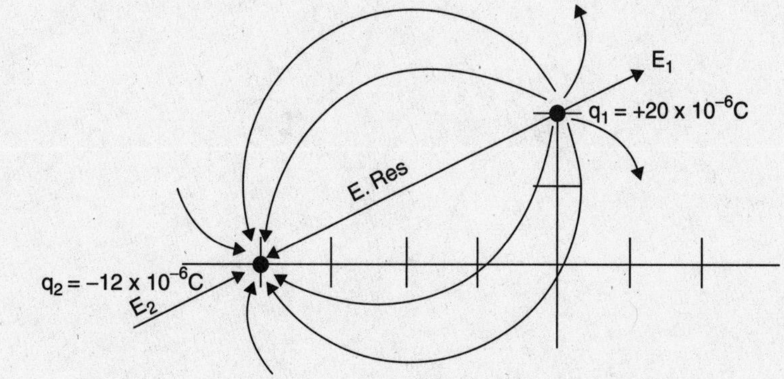

(b) $$\mathbf{V} = \frac{\mathbf{\Sigma kq}}{\mathbf{r}} = k\left(\frac{q_1}{\mathbf{r}} + \frac{q_2}{\mathbf{r}}\right) = 9 \times 10^9\,\text{Nm}^2/\text{C}^2\left[\left(\frac{+20 \times 10^{-6}\,\text{C}}{2\,\text{m}}\right) - \left(\frac{12 \times 10^{-6}\,\text{C}}{4\,\text{m}}\right)\right]$$

$$= \mathbf{6.2 \times 10^4\ Nm/C\ or\ J/C}$$

(c) Determining the force on the test charge q_3 involves the vector addition of the forces exerted on q_3 by the other two charges, q_1 and q_2.

$$\mathbf{F_{1\text{-}3}} = \frac{kq_1q_3}{(r_{1-3})^2} = (9 \times 10^9\,\text{Nm}^2/\text{C}^2)\left[\frac{(+20 \times 10^{-6}\,\text{C})(+6 \times 10^{-6}\,\text{C})}{(2\,\text{m})^2}\right] = 0.27\ \text{N (repelling)}$$

$$\mathbf{F_{2\text{-}3}} = \frac{kq_1q_3}{(r_{2-3})^2} = (9 \times 10^9\,\text{Nm}^2/\text{C}^2)\left[\frac{(-12 \times 10^{-6}\,\text{C})(+6 \times 10^{-6}\,\text{C})}{(4\,\text{m})^2}\right] = -0.041\ \text{N (attracting)}$$

Taking the sum of these components: $\mathbf{F_{1\text{-}3}} = [(0.27\ \text{N})^2 + (0.041\ \text{N})^2]^{\frac{1}{2}} = 0.27\ \text{N}$

For the angle: $\tan\Theta = \frac{0.041}{0.27} = 0.152$ and $\Theta = 8.6°$

The magnitude of the resulting force on q_3 is 0.27 N at an angle of 8.6°.

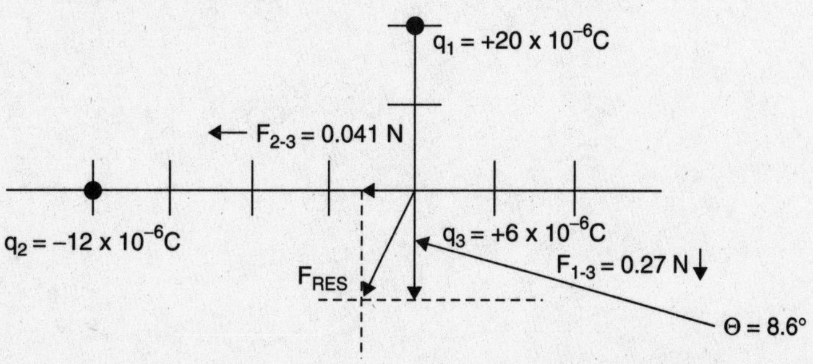

(d) Using **W = Fd**,

$$\mathbf{U = qV} = (+6 \times 10^{-6}\text{C})(6.2 \times 10^4\text{Nm/C}) = \mathbf{0.37\ J}$$

CliffsAP Physics C Practice Exam 1

E & M. 2. **(a)** Since the total initial charge of $-q$ is shared between the two spheres, each has a charge of $-\frac{q}{2}$.

(b) Using $\mathbf{F} = \frac{\mathbf{kq_1q_2}}{\mathbf{r^2}}$, (k is substituted for $\frac{1}{4\pi\varepsilon_0}$ for brevity)

Final separation of centers = (x – d).

$$F = \frac{k(-q/2)(-q/2)}{(x-d)^2} = \frac{kq^2}{(4)(x-d)^2}$$

(c) The electric field is 0 at the center.

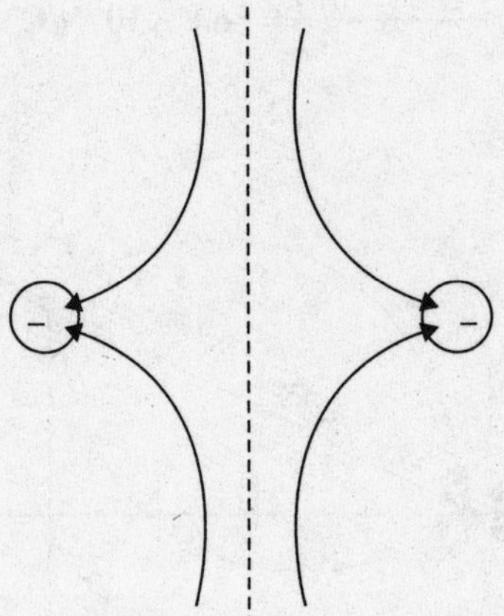

(d) Using $\mathbf{V} = \frac{\mathbf{\Sigma Fd}}{\mathbf{q}} = \frac{\Sigma(kq_1q_2/r^2)(r)}{q} = \frac{\Sigma kq}{r} = k\frac{(-q/2)}{(x-d)} + k\frac{(-q/2)}{(x-d)} = \frac{-kq}{(x-d)}$

(e)

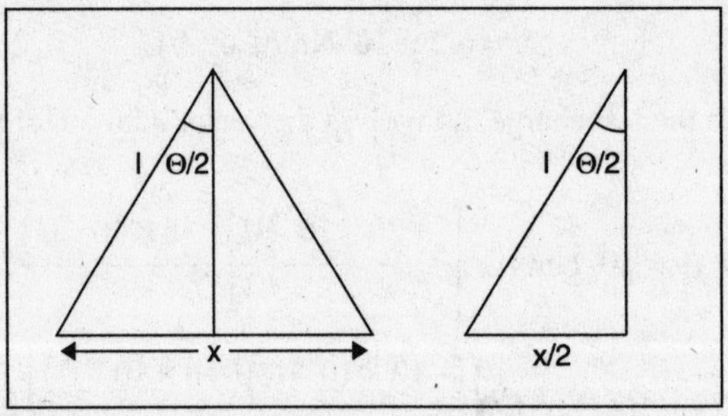

$$\sin\Theta/2 = \frac{x/2}{l} \text{ and } \mathbf{x = 2l\,(sin\,\Theta/2)}$$

(f) Using $\mathbf{V} = \frac{\mathbf{\Sigma kq}}{\mathbf{r}} = k\left[\frac{-q/4}{(x'-d)} + \frac{-q/4}{(x'-d)}\right] = \frac{-kq/2}{x'-d}$

E & M. 3. **(a)** Resistances of the circuit are as follows:

Internal resistance: r = 0.10 Ω

Top branch with R_1 in parallel with R_2 and R_3: 3.0 Ω

Side branch with R_4, R_5, and R_6 all in parallel: 2.0 Ω

Bottom of circuit with R_7 and R_8 in series: 6.0 Ω

Total equivalent resistance: **11.1 Ω**

(b) Using **V = IR**,

$$I = \frac{V}{R} = \frac{24.0\ \text{V}}{11.1\ \Omega} = \mathbf{2.16\ A}$$

(c) Since the branch side with R_2 and R_3 is 12 Ω, or 3 times the resistance of R_1, it gets $\frac{1}{4}$ of the current, or 0.54 A.

Using **V = IR**, V(for R_3) = (0.54 A)(6.0 Ω) = **3.24 V**.

(d) The 4 Ω R_4 gets $\frac{1}{2}$ the current in the right branch, while both of the 8 Ω resistors R_5 and R_6 get $\frac{1}{4}$ of the current in the branch (2.16 / 4 or 0.54 A).

Using **V = IR**, V(for R_5) = (0.54 A)(8.0 Ω) = **4.32 V**.

(e) The current through the 2.0 Ω resistor R_8 is the main line current 2.16 A.

Using **V = IR**, V(for R_8) = (2.16 A)(2.0 Ω) = **4.32 V**.

CliffsAP Physics C Practice Exam 2

Physics C

You may take the entire C exam or Mechanics only or Electricity and Magnetism only as follows:

	Entire C Exam	*Mechanics Only*	*Electricity and Magnetism Only*
1st 45 min.	Sect. I—Mechanics	Sect. I—Mechanics	Sect. I—Electricity and Magnetism
	35 questions	35 questions	35 questions
	Multiple Choice	Multiple Choice	Multiple Choice
2nd 45 min.	Sect. I—Electricity and Magnetism	Sect. II—Mechanics	Sect. II—Electricity and Magnetism
	35 questions	3 questions	3 questions
	Multiple Choice	Free Response	Free Response
3rd 45 min.	Sect. II—Mechanics		
	3 questions		
	Free Response		
4th 45 min.	Sect. II—Electricity and Magnetism		
	3 questions		
	Free Response		

Separate grades are reported for Mechanics and for Electricity and Magnetism. Each section of each examination is 50 percent of the total grade.

Each question in a section has equal weight. Calculators are *not permitted* on Section I of the exam but are allowed on Section II. However, calculators cannot be shared with other students, and calculators with typewriter-style (QWERTY) keyboards will not be permitted. A table of information that may be helpful is found on the following page.

The Physics C examination contains a total of 70 multiple-choice questions. If you are taking

Mechanics only, please be careful to answer multiple choice questions 1–35.

Electricity and Magnetism only, please be careful to answer multiple choice questions 36–70.

Mechanics *and* Electricity and Magnetism (the entire examination), answer multiple-choice questions 1–70.

Information Table

Constants and Conversions		*Units*	
1 Atomic mass unit	$u = 1.66 \times 10^{-27}$ kg	meter	m
	$= 931$ MeV/ c^2	kilogram	kg
Proton mass	$m_p = 1.67 \times 10^{-27}$ kg	second	s
Neutron mass	$m_n = 1.67 \times 10^{-27}$ kg	ampere	A
Electron mass	$m_e = 9.11 \times 10^{-31}$ kg	kelvin	K
Electron charge	$e = 1.60 \times 10^{-19}$ C	mole	mol
Avogadro's number	$N_0 = 6.02 \times 10^{23}$ mol^{-1}	hertz	Hz
Universal gas constant	$R = 8.31$ J / (mol · K)	newton	N
Boltzmann's constant	$k_B = 1.38 \times 10^{-23}$ J / K	pascal	Pa
Speed of light	$c = 3.00 \times 10^8$ m / s	joule	J
Planck's constant	$h = 6.63 \times 10^{-34}$ J · s	watt	W
	$= 4.14 \times 10^{-15}$ eV · s	coulomb	C
	$hc = 1.99 \times 10^{-25}$ J · m	volt	V
	$= 1.24 \times 10^{3}$ eV · nm	ohm	Ω
Vacuum permittivity	$\varepsilon_0 = 8.85 \times 10^{-12}$ C^2 / N · m^2	henry	H
Coulomb's law constant	$k = 1/4\pi\varepsilon_0 = 9.0 \times 10^9$ N · m^2 / C^2	farad	F
Vacuum permeability	$\mu_0 = 4\pi \times 10^{-7}$ (T · m) / A	tesla	T
Magnetic constant	$k' = \mu_0 / 4\pi = 10^{-7}$ (T · m) / A	degree Celsius	°C
Universal gravitation constant	$G = 6.67 \times 10^{-11}$ Nm^2/kg^2	electron Volt	eV
Acceleration due to gravity at the earth's surface	$g = 9.8$ m/s^2		
1 atmosphere pressure	1 atm $= 1.0 \times 10^5$ Pa $= 1.0 \times 10^5$ N / m^2		
1 electron volt	1 eV $= 1.60 \times 10^{-19}$ J		

Prefixes			*Trigonometric Function Values for Common Angles*			
Factor	Prefix	Symbol	Θ	sin Θ	cos Θ	tan Θ
10^9	giga	G	0°	0	1	0
10^6	mega	M	30°	$\frac{1}{2}$	$\frac{\sqrt{3}}{2}$	$\frac{\sqrt{3}}{3}$
10^3	kilo	k	37°	$\frac{3}{5}$	$\frac{4}{5}$	$\frac{3}{4}$
10^{-2}	centi	c	45°	$\frac{\sqrt{2}}{2}$	$\frac{\sqrt{2}}{2}$	1
10^{-3}	milli	m	53°	$\frac{4}{5}$	$\frac{3}{5}$	$\frac{4}{3}$

Prefixes			***Trigonometric Function Values for Common Angles***			
10^{-6}	micro	μ	60°	$\frac{\sqrt{3}}{2}$	$\frac{1}{2}$	$\sqrt{3}$
10^{-9}	nano	n	90°	1	0	∞
10^{-12}	pico	p				

These conventions are used in the examination:

1. *Unless stated otherwise, inertial frames of reference are used.*
2. *The direction of electric current flow is the conventional direction of positive charge.*
3. *For an isolated electric charge, the electric potential at an infinite distance is zero.*
4. *In mechanics and thermodynamics equations, W represents work done* ***on*** *a system.*

GO ON TO THE NEXT PAGE

Physics C
Sample Exam #2
Section I, Mechanics

Time — 45 Minutes

35 Questions

Directions: Each of the following questions or incomplete statements is followed by five possible answers. Select the best answer and then fill in the corresponding oval on the answer sheet.

Note: You may use $\mathbf{g} = 10\ \text{m/s}^2$ to simplify calculations.

1. The motion of a particle having non-constant acceleration along the **x**-axis is given as $\mathbf{a} = 14t$, where **a** is in m/s^2 and t is in sec. The particle starts from rest and at t = 0 its speed and position are zero. When t = 3 seconds, the particle is located at

(A) x = 7 m
(B) x = 21 m
(C) x = 42 m
(D) x = 63 m
(E) x = 84 m

2. A projectile is launched with velocity 100 m/s at an angle of 30° with the ground. The projectile reaches a maximum height of

(A) 25 m
(B) 75 m
(C) 125 m
(D) 250 m
(E) 500 m

3. A mass M is hung on a string and is made to oscillate. Its period of oscillation is directly proportional to

(A) $(g/s)^{\frac{1}{2}}$
(B) $(s/g)^{\frac{1}{2}}$
(C) $(s/m)^{\frac{1}{2}}$
(D) $(s^2g/m)^{\frac{1}{2}}$
(E) $(mg/s)^{\frac{1}{2}}$

4. A satellite is in a circular orbit around the earth a distance D from the earth's center and has a speed of $\mathbf{v_D}$. Its speed needed to orbit at twice that distance (2D) would be

(A) $\frac{(2\mathbf{v_D})^{\frac{1}{2}}}{2}$
(B) $\frac{\mathbf{v_D}}{2}$
(C) $\frac{\mathbf{v_D}}{(2)^{\frac{1}{2}}}$
(D) $2\mathbf{v_D}$
(E) $\frac{2\mathbf{v_D}}{(2)^{\frac{1}{2}}}$

Questions 5–6 refer to the following figure.

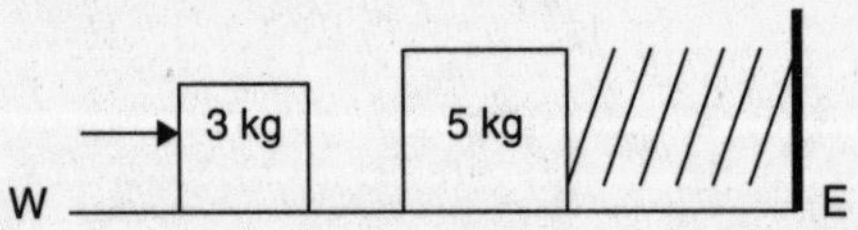

A 3 kg mass with initial velocity 2 m/s slides east on a frictionless surface and collides with and sticks to a stationary 5 kg mass. An unstretched, massless spring connects the 5 kg mass to an abutment as shown.

5. The speed of the system immediately after the 3 kg mass collides and sticks is closest to

(A) 0.2 m/s
(B) 0.4 m/s
(C) 0.6 m/s
(D) 0.8 m/s
(E) 1.0 m/s

GO ON TO THE NEXT PAGE

CliffsAP Physics C Practice Exam 2

6. If the spring constant is 8 N/m, the blocks stop after sliding approximately

(A) 0.25 m
(B) 0.50 m
(C) 0.75 m
(D) 1.0 m
(E) 1.25 m

7. A solid disk of mass M and radius R rolls, without slipping, along a horizontal surface at a speed $\mathbf{v_0}$. If the rotational inertia about its center of mass is $\left(\frac{1}{2}\right)mr^2$, the maximum vertical height to which it can roll up an incline is

(A) $\frac{3v^2}{4g}$
(B) $\frac{3v^2}{2g}$
(C) $3v^2g$
(D) $\frac{4v^2}{3g}$
(E) $\frac{4v^3}{3g}$

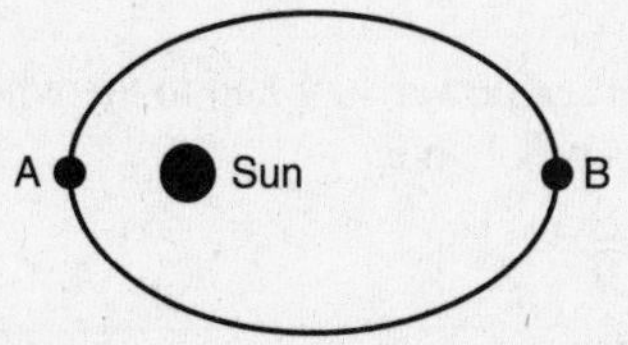

8. As a planet orbits the Sun in its elliptical path, it travels from position A to position B. Which statement relates its angular momentum and speed?

	Angular Momentum	*Speed*
(A)	Increases	Decreases
(B)	Remains constant	Increases
(C)	Remains constant	Remains constant
(D)	Decreases	Remains constant
(E)	Remains constant	Decreases

9. A nonuniform spring is stretched by an amount **x**, requiring a force **F** represented by $\mathbf{F} = 20\mathbf{x} - 8\mathbf{x}^2$, where **F** is in Newtons and **x** is in meters. When the spring is stretched 3 m from its equilibrium position, the change in potential energy is

(A) 6 J
(B) 8 J
(C) 10 J
(D) 18 J
(E) 28 J

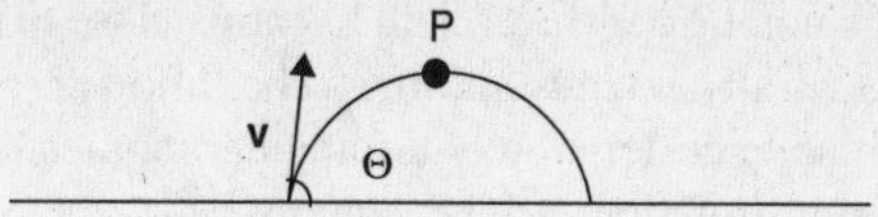

10. A projectile is fired into the air at an angle Θ with the ground. At its highest point P, its velocity and acceleration are given by

	v	*a*
(A)	↑	→
(B)	→	→
(C)	↑	↑
(D)	→	↓
(E)	↑	↓

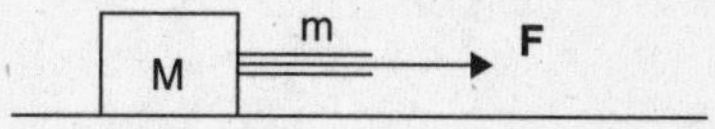

11. A rope of mass m pulls a block of mass M across a frictionless floor. A force **F** is applied at the end of the rope. If **a** represents the acceleration of the block, the force applied directly to the block is

(A) F – ma
(B) F – Ma
(C) F + M/a
(D) F + ma
(E) F + Ma

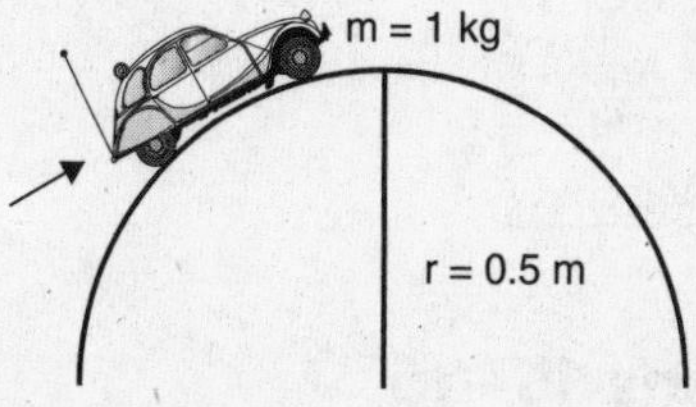

12. A 1 kg remote-controlled car is driven over a semicircular bump of radius 0.5 m. The maximum speed it can reach without leaving the surface is

(A) 1 m/s
(B) 2 m/s
(C) 3 m/s
(D) 4 m/s
(E) 5 m/s

GO ON TO THE NEXT PAGE

13. A spinning ice skater with his arms close to his body fully extends his arms. The following is an accurate description of his angular momentum **L** and his rotational kinetic energy $E_{K\,ROT}$.

	L	$E_{K\,ROT}$
(A)	Increases	Decreases
(B)	Remains constant	Increases
(C)	Remains constant	Remains constant
(D)	Decreases	Remains constant
(E)	Remains constant	Decreases

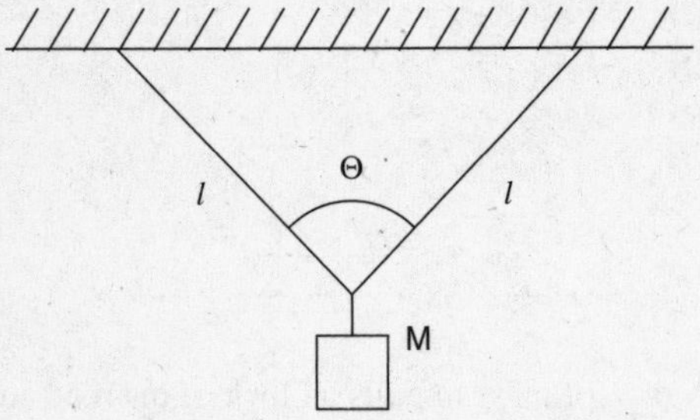

14. A mass M is hung from a string of negligible mass as shown. The tension T anywhere along the string is

(A) $\dfrac{Mg}{2\cos\Theta/2}$

(B) $\dfrac{Mg}{\cos\Theta}$

(C) $\dfrac{2\,Mg}{\cos\Theta}$

(D) $\dfrac{4\,Mg}{\cos\Theta}$

(E) $\dfrac{Mg\cos\Theta}{2}$

Questions 15–16

The motion of a particle moving in a circle is given by:

$$x = 2\cos(4t)$$

$$y = 2\sin(4t)$$

where t is in seconds.

15. The period of revolution of the particle is

(A) $(\pi/4)$ seconds.
(B) $(\pi/2)$ seconds.
(C) π seconds.
(D) 2π seconds.
(E) 4π seconds.

16. The speed of the particle is

(A) 2 m/s
(B) 4 m/s
(C) 6 m/s
(D) 8 m/s
(E) 10 m/s

Questions 17–18

A flywheel having rotational inertia I rotates with angular velocity ω on a fixed frictionless axle. It then slows uniformly to rest in time t, when an outside torque is applied.

17. The net torque on the flywheel is

(A) $\dfrac{-I\omega}{t^2}$

(B) $\dfrac{-I\omega^2}{t}$

(C) $\dfrac{-\omega}{It}$

(D) $\dfrac{-\omega^2}{It}$

(E) $\dfrac{-I\omega}{t}$

18. The average power supplied to the wheel in the previous problem is

(A) $\dfrac{-2I\omega}{3t}$

(B) $\dfrac{-I\omega^2}{2t}$

(C) $\dfrac{2I\omega^2}{3t}$

(D) $\dfrac{3I\omega^2}{2t}$

(E) $\dfrac{I\omega}{2t^2}$

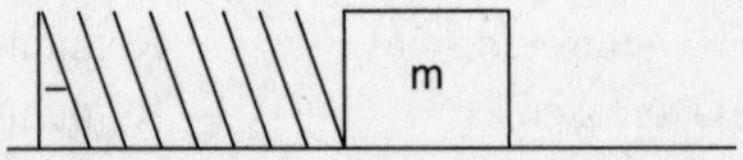

19. An ideal spring of negligible mass and spring constant k is attached to a wall and a mass m as shown. The block is on a frictionless surface. If the mass is set oscillating with maximum velocity **v**, its maximum amplitude **x** is

(A) $\left(\frac{1}{2}\right)kx^2$

(B) $v^2 / 2m$

(C) $(mv / k^2)^{\frac{1}{2}}$

(D) $(mv^2 / k)^{\frac{1}{2}}$

(E) $(km / v^2)^{\frac{1}{2}}$

GO ON TO THE NEXT PAGE

Questions 20–21 refer to the following figure.

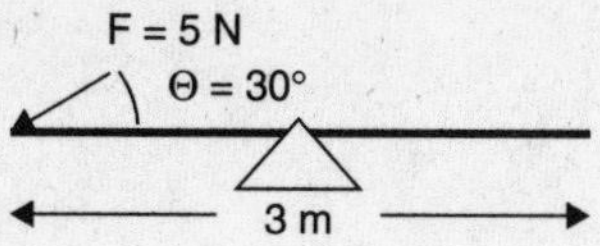

20. A see-saw of length 3 m and mass of 50 N is supported by the fulcrum at the center. A 5 N force F is applied at one end at an angle of 30° as shown. Where must a 10 N weight be placed to maintain rotational equilibrium?

(A) $\left(\frac{3}{8}\right)$ m left of center

(B) $\left(\frac{3}{8}\right)$ m right of center

(C) $\left(\frac{5}{8}\right)$ m left of center

(D) $\left(\frac{5}{8}\right)$ m right of center

(E) $\left(\frac{7}{8}\right)$ m right of center

21. If the system is in translational and rotational equilibrium with the 10 N weight applied, the upward force provided by the fulcrum at the center is most nearly

(A) 13 N
(B) 33 N
(C) 50 N
(D) 63 N
(E) 80 N

22. Doing 2000 J of work on a 100 kg mass may cause it to be raised to a ledge of height

(A) 1 m
(B) 2 m
(C) 3 m
(D) 4 m
(E) 5 m

23. An object of mass m that orbits a planet of mass M and radius R at a distance R above the surface has kinetic energy

(A) 0

(B) $\frac{GMm}{4R}$

(C) $\frac{GMm}{2R}$

(D) $\frac{GM}{2R^2}$

(E) $\frac{GMm}{R^2}$

Questions 24–26 refer to the following figure.

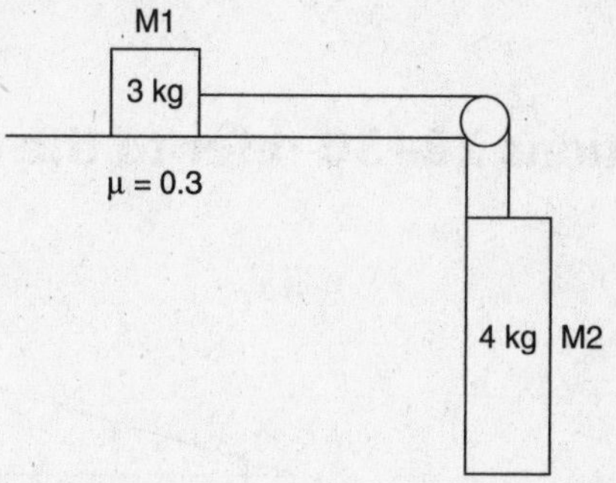

24. In the above illustration, a 3 kg mass M1 rests on a horizontal table. A second mass of 4 kg M2 hangs over the side of the table, connected to M1 by a string that passes over a frictionless, massless pulley. The coefficient of friction between M1 and the table is 0.3.

The acceleration of M1 is most nearly

(A) 2.5 m/s^2
(B) 4.5 m/s^2
(C) 8.5 m/s^2
(D) 12.5 m/s^2
(E) 15.5 m/s^2

25. The net force on M1 is

(A) 6 N
(B) 13 N
(C) 18 N
(D) 24 N
(E) 31 N

GO ON TO THE NEXT PAGE

26. The coefficient of friction between M1 and the table is doubled. The new acceleration of the system is nearest to

(A) 2 m/s^2
(B) 3 m/s^2
(C) 4 m/s^2
(D) 5 m/s^2
(E) 6 m/s^2

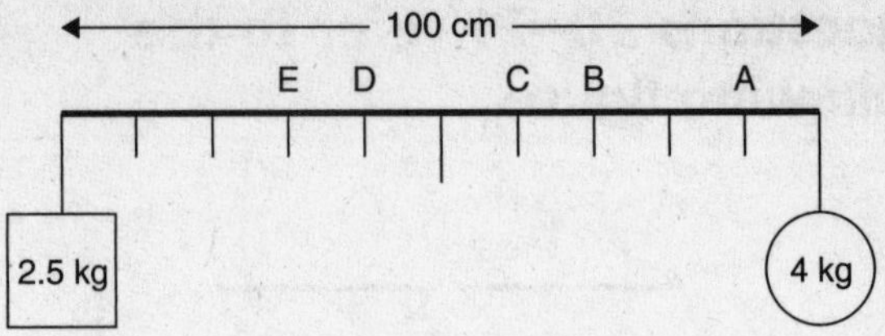

27. Two masses, 2.5 and 4 kg, hang from opposite ends of a meter stick of negligible mass. At which point should a string be attached to suspend this system and maintain rotational equilibrium?

(A) A
(B) B
(C) C
(D) D
(E) E

Questions 28–30 refer to the following figure.

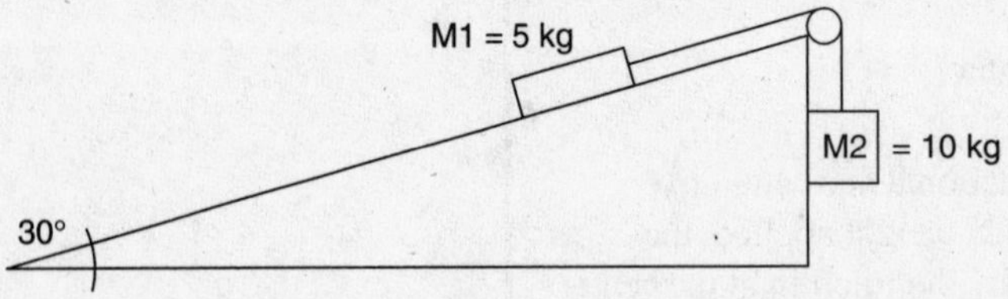

Two masses, M1 (5 kg) and M2 (10 kg) are attached by a string placed over a frictionless pulley as shown. The surface is frictionless and the incline is 30°.

28. The net force on the system is

(A) 5 N
(B) 20 N
(C) 25 N
(D) 50 N
(E) 75 N

29. The acceleration of mass M2 is

(A) 2 m/s^2
(B) 5 m/s^2
(C) 8 m/s^2
(D) 10 m/s^2
(E) 12 m/s^2

30. The 5 kg mass (M1) is replaced by a 10 kg mass. The new acceleration of M2 is

(A) 1.5 m/s^2
(B) 2 m/s^2
(C) 2.5 m/s^2
(D) 3 m/s^2
(E) 3.5 m/s^2

GO ON TO THE NEXT PAGE

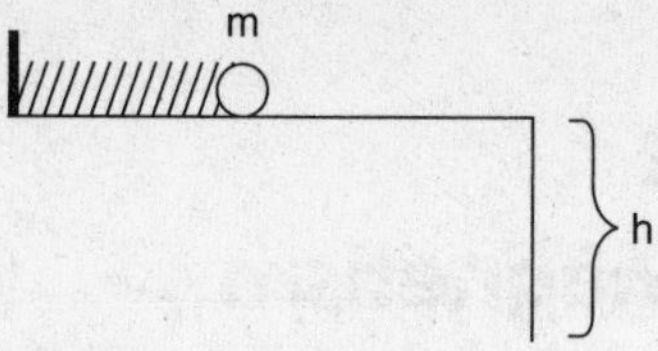

31. A ball of mass m is launched from a horizontal table of height h by a spring having a spring constant k. After the ball leaves the spring, it hits the floor after a time of

(A) $2h / g$
(B) $\frac{2h}{\sqrt{g}}$
(C) $\sqrt{gh}$
(D) $\frac{h}{\sqrt{g}}$
(E) $\sqrt{2gh}$

Questions 32–33 refer to the following figure.

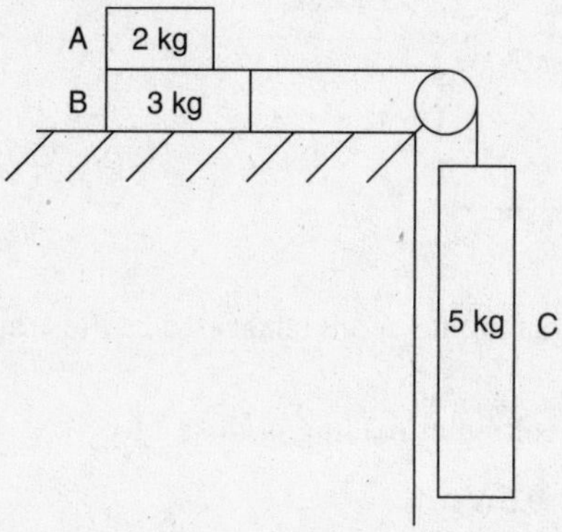

A 2 kg mass A is placed and secured atop a 3 kg mass B. Both rest on a horizontal surface having a coefficient of friction 0.6 with the bottom of the masses. A mass of 5 kg hangs by a cord attached to a frictionless pulley as shown.

32. The acceleration of the 3 kg mass is most nearly

(A) 1 m/s^2
(B) 2 m/s^2
(C) 3 m/s^2
(D) 4 m/s^2
(E) 5 m/s^2

33. Block A is removed and the experiment is repeated. The acceleration of the 5 kg mass is now

(A) 1 m/s^2
(B) 2 m/s^2
(C) 3 m/s^2
(D) 4 m/s^2
(E) 5 m/s^2

Questions 34–35 refer to the following figure.

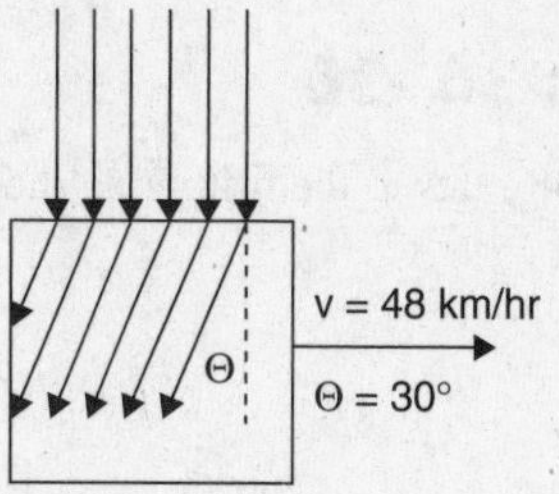

Falling raindrops make an angle of 30° with the vertical when they hit the side window of a car traveling at a constant speed of 48 km/hr.

34. The vertical speed of the raindrops hitting the window is nearest to

(A) $\frac{9}{\sqrt{3}}$ m/s
(B) $\frac{13}{\sqrt{3}}$ m/s
(C) $\frac{24}{\sqrt{3}}$ m/s
(D) $\frac{29}{\sqrt{3}}$ m/s
(E) $\frac{40}{\sqrt{3}}$ m/s

35. Two equal masses oscillate vertically on two different springs with differing values for k. The spring of lesser spring constant has the

(A) larger oscillation frequency.
(B) smaller oscillation amplitude.
(C) larger oscillation amplitude.
(D) shorter oscillation period.
(E) larger oscillation period.

END OF SECTION I, MECHANICS. IF YOU FINISH BEFORE TIME IS CALLED, YOU MAY CHECK YOUR WORK ON THIS SECTION ONLY. DO NOT TURN TO ANY OTHER TEST MATERIALS.

STOP

Physics C
Sample Exam #2
Section II, Electricity and Magnetism

Time – 45 Minutes
35 Questions

Directions: Each of the following questions or incomplete statements is followed by five possible answers. Select the best answer and then fill in the corresponding oval on the answer sheet.

Questions 36–38

Questions 36–38 refer to the following diagram of a DC circuit.

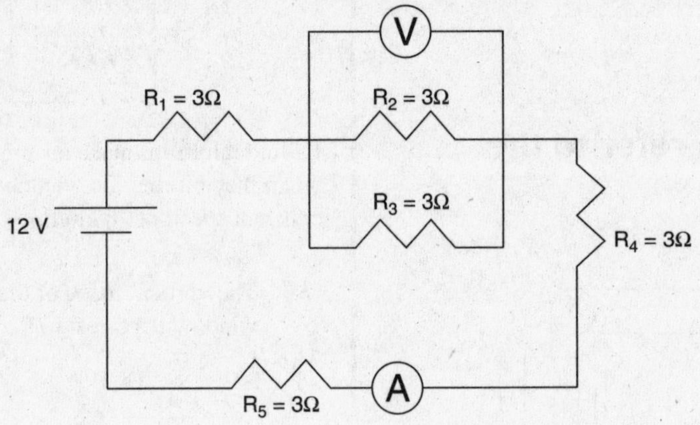

A DC circuit containing five 3 Ω resistors, a 12 V source, a voltmeter and an ammeter are connected as illustrated.

36. The equivalent resistance of the circuit is

(A) 5 Ω
(B) 7 Ω
(C) 9 Ω
(D) 11 Ω
(E) 13 Ω

37. The ammeter reading is closest to

(A) 1 A
(B) 2 A
(C) 3 A
(D) 4 A
(E) 5 A

38. The voltmeter reading is closest to

(A) 0.5 V
(B) 1.0 V
(C) 1.5 V
(D) 2.0 V
(E) 2.5 V

39. A capacitor with its plates separated by a thin layer of insulating material is discharged and the insulating material is removed. Compared to its original capacitance its capacitance with air separation

(A) is zero.
(B) is unchanged.
(C) is less now.
(D) is greater now.
(E) cannot be determined.

GO ON TO THE NEXT PAGE

CliffsAP Physics C Practice Exam 2

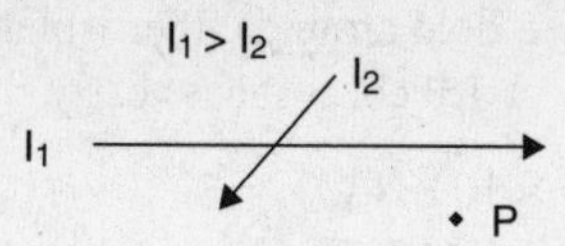

40. The magnetic field at point P is directed

(A) toward the top of the page.
(B) toward the bottom of the page.
(C) into the page.
(D) out of the page.
(E) to the right.

Questions 41–42 refer to the following figure.

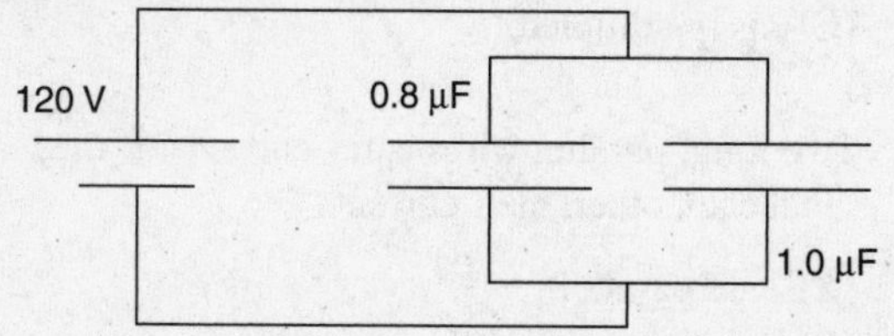

Two capacitors, 0.8 μF and 1.0 μF are connected in parallel and charged to a 120 V potential difference.

41. The total charge acquired by the capacitors is most nearly

(A) 0.2 μC
(B) 2 μC
(C) 22 μC
(D) 216 μC
(E) 2160 μC

42. The capacitors are now discharged and reconnected in series with the same power source. The new total charge acquired is nearest to

(A) 0.5 μC
(B) 5.0 μC
(C) 50 μC
(D) 500 μC
(E) 5000 μC

43. A charge of +20 nC is moved from a point where the potential is 0.50 V to another point where the potential is 0.75 V. The energy change in Joules is

(A) 2.5×10^{-9} J
(B) 5.0×10^{-9} J
(C) 7.5×10^{-9} J
(D) 5.0×10^{-8} J
(E) 7.5×10^{-8} J

Questions 44–45 refer to the following figure.

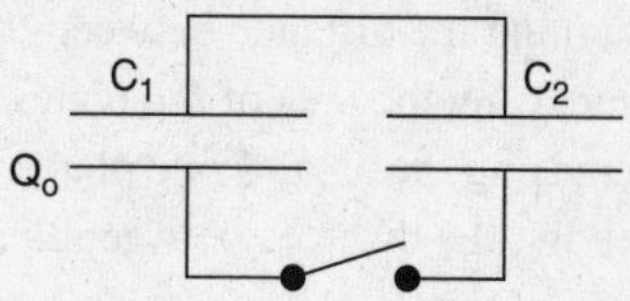

A 5.0 μF capacitor C_1 is charged to an initial potential difference of 12.0 V and connected to an initially *uncharged* 13.0 μF capacitor C_2 and an open switch.

44. When the switch is closed, the final potential difference V_{FINAL} across both C_1 and C_2 is most nearly

(A) 1 V
(B) 3 V
(C) 5 V
(D) 7 V
(E) 9 V

45. The total potential energy stored in capacitors C_1 and C_2 *before* and *after* the switch S is closed is most nearly

	$U_{INITIAL}$	U_{FINAL}
(A)	99 μJ	13 μJ
(B)	140 μJ	27 μJ
(C)	200 μJ	48 μJ
(D)	300 μJ	73 μJ
(E)	360 μJ	99 μJ

GO ON TO THE NEXT PAGE

46. The work necessary to move a point charge of 2.0×10^{-3} C from the origin to another point 1.5 m away is 3.0 J. The potential difference between the two points is

(A) 4.5×10^{-4} V
(B) 1.5×10^{-3} V
(C) 3.0×10^{-2} V
(D) 10^{3} V
(E) 1.5×10^{3} V

47. In a parallel-plate capacitor, the following will double the charge:

(A) Doubling the distance between the plates
(B) Decreasing the area of the plates
(C) Increasing the area of one plate
(D) Halving the distance between the plates
(E) None of these

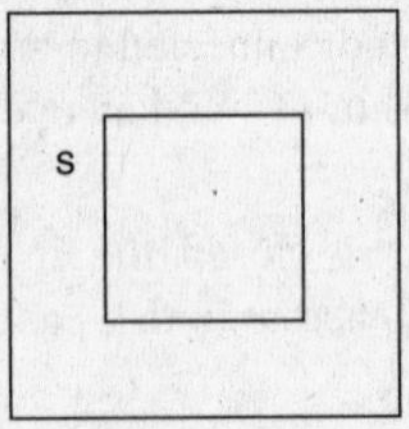

48. Two concentric squares of sides s and 2s, respectively, are made of the same type and thickness of wire. They lie in the same plane, as shown. If the resistance of the larger square is R, the resistance of the smaller square is

(A) R/4
(B) R/2
(C) R
(D) 2R
(E) 4R

Questions 49–50

Two electrons are separated by a distance of 1.0 cm and are separated by air.

49. The electrostatic force between them is most nearly

(A) -2.3×10^{-25} N
(B) -2.3×10^{-24} N
(C) -1.6×10^{-24} N
(D) $+1.6 \times 10^{-24}$ N
(E) $+2.3 \times 10^{-24}$ N

50. The electric field around either isolated electron at a distance of 1.0 cm is most nearly

(A) -1.4×10^{-5} N/C
(B) -1.4×10^{-4} N/C
(C) -1.4×10^{-3} N/C
(D) $+1.4 \times 10^{-4}$ N/C
(E) $+1.4 \times 10^{-5}$ N/C

51. A parallel plate capacitor is connected to an emf ε with plate separation distance s. If the plate separation becomes s/2, the charge on the plates

(A) is quartered.
(B) is halved.
(C) is unchanged.
(D) is doubled.
(E) is quadrupled.

52. Two long, parallel wires carry currents. If they repel each other, their currents

(A) are parallel.
(B) are antiparallel.
(C) are equal.
(D) are unequal.
(E) draw from the same source.

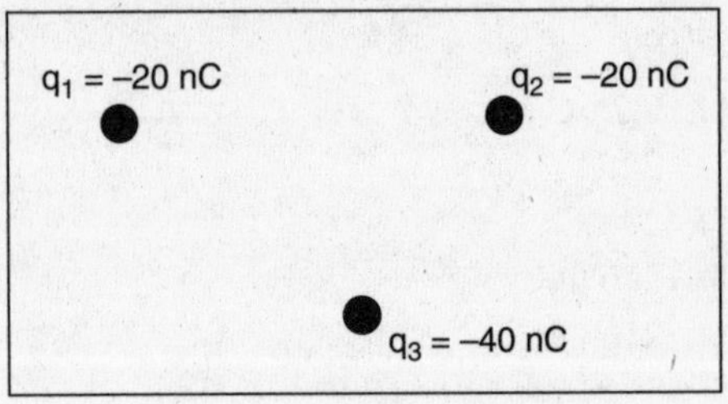

53. Three electric point charges, $q_1 = -20$ nC, $q_2 = -20$ nC, and $q_3 = -40$ nC are arranged as shown. The direction of the force on charge q_3 is closest to

(A) ↑
(B) ↓
(C) →
(D) ←
(E) ↖

GO ON TO THE NEXT PAGE

CliffsAP Physics C Practice Exam 2

Questions 54–55

Questions 54–55 pertain to the following arrangements of electric charges located at the vertices of an equilateral triangle. Point **X** is equidistant from the charges.

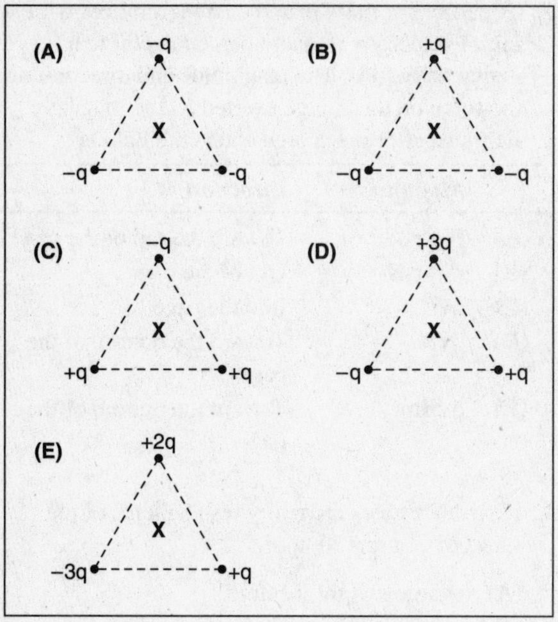

54. In which arrangement is the electric field at point *X* equal to zero?

(A) A
(B) B
(C) C
(D) D
(E) E

55. In which arrangement is the electric field at point *X* located midway between two of the charges?

(A) A
(B) B
(C) C
(D) D
(E) E

Questions 56–57 refer to the following figure.

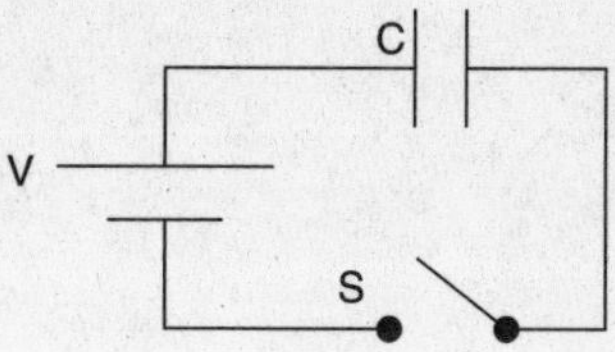

At time t = 0 the switch in the above circuit is closed.

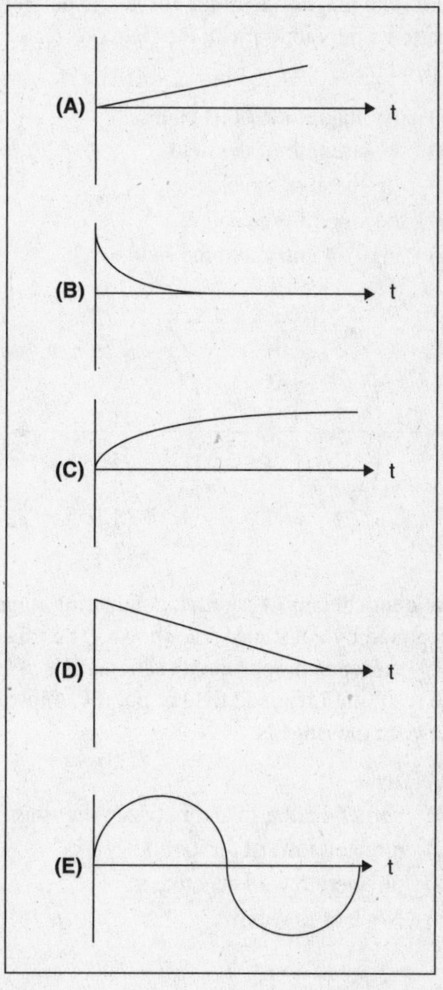

56. Of the preceding five graphs, which best represents the potential difference across the capacitor C as a function of time?

(A) A
(B) B
(C) C
(D) D
(E) E

GO ON TO THE NEXT PAGE

57. Of the preceding five graphs, which best represents the current through the capacitor C as a function of time?

(A) A
(B) B
(C) C
(D) D
(E) E

58. A charged particle moves through a magnetic field. The magnitude of the force it experiences due to its interaction with the field *does not* depend on

(A) the magnitude of its charge.
(B) the strength of the field.
(C) the speed of the charge.
(D) the sign of the charge.
(E) angle of entry into the field.

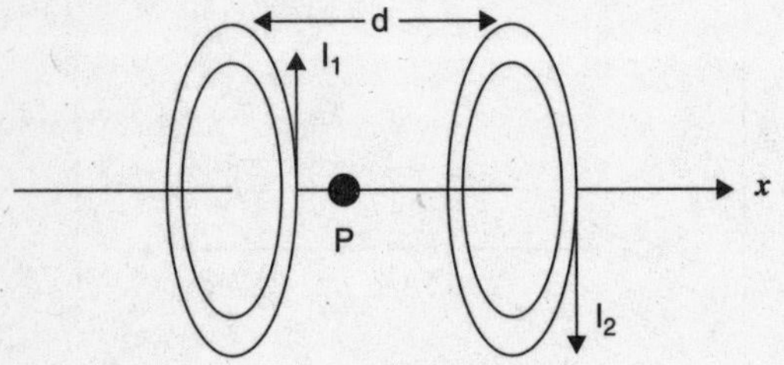

59. An identical pair of parallel, conducting rings is separated by a distance d as shown. The rings carry currents in opposite directions. If $I_1 > I_2$, the resulting magnetic field at point P, midway between the rings is

(A) zero.
(B) perpendicular to and between the rings.
(C) directed toward positive **x**.
(D) directed toward negative **x**.
(E) None of the above.

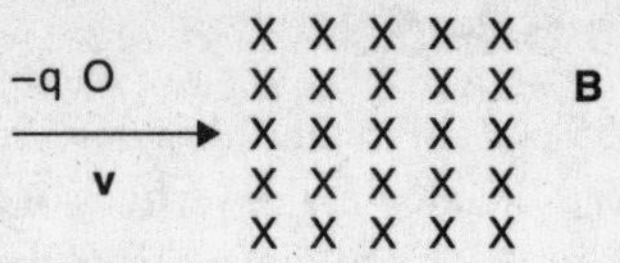

60. A particle of mass m and charge -q moves with initial velocity **v** into and perpendicular to a magnetic field **B**. The magnitude and direction of the force on the charge exerted by the magnetic field just after the charge enters the field is

	Magnitude	*Direction*
(A)	(kq/r^2)	Toward the top of the page
(B)	(kq/r)**B**	Out of the page
(C)	qv**B**	Into the page
(D)	qv**B**	Toward the bottom of the page
(E)	qvBm	Toward the bottom of the page

61. Kirchoff's Laws are really restatements of the Laws of Conservation of

(A) charge and momentum.
(B) energy and momentum.
(C) energy and power.
(D) resistance and charge.
(E) charge and energy.

Questions 62–64

A current of 1.5 A is supplied to a coil of wire having 400 turns. A flux of 10^{-2} Wb develops through the center of the coil in 0.20 sec.

62. The average emf induced in the coil is most nearly

(A) –20 V
(B) –2 V
(C) 0 V
(D) 2 V
(E) 20 V

GO ON TO THE NEXT PAGE

CliffsAP Physics C Practice Exam 2

63. The coil's inductance is most nearly

(A) 2 H
(B) 3 H
(C) 4 H
(D) 5 H
(E) 6 H

64. The energy stored in the coil's magnetic field is

(A) 2 J
(B) 3 J
(C) 4 J
(D) 5 J
(E) 6 J

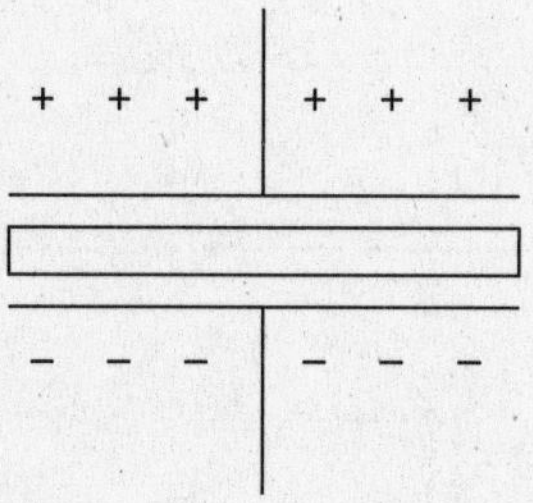

65. The direction of the electric field inside the dielectric is

(A) out of the page.
(B) into the page.
(C) toward the top of the page.
(D) toward the bottom of the page.
(E) nonexistent.

66. A resistor and an inductor are connected to a source of emf. When the switch is closed, the inductor

(A) reaches maximum potential immediately.
(B) reaches maximum potential after awhile.
(C) conducts but does not store energy.
(D) does not oppose the current through it.
(E) reaches zero after awhile.

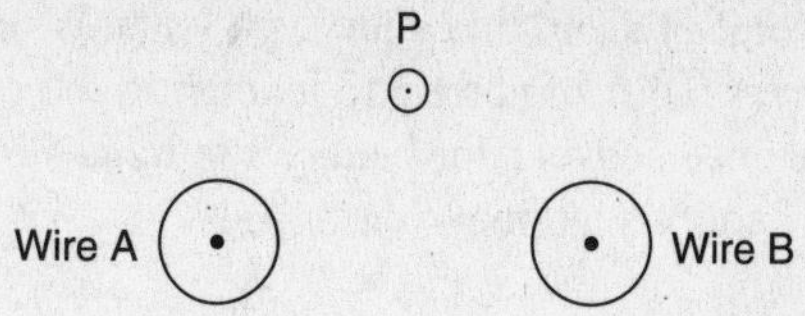

67. Two long parallel wires, A and B, both carry identical currents, directed out of the page. At point P, equidistant from both wires, the direction of the resultant magnetic field is

(A) nonexistent.
(B) to the left of the page.
(C) to the right of the page.
(D) into the page.
(E) out of the page.

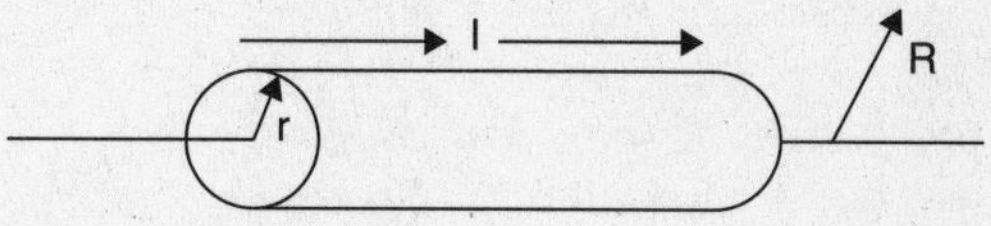

68. A long wire of radius **r** carries a current **I** as shown. The graph that best describes the magnetic field intensity **B** as a function of the distance **R** from the central axis of the wire is

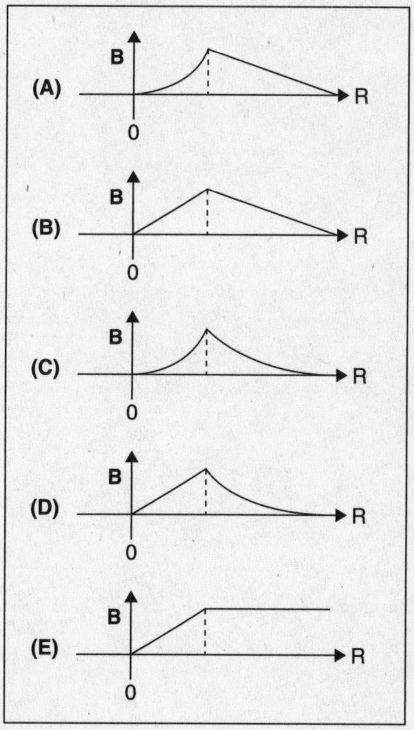

GO ON TO THE NEXT PAGE

69. The emf of a certain battery is 6 volts. When it delivers 0.4 A of current to its circuit, the potential difference between the battery's terminals is 5 volts. The battery's internal resistance is

(A) 0.5 Ω
(B) 1.5 Ω
(C) 2.5 Ω
(D) 3.5 Ω
(E) 4.5 Ω

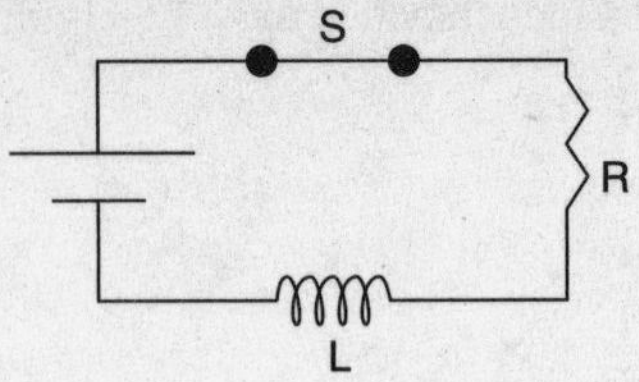

70. A battery is connected to a resistor R, an inductor L and a switch S. The graph that best depicts the circuit current when the switch is opened after a long time is

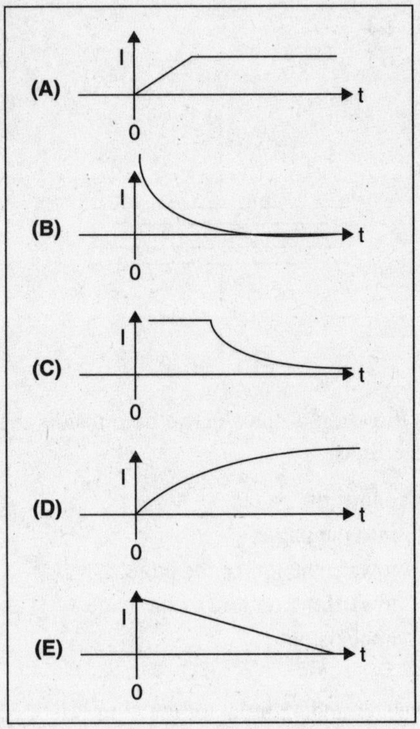

END OF SECTION I, ELECTRICITY AND MAGNETISM. IF YOU FINISH BEFORE TIME IS CALLED, YOU MAY CHECK YOUR WORK ON THIS SECTION ONLY. DO NOT GO ON TO SECTION II UNTIL YOU ARE TOLD TO DO SO.

Physics C
Section II
Free-Response Questions

Mechanics	45 Minutes	3 required questions of equal weight
Electricity and Magnetism	45 Minutes	3 required questions of equal weight

Section II is 50 percent of the total grade for each of the two examinations.

Mark one of the three boxes to indicate which questions you will answer.

❑ Mechanics only
❑ Electricity and Magnetism only
❑ Both Mechanics and Electricity and Magnetism

GO ON TO THE NEXT PAGE

Information Table

Constants and Conversions		**Units**	
1 Atomic mass unit	$u = 1.66 \times 10^{-27}$ kg	meter	m
	$= 931$ MeV/ c^2	kilogram	kg
Proton mass	$m_p = 1.67 \times 10^{-27}$ kg	second	s
Neutron mass	$m_n = 1.67 \times 10^{-27}$ kg	ampere	A
Electron mass	$m_e = 9.11 \times 10^{-31}$ kg	kelvin	K
Electron charge	$e = 1.60 \times 10^{-19}$ C	mole	mol
Avogadro's number	$N_0 = 6.02 \times 10^{23}$ mol^{-1}	hertz	Hz
Universal gas constant	$R = 8.31$ J / (mol · K)	newton	N
Boltzmann's constant	$k_B = 1.38 \times 10^{-23}$ J / K	pascal	Pa
Speed of light	$c = 3.00 \times 10^8$ m / s	joule	J
Planck's constant	$h = 6.63 \times 10^{-34}$ J · s	watt	W
	$= 4.14 \times 10^{-15}$ eV · s	coulomb	C
	$hc = 1.99 \times 10^{-25}$ J · m	volt	V
	$= 1.24 \times 10^3$ eV · nm	ohm	Ω
Vacuum permittivity	$\varepsilon_0 = 8.85 \times 10^{-12}$ C^2 / N · m^2	henry	H
Coulomb's law constant	$k = 1/4\pi\varepsilon_0 = 9.0 \times 10^9$ N · m^2 / C^2	farad	F
Vacuum permeability	$\mu_0 = 4\pi \times 10^{-7}$ (T · m) / A	tesla	T
Magnetic constant	$k' = \mu_0 / 4\pi = 10^{-7}$ (T · m) / A	degree Celsius	°C
Universal gravitation constant	$G = 6.67 \times 10^{-11}$ Nm^2/kg^2	electron Volt	eV
Acceleration due to gravity at the earth's surface	$\mathbf{g} = 9.8$ m/s^2		
1 atmosphere pressure	1 atm $= 1.0 \times 10^5$ Pa		
	$= 1.0 \times 10^5$ N / m^2		
1 electron volt	1 eV $= 1.60 \times 10^{-19}$ J		

Prefixes

Factor	Prefix	Symbol
10^9	giga	G
10^6	mega	M
10^3	kilo	k
10^{-2}	centi	c
10^{-3}	milli	m
10^{-6}	micro	μ
10^{-9}	nano	n
10^{-12}	pico	p

Trigonometric Function Values for Common Angles

Θ	sin Θ	cos Θ	tan Θ
0°	0	1	0
30°	$\frac{1}{2}$	$\frac{\sqrt{3}}{2}$	$\frac{\sqrt{3}}{3}$
37°	$\frac{3}{5}$	$\frac{4}{5}$	$\frac{3}{4}$
45°	$\frac{\sqrt{2}}{2}$	$\frac{\sqrt{2}}{2}$	1
53°	$\frac{4}{5}$	$\frac{3}{5}$	$\frac{4}{3}$
60°	$\frac{\sqrt{3}}{2}$	$\frac{1}{2}$	$\sqrt{3}$
90°	1	0	∞

These conventions are used in the examination:

1. *Unless stated otherwise, inertial frames of reference are used.*
2. *The direction of electric current flow is the conventional direction of positive charge.*
3. *For an isolated electric charge, the electric potential at an infinite distance is zero.*
4. *In mechanics and thermodynamics equations, W represents work done* ***on*** *a system.*

Advanced Placement Physics C Equations

Newtonian Mechanics

$\mathbf{v} = \mathbf{v}_0 + \mathbf{a}t$	**a** = acceleration
$\mathbf{x} = \mathbf{x}_0 + \mathbf{v}_0 t + \frac{1}{2}\mathbf{a}t^2$	**F** = force
$\mathbf{v}^2 = \mathbf{v}_0^2 + 2\mathbf{a}(\mathbf{x}-\mathbf{x}_0)$	f = frequency
$\Sigma\mathbf{F} = \mathbf{F}_{NET} = m\mathbf{a}$	h = height
$\mathbf{F} = \frac{d\mathbf{p}}{dt}$	I = rotational inertia
$\mathbf{J} = \int \mathbf{F}\,dt = \Delta\mathbf{p}$	**J** = impulse
$\mathbf{p} = m\mathbf{v}$	K = kinetic energy
$\mathbf{F}_{fric} \leq \mu\mathbf{N}$	k = spring constant
$\mathbf{W} = \int \mathbf{F} \cdot d\mathbf{r}$	l = length
$K = \frac{1}{2}m\mathbf{v}^2$	**L** = angular momentum
$P = \frac{d\mathbf{W}}{dt}$	m = mass
$P = \mathbf{F} \cdot \mathbf{v}$	**N** = normal force
$\Delta U_g = m\mathbf{gh}$	P = power
$\mathbf{a}_c = \frac{\mathbf{v}^2}{\mathbf{r}} = \omega^2\mathbf{r}$	**p** = momentum
$\boldsymbol{\tau} = \mathbf{r} \times \mathbf{F} = \mathbf{rF}\sin\Theta$	r = radius, distance
$\Sigma\boldsymbol{\tau} = \boldsymbol{\tau}_{NET} = I\alpha$	**r** = position vector
$I = \int \mathbf{r}^2\,dm = \Sigma m\mathbf{r}^2$	T = period
$\mathbf{r}_{CM} = \Sigma m\mathbf{r}/\Sigma m$	t = time
$\mathbf{v} = r\boldsymbol{\omega}$	U = potential energy
$\mathbf{L} = \mathbf{r} \times \mathbf{p} = I\boldsymbol{\omega}$	v = velocity
$K = \left(\frac{1}{2}\right)I\omega^2$	v = speed
$\boldsymbol{\omega} = \boldsymbol{\omega}_O + \boldsymbol{\alpha}t$	W = work done on a system
$\boldsymbol{\Theta} = \boldsymbol{\Theta}_O + \boldsymbol{\omega}_O t + \left(\frac{1}{2}\right)\alpha t^2$	**x** = position
$\mathbf{F}_s = -k\mathbf{x}$	μ = coefficient of friction
$U_s = \frac{1}{2}k\mathbf{x}^2$	**Θ** = angle, displacement
$T = (2\pi/\omega) = 1/f$	**τ** = torque
$T_P = 2\pi(l/\mathbf{g})^{\frac{1}{2}}$	**ω** = angular speed
$T_S = 2\pi(m/k)^{\frac{1}{2}}$	**α** = angular acceleration
$\mathbf{F}_G = -\frac{Gm_1 m_2}{\mathbf{r}^2}$	
$U_G = -\frac{Gm_1 m_2}{\mathbf{r}}$	

Electricity and Magnetism

$\mathbf{F} = \frac{1}{4\pi\varepsilon_0}\frac{q_1 q_2}{\mathbf{r}^2}$	A = area
$\mathbf{E} = \frac{\mathbf{F}}{q}$	**B** = magnetic field
$\oint \mathbf{E} \cdot d\mathbf{A} = Q/\varepsilon_0$	C = capacitance
$U_E = qV = \frac{1}{4\pi\varepsilon_0}\frac{q_1 q_2}{r}$	**d** = distance
$\mathbf{E} = -\frac{dV}{d\mathbf{r}}$	**E** = electric field
$V = \frac{1}{4\pi\varepsilon_0}\Sigma\frac{(q_1)}{\mathbf{r}_1}$	ε = emf
$C = \frac{Q}{V}$	**F** = force
$C = \frac{\kappa\varepsilon_0 A}{\mathbf{d}}$	I = current
$C_{PAR} = C_1 + C_2 + \ldots C_n$	L = inductance
$\frac{1}{C_{SER}} = \frac{1}{C_1} + \frac{1}{C_2} + \ldots\frac{1}{C_n}$	l = length
$I = \frac{dQ}{dt}$	n = number of turns
$UC = \frac{1}{2}QV = \frac{1}{2}CV^2$	P = power
$R = \rho\,(l/A)$	Q = charge
$V = IR$	q = point charge
$R_{SER} = R_1 + R_2 + \ldots R_n$	R = resistance
$\frac{1}{R_{PAR}} = \frac{1}{R_1} + \frac{1}{R_2} + \ldots\frac{1}{R_n}$	**r** = distance, displacement
$P = IV$	t = time
$\mathbf{F}_B = q\mathbf{v} \times \mathbf{B} = q\mathbf{vB}\sin\Theta$	U = potential (stored) energy
$\Phi\mathbf{B} \cdot d\mathbf{l} = \mu_0 I$	V = electric potential difference
$F = \int I\,dl \times B$	**v** = velocity
$\mathbf{B}_s = \mu_0 nI$	v = speed
$\phi_M = \int \mathbf{B} \cdot dA$	ρ = resistivity
$\varepsilon_{AVE} = \frac{d\phi_M}{dt}$	ϕ_M = magnetic flux
$\varepsilon = -L\frac{dI}{dt}$	κ = dielectric constant
$U_L = \frac{1}{2}LI^2$	

Geometry and Trigonometry

Rectangle

$A = bh$

Triangle

$A = \frac{1}{2}bh$

Circle

$A = \pi r^2$

$C = 2\pi r$

Paralelopiped

$V = lwh$

Cylinder

$V = \pi r^2 h$

$S = 2\pi rh + 2\pi r^2$

Sphere

$V = \frac{4}{3}\pi r^3$

$S = 4\pi r^2$

Right Triangle

$a^2 + b^2 = c^2$

$\sin \Theta = a/c$

$\cos \Theta = b/c$

$\tan \Theta = a/b$

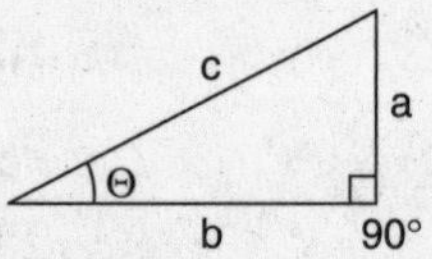

A = area
C = circumference
V = volume
S = surface area
b = base
h = height
l = length
w = width
r = radius

Calculus

$$\frac{df}{dx} = \frac{df}{du}\frac{du}{dx}$$

$$\frac{d}{dx}\left(x^n\right) = nx^{n-1}$$

$$\frac{d}{dx}\left(e^x\right) = e^x$$

$$\frac{d}{dx}(\ln x) = \frac{1}{x}$$

$$\frac{d}{dx}(\sin x) = \cos x$$

$$\frac{d}{dx}(\cos x) = -\sin x$$

$$\int x^n\,dx = (1/n+1)\,x^{n+1},\ n \neq -1$$

$$\int e^x\,dx = e^x$$

$$\int (dx/x) = \ln|x|$$

$$\int \cos x\,dx = \sin x$$

$$\int \sin x = -\cos x$$

GO ON TO THE NEXT PAGE

CliffsAP Physics C Practice Exam 2

Physics C
Section II, Mechanics

Time – 45 Minutes

3 Questions

Directions: Answer all three questions. The suggested time is approximately 15 minutes for answering each question, each of which is worth 15 points. The parts within a question may *not* have equal weight.

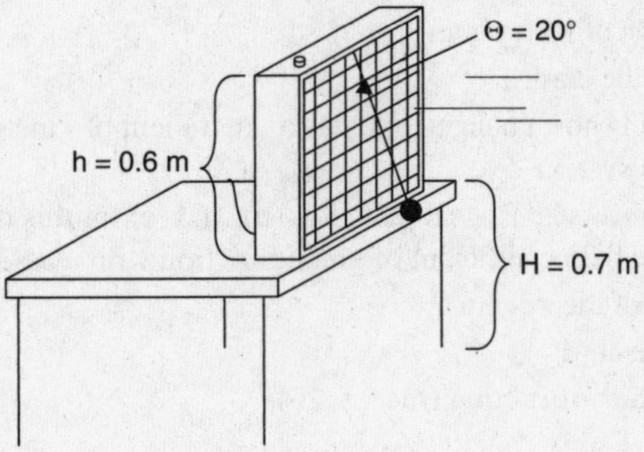

Mech. 1. A small ball of mass 0.15 kg is attached to a 0.6 m long string of negligible mass. The string and ball are attached to a fan 0.6 m in height. The fan sits on a table that is 0.7 m high. The fan is turned on to medium speed and the air from the fan blows the ball outward to an angle Θ as shown.

(a) Draw and label all forces acting on the ball.

(b) Determine the force of air on the ball if $\Theta = 20°$.

(c) How far from the fan is the ball?

(d) The string breaks. The fan exerts a horizontal force on the ball for 0.39 seconds. How far from the bottom of the table does the ball land?

(e) What is the velocity of the ball as it hits the floor?

GO ON TO THE NEXT PAGE

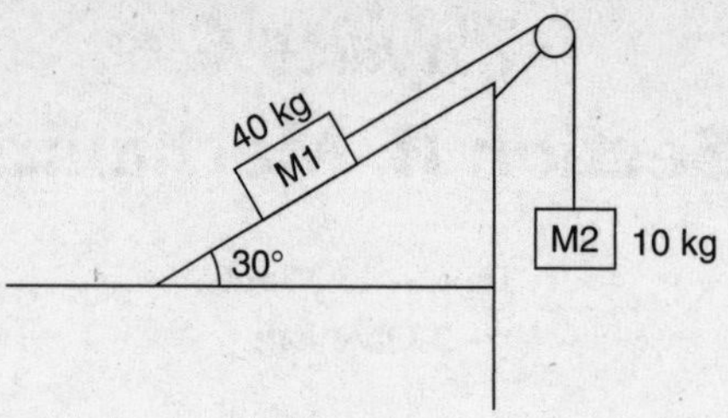

Mech. 2. A 40 kg block M1 sits on a 30° incline. A massless cord attaches it to a 10 kg mass M2, which hangs as shown. The cord passes over a frictionless pulley and the surface of the incline is frictionless as well.

(a) Determine the acceleration of the system.

(b) Determine the tension in the cord.

(c) The surface of the incline is now changed and has a coefficient of kinetic friction of 0.25 with mass M1. Describe the result on the system.

(d) The two masses are now reversed. Repeat parts (a), (b), and (c) for this new situation, assuming a frictionless surface in A and B and the 0.25 coefficient of kinetic friction with mass M2 in part (c). These will include

i. The new acceleration of the system

ii. The new tension in the cord

iii. The result of the addition of friction ($m_K = 0.25$)

GO ON TO THE NEXT PAGE

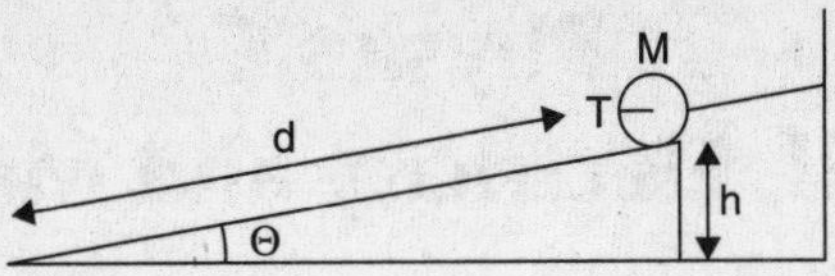

Mech. 3. A solid sphere of mass M and radius r is held motionless by a string on an incline that makes an angle Θ with the horizontal. The sphere is a vertical distance h from the ground.

Determine all answers in terms of M, r, h, d, **g**, and Θ. (Units are kg, m, m, m, m/s^2 and °, respectively.)

(a) Draw and label all forces acting on the sphere.

(b) What is the tension in the string?

The string is cut and the sphere rolls without slipping to the end of the incline, a distance d.

(c) If the rotational inertia I of the sphere is $\left(\frac{2}{5}\right)mr^2$, determine the linear velocity of the sphere at the bottom of the incline.

(d) Determine the sphere's rotational acceleration.

(e) Give an expression for the amount of time it takes for the sphere to reach the bottom of the incline.

GO ON TO THE NEXT PAGE

Physics C
Section II, Electricity and Magnetism

Time – 45 Minutes
3 Questions

Directions: Answer all three questions. The suggested time is approximately 15 minutes for answering each question, each of which is worth 15 points. The parts within a question may NOT have equal weight.

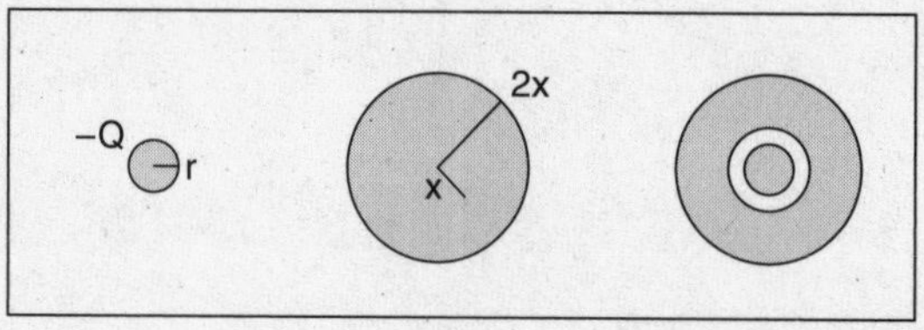

E&M. 1. A charged, solid metallic sphere of radius r and total charge –Q is placed inside an uncharged, solid metallic spherical shell having an inner radius x and an outer radius 2x.

(a) In terms of fundamental constants and the given quantities, give an expression for the potential outside the small sphere before being placed inside the shell.

(b) After the small sphere is placed inside the metallic shell, indicate the induced charge on the shell's inside and outside surfaces.

(c) Give an expression for the electric field at some distance 2x–r between the inner sphere and metallic shell.

(d) Determine the potential of the inner sphere relative to the ground, or at a distance >2x.

GO ON TO THE NEXT PAGE

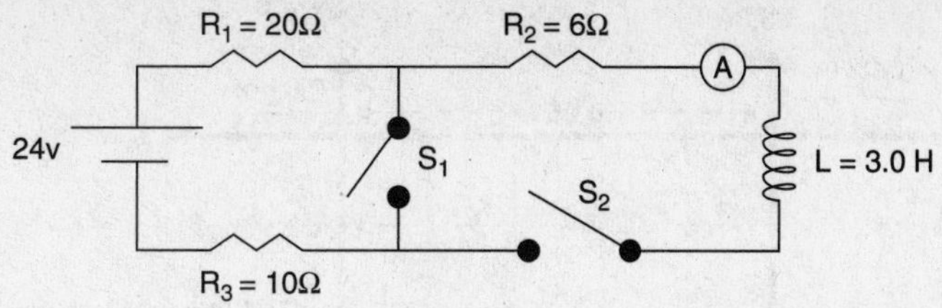

E&M. 2. A 24 V source of emf is connected to three resistors, an inductor, and two switches as shown. Switch S_1 separates part of the circuit.

(a) With switch S_1 closed and with switch S_2 in the open position, determine the
   i. current in the circuit.
   ii. power dissipated by the circuit.

(b) Switch S_1 is now opened, and switch S_2 is closed. At a time when there is maximum current through the circuit, determine
   i. the reading in the ammeter.
   ii. the potential drop across the 10 Ω resistor.
   iii. the energy stored in the magnetic field of the inductor.

GO ON TO THE NEXT PAGE

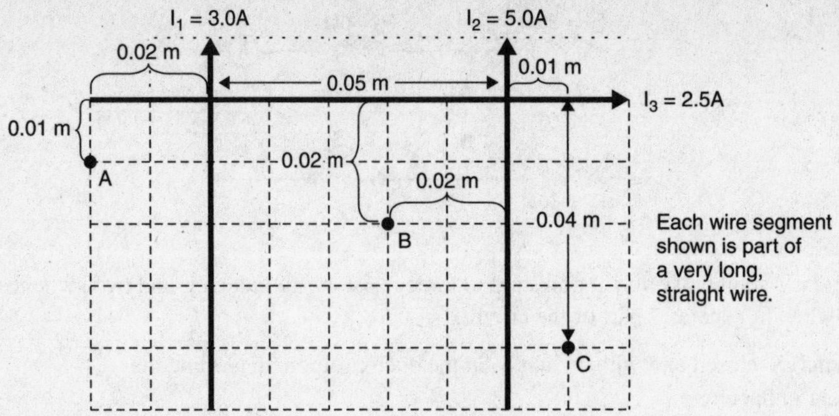

E&M. 3. Three long, straight, coplanar current-carrying wires are also coplanar with points A, B, and C. $I_1 = 3.0A$, $I_2 = 5.0A$, and $I_3 = 2.5A$.

Point A lies 0.02 m from I_1, 0.07 m from I_2, and 0.01 m from I_3.

Point B lies 0.03 m from I_1, 0.02 m from I_2, and 0.02 m from I_3.

Point C lies 0.06 m from I_1, 0.04 m from I_2, and 0.01 m from I_3.

Air separates the wires.

(a) Determine the strength and the direction of the magnetic field at
  i. Point A
  ii. Point B
  iii. Point C

(b) Determine the magnitude and the direction of the force between the wires carrying I_1 and I_2 on a 0.06 m wire segment.

(c) Determine the magnitude and direction of the force between the wires carrying I_1 and I_3.

End of Examination

CliffsAP Physics C Practice Exam 2

Sample C Exam #2 Answers and Comments

Mechanics

Multiple Choice

1. **(D)** $\mathbf{a} = \frac{dv}{dt} = 14t$

   Taking the antiderivative: $\int 14t\,dt = (14)\left(\frac{1}{2}\right)(t^2) = 7\,t^2$ (This is the velocity.)

   Taking the antiderivative of the velocity to get the displacement:

   $$\int 7t^2\,dt = (7)\left(\frac{1}{3}\right)(t^2) = \left(\frac{7}{3}\right)t^3$$

   Substituting 3 seconds for t results in the displacement of **63 m.**

2. **(C)** $\sin 30° = \frac{\mathbf{v_{O\,VERTICAL}}}{100\text{ m/s}}$

   $$\mathbf{v_{O\,VERTICAL}} = 50\text{ m/s}$$

   Using $\mathbf{v_F^2 = v_O^2 + 2ad}$, $(50\text{ m/s})^2 = 0 + (2)(10\text{ m/s}^2)(\mathbf{h})$

   $$\mathbf{h = 125\text{ m}}$$

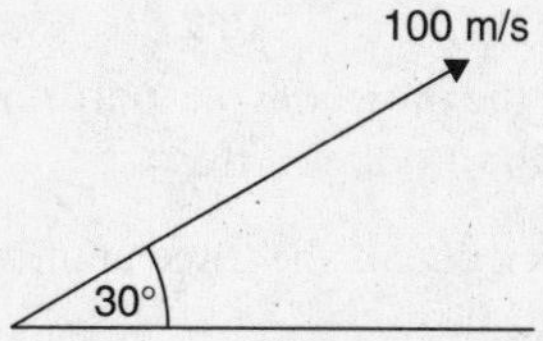

3. **(B)** Using $\mathbf{T = 2\pi(1 / g)^{\frac{1}{2}}}$,

   $$T = 2\pi(s\,/g)^{\frac{1}{2}}$$

4. **(C)** For any object in a circular orbit around a planet, its weight equals the centripetal force exerted on it by the planet. $\mathbf{F} = mg = \frac{mv^2}{r}$ and $mg_D = \frac{mv_D^2}{D}$

   Since $\mathbf{a} = \frac{F}{m}$, for gravitational acceleration: $\mathbf{g} = \frac{GM_E}{D^2}$

   $\frac{GM_E}{D^2} = \frac{v_D^2}{D}$ can now be used in a ratio for D to 2D:

   $$\frac{GM_E\,/D = v_D^2}{GM_E\,/2D = v_{2D}^2} \text{ yields } v_D^2 = \frac{v_D}{(2)^{\frac{1}{2}}}$$

5. **(D)** Using $\mathbf{\Sigma mv_O = \Sigma m_{FINAL}v_{FINAL}}$,

   $$(3\text{ kg})(2\text{ m/s E}) + 0 = (8\text{ kg})(\mathbf{v_{FINAL}})$$

   $$v_{FINAL} = 6/8\text{ m/s or } \mathbf{0.75\text{ m/s}}$$

6. **(C)** The kinetic energy of the blocks is completely transferred to the spring at the point where the system stops. $\mathbf{E_K = E_{P\,SPRING}}$

   $$\left(\frac{1}{2}\right)mv^2 = \left(\frac{1}{2}\right)kx^2$$

   $$\mathbf{x} = (mv^2/k)^{\frac{1}{2}} = [(8\text{ kg})(0.75\text{ m/s})^2 / 8\text{ N / m})]^{\frac{1}{2}} = \mathbf{0.75\text{ m}}$$

7. **(A)** The total kinetic energy of the disk will be completely transformed into gravitational potential when it comes to a momentary stop up the incline.

$$E_{K\ ROTATIONAL} + E_{K\ LINEAR} = E_{P\ GRAVITATIONAL}$$

$$\left(\frac{1}{2}\right)(I)(\omega^2) + \left(\frac{1}{2}\right)mv^2 = mgh$$

$$\left(\frac{1}{2}\right)\left(\frac{1}{2}\right)(mr^2) + \left(\frac{1}{2}\right)mv^2 = mgh$$

$$\left(\frac{1}{4}\right)v^2 + \left(\frac{1}{2}\right)v^2 = gh$$

$$h = \frac{3v^2}{4g}$$

8. **(E)** The planet's angular momentum remains constant at all points in its orbit because although its linear speed may change, its angular speed does not. As the planet moves away from the Sun, its linear speed decreases due to decreased gravitational attraction.

9. **(D)** Since $F(x) = -\frac{dU(x)}{dx}$,

$$-dU(x) = F(x)dx$$

$$\int -dU(x) = \int F(x)dx$$

$$-U(x) = \int (20x - 8x^2\,dx)$$

$$U(3m) = -(90 - 72)\ J = -18\ J \text{ or a stretched difference of } 18\ J$$

10. **(D)** At point P, the apex of its trajectory, the projectile has only horizontal velocity and, "since gravity is never turned off," the particle's acceleration is toward the earth.

11. **(A)** The force applied directly to the block equals the force applied to the rope **F minus** the force needed to accelerate the rope of mass m. This results in

$F_{BLOCK} = F - ma$

12. **(B)** At the maximum speed at the top of the bump, the car's weight would equal the centripetal force. $mg = \frac{mv^2}{r}$

$$v = (gr)^{\frac{1}{2}} = [(10\ m/s^2)(0.5\ m)]^{\frac{1}{2}} = 5^{\frac{1}{2}}\ m/s \text{ (closest to 2 m/s)}$$

13. **(E)** By the Law of Conservation of Momentum, both linear and angular momentum are always conserved. This means that in the absence of any outside applied force or torque, the total momentum does not change. Since $E_{K\ ROTATIONAL} = \left(\frac{1}{2}\right)I\omega^2$, changing the mass distribution by changing arm position only changes I, which in turn changes the angular velocity ω.

14. **(A)** $\cos \Theta/2 = \frac{Mg/2}{T}$

$$T = \frac{Mg}{2\cos\Theta/2}$$

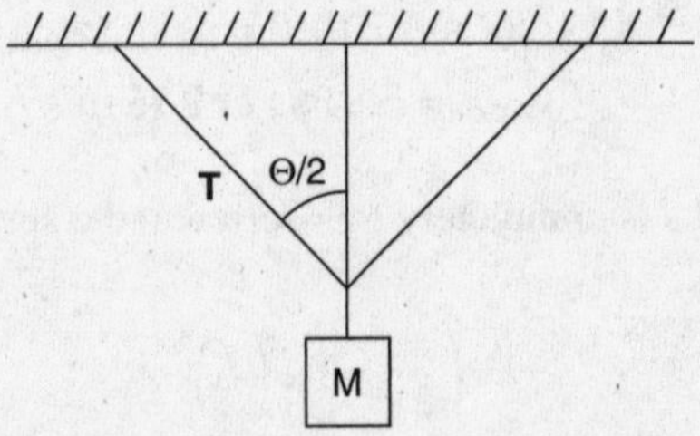

15. (B) $T = \frac{2\pi}{\omega}$, $\omega = 4$

$$T = \frac{2\pi}{4} = \frac{\pi}{2} \text{ seconds}$$

16. (D) Using $\mathbf{v} = \boldsymbol{\omega}\mathbf{r}$

$$v = (4 \text{ rad/sec})(2 \text{ m/rad}) = 8 \text{ m/s}$$

17. (E) Using $\boldsymbol{\tau} = \mathbf{I}\boldsymbol{\alpha} = \mathbf{Fr}$

$$\tau = I\frac{(0-\omega)}{t} = \frac{-I\omega}{t}$$

18. (B) Using $P = \frac{W}{t} = \frac{Fd}{t} = \frac{\tau\Theta}{t}$

$$= \left(\frac{-I\omega}{t}/t\right)\left(\omega t + \frac{1}{2}\alpha t^2\right)$$

$$= \frac{-I\omega}{t^2}\left(\omega t + \left(\frac{1}{2}\right)\left(0\frac{-\omega}{t}\right)\left(t^2\right)\right)$$

$$= \frac{-I\omega}{t^2}\left(\omega t - \frac{\omega t}{2}\right)$$

$$= \frac{-I\omega}{t^2}\left(\frac{\omega t}{2}\right) = \frac{-I\omega^2}{2t}$$

19. (D) Setting the compressed energy in the spring equal to the kinetic energy:

$$\left(\frac{1}{2}\right)\mathbf{kx^2} = \left(\frac{1}{2}\right)\mathbf{mv^2}$$

$$\mathbf{kx^2 = mv^2}$$

$$\mathbf{x = (mv^2 / k)^{\frac{1}{2}}}$$

20. (B) Since the see-saw must be in rotational equilibrium, the sum of the clockwise torques must equal the sum of the counterclockwise torques:

$$\Sigma\tau_{CW} = \Sigma\tau_{CCW}$$

$$\Sigma\tau_{CW} = \mathbf{r} \times \mathbf{F}$$

$$\Sigma\tau_{CW} = \left(\frac{3}{2}\right)(5 \text{ N})(\sin 30°)$$

$$(10 \text{ N})(x) = \left(\frac{15}{4}\right) \text{m}$$

$$x = \left(\frac{3}{8}\right) \text{m right of center}$$

21. (D) For the system to be in translational equilibrium, the sum of the upward (supporting) forces must equal the sum of the downward forces.

$$\Sigma F_{UP} = \Sigma F_{DOWN}$$

$$\Sigma F_{UP} = (50 \text{ N}) + (10 \text{ N}) + (5 \text{ N})(\sin 30°) = \mathbf{62.5 \text{ N}}$$

22. (B) Using $\mathbf{W = F \cdot d}$,

$$d = \frac{W}{F} = \frac{2000 \text{ J}}{(100 \text{ kg})(10 \text{ m/s}^2)} = 2 \text{ m}$$

23. (B) Setting $F_C = F_G$, $\frac{mv^2}{2R} = \frac{GMm}{4R^2}$ yields $v^2 = \frac{GM}{2R}$

$$E_K = \frac{1}{2}mv^2 = \frac{1}{2}m\left(\frac{GM}{2R}\right) = \frac{GMm}{4R}$$

24. (B) Using ($\mathbf{F_{NET} = F_{WT\,4KG} - F_F}$)

$$= (4\text{ kg})(10\text{ m/s}) - \mu \mathbf{F_N}$$
$$= (40\text{ N}) - (0.3)(3\text{ kg})(10\text{ m/s}^2) = 40\text{ N} - 9\text{ N} = 31\text{ N of net force}$$

Now using $\mathbf{F_{NET} = ma}$ to find the acceleration of the system, $\mathbf{a = F_{NET} / m}$. (Here, m is the sum of the masses in the system.)

$$\mathbf{a} = (31\text{ N}) / (3\text{ kg} + 4\text{ kg}) = 4.4\text{ m/s}^2 \text{ (closest to } 4.5\text{ m/s}^2)$$

25. (B) The net force on M1 is what causes it to accelerate:

$$\mathbf{F_{NET}} = M_1\mathbf{a_1} = (3\text{ kg})(4.4\text{ m/s}^2) = 13.2\text{ N}$$

26. (B) Doubling the friction coefficient to 0.6 means that the frictional force is doubled:

$$\mathbf{F_{NET} = F_{WT\,M2} - F_{F\,M1}} = 40\text{ N} - (0.6)(3\text{ kg})(10\text{ m/s}^2) = 40\text{ N} - 18\text{ N} = 22\text{ N}$$

Now calculating the new acceleration:

$$\mathbf{a = F_{NET} / m_{TOT}} = 22\text{ N} / 7\text{ kg} = 3.1\text{ m/s}^2 \text{ (closest is } 3\text{ m/s}^2)$$

27. (C) Using CW and CCW torques: (let x be the distance from the 8 kg mass)

$$\text{CCW torque} = \text{CW torque}$$
$$(2.5\text{kg})(10\text{ m/s}^2)(100\text{ cm} - x) = (4\text{ kg})(10\text{ m/s}^2)(x)$$
$$\mathbf{x = 38\text{ cm}}\textbf{, position C}$$

28. (E) The net force on the system is the weight of the 10 kg mass MINUS the parallel force of the 5 kg mass. $\mathbf{F_{NET}} = (M2)(g) - \mathbf{F_{P\,M1}}$

$$= (10\text{ kg})(10\text{ m/s}^2) - (5\text{ kg})(10\text{ m/s}^2)(\sin 30°)$$
$$= 100\text{ N} - 25\text{ N} = \mathbf{75\text{ N}}$$

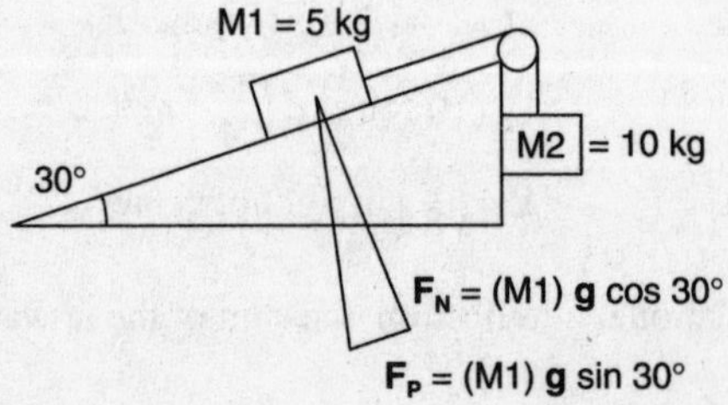

29. (B) Using $\mathbf{F_{NET} = ma}$,

$$\mathbf{a} = \frac{\mathbf{F_{NET}}}{m_{TOTAL}} \text{ (Both masses will accelerate at the same rate.)}$$
$$\mathbf{a} = (75\text{ N}) / (15\text{ kg}) = 5\text{ m/s}^2$$

30. (C) Again, using $\mathbf{F_{NET} = ma}$,

$$\mathbf{a} = \frac{\mathbf{F_{NET}}}{m_{TOTAL}} \text{ (Both masses will accelerate at the same rate.)}$$
$$\mathbf{a} = \frac{(10\text{ kg})(10\text{ m/s}^2) - (10\text{ kg})(10\text{ m/s}^2)(\sin 30°)}{20\text{ kg}}$$
$$= 50\text{ N} / 20\text{ N} = 2.5\text{ m/s}^2$$

31. (B) Using $\mathbf{d_{VERT}} = (\mathbf{v_{O\ VERT}})(\mathbf{t}) + \left(\frac{1}{2}\right)\mathbf{at^2}$,

$$\mathbf{h} = (0) + (\mathbf{g}/2)(\mathbf{t}^2) \text{ and } \mathbf{t} = (\mathbf{2h} / \mathbf{g})^{\frac{1}{2}}$$

32. (B) First, finding the net force on the system:

$$\begin{aligned}\mathbf{F_{NET}} &= \mathbf{mg_{5\,kg}} - \mathbf{F_{F\,2+3\,kg}} \\ &= (5\text{ kg})(10\text{ m/s}^2) - \mu\mathbf{F_{N\,2+3\,kg}} \\ &= (50\text{ N}) - (0.6)(5\text{ kg})(10\text{ m/s}^2) \\ &= 20\text{ N}\end{aligned}$$

Using $\mathbf{F_{NET}} = \mathbf{ma}$, (The entire system will be accelerated together.)

$$\mathbf{a} = \frac{\mathbf{F_{NET}}}{\mathbf{m}} = \frac{20\text{ N}}{10\text{ kg}} = 2\text{ m/s}^2$$

33. (D) $\mathbf{F_{NET}} = \mathbf{mg_{5\,kg}} - \mathbf{F_{F\,3\,kg}}$

$$= (50\text{ N}) - (0.6)(3\text{ kg})(10\text{ m/s}^2) = 32\text{ N}$$

Using $\mathbf{F_{NET}} = \mathbf{ma}$, (The entire system will be accelerated together.)

$$\mathbf{a} = \frac{\mathbf{F_{NET}}}{\mathbf{m}} = \frac{32\text{ N}}{8\text{ kg}} = 4\text{ m/s}^2$$

34. (E) $\tan 30° = \dfrac{48\text{ km/hr}}{\mathbf{v_{VERT}}}$

$$\mathbf{v_{VERT}} = \frac{(48\text{ km/hr})}{\left(3^{\frac{1}{2}}\big/3\right)} \cdot \frac{(1\text{ hr})}{(3600\text{ sec})} \cdot \frac{(1000\text{ m})}{(\text{km})} = \frac{40}{(3)^{\frac{1}{2}}} \text{ or } \frac{40}{\sqrt{3}}$$

35. (E) A smaller spring constant means that the spring exerts less force per stretch or compression distance than one of greater k. This translates into fewer Newtons of force per meter of stretch or compression. The result is a spring being "looser" or less stiff, which allows for a longer oscillation period.

Electricity and Magnetism

Multiple Choice

36. (D) Since the **series** combination of 3 Ω resistors is in series with the parallel 3 Ω resistors, the equivalent resistance of the circuit equals 9 Ω + the parallel resistors:

$$\frac{1}{R_{2,3}} = \frac{1}{3\Omega} + \frac{1}{3\Omega} = \frac{2}{3\Omega} \text{ and } R_{2,3} = 1.5\ \Omega$$

The equivalent resistance of the circuit = (9 + 1.5) Ω = 10.5 Ω

37. (A) Using $\mathbf{V} = \mathbf{IR}$,

$$I = V/R = (12\text{ V}) / (10.5\ \Omega) = 1.1\text{ A}$$

38. (C) The current of about 1 A splits evenly to both R_2 and R_3, which both receive about 0.5 A. Therefore, the voltmeter reads:

$$V = IR = (0.5\text{ A})(3\ \Omega) = \text{about } 1.5\text{ V}$$

39. **(C)** Capacitance is governed by the amount of charge that can be stored on parallel plates before electrons jump across and discharge the capacitor. Less dense insulating material between the plates allows fewer charges to build up before passing through to discharge.

40. **(C)** Using the right hand, **pointing** the thumb in the direction of current I_1, the fingers point **into the page**, indicating that the direction of the magnetic field at point P for current I_1 is into the page. **Pointing** the thumb in the direction of current I_2, the fingers point **out of the page**, indicating that the direction of the magnetic field at point P for current I_2 is out of the page. Since $I_1 > I_2$, the resulting magnetic field is **into the page.**

41. **(D)** Using **Q = VC,** plus the fact that the capacitances, being in parallel, are added: Q = (120 V)(1.8 μC) = 216 μC

42. **(C)** Using **Q = VC,** plus the fact that the capacitances, being in series, are added inversely: $\frac{1}{C_{TOT}} = \frac{1}{0.8\,\mu F} + \frac{1}{1.0\,\mu F}$ and $C_{TOT} = \frac{4}{9}\,\mu F$

$$Q = (120\text{ V})\left(\frac{4}{9}\,\mu F\right) = 53.3\,\mu C$$

43. **(B)** Using **U = Vq,**

$$\Delta U = (\Delta V)(q) = q\,(V_2 - V_1) = (+20 \times 10^{-9}\text{C})(0.75\text{ V} - 0.50\text{ V})$$
$$= (+20 \times 10^{-9}\text{C})(0.25\text{ V}) = 5.0 \times 10^{-9}\text{ J}$$

44. **(B)** Since the original charge is now shared, $Q_O = Q_1 + Q_2$

Using **Q = VC,**

$$V_O C_1 = V_{FINAL} C_1 + V_{FINAL} C_2$$
$$V_{FINAL} = V_O \frac{(C_1)}{(C_1 + C_2)} = (12.0\text{ V})\frac{(5.0\,\mu F)}{(5.0\,\mu F + 13.0\,\mu F)} = 3.3\text{ V}$$

45. **(E)** Using $U_C = \left(\frac{1}{2}\right)CV^2$,

$$U_{INITIAL} = \left(\frac{1}{2}\right)(5.0\,\mu F)(12.0\text{ V})^2 = \left(\frac{144}{2}\right)(5.0\,\mu F) = 360\,\mu J$$
$$U_{FINAL} = \left(\frac{1}{2}\right)(5.0\,\mu F + 13.0\,\mu F)(3.3\text{ V})^2 = (9)(11)\,\mu J = 99\,\mu J$$

46. **(E)** Using **U = qV,**

$$V = \frac{U}{q} = \frac{3.0\text{ J}}{2.0 \times 10^{-3}\text{ C}} = \mathbf{1.5 \times 10^3\ V}$$

47. **(D)** Using $\mathbf{C = \frac{\kappa\varepsilon_0 A}{d} = \frac{Q}{V}}$,

Charges increase with plate area increase and/or separation distance increase. Therefore, $\left(\frac{1}{2}\right)$ d doubles charge.

48. **(B)** Using $\mathbf{R = \frac{\rho l}{A}}$ and assigning **r** to the small square and **R** to the large one:

$$\frac{r}{R} = \frac{1}{L} \text{ and } \mathbf{r} = \frac{1R}{L} \text{ (the } \rho \text{ and A for both wires are the same)}$$
$$\mathbf{r = 4s\ R / 8s = \left(\frac{1}{2}\right)R}$$

49. **(E)** Using $\mathbf{F = \frac{kq_1 q_2}{r^2}}$,

$$F = \frac{\left(9 \times 10^9\,\text{Nm}^2/\text{C}^2\right)\left(-1.6 \times 10^{-19}\,\text{C}\right)\left(-1.6 \times 10^{-19}\,\text{C}\right)}{(0.01\,\text{m})^2} = +2.3 \times 10^{-24}\text{ N}$$

50. **(A)** Using $\mathbf{E = \frac{F}{q}}$,

$$E = \frac{+2.3 \times 10^{-24}\text{ N}}{1.60 \times 10^{-19}\text{ C}} = 1.4 \times 10^{-5}\text{ N/C}$$

51. **(D)** Using $C = \frac{\kappa \varepsilon_0 A}{d} = \frac{Q}{V}$

Charges increase with plate area increase and/or separation distance increase. Therefore, $\left(\frac{1}{2}\right)$ d doubles charge.

52. **(B)** Antiparallel currents repel each other due to their parallel and mutually repulsive magnetic fields and their resulting repulsive forces.

53. **(B)**

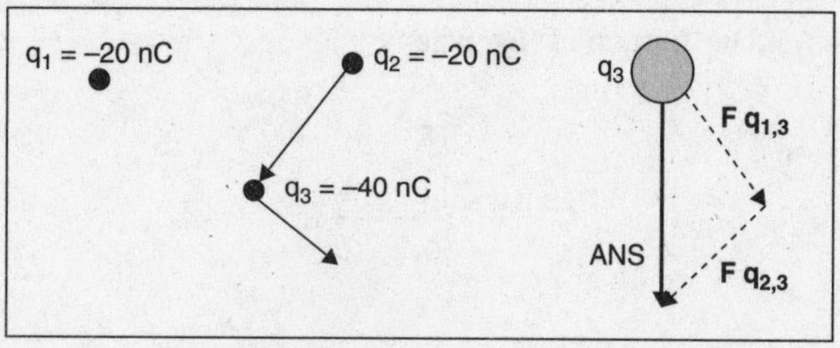

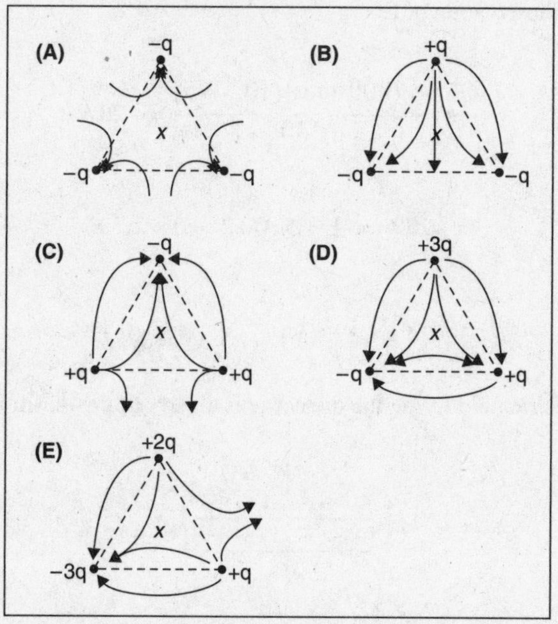

54. **(A)** There are no field lines at the center of A due to symmetrical repulsion.

55. **(B)** Field lines in B point away from the positive on top toward the two negatives on the bottom, through their center as well.

56. **(C)** When the switch is closed, current flows to the capacitor and the maximum current flows out immediately. After awhile, the charge buildup on the incoming plate allows no more charge to be repelled from the opposite plate. As a result, current ceases to flow and maximum potential difference is reached.

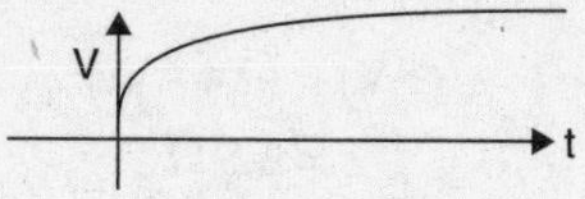

57. **(B)** **(Refer to the previous answer.)**

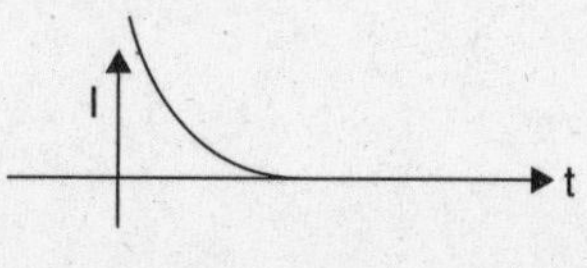

58. **(D)** Using $\mathbf{F_B = qvXB}$, the charge magnitude, speed, entry angle into the field and the magnetic field strength determine the resulting force on the charge.

59. **(D)** Using the right hand and pointing the thumb in the direction of each current, the fingers point in the direction of the magnetic field. For $\mathbf{I_1}$ they point in the negative **x** direction. For $\mathbf{I_2}$ they point in the positive **x** direction. Since $\mathbf{I_1 > I_2}$, the resultant is in the **negative x direction.**

60. **(D)** Using $\mathbf{F = qvXB}$ and $\mathbf{F = I\,X\,B}$

For negative q or **I, F** is toward the **bottom of the page.**

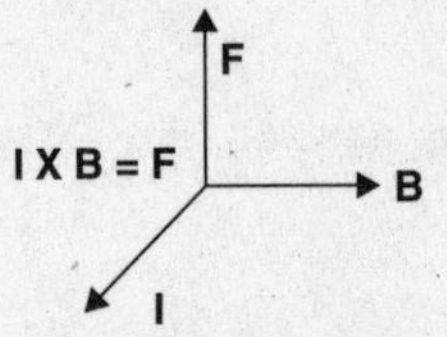

61. **(E)** The JUNCTION LAW is a restatement of the Law of Conservation of Charge. The POTENTIAL LAW is a restatement of the Law of Conservation of Energy.

62. **(A)** Using $\varepsilon = -N\frac{d\Phi_B}{dt}$,

$$\varepsilon = -\frac{(400\text{ turns})(10^{-2}\text{ Wb})}{(0.20\text{ sec})} = \mathbf{-20\ V}$$

63. **(B)** Using $\varepsilon = -\frac{LdI}{dt}$,

$$-20\text{ V} = -L(1.5\text{ A}/0.2\text{ sec}) = 2.7\text{ H}$$

64. **(B)** Using $U_L = \frac{1}{2}LI^2$,

$$U_L = \left(\frac{1}{2}\right)(2.7\text{ H})(1.5\text{ A})^2 = \mathbf{3.0\ J}$$

65. **(C)** The direction of the electric field inside the dielectric is always **opposite the external field.**

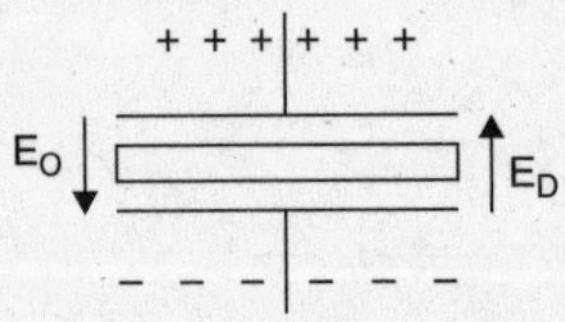

66. **(B)** The inductor begins to transform electrical energy into radiant energy and builds up its magnetic field slowly, while opposing the current through it. When the magnetic field is at its maximum, so is the potential across it.

67. **(B)** Applying the right hand rule, pointing the thumb in the direction of of each current, the fingers point **to the left for both currents.**

68. **(D)** Inside the wire, the magnetic field intensity **B** rises directly with distance from the central axis to its surface, where $\mathbf{R \le r}$. Outside the wire where $\mathbf{R} \ge r$, **B** diminishes inversely with radial distance.

69. **(C)** Using **V = IR**, and the fact that there is a difference of 1V in the battery when in operation,

$$(1\text{ V}) = (0.4\text{ A})(r)$$

$$\mathbf{r = 2.5\ \Omega}$$

70. **(B)** Since $\mathbf{I_{MAX} = V / R}$, the current decay is exponential.

Mechanics

Free Response

Mech. 1. **(a)**

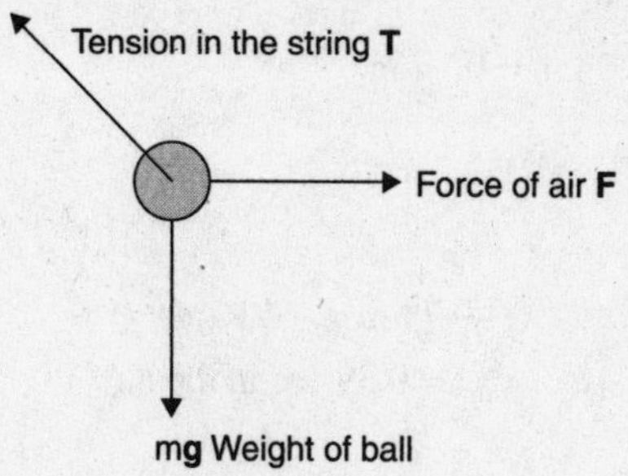

(b) $\tan 20° = \frac{F_{AIR}}{mg}$

$$F_{AIR} = (0.364)(0.15\text{ kg})(9.8\text{ m/s}^2)$$
$$= \mathbf{0.54\ N}$$

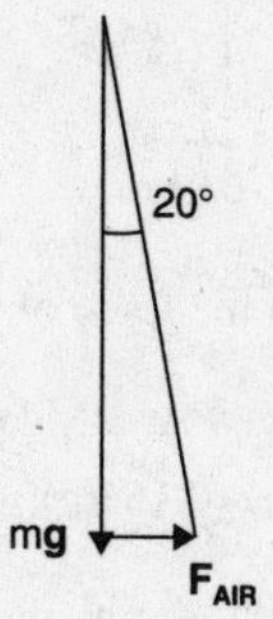

(c) $\sin 20° = \frac{x}{0.6\text{ m}}$

$$x = (0.342)(0.6\text{ m}) = \mathbf{0.21\ m}$$

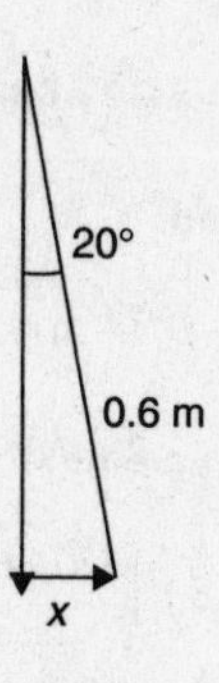

(d) First, find the total height of the ball:

$$\mathbf{y} = (0.6)(\cos 20°) = 0.56\,\text{m}$$

The table's height = 0.7 m.

The ball rises 0.7 m – 0.56 m = 0.04 m

The total height of the ball = 0.7 + 0.04 m = 0.74 m

Find the time the ball is in the air:

Using $\mathbf{d} = \mathbf{v_O}t + \left(\frac{1}{2}\right)\mathbf{a}t^2$

$$-0.74\text{ m} = -4.9\text{ m/s}^2\,(t^2)$$

$$t = 0.39\text{ sec in the air}$$

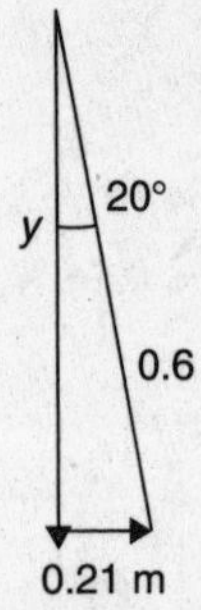

To find the velocity of the ball horizontally from the force of the fan:

$$\mathbf{F}\Delta t = \mathbf{M}\Delta\mathbf{v}$$

$$(0.54\text{ N})(0.39\text{ sec}) = (0.15\text{ kg})(\Delta v)\text{ yields } v_H = 1.4\text{ m/s}$$

Using $\mathbf{d} = \mathbf{v_O}t + \left(\frac{1}{2}\right)\mathbf{a}t^2$ (for the horizontal distance traveled)

$$\mathbf{d_H} = (1.4\text{ m/s})(0.39\text{ sec}) + 0 = 0.55\text{ m}$$

(e) First, for the final horizontal velocity:

$$= 1.4\text{ m/s} = \mathbf{v_F}\text{ horizontally.}$$

For the final vertical velocity: Using $\mathbf{v_F}^2 = \mathbf{v_O}^2 + \mathbf{2ad}$:

$$\mathbf{v_F}^2 = 0 + (2)(-9.8\text{ m/s}^2)(-0.74\text{ m})\text{ and }\mathbf{v_F} = -3.8\text{ m/s vertically}$$

The ball lands with a final speed of $[(1.4\text{ m/s})^2 + (-3.8\text{ m/s})^2]^{\frac{1}{2}} = 4.0\text{ m/s}$.

$$\Theta\text{: } \tan\Theta = \left|\frac{1.4}{3.8}\right| = 0.368\text{ and }\Theta = 20°$$

The ball lands with a speed of 4.0 m/s at an angle of 20° with the vertical.

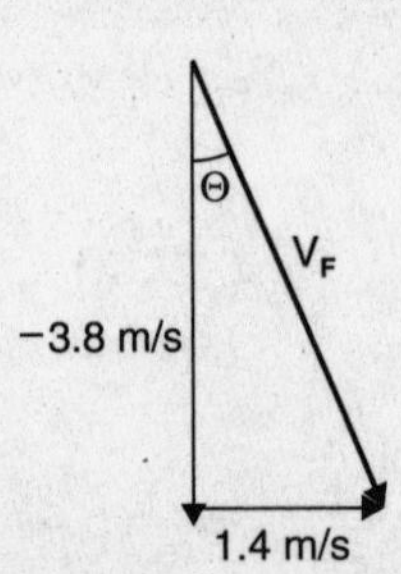

Mech. 2. **(a)** Acceleration is due to the net force acting on the sum of the masses. The net force is the difference between the **PARALLEL FORCE** pulling the M1 block **DOWN THE RAMP (to the left)** and the **WEIGHT** of the M2 block pulling the system **to the right.**

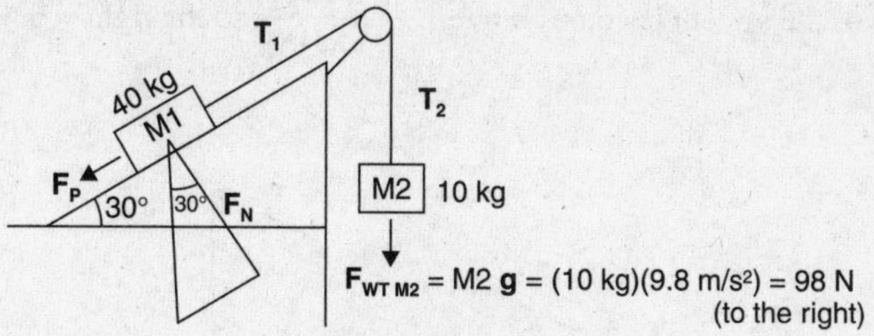

$$\mathbf{F_{P\,M1} = (M1)\ g\ \sin 30°}$$
$$= (40\text{ kg})(9.8\text{ m/s}^2)(0.5)$$
$$= 196\text{ N } \textbf{(to the left)}$$

The NET FORCE is 196 N – 98 N = 98 N **to the left**

Using $\mathbf{F_{NET} = ma}$,

$$\mathbf{a} = \frac{\mathbf{F_{NET}}}{\text{m}} = \frac{(98\text{ N})}{(40+10)\text{ kg}} \text{ to the left} = \mathbf{1.96\ m/s^2 \text{ or } 2.0\ m/s^2} \textbf{ to the left or DOWN THE INCLINE}$$

(b) The tension in the string is the weight plus the accelerating force on M2:

$$\mathbf{T} = (10\text{ kg})(9.8\text{ m/s}^2) + (10\text{ kg})(2.0\text{ m/s}^2) = 98\text{N} + 20\text{N} = 118\text{ N} \approx 120\text{ N}$$

(c) Adding friction to the incline adds a force that retards any motion. Already, there are forces of 196 N **to the left** and 98 N **to the right.** Finding the frictional force: Using $\mu = \mathbf{F_F / F_N}$,

$$\mathbf{F_F} = \mu\ \mathbf{F_N} = (0.25)(40\text{ kg})(9.8\text{ m/s}^2)(\cos 30°) = 85\text{ N}$$

This opposes motion down the incline. The net force now becomes

$$\frac{\mathbf{F_{NET}}}{\text{m}} = \mathbf{F_{P\,M1} - (F_{WT\,M2} + F_F)}$$
$$\mathbf{= 196\ N - 98\ N - 85\ N = 13\ N} \textbf{ to the left}$$

The new acceleration of the system becomes

$$\mathbf{a} = \frac{\mathbf{F_{NET}}}{\text{m}} = \frac{(13\text{ N})}{(40+10)\text{ kg}} \text{ to the left} = \mathbf{0.26\ m/s\ 2} \textbf{ to the left or DOWN THE INCLINE}$$

(d) i. $\mathbf{F_P} = (10\text{ kg})(9.8\text{ m/s}^2)(\sin 30°) = 49\text{ N}$ **(to the left)**

$$\mathbf{F_{WT\,M2} = M2\ g} = (40\text{ kg})(9.8\text{ m/s}^2) = 392\text{ N } \textbf{(to the right)}$$

$$\mathbf{a} = \frac{\mathbf{F_{NET}}}{\text{m}} = \frac{(392-49)\text{ N}}{50\text{ kg}} = \mathbf{6.9\ m/s^2} \textbf{ to the right.}$$

ii. $\mathbf{T_1 = T_2} = 40\text{ kg } \mathbf{(g - a)} = 40\text{ kg } (9.8\text{ m/s}^2 - 6.9\text{ m/s}^2)$

$$= 40\text{ kg } (2.9\text{ m/s}^2)$$
$$= 116\text{ N}$$

iii. $\mathbf{F_F} = \mu\ \mathbf{F_N} = (0.25)(10\text{ kg})(9.8\text{ m/s}^2)(\cos 30°) = 21\text{ N}$

$$\mathbf{F_P\ 10\ KG} = (10\text{ kg})\ \mathbf{g}\ \sin 30° = 49\text{ N } \textbf{to the left}$$
$$\mathbf{F_{WT}\ 40\ KG} = (40\text{ kg})(9.8\text{ m/s}^2) = 392\text{ N } \textbf{to the right}$$

Since the much larger force is now to the right, friction acts to the left:

$$\frac{F_{NET}}{m} = \mathbf{392\ N - 49\ N - 21\ N = 322\ N\ to\ the\ right}$$

The new acceleration of the system becomes $\mathbf{a} = \frac{F_{NET}}{m} = \frac{(322\,N)}{(50)\,kg}$ to the right = **6.4 m/s² to the right**

Mech. 3. **(a)**

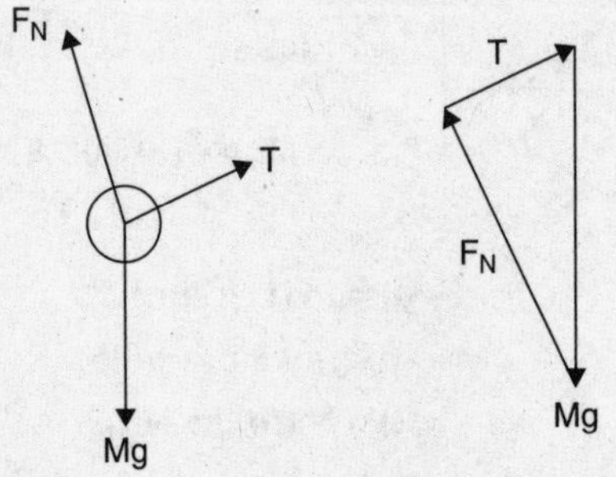

(b) $\mathbf{T = F_P = Mg \sin \Theta}$

(c) The kinetic energy the sphere has at the bottom of the incline must be equal to the potential energy it had at the top.

$$\mathbf{E_{K\ TOTAL} = E_P}$$

$$\mathbf{E_{K\ LINEAR} + E_{K\ ROTATIONAL} = E_P}$$

$$\mathbf{\left(\frac{1}{2}\right)Mv^2 + \left(\frac{1}{2}\right)I\omega^2 = Mgh}$$

$$\left(\frac{1}{2}\right)Mv^2 + \left(\frac{1}{2}\right)\left(\frac{2}{5}\right)(Mr^2)\omega^2 = Mgh$$

$$\frac{v^2}{2} + \frac{r^2}{5}(v / r)^2 = gh$$

$$\left(\frac{7}{10}\right)v^2 = gh \text{ and } \mathbf{v = (10\ gh / 7)^{\frac{1}{2}}}$$

(d) Using $\mathbf{V_f^2 = V_o^2 + 2ad}$:

$$\frac{10}{7}gh = 2(a)(d)$$

The sphere's linear acceleration is $\mathbf{a_{LINEAR}} = \frac{5}{7}$ gh/d and d = h/sin Θ

$$\mathbf{a_{LINEAR}} = \frac{5}{7} g \sin \Theta$$

Since $\boldsymbol{\alpha} = \frac{\mathbf{a}}{r}$, $\boldsymbol{\alpha} = \frac{5g \sin \Theta}{7r}$

(e) Using $\mathbf{d = v_o t} + \left(\frac{1}{2}\right)\mathbf{at^2}$,

$$\mathbf{d} = \left(\frac{1}{2}\right)\mathbf{(a)(t^2)}$$

$$t = (\mathbf{2d/a})^{\frac{1}{2}} = \left(\frac{\mathbf{2d}}{\frac{5}{7}\mathbf{g}\sin\Theta}\right)^{\frac{1}{2}}$$

$$\left(\frac{\mathbf{14d}}{5\,\mathbf{g}\sin\Theta}\right)^{\frac{1}{2}}$$

Electricity and Magnetism

Free Response

E&M. 1. **(a)** Since the sphere is metallic, all the charge is distributed on its outside surface and behaves as if it were a point charge.

Using $\mathbf{V} = \frac{1}{4\pi\varepsilon_0}\frac{q}{r}$,

$$\mathbf{V} = \frac{1}{4\pi\varepsilon_0}\frac{-Q}{r} \text{ or } \frac{-kQ}{r}$$

(b) Since the shell was initially uncharged, the total sum of any induced charges must still equal zero. Since the negative core will repel electrons in the shell, its **inside surface will have an induced charge of +Q and its outside surface will have an induced charge of –Q.**

(c) $\mathbf{E} = \frac{1}{4\pi\varepsilon_0}\frac{Q}{s^2}$ or $\frac{kQ}{s^2}$ or $\frac{kQ}{(2x-r)^2}$

(d) The total potential of the inner sphere, relative to the ground, when surrounded by the shell, is the sphere's potential at its surface minus the potential of the inner surface of the enclosing shell:

$$\mathbf{V_{SPHERE} = V_r - V_x = -\Sigma Es}$$

$$-\int_x^r E \cdot ds = -\int_x^r \frac{kQ}{s^2} ds \text{ and}$$

$$\int \frac{1}{s^2} ds = \frac{-1}{s}$$

$$\mathbf{\Delta V = kQ\,(1/r - 1/x)}$$

$$\text{or } \mathbf{\Delta V} = \frac{V(x-r)}{x}$$

E&M. 2. **(a)** i. Using $\mathbf{V = IR}$,

$$\mathbf{I} = \frac{V}{R} = \frac{24\text{ V}}{30\ \Omega} = \mathbf{0.80\ A}$$

ii. Using $\mathbf{P = IV}$,

$$P = (0.80\text{ A})(24\text{ V}) = \mathbf{19.2\ W}$$

(b) i. Using $\mathbf{V = IR}$,

$$I = \frac{V}{R} = \frac{24\text{ V}}{36\ \Omega} = 0.67\text{ A}$$

ii. Using V = IR,

$$V_{10\Omega} = IR_{10\Omega} = (0.67\text{ A})(10\ \Omega) = 6.7\text{ V}$$

iii. Using $U_L = \frac{1}{2}LI^2$

$$= \left(\frac{1}{2}\right)(1.5\text{ H})(0.67\text{ A})^2 = 0.68\text{ J}$$

E&M. 3. **(a)** Using the Biot-Savart Law application for long, straight wires: $\mathbf{B} = \frac{\mu_0 I}{2\pi r} = \frac{(4\pi \times 10^{7} \text{T}\cdot\text{m/A})(\text{I})}{2\pi r} = \frac{(2\times 10^{-7}\text{T}\cdot\text{m/A})(\text{I})}{r}$ and the right thumb pointed in the current's direction:

i. For point A:

$$\mathbf{B}_{I_1} = \frac{(2\times 10^{7}\text{T}\cdot\text{m/A})(3.0\text{A})}{(0.02\,\text{m})} = 300 \times 10^{-7}\,\text{T} = 3.0\times 10^{-5}\,\text{T out of the page}$$

$$\mathbf{B}_{I_1} = \frac{(2\times 10^{7}\text{T}\cdot\text{m/A})(5.0\text{A})}{(0.07\,\text{m})} = 143 \times 10^{-7}\,\text{T} = 1.4\times 10^{-5}\,\text{T out of the page}$$

$$\mathbf{B}_{I_2} = \frac{(2\times 10^{7}\text{T}\cdot\text{m/A})(2.5\text{A})}{(0.01\,\text{m})} = 500 \times 10^{-7}\,\text{T} = 5.0\times 10^{-5}\,\text{T into the page}$$

The sum of the values and directions for **B** at point A due to all three currents is **6.0 × 10⁻⁶ T into the page.**

ii. For point B:

$$\mathbf{B}_{I_1} = \frac{(2\times 10^{7}\text{T}\cdot\text{m/A})(3.0\text{A})}{(0.03\,\text{m})} = 200 \times 10^{-7}\,\text{T} = 2.0\times 10^{-5}\,\text{T into the page}$$

$$\mathbf{B}_{I_2} = \frac{(2\times 10^{7}\text{T}\cdot\text{m/A})(5.0\text{A})}{(0.02\,\text{m})} = 500 \times 10^{-7}\,\text{T} = 5.0\times 10^{-5}\,\text{T out of the page}$$

$$\mathbf{B}_{I_3} = \frac{(2\times 10^{7}\text{T}\cdot\text{m/A})(2.5\text{A})}{(0.02\,\text{m})} = 250 \times 10^{-7}\,\text{T} = 2.5\times 10^{-5}\,\text{T into the page}$$

The sum gives, for point B: 50×10^{-7} T = **5.0×10^{-6} T out of the page.**

iii. For point C:

$$\mathbf{B}_{I_1} = \frac{(2\times 10^{7}\text{T}\cdot\text{m/A})(3.0\text{A})}{(0.06\,\text{m})} = 100 \times 10^{-7}\,\text{T} = 1.0\times 10^{-5}\,\text{T into the page}$$

$$\mathbf{B}_{I_2} = \frac{(2\times 10^{7}\text{T}\cdot\text{m/A})(5.0\text{A})}{(0.01\,\text{m})} = 1000 \times 10^{-7}\,\text{T} = 10.0\times 10^{-5}\,\text{T into the page}$$

$$\mathbf{B}_{I_2} = \frac{(2\times 10^{7}\text{T}\cdot\text{m/A})(2.5\text{A})}{(0.04\,\text{m})} = 125 \times 10^{-7}\,\text{T} = 1.3\times 10^{-5}\,\text{T into the page}$$

The sum gives, for point C: 12.3×10^{-5} T = 1.2×10^{-4} T into the page.

(b) For the magnitude of the force between the wires carrying I_1 and I_2:

Using $\mathbf{F} = \frac{\mu I_1 I_2 1}{2\pi r}$

$$\mathbf{F} = \frac{(2\times 10^{-7}\text{T}\cdot\text{m/A})(3.0\text{A})(5.0\text{A})(0.06\,\text{m})}{(2)(3.1416)(0.05\,\text{m})} = \mathbf{5.7 \times 10^{-7}\,N}$$

Since the currents are parallel, the magnetic fields between them attract. The resultant force is **toward one another.**

(c) Since the two currents I_1 and I_3 are perpendicular to each other, there is no resulting force between them.